Mitchell M.Samuelian is the CEO and Founder of NuConsult Services, LLC, a renewable energy consulting firm that supports asset owners in construction management, O&M, operational improvement, compliance, policies, procedures, training, and safety across large wind, solar, and battery storage portfolios. He brings more than 40 years of electric utility and renewable energy experience, with a career spanning corporate asset management, environmental health and safety, plant management, and strategic planning in both traditional and renewable power sectors.

Mitch has held several senior leadership roles that helped shape modern utility-scale solar operations. He previously served as CEO of Tonian Renewables, Vice President of O&M for NRG/Clearway Energy, Director of Plant Operations for First Solar, and General Manager of the 392 MW Ivanpah Solar Electric Generating System, one of the world's largest concentrated solar thermal plants. Over his career, he has been involved in the development, construction, and operations of more than 5,000 MW of utility-scale solar projects. Earlier, he led the Tennessee Valley Authority's Hydro and Renewable Operations and Maintenance program, overseeing more than 5,200 MW of pumped storage, hydroelectric, solar, wind, and methane assets across five southeastern states.

Mitch's technical foundation was forged in the nuclear power sector, where he earned both Reactor Operator and Senior Reactor Operator licenses from the U.S. Nuclear Regulatory Commission. He holds an MBA with a minor in Human Resources from the University of Phoenix and a BS in Nuclear Engineering from Thomas Edison State College, combining engineering rigor with a strong focus on human performance and organizational leadership.

Today, through NuConsult, he leads owner's representative services on major renewable energy construction programs and supports the transition of large solar portfolios to self-operations and optimized O&M strategies, emphasizing disciplined procedures, safety, and reliability that align with the principles of the Industrial Operators Handbook.

# CONSTRUCTION MANAGER'S HANDBOOK

FIRST EDITION

Paperback ISBN: 979-8-9955861-0-4

# Contents

# How to Use This Book

This book was written to function in two ways at the same time. It is a teaching text that explains why failures happen, how construction systems drift, and how leadership, planning, and procedural discipline determine the difference between a stable project and a preventable loss. It is also a field manual for working professionals who need practical systems that can be used on active jobsites, under real production pressure, with incomplete information, multiple stakeholders, and constant change.

That dual purpose matters. This is not a book to be read once from front to back and then shelved. It is designed to be studied, referenced, carried into meetings, used in pre-task planning, and returned to whenever a team needs to reset its thinking. Some readers will move through it sequentially, starting with the foundations of failure and learning, then progressing through leadership, regulation, team systems, adaptability, and operational tools. Others will use it more like a handbook, turning directly to the chapter that matches the issue in front of them, whether that issue is procedural discipline, risk management, human factors, independent verification, continuous improvement, or emergency response.

The first way to use this book is as a systems text. Read it in order if you want to understand the full architecture of the argument. The early chapters establish the moral and operational foundation of the work. They explain that major failures are rarely random, and that collapse is usually the final visible result of hidden drift in design, communication, oversight, sequencing, leadership, and organizational culture. If you are a student, a new project engineer, a foreman moving into leadership, or a manager trying to sharpen your larger framework for decision-making, that full progression matters. It shows how technical, procedural, financial, and ethical responsibilities connect.

The second way to use this book is as a field reference. Each chapter is meant to stand on its own well enough that a superintendent, project manager, safety lead, or engineer can pull from it in a targeted way. A team preparing for a difficult pour can go directly to the planning, temporary states, independent verification, and manager's toolkit chapters. A leader dealing with recurring quality issues can work from the chapters on procedural discipline, quality control, continuous improvement, and accountability. A team confronting weather exposure, unstable ground conditions, or restart decisions can move directly into the chapters on emergency response, resilience, sustainability, and site recovery. The value of the book is not only in being read, but in being used.

A third way to use this book is as a training framework. The case studies, chapter takeaways, prompts, and templates can be turned into short teaching modules for crews, supervision teams, or internal leadership development. A chapter does not need to become a lecture. It can become a 15-minute conversation, a weekly huddle topic, a foreman development session, a planning workshop, or a post-incident review guide. In that sense, the book is meant to support organizational learning. It gives companies and field leaders a common language for discussing failure, systems, drift, risk, communication, and correction without reducing those ideas to slogans.

To get the most value from this book, read each chapter with two questions in mind. First, what is this chapter saying about how work actually fails or succeeds under pressure. Second, what changes would this require in how I plan, verify, communicate, supervise, or lead. That second question is the more important one. The purpose of the book is not simply to improve understanding. It is to improve judgment. Every chapter should move the reader toward clearer decisions, stronger systems, and more disciplined execution.

The case studies should be used carefully. They are not included for spectacle. They are included because every major failure teaches something about ordinary work. It is easy to distance yourself from a catastrophe and assume it belongs to another time, another company, another state, or another sector. That is almost always the wrong response. The discipline of studying failure is learning to see familiar patterns in unfamiliar tragedies. A bridge collapse, a crane failure, a balcony drop, a demolition accident, or a process explosion all reveal human and organizational conditions that still exist today. When you read the case studies, the right question is not "How could they let that happen?" The better question is "Where do these same pressures show up in our work, and what systems do we have to stop them?"

The operational takeaways at the end of chapters should be treated as field prompts, not as decoration. They are there to help readers convert ideas into action. A reader can use them to structure a morning huddle, a weekly superintendent meeting, an onboarding conversation, or a lessons-learned review. If the chapter itself provides the reasoning, the takeaways provide the handle for immediate use.

The appendices are meant to make the book practical. They are not filler at the back. They are the working layer that supports the chapters. The glossary gives shared language to complex ideas. The flowcharts make recurring processes visible. The templates turn principles into usable tools. The assignments and review questions make the book teachable in classrooms, internal programs, and leadership cohorts. The reading and standards list gives readers a path to go deeper. Used together, these back-of-book materials make it easier for a team or institution to turn the book into a living system rather than a static text.

This book should also be read with humility. Construction is too complex for anyone to master through confidence alone. The people who get the most from this material will usually be those willing to pause, check assumptions, revisit familiar practices, and question whether what has become normal is still safe, sound, and defensible. The strongest readers of this book will not simply agree with it. They will test it against their own projects, improve their own systems through it, and use it to challenge weak habits that production pressure often hides.

In practical terms, a student may read this book for understanding. A superintendent may use it to sharpen field systems. A project manager may use it to structure accountability, planning, and communication. A safety leader may use it to deepen pre-task planning, independent verification, and learning loops. An owner or executive may use it to evaluate whether the organization's systems reward real control or just the appearance of progress. All of those uses are valid because the book was written for the whole construction chain of responsibility, not just one role.

The simplest way to use this book is this: read the chapter that matches the risk in front of you, use the takeaways to force a better conversation, use the templates to structure action, and return to the case studies whenever your team begins to believe that familiarity is the same thing as control. It is not. That confusion has cost the industry too much already.

If this book does its job well, it should leave the reader with more than information. It should create a habit of seeing systems more clearly, questioning shortcuts sooner, and building teams that can think, communicate, and recover under pressure. That is how this book should be used. Not as a collection of ideas, but as a working discipline.

# Introduction

## Creating Systems that Last

Construction is more than manipulating steel and pouring concrete. It is the intentional orchestration of human intellect, mechanical precision, and procedural discipline that together give physical form to a vision. Each beam, column, and connection is an agreement based on trust in design, process, leadership, and one another. While the strength of a structure matters, its true longevity lies in the integrity of the system that harmoniously brings it together.

This handbook explores those systems. It is not merely a technical manual or a construction management guide; it is a call to reexamine our purpose as builders. It challenges both seasoned professionals and emerging field leaders to define what a truly resilient project system looks like under real-world conditions. A successful project is not just the product of workmanship; it is the result of how well we plan, communicate, coordinate, and lead. The line that divides a project remembered as a triumph from one remembered as a failure is drawn long before any concrete is poured. It begins with the invisible architecture of management itself, those frameworks and decisions that govern safety, efficiency, and durability under pressure.

In this environment, where climate, economy, regulation, and human behavior intersect daily, we must design our systems to protect progress rather than impede it. Codes, inspections, and safety protocols are not red tape; they are the guardrails that keep ambition from becoming harm. These layers of accountability, technical, managerial, and ethical, define the boundaries within which innovation can thrive responsibly. Planning, risk management, compliance, and communication are not bureaucratic hoops to jump through; they are the reinforcements that sustain the project when pressures mount.

This book is the construction-sector analog of the Industrial Operator's Handbook: a practical, field-facing operations manual for those leading work on-site. It is written specifically for superintendents, foremen, project managers, field engineers, and safety professionals—those who carry the daily burden of decision-making, coordination, and frontline execution. Its pages are filled not with abstract theories or generic management advice, but with the operating principles, practical systems, and repeatable disciplines that enable projects to consistently succeed in the real world.

The chapters that follow will not explain how to pour concrete or set steel. Instead, they address the systems that control those outcomes, the human and organizational frameworks that determine how well a project functions when it is under strain. A successful build is a "system of systems," each one dependent on the other. When any link falters, in design, in communication, or in execution, the entire structure is at risk. This book's purpose is to ensure those links hold.

The construction industry is driven by profit margins, efficiency, and schedule demands. These are not distractions from the work—they are the work. We build to achieve deadlines, control costs, and deliver value. But we must also remember why those outcomes matter. Behind every budget line and every production target are people: the workers who must return home safely, the families who rely on them, the communities that live beside our projects, and the environment that supports us all.

Every decision made in construction should be anchored by four drivers: people, environment, purpose, and quality. These are not abstract values; they are the compass points that keep daily decisions aligned with long-term integrity. They remind us that every schedule change, design modification, and cost adjustment carries consequences that extend far beyond paperwork. To lead in construction is to make those decisions with full awareness of their ripple effects—and to lead in a way that builds not just structures, but systems that last.

Every decision made in construction should be anchored by four drivers: people, environment, purpose, and quality. These are not abstract principles; they are the compass points that keep daily decisions aligned with long-term integrity. They remind us that every schedule change, design modification, and cost adjustment carries consequences that extend far beyond paperwork.

## People

Everyone involved in a project shall stay safe and go home safely. It sounds simple, yet it is the first value to slip when pressure builds. Construction is full of constant adjustments — shifting schedules, cost of turnover, design revisions, budget changes, material substitutions — each small on its own but powerful in the aggregate. In the tension between time and cost, people can fade from the immediate conversation. When that happens, they become the ones most affected by the outcomes of poor decisions. Protecting people means keeping them at the center of every operational choice, not as an afterthought but as the first condition of success. The systems we build to safeguard workers are not separate from productivity; they are what make productivity sustainable.

## Environment

Our work must minimize its footprint and preserve what it touches. This is not only an ethical responsibility but a measure of foresight. Each project interacts with its surroundings long after we leave the site — in the materials it consumes, the resources it displaces, and the resilience it either strengthens or weakens. Environmental stewardship in construction means thinking beyond project completion and considering the full life cycle of what we build. If we design with care for the land and the systems that sustain it, our projects inherit that same endurance. A structure that respects its environment will serve longer, require less, and contribute more to the communities that depend on it.

## Purpose

We must focus on meaningful, non-urgent work that sustains long-term value rather than chasing short-term urgency. True efficiency is not the ability to do more faster; it is the ability to do what matters most and make it last.

When projects operate in a constant state of reaction — racing to meet compressed schedules or fill gaps created by oversight — the result is often rework, exhaustion, and preventable failure. Purposeful work, on the other hand, creates stability. It allows space for verification, communication, and craftsmanship. When we slow down enough to make deliberate decisions, we accelerate the reliability of everything that follows. Purposeful work is not a delay; it is discipline. It separates productive urgency from destructive haste.

## Quality

Every outcome must reflect craftsmanship, care, and professional pride. Quality is not the responsibility of one department; it belongs to everyone involved, from the engineer drafting a detail to the contractor setting the final bolt. It is the thread that connects vision to execution. When quality is sacrificed for convenience or ego, the entire system suffers — errors multiply, systems weaken, and reputations erode. True quality is a form of respect: respect for the profession, for the client, and for the people who will live or work within what we build. It reflects character as much as competence. Across the case studies explored in this book, when quality was compromised, the result was predictable — complexity turned into chaos, and progress turned into loss.

Together, these four drivers form a practical framework for every decision we make. They balance the demands of business with the responsibilities of stewardship. When people, environment, purpose, and quality remain visible in our process, profit and performance follow naturally — because the work is sound, the systems hold.

These drivers are not theoretical. They are practical tools for judgment, benchmarks for decision-making that keep work aligned with both purpose and performance. When we allow these values to guide us, profit becomes sustainable, efficiency becomes humane, and schedules become achievable without compromise.

We build to meet our timelines and our budgets — that is the truth of business. But we also build to create something that will last beyond the transaction. Financial success and performance are necessary measures of a project, but they are not the only ones. The enduring test is whether what we build strengthens the world around it. Whether it is a simple community bridge or a multi-story tower that reshapes a skyline, each project contributes to a shared human story.

What we build will outlive us. It will become part of someone's daily routine, their safety, and their sense of belonging. Our work is not just efficient execution or technical precision; it is an act of legacy. Each project carries both the evidence of our skill and the reflection of our values. We build to sustain economies, but also to sustain trust — to show that progress and care can coexist, and that what we leave behind is worthy of those who will inherit it.

This handbook will ask you to confront both failure and opportunity. It will challenge assumptions about confidence, risk, and the hidden costs of neglect. It will show how systems transform complexity into control, and how leadership turns aspiration into endurance. It will remind you that every technical decision carries ethical weight, and that the truest form of leadership is accepting responsibility for it.

Throughout these chapters, you will see construction not as a sequence of tasks, but as a living system of coordination, accountability, and learning. You will see how the organization translates vision into order, how risk management clarifies uncertainty, and how procedural discipline preserves institutional memory. You will see how technology accelerates progress, but cannot replace judgment or a moral compass.

This work is both an examination and a call to action. It invites every professional who builds, designs, or manages to look inward before looking outward. The future of construction will not be measured by how many structures we complete, but by how we construct the systems that make those structures possible. Quantity will not define our legacy; resilience will.

What follows is a call to think differently — to build organizational and governmental systems as durable and principled as the structures they support. To lead with accountability. To design processes that protect human life and conserve collective wisdom. To measure success not only by what we build, but by how safely, ethically, and intelligently we build it.

**Case Study 1**

# The Kleen Energy Explosion

## Middletown, Connecticut, 2010

The Kleen Energy Explosion (2010)

On a cold Sunday morning in February 2010, a nearly finished power plant in Middletown, Connecticut, carried out what many in the industry then considered routine. Crews set up a high-pressure fuel gas blow to clear debris from new piping as part of commissioning. Minutes later, a dense gas cloud that had collected in a tight courtyard found an ignition source and detonated. Six workers lost their lives, more than fifty were injured, and a billion-dollar project was left in ruins. Nothing about the materials was mysterious. What failed was judgment, planning, and ownership in the face of real schedule pressure.

The operation was simple in concept. Natural gas would be forced through the lines and vented to the atmosphere to carry construction debris out of the system. In practice, the venting occurred between large plant structures where air movement was limited, and work was still active. Gas pooled quietly at ground level and drifted through adjacent work areas while welding, energized circuits, heaters, and engines remained in service. An electrical arc or hot work spark was enough to ignite the invisible cloud. The blast that followed tore through the power block and scattered evidence of a deeper problem that had been building for months.

A high-energy task of this magnitude should have been planned like a critical lift. There was no focused job hazard analysis for the blow, no written procedure defining the footprint, and no brief assigning stop-work authority. Vent routing and stack height were not engineered to carry gas away from people. There was no coordinated evacuation of non-essential personnel, no systematic shutdown of ignition sources, and no staged emergency response for a fuel-air event. The site had many companies working in parallel, and everyone assumed someone else had the hazard covered. In reality, no one person owned the risk.

The human pattern is familiar. What began as a shortcut years earlier had hardened into accepted practice. Using fuel gas to clean lines felt normal because it had been done before without consequence. That success was misread as proof of safety. Schedule pressure added urgency; it was Superbowl Sunday, and the team was behind, and the window of opportunity was convenient. Prior blows that had not ignited were treated as experience rather than warnings. The result was a textbook case of normalization of deviation and diffusion of responsibility. When many firms share a site, and no one is clearly in charge of a hazardous operation, decisions default to habit.

The aftermath made the lessons plain. Codes and standards were updated to prohibit the use of flammable gas for cleaning or purging and to require safer alternatives such as air, nitrogen, or mechanical methods. Enforcement actions and industry alerts underscored a simple truth that should have been obvious all along. No deadline, no budget, and no commissioning target justifies creating a blast cloud in the middle of a working site. Connecticut banned gas blowouts on power projects, and many jurisdictions and owners adopted the new standard of care in their specifications.

What would a competent construction manager have done differently? First, reject the method and require a non-flammable medium. Second, if any release is unavoidable, design the vent path and the space so that gas is routed high and remote, with dispersion calculated and verified. Third, own the footprint. Barricade the zone, evacuate non-essential personnel, shut engines down, de-energize circuits, suspend hot work, and verify controls before any flow begins. Fourth, name one qualified person to coordinate the operation across every contractor, hold the permit, and carry stop work authority on the radio. Finally, run a short, specific briefing, monitor conditions, and stop at the first sign of drift.

This case belongs at the front of the book because it is not about one bad decision. It is about a sequence that construction teams know too well. A familiar practice that saves time. A congested space that no one maps. Parallel contractors who all assume the others have it handled. Real pressure to deliver. When high-energy work is not planned with discipline, the job organizes itself around convenience. People are always the ones who pay for that convenience. Our responsibility as builders is to make the safe way the normal way and to place clear, accountable leadership over any task that can release uncontrolled energy.

In the coming chapters, we will break down other incidents, structural failures, collapses, and process events, all of which will reveal unique patterns but similar human causes. We will learn about communication breakdowns, procedural drifts, leadership, and culture, and how they neither add to nor subtract from risk but rather build the discipline needed to manage complex risk. We will learn about not just what failed, but why we failed and how our industry can build systems resilient enough to mitigate the preventable.

## Case Study 2

# The Vdara Hotel "Death Ray" Incident

### Las Vegas, Nevada, 2010

Amid the desert sun in Las Vegas, where reflective, glass-shrouded high-rises serve as markers of opulence and creativity, one building curved the light in a way no one had imagined. The Vdara Hotel, a 57-story, curved tower designed by one of the highly acclaimed architects of our time, became famous not for its architectural accolades but for an unexpected side effect that emerged in the first few months of opening in 2010, which the locals named "the death ray." Within months of opening, visitors at the hotel using the pool reported hair blowing up, plastic melting, and the temperature sizzling off the deck below, as concentrated beams of sunlight smeared on the deck, scorching intermittent spots.

The issue arose from the building's most distinctive design component, a concave façade clad in reflective glass. The curvature, intended to create elegance and flow, acted as a gigantic mirror, intensifying sunlight on the pool area. The building had been carefully modeled for energy efficiency and selected the glass coating to minimize solar heat gain in the building's rooms; in a paradoxical conflict, the protective system that meant to protect the building from the generation of heat by sunlight now emitted it outward into space, creating a hot zone, "Death-Ray".

The designers anticipated glare but attempted to address it by applying a low-emissivity film to the glass to reduce reflection. But their analysis underestimated the amount of sunlight, both in quantity and intensity, which would be experienced in the hot, dry climate of Las Vegas, and how the geometry of the façade would focus and amplify this sunlight. Once the phenomenon became public, guests and the service staff reported in vivid detail incidents of newspapers curling, plastic drink covers warping, and people rushing from their deckchairs to escape the invisible magnifying heat.

The event resulted in no physical harm or structural damage, but it became a public example of what can happen when aesthetic form and effect prevail over thought and foresight. The "death ray" event was not a structural failure, a building collapse, nor an incorrect calculation of building materials. It was the failure of a systems approach, a recognition that no design choice can be made without consideration and examination, and that every design choice carries a series of environmental and human consequences. The designer not only knew about the potential phenomenon and studied it, but also turned a blind eye to the consequences because it was their signature style.

From a construction management perspective, the Vdara case demonstrates that design intent, materials selection, and operational experience must be interconnected with thoughtful pre-analysis and resulting feedback. In complex projects, such as many in healthcare, engineers and managers often use various forms of digital modeling and the endorsements and assurances of consultants to validate their design decisions. However, the visible response of the glass's structural ability to sunlight is regarded as an output pigeonhole design intent, as well as discipline norms and values. It demonstrates how to test your assumptions respectfully under real conditions rather than optimistically in your previous modeling. The designers anticipated glare on the patio but considered it a reasonable, alterable nuisance. They predicted a hotspot around 36°C (97°F), but it turned out to be double that, around 72°C (162°F). The mitigation being only partially successful ultimately resulted in a series of embarrassing, costly, and reputationally damaging situations across all stakeholders involved.

The other piece of the picture is that failures do not always occur as collapses and explosions. They begin as small discomforts, anomalous experiences, or things that are often overlooked, while there appear to be inadequacies in how these issues are addressed in the process and communication. The Vdara incident reminds us that safety is not just about structural stability and integrity but also about environmental behavior and user experience. Projects that embrace aesthetic innovation need to balance creativity with rigorous technical review.

The recurring dynamics of the case at Vdara included: communication isolation between design and construction teams; an opportunity for inquiry in risk assessment rather than a simple model of compliance; and reliance on precedent whenever projects rest on the edge of the norm, rather than proposing something unique. These team dynamics become systemic issues, regardless of whether the failure is dramatic or quiet. Hopefully, by closely reviewing a range of strategic cases, this book will help professionals develop a habit of looking more deeply into each example, beyond cherished design intent, and understanding what really matters: human impact. An understanding of mechanisms for failure, even in moments of unintentional absurdity like the "death ray" of the Vdara incident, strengthens the discipline of construction management and its recognized construct to achieve the highest purpose: build buildings that will respect the safety and intelligence of its users, occupants, and the responsibility of its stewardship.

Case Study 3

# The Taum Sauk Upper Reservoir Dam Failure

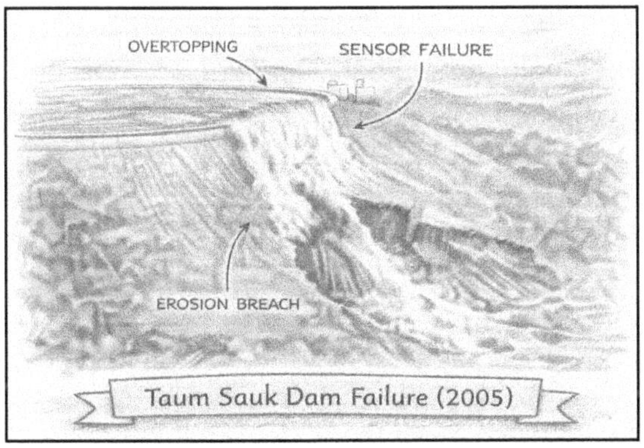


On the morning of December 14, 2005, before the sun rose in Missouri, a relatively quiet mountain was awakened by a sound like bursting water. The Taum Sauk Upper Reservoir Dam, part of a pumped-storage hydroelectric facility, catastrophically failed. In less than ten minutes, over a billion gallons of water cascaded down Proffit Mountain, taking whatever was in its path; massive amounts of water and debris flowed into Johnson's Shut-Ins State Park below. The surge left destruction in its path, taking down trees, cars, and houses. In the rush, the park superintendent's family sat in their home, which stood directly in the flood's path. Fortunately, the family survived, but not without injuries to the family and property damage, leaving a scar upon the landscape and state.

The Taum Sauk disaster was not caused by natural forces or an act of God. It was an overreliance on technology, poor human monitoring, and a cascade of judgment failures. The dam, constructed in the 1960's, was a ring-shaped embankment dam on a mountain that stored water from a lower basin for energy storage. The dam did not have a spillway, relying entirely on sensors and automated controls to monitor and alert to the risk of overfilling. For decades, none of the systems failed or caused a threat to anyone's safety. Over time, the natural settlement of the dam structure, sensor drift, and the loss of institutional knowledge led to this incident. The operators of the facility had more faith in their instruments than in their own perception of events, and when the instrumentation failed, the dam was left a sitting duck.

In the months prior to the failure, there had been some warning signs. In September 2005, the reservoir briefly overtopped while the system was pumping, an event that should have instigated a shutdown and immediate investigation. Instead, the engineering team misread the situation, changed the supervisory control logic to stop pumping slightly earlier, and scheduled the physical repair work for the next maintenance cycle. This was a convenient, but dangerous, course of action. The sensor readouts continued to falsely indicate the water level was lower than it actually was, and the backup probes were placed too low to detect an unsafe condition. During an overnight pumping operation in December, the sensor read "safe" even as water crept over the reservoir's lowest rim. Once overtopping began, erosion progressed rapidly through the earthen embankment to the overtopping, causing the dam wall to fail.

In the end, it turned out to be a combination of human and instrumentation errors. During the recent liner replacement project, the instrumentation was poorly secured, resulting in inaccurate measurements. The crest of the dam had also settled unevenly over time, leading to freeboard margins that the operators assumed were existing. Most importantly, the dam operator (AmerenUE) failed to notify federal regulators of the previous overtopping event. The Federal Energy Regulatory Commission (FERC) later concluded that the company had both underestimated the severity of the September incident and had proceeded to operate fully aware of the existing safety issues within the dam. The organizational culture exhibited an inappropriate degree of trust in automation and a level of comfort with temporary fixes for operational safety issues.

This incident starkly demonstrates how modern infrastructure can catastrophically fail when human attentiveness to risks is relinquished to complacency in reliance on technology. The control system did exactly what it was programmed to do, but it was programmed based on false premises. The operators of the dam did not commit an egregious act of ownership neglect; they were simply constrained by an organizational system that had normalized the risk. The harmful effects came on subtly. Each incremental concession, a delayed maintenance repair, software adjustment, or assumption that "surely it will be ok," simply adds an additional layer to an already hidden susceptibility. Over time, those layers become indistinguishable from routine practice until the system breaks.

The financial and environmental consequences were dramatic. The flood inundated public land, destroyed state infrastructure, and cost Ameren Bank nearly $200 million in settlements and fines, as well as additional construction costs. The dam was completely rebuilt as a roller-compacted concrete dam with an appropriate spillway and enhancements to the dam monitoring system. The company's reputation was damaged, leading to sweeping changes to dam safety oversight nationwide. FERC created new requirements for chief dam safety engineers, routine physical inspections, and formal reporting for abnormal events. What had begun as "the advanced automation we employed" quickly became a cautionary tale about the dangers of remote operations without human oversight.

The Taum Sauk failure presents an opportunity for reflection on a constructible truth of design and operations: systems do not fail suddenly; they fail slowly and quickly, all at once. We observe cracks in the process before we observe cracks in the material. There is an opportunity for construction management students and professionals to see firsthand, in this example, the consequences of avoided maintenance, unreported anomalies, and misplaced faith in technology. The current construction industry faces similar pressures of short timeframes, complex systems, and an increasing reliance on data and automation.

In this book, we will return to the themes found here: the interplay between human thought and mechanical precision; the degradation of safety culture in the face of operational pressures; and organizational silos, which, intentionally or unintentionally, delay or dilute corrective action.

By examining case studies like Taum Sauk, we could reflect on what it looks like in anatomy when failure occurs, and how technical precision, though error-resistant, with limited scrutiny during practice, can become a mask for fragility. The aim of this work is not dissimilar to telling the story of a failure; it is to end with failure transformed into prevention, foresight, and resilience. The Taum Sauk dam did not simply "break a wall of concrete and stone," it broke the illusion that technology alone guarantees safety. When the dam was rebuilt, part of that rebuilding was cultural, not simply structural; the future of construction management depends on our learning from the past, and on doing so with a degree of humility and discipline.

## *Part I – Foundations of Failure and Learning*

**Roles:** *All construction professionals,* especially project engineers, new managers, and seasoned superintendents.

**Focus:** Establishes a shared understanding of why failures happen and how we learn from them. By examining real-world construction disasters and their root causes, Part I lays the ethical and practical foundation for what follows.

**Operational Decisions Informed:** Big-picture thinking and moral responsibility. After this section, a superintendent or project engineer will approach decisions with a sharper eye for early warning signs and a deeper commitment to safety and quality – ensuring that past mistakes inform future actions.

Chapter 1

# Why Projects Fail - Systemic vs. Isolated Failure

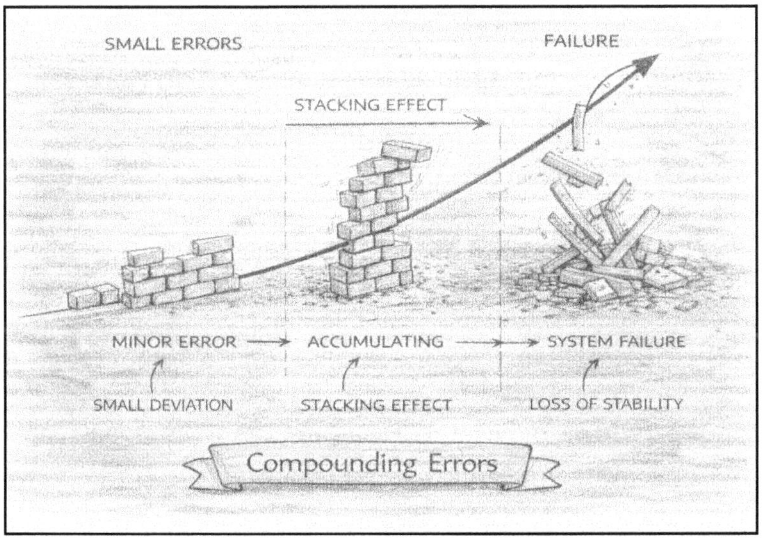


No project is conceived with the goal of failure. Every set of drawings, every schedule, and every safety plan is a manifestation of human ambition shaped into order. History reminds us, however, that failure is not contrary to construction but one of its outcomes. The question is not whether a project will fail, but how well the systems supporting it are built to absorb and adapt to failure.

A single act of negligence or an isolated mistake does not cause most project failures. Most failures result from systems that are, over time, gently nudged away from their intended design. A structure collapsing, a significant safety event, or a financial meltdown might be the last link in a lengthy yet invisible chain. Each link may appear insignificant, a rushed decision, a lack of checking, a delayed conversation, but in unison, they comprise a chain of events that no one person can read until it is completed.

To understand the act of failure in construction, we should first recognize that projects function like living systems. They are dynamic organisms made up of interdependent parts: design, procurement, execution of work on-site, quality oversight, and leadership—all existing in a stressed state. When one element weakens, it transmits that stress to other elements. If the system lacks elasticity (as it relates to construction projects) or feedback loops, stress builds until it exceeds capacity. What looks, in hindsight, like an abrupt failure is, in fact, almost always the slow depletion of invisible guards.

Isolated failures are easy to explain and, therefore, comforting. They allow us to point fingers and move on, thinking we have solved the problem. A worker didn't comply with a rule. A drawing was wrong. A safety check was not made. But systemic failures do not provide simple explanations because they confer joint responsibility, indicating that the conditions allowing for an error were built into the project. The issue is not who failed; the issue is what the system allowed.

The construction industry, by nature, lives at the intersection of complexity and uncertainty. Each project is singular and tends to embody thousands of tasks and hundreds of contributors, all working under multiple contractual regimes. This fragmentation creates invisible fault lines. Communication across trades, between layers of management, and with oversight authorities is easily muddied, with assumptions often replacing facts. A late permit, a subcontractor's interpretation, or a missing inspection can move through a project just like an unseen current would. By the time the symptom manifests itself, the cause is buried under documents and points of accountability.

One of the ongoing obstacles to construction management is that our systems reward production rather than reflection. Success is gauged by calendar adherence and cost-effectiveness—measures which show momentum but not vulnerability. Rarely will the organization stop mid-project to ask whether the control systems are actually working. The industry's pace promotes the belief that momentum fosters confidence, which in turn masks error. The result is a form of procedural inertia: the work continues because stopping feels more dangerous than moving forward.

When our failures occur, they often expose patterns rather than surprises. They reveal that many people could see pieces of the problem but had no process for synthesis. The engineer saw the technical inconsistency. The foreman felt the unsafe shortcut. The manager sensed the schedule compression. Each insight was held in isolation, and no one system linked them. A systemic failure is not a failure of knowledge; it is a failure of integration.

To prevent recurrence, we must learn to observe construction projects not as collections of tasks to be punting, but as interdependent ecosystems where each decision has consequences. The value of a project is not in the ability to make quality parts; it lies in the quality of the relationships among the parts. Communication loops, feedback loops, and disciplined checking are the load path of a successful project. As connections weaken, the signs become clearer: rework is simply accepted as the status quo, small deviations are tolerated, and paperwork is signed without assessing its value or ensuring the work is done. Leadership is more interested in completing the task than understanding it. These experiences are precursors to failure, not something dramatic, but subtle cultural changes.

This book begins with the perspective that failure is not a mystery to be solved, but a process to be analyzed, with each failure representing its own anatomy and a system. To understand how projects fail, we must learn to trace those backward from the point of failure to the first unnoticed drift. Only then can we create organizations resilient enough to withstand the same pressures that caused others to fail.

## Method of Studying Failures (Root Cause, RCA, Forensics)

We study failure not to blame, but because work demands that we understand. Every failure, accident, or near miss tells a story about how people, procedures, and systems act under pressure. To reveal the story, we must balance curiosity with discipline, evidence with empathy, and science with humility. Root cause and forensic reviews are not rituals performed simply to meet standards after the fact; they are rituals performed in respect for the work itself — to ensure that every piece of knowledge lost, whether paid in cost or suffering, can be turned into knowledge and progress.

The inquiry starts with humility. The aim is not to show who was wrong, but simply what went wrong. Any process you pursue must be guided by several principles: protect life, preserve evidence, distinguish fact from inference, respect context, and share the lessons learned with those who need them most. A failure that we reconstruct honestly usually indicates that those involved behaved reasonably within the system available to them. The question is seldom, "Who made the mistake?" but "Why did the system allow for the mistake to occur at all?"

The first step in any loss is securing the scene and preserving what it might still have to teach. The urgency that follows a loss can easily lead to the loss of evidence that could help prevent a similar loss in the future. The scene must be stabilized (as much as possible) to rule out any further harm or loss; however, disturbance must be minimized. Evidence in construction is often found in the most transient of things: formwork, scaffolding, rigging, lifts, and temporary electrical feeds. Before those are removed/deconstructed, there needs to be, at a minimum, careful documentation of the scene-wide images for context, close-up images of mechanisms, and notes describing location, weather, and conditions. Digital copies of records, drawings, and instruments (if needed) need to be captured before power is restored or the system is wiped. Even small observations related to bolt placements, weld conditions, etc., can help confirm or refute a theory later.

In addition to preserving the physical record, it is important to preserve human evidence. Identifying and interviewing witnesses quickly may yield information that would otherwise fade from memory over time or with prompting from colleagues. Interviews should be conversations, not interrogations. The investigator initially listens with the intent to understand what each person witnessed, what they believed to be true, and what sorts of pressures or preconceived notions influenced their actions. Most human error occurs in good faith and stems from a variety of competing demands. When individuals are allowed to explain their logic and actions, patterns emerge that suggest not carelessness but normal behavior operating under duress.

From these descriptions and materials, the timeline begins to take shape.

Ideally, it will start with the last point where everything was stable, then progress the event forward in small sublations from that point to the failure point, and finally delve into the immediate response. Veracity must be substantiated by credible sources: inspection logs, radio calls, daily reports, photographs, telemetry, etc. Gaps are to be expected and identified when that is prudently done, rather than filling them in with supposition or speculation about what might have happened. When multiple trades or crews are working side by side, mapping their work to one another often reveals the point where miscommunication or the way a task was accomplished led to conflicting assumptions. The goal of the timeline is not to develop a single story but to retell how a set of normal events converged to create one abnormal event.

After the timeline is accurately formed, the analysis begins. No single tool can illustrate the complete truth, so a wise investigator utilizes several tools. Causal factor mapping demonstrates how conditions and their actions are linked. Change analysis shows what differed on the day of the event from the baseline of safe operation. Is it equipment, people, environment, or timing? Barrier analysis examines the safeguards that were in place and whether they were missing, ineffective, or ignored. Both event trees and fault trees analyze how a set of small influences can combine to create a major failure. Human factors analysis examines the cognitive environment of the people involved with the task- workload, fatigue, distractions, available information, etc. Tools such as the Five Whys or fishbone diagrams can certainly have their place, but it is important that the analysis does not stop at "human error" and is supported by an attempt to find a systemic cause.

Construction presents a unique challenge because the state of construction work is constantly changing. Construction involves many elements that remain fixed, such as the state of the work at the end of the day. However, the project evolves daily, passing through dozens of temporary conditions that will never recur once that section of the structure is finished. Concrete cures, shoring is struck, cranes are configured for specific lifts, and scaffolding is moved, each representing a transitional moment when the risk was at its highest. Each time a state changed, new risks were introduced that might not have been considered when developing the work plan or verifying its adequacy to safely complete the work. Each should be treated as an engineered condition with a design and specified inspection criteria, even if the state is temporary. In retrospect, many of the worst failures in the history of the construction industry occurred during these in-between states, where the work was neither temporary nor permanent, and no one was accountable. Contextually, if you were to try to understand why control was lost, mapping out how that condition was designed and verified often explains it. Each finding should be valid and seen as true. If possible, independent verification of findings provides an objective lens and challenges assumptions that might otherwise go unchecked, as they could be overlooked by someone too involved in the work. There should be no shortage of a qualified peer to verify calculations, examine the evidence, and assess the reasoning in the findings. Counterfactual testing yields stronger conclusions simply by asking "what if." If we could have averted the incident by doing something differently, and/or if the safety and containment controls had functioned, would the result have been any different? If we had delayed entering the plan, would the conditions have been different? These tests assess the durability of conclusions and identify which elements or points of leverage were critical in the incident, potentially preventing loss of control.

Documentation is the thread that links all of this together. The documentation should be detailed and comprehensive enough for someone unfamiliar with the project to trace the logic from reconnaissance evidence to the conclusions. The documentation should label photographs, drawings, and witness statements. The documentation should indicate uncertainties, not just facts. Keep a version history of the analysis of drafts so that the changes in the analysis of thought are apparent. Good documentation not only validates the findings currently evident but also serves as a repository of future lessons to review when similar patterns of hazards emerge. The clarity of the documentation will ultimately determine whether the insights gained from the investigation become a living resource or a forgotten file.

Then the analysis turns to "cause" and the determination of the critical point. Contributing or contributory factors are those that made something more likely or made the consequence more robust. Root cause or root causes are the systemic frailties that permitted circumstances to continue unchecked (inappropriately). Weather, fatigue, or design change might be contributing factors; however, the deeper cause is usually an organizational systemic weakness—poor planning, incomplete integration, or leadership inappropriately preferred speed over certainty. Referring to an event as "human error" indicates that searching has concluded too early. The more correct question would be "what made it matter to the system?" The lesson holds value when a reckless choice is actually a reasonable choice to act on, despite constraints that were inappropriate to the process.

Corrective actions should change the system, not just behavior. They should include simplified procedures, remove hazards (where practicable), and add redundancy in energy sources where hazards/constraints cannot be eliminated. For example, substituting a cleaning process where staff would have used hazardous removal and operating an inert purge to remove operational hazards while driving the removed value. Provide effective management of the lift used in the lifting process by including an independent person who performs a lifting verification. Moving the hold point from the worker, who waited for a shift/decision from management upon completing a predetermined deliverable, ensured that members did not alter the loading of structural supports or hold during the decision-making process. Each corrective action should have an assigned responsible party, a defined outcome, and a time frame for validation. For the incident to be meaningful, it must be closed in the field, with feedback from the affected unit confirming that they have, in fact, changed the system. And this is what becomes trust in leadership learning built.

Finally, is the learning capture-report that is never read, created for the next crew or project. Wrap-up and share the modifications made, learnings in near misses, and make the modifications tangible. Technological tools, "lessons learned", and more conventional tools of training modules, checklists, standards, and planning prompts. It might be beneficial to the team if problems are documented in a centralized database and categorized by hazard, project phase, trade discipline, etc. Additionally, referring to the learning tool during preconstruction planning should take 5 minutes at the tailgate meeting. Did they discuss a near miss? Encourage teams to discuss 'reluctantly' what they experienced - telling their near misses with real context. Only if an organization views transparency as a strength and builds institutional memory will this become the most valuable form of risk prevention or examination, one that evaluates all variables.

The integrity of an inquiry relies not only on how it is conducted but also on the orientation of the researchers who conduct it. When the preparation of a project begins with curiosity, it drives questions deeper into the problem. When an investigation is grounded in humility, it will, in its own way, ensure the outcome is reflected honestly. When an investigation is grounded in a commitment to accuracy, it enables the work to be trusted and gives credibility to the findings. Ultimately, when those three ideals steer an investigation, they transcend the administrative requirements of an inquiry process and become a vehicle for improvement. In a formal investigation, the lessons learned from failure transform loss into knowledge and into resilient systems.

## Moral Imperative: Human Life and Cost of Harm

Every construction project implicates a moral agreement beyond any legal contract or professional seal. It is a silent agreement to map a process for professionals to follow, ensuring that no one will be harmed in the act of creation. Every decision implies a moral position, even the act of changing a detail in a serial number or breaking a string of a schedule. Every decision can impact someone's life (or everyone). Construction is a physical responsibility. The force on one beam or the tension in one cable could alter a person's life and future. A profession as responsible as construction requires that knowledge, starting with accountability.

When accidents happen, they are not only mechanical or procedural failures, as we identify them; they are also ethical failures. They constantly remind us that safety is not a department, a checklist, or a compliance audit; it is simply respect for life. Every regulation and every code point to a story that was written in loss (blood, injury, or grief). Every clause of an OSHA compliance standard and every section of a building standard points to an ethical dimension of construction that grows from remembering that cost. Engaging in a discussion of workplace ethics is not a nebulous concept. Ethics is an inherent part of decisions made each day: do we stop a task when the next step is unclear, challenge an engineer's assumptions about the work story, or voice concerns about a subcontractor who has taken a shortcut? Ethics occur in the quiet, fleeting moments between instruction and action. The moral imperative of leadership is to make the hidden visible and safe. Authentic leadership does not celebrate risk tolerance, but rather builds risk awareness. It teaches that stopping a job is not a sign of weakness but wisdom, and that raising a concern is not an act of insubordination but dedication.

The costs of an injury extend past the incident itself. When a worker is injured, there are impacts on family, crew, and beyond to the community. The emotional toll of the incident can last for years for those who witnessed it or took action afterward. The organization's reputation suffers not just in the eyes of clients or regulators, but also in the consciousness of its own people. Projects stall, teams fracture, and shared purpose is diminished. Human costs become economic costs, and economic costs become cultural.

It is easy to consider workplace accidents as singular events, though they rarely are. Accidents reveal the gradual erosion of ethical attention that takes place when production takes precedence over protection. The pressure to deliver a product faster or cheaper takes the form of efficiency. Efficiency without ethics is tenuous. The appearance of old common wisdom of "progress" (sometimes called production) can hide the gradual erosion of the values that keep people secure. As the moral framework deteriorates, so will every technical system that relies on it.

The ethical mandate of construction management requires project leaders to view every schedule, budget, and plan through the lens of human consequences. A schedule that has been compressed beyond reason is not a scheduling issue; it is an ethical issue. A step missing from verification and inspection is not merely a technical omission from a plan of work; it is a moral failure. When the pressure to proceed replaces the responsibility to stop, harm is the inevitable result. Ethical management begins where courage replaces convenience.

The history of construction provides numerous examples in which moral clarity became apparent only after disaster. Bridges collapsed because engineers dismissed signs of distress. Towers failed because safety inspections were treated as sign-offs. Workers died because the chain of communication unraveled under stress. Each of these tragic events has led to new codes and regulations, but regulatory compliance is not an ethical obligation. Ethical obligation exists upstream of regulatory compliance; it begins long before a code check, inspection, or citation. It begins in design meetings, preconstruction meetings, and everyday planning conversations about decision-making, where decisions about risk first emerge.

Ethics in construction cannot depend on individual moral fortitude and professional judgment alone. Ethics must be embedded in systems that make the right decision the easier decision. When policies encourage employees to report near misses in the field without fear of repercussion, when owners penalize workers for unsafe conditions and reward safety performance as seriously as schedule performance, and when engineers can stop work under unsafe conditions without being punished, ethics becomes the standard. Ethics goes from principle to practice. Everyone connected to the fields of construction, design, contracting, inspection, or management should consider themselves a steward of human life. The buildings, bridges, and roads that we construct will be in place regardless of how long it takes for our signatures to fade from the drawings. The decisions we make will be there for years to come. Our responsibility is not solely to carry a project through completion, but to do so in a way that honors all who constructed it. Success is not measured simply by the physical structure remaining, but rather by the fact that each individual who signed the invoices stands as well.

The ethical responsibility of this vocation is not optional, nor is it ceremonial. It is the bedrock that all systems rest upon. The codes and standards provide the structure, but ethics provide the rationale. Without that rationale, no level of advanced safety, technology, or management system will survive the test of time. Ethics is the unseen bulletin that quietly provides structure to the entire industry. Ethics is the invisible steel that keeps competence fused to trust, and trust fused to care.

To build with ethics, therefore, is to build with foresight. To recognize that every decision becomes part of a legacy that is not measured in profit or reputation, but in lives saved. The ultimate test of any project will not be whether it was built to specification, but whether it served to protect those who made it possible. Construction, at its essence, is stewardship of land, materials, energy, and, importantly, of human life. Understanding the cost of harm involves recognizing the burden of that stewardship and moving forward with discipline, empathy, and resolve.

## Industrial Parallels: Lessons from Operators and Other High-Risk Industries

Construction does not exist alone. Construction intersects with other industries that also share the equilibrium of human judgment, energy, and dire consequences. Aviation, nuclear power, chemical processing, and offshore oil have all found themselves at moments when a single event has revealed the fragility of their systems. These industries did not change systems based on technologies; they changed systems through disciplined learning. The evolution of these industries provides construction as a living laboratory for considering maturity in risk management.

In the 1970s and 1980s, aviation and nuclear power, among other industries, experienced their defining disasters. In the wake of fatal collisions and midair events, the aviation

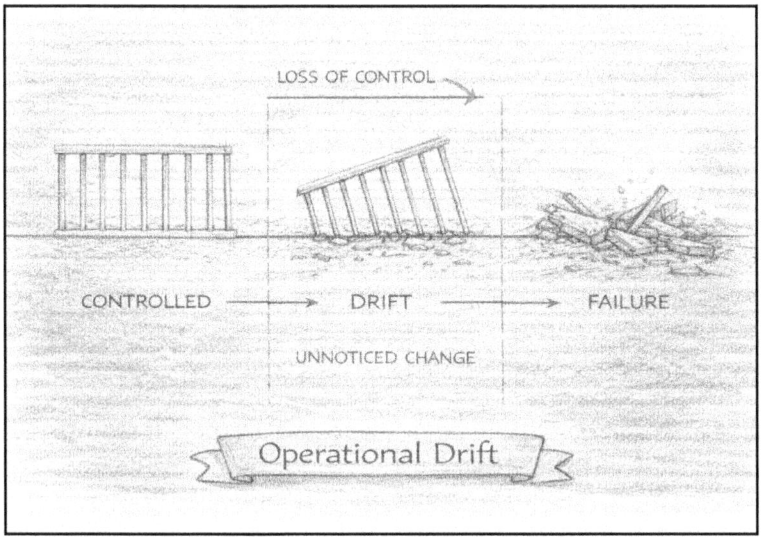


industry established independent investigation boards, standardized flight operations manuals, and implemented crew resource management. This training emphasized (1) communications; (2) shared decision-making; and (3) that any crew member could call attention to danger, regardless of rank. The change was not only procedural but cultural. Over time, this resulted in one of the world's safest complex systems. Similar to aviation, the nuclear industry experienced its own evolution following the Three Mile Island incident.

This failure in safety was not due to insufficient technology but to "confusion and miscommunication...." In the aftermath of that incident, the nuclear industry developed a network of peer-review organizations, independent oversight entities, and incident-transparency reporting systems. These developments created an environment in which operators learned collectively rather than independently. Each abnormal event became a mechanism for shared learning and improvement. Similarly, after the tragedy, the chemical and energy sector constructed systemic resilience. The Bhopal disaster in India in 1984 and the 2005 Texas City refinery explosion illustrated how insignificant deviations, normalized over time, could combine with poor communication to release catastrophic energy. These events served as catalysts for global responses in hazard analysis, emergency preparedness, and process safety management. The focus shifted from technical perfection to the reliability of personnel—how personnel interpret, adhere to, and possibly ignore procedures. The lesson was simple to articulate but profound and powerful in delivery: technology can only be as safe as the culture from which it operates.

In construction, similar energy pathways and human pressures exist and perpetuate the failures seen in aviation, nuclear, chemical, and energy sectors. A crane operator balancing wind conditions on a jobsite, a concrete team racing against the temperature, and a project manager scaling back scope to accommodate client pressures all perform their work in complex environments that require rapid decisions based on incomplete information. While these decision pressures are present in control rooms and cockpits, the construction sector has been slow to adopt a systems-based lens. Many of the organizational safeguards we now expect in other sectors—sealed, formalized communication loops; structured turnover briefings; independent verification of work scope—are not consistently used in construction.

The aviation practice of crew resource management provides a valid parallel to construction. In crew resource management, the authority is shared when 'safety' is involved. A laborer, foreman, or engineer must be confident in sharing safety concerns, and the leaders must be humble enough to listen and act. In aviation, the cultural change saved lives—in construction, it's an opportunity to stop the quiet normalization of unsafe acts that precedes tragedies.

The nuclear sector's practice of peer review also gives insight. Operators routinely solicit independent teams to visit, review their procedures, maintenance, and event history. These visits are not audits with punishment as the endpoint, but rather learning opportunities that lead to improvement. In construction, a similar model can be used before critical construction milestones – invite outside superintendents, engineers, or safety professionals to challenge assumptions, validate risk, and review readiness for critical lifts of temporary works. This practice shifts accountability from compliance to collaboration.

The chemical industry understands the need for redundancy in its systems and the ability to identify hazards. Many of their significant accidents stemmed from teams placing too much reliance on automation or thinking that the instruments would recognize deviations. Construction typically shares this theme of overreliance on paperwork or technology. Checklists, sensors, and digital systems can be effective tools; however, they do not reduce the need for disciplined attention to the process. The message is not to replace innovation, but to integrate it with human situational awareness. The best system is one in which technology complements human intellect rather than displaces it.

Across all high-risk industries, one common theme emerges: they treat failure as an opportunity for information, not as an embarrassment. They track, share, and discuss failures without the fear of embarrassing someone. This open failure culture is what we know as a learning culture—a condition where organizations constantly investigate how the work is done, not how the work is written to be done. Construction may still be maturing to that type of learning culture; however, we can quickly learn from these acknowledgment systems.

The future of our profession will depend on how we translate these systems into our own environment. We need to normalize near-miss reporting without punishment, establish structured pre-task briefings similar to having structured aviation flight checks, and incorporate independent verification of at-risk work. We need to move communication forward and pride backward.

The industries that experienced their worst failure chose to face them openly and systematically. They did not just rebuild their facilities; they rebuilt how they thought about it. Construction stands at the same crossroad. We have the benefit of hindsight without yet having a catastrophe define us on a global scale. The question before us is whether we will learn in voluntary circumstances or learn through tragedy. The decision will define not only our safety reputation, but our industry character for decades to come.

## Independent Verification as Cultural Practice

All strong systems have verification as a factor. Verification is why confidence does not become assumption, or the intention leads to error. Verification in construction ensures that what was designed, planned, and communicated has been understood and accomplished as needed. However, verification is not simply a technical requirement. It is a cultural practice—a collective obligation to practice humility together, discipline together, and hold each other accountable together.

Independent verification starts with accepting that no one person, regardless of their experience, is exempt from oversight. Complex projects have thousands of moving parts, making an entirely accurate interpretation of drawings, specifications, and procedures critical. Flaws in your work can creep in so quietly, through a drawing revision, a misread tolerance, or a change made in the field that no one else reviewed. Independent verification allows another participant, who is trusted and not hired for the project, to evaluate witnessed assumptions and confirm that what appears correct is indeed correct.

There is no disagreement with verification, which hints at distrust; rather, it is simply a realization of human limitation. Verifying work should not be done spontaneously; it should be organized in advance. Verification should be placed on the program schedule and delineated within the scopes of work. For example, critical lifts, formwork designs, post-tensioning work, or any work that entails a large amount of energy and/or irreversible consequences generally warrant a second verification before work begins. The review should be independent of the work, meaning it should be initiated by someone with no stake in the outcome. Independence is necessary for the integrity of the review; an outside observer will find patterns and risks that familiarity obscures.

Verification exists beyond drawings and calculations. It exists on the job site, in the actual work. Inspections, pre-task briefings, and walk-downs provide opportunities for independent verification or observation. When verification becomes routine, it changes the nature of the project. Teams begin to embrace questions rather than resent them. Supervisors are beginning to view verification as a partnership rather than an interruption. A culture of verification promotes the confidence of the system in which we work and think.

The collapses that shaped our industry are often rooted in missed historical verification stories. For example, the Hyatt Regency walkway collapse, the Willow Island cooling tower incident, and the L'Ambiance Plaza emergency all had early signs that went unverified or signed off by an uninformed or uninspired peer. Each incident had a point at which a second verification could have broken the chain of events that led to failure. Each event serves as a reminder that verification is not a box to check. Verification is an opportunity that allows for enough time to see clearly for ourselves. This brief moment of confirmation is often all that stands between a near miss and disaster.

Systems of verification should provide for fully effective two-way communications between the design process, construction, and quality control. A review is not limited to approvals alone but should also produce feedback loops. When an engineer or inspector raises a concern, the finding must be documented and resolved. Closed-loop verifications can turn insight into learning and build a record that improves future performance. After a while, these processes become self-sustaining: the more verifications are practiced, the more the organization trusts that it is doing it right the first time.

Leaders have a bigger role in shaping this structure. When leaders see verification as an impediment to productivity, the workforce does too. When managers reward teams for identifying their own errors rather than keeping the problems in their own pockets, the organization matures. Independent verification needs to be a form of protection, not punishment—protection for the project, the client, and all safety obligations for every worker who relies on what was built.

To develop this discipline, we need to be consistent. Verification must be present at every stage of the project life cycle, from design and procurement reviews to field inspections and final acceptance testing. Each of these stages must have clearly defined criteria for what must be verified independently, and by whom. The goal is not to repeat the verification process. The goal should be to layer assurance. When multiple qualified people review a common and critical element, confidence becomes verifiable rather than assumed. When independent verification is at its most mature, it becomes a mindset rather than a step. It is a reflex to look again, to ask another question, and to check your own certainty. It embodies a culture that understands that truth in construction is never proven once; truth is constantly being validated. A strong verification culture acts against complacency and becomes a habit. It reminds everyone involved—engineer, manager, craft professional—that quality and safety are not sustained by trust alone but by the validation of quality and safety through intentional review.

Verification does not slow progress; it illustrates progress. The most trustworthy projects are not the ones that work most rapidly, but the ones that stop as necessary. Independent verification creates momentary pauses amid the voice of production in a fast-paced industry, where noise gives way to the stillness of verification. In that pause, errors are caught, understanding is aligned, and systems are strengthened. That is how a project, and the people building it, learn to endure.

## Leadership Under Stress

Leadership in construction is often tested at the moments when the plan is starting to come apart. Schedules begin to tighten, the weather shifts, and competing influences pull people in seemingly opposite directions. It is often at these moments when both the outcome of a project and the character of those leading it are established. Calm, courage, and clarity

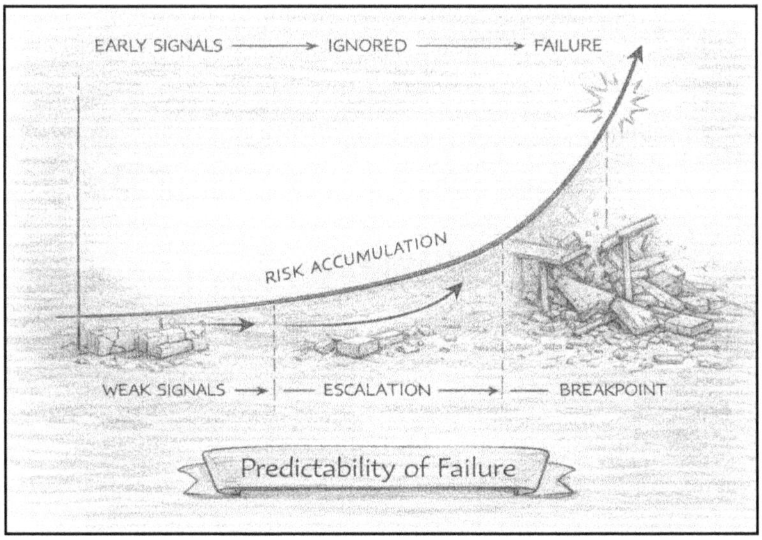


become the tools that keep the system together when everything else seems to want to fall apart.

At most jobs, stress is not an exception; it is a preferred partner. Teams are working against the clock, resources are in flux, and decisions need to be made quickly. What distinguishes sustainable success from catastrophe is not the lack of stress, but how it is handled. A good leader takes pressure and transforms it into focus, aligning safety and progress rather than pitting them against each other.

History is not short of lessons. During the Big Blue crane accident in Milwaukee, a massive crane tipped over while being moved in high winds. The decision to proceed, made under the duress of an intense schedule, ignored all of the warnings. We started with confidence, then transitioned to urgency, and finally to an incident. In the case of the Willow Island cooling tower collapse, workers placed concrete on lifts that had not yet cured to meet their accelerated timeline. The leader confused momentum for control, and 51 people died.

These occurrences reveal a basic truth about leadership. Leadership does not fail when it makes decisions; it fails when it does not stop. The courage to pause work, consider the next action, or delay to confirm is often the distinction between recovering and regretting. The job site at the L'Ambiance Plaza collapse removed shoring systems even while the structure's integrity was in question; progress was deemed more important than the lack of confirmation, and the building fell. The pattern is clear—pressure, confidence, silence, collapse.

When conditions start to deteriorate, maintaining composure is key to control. After an incident, the leader who moves calmly through chaos, listens, and acts restores order far sooner than the one who demands it. The crew does not draw strength from the leader's authority alone; it draws strength from the leader's disposition. The capacity to maintain equanimity when others are apprehensive is not a function of personality; it is preparation. Those who know their systems well can focus on solutions rather than emotion.

Stress also reveals culture. In high-functioning teams, communication stays open, even under stress. People voice concerns without trepidation, and information flows upward as readily as it flows downward. In weak cultures, silence permeates. Workers sense risk but say nothing; supervisors see a drift but presume someone else will say something; and managers keep to a schedule even as warnings accrue. The Hyatt Regency walkway collapse is a lasting example of what happens when communication collapses under authority. A manageable connection detail was changed without complete inspection, since questioning it would have challenged time and authority. When communication collapses, systems collapse.

Leadership in construction isn't about reducing stress, but preparing for it. There should be anticipation of decision points where they will become more constraining—high-risk lifts, critical pours, shift changes, etc.—and the identification of clear pathways for input and validation. Leaders who are prepared for these moments of high stress manage them effectively. They create a procedure to protect decision-making when time and external pressures obstruct perception.

Good leadership is also aware of the contagion of calm. A foreman who communicates uncertainty calmly teaches a crew to consider rather than react. A project manager who acknowledges uncertainty sets the bar for realistic assessment. An executive who supports an employee's safety check sends a clear message that prioritizing safety is never negotiable. Behavior flows throughout every level of the organization.

One thoughtful, composed decision in a crisis can change how hundreds will react when the next crisis occurs. Across many of the case studies in the book, which include examples of crisis failures, the difference between failure and recovery was just seconds and a single decision. The Formosa Plastics explosion, the Hard Rock Hotel collapse, and many others shared the same DNA: leaders knew there were risks but couldn't act because of competing priorities. Each decade contains the same lesson. Stress without structure becomes chaos. Authority without humility becomes blindness. Confidence without fact becomes danger.

When tensions increase, leaders need to return to the basics of crisis response: clear communication, factual assessment, and a disciplined pause. They are not ideals; they are habits developed through repetition and reinforcement by culture. The organizations that train their people to slow down at the first sign of uncertainty are the ones that will avoid escalation. Stopping to think is a form of leadership.

The strength of a construction project is not just defined by design or engineering, but also by how people respond when things go wrong. Systems and procedures matter, but systems rely on judgment, and judgment relies on being steady in a moment of confusion. The leader who can think sharply when everything else is fading protects not just the project but the people working in it.

The future of this industry will be defined by how we can support leaders who maintain moral clarity, operational focus, and human empathy in the most difficult moments. All projects will involve stress, but it is how we respond to that stress that makes the difference between failure and growth. The target is not to groom leaders to never feel pressure, but rather to provide leaders with the capability to transition that pressure into alignment, with people, systems, and purpose intact when it matters most.

## Building the Roadmap

The chapters that follow are specifically aimed at guiding a transition from acknowledging failure to designing systems to eliminate it. The journey begins with accepting that failure is not an event; it is a process that gradually shifts over time in both intent and outcome. In studying how drift occurs, we incorporate mechanisms of reorientation into intent and maintain the project framework with purpose, clarity, and safety. The new path of this text is designed to transform explicit and tacit attention from reflection to mechanism, from theory to practice, and from individual accountability to collective reliability.

---

**Part I** will lay that foundation. It starts with what we have examined here: what failure is, the ethics of obligation, and the social/economic weight of our choices. Part I will take us through case studies to examine how patterns emerge in decades of construction history. Each case shows how familiar pressures — time, cost, pride, and exhaustion — interacted with weak systems to produce outcomes no one wanted, nor intended. Nothing is intended here to rehash the tragedy for tragedy's sake. Rather, we recognize that each incident determined our codes, standards, and expectations. The case studies serve as our framework to understand the anatomy of failure that modern construction still faces.

**Part II** will transition from reflection in the process to reflection in the structure. The text analyzes the processes of planning, risk management, and leadership—the unseen architecture that determines whether a project succeeds or fails when pressured. In this section, we focus specifically on the systematic tools that turn philosophy into action: work breakdown structures, risk registers, standard operating procedures, and communication systems to keep teams aligned. Our goal is to give the manager and engineer the ability to see these patterns before they occur as incidents. While sound systems take into account many sources of uncertainty, they do not eliminate them; they organize them into manageable forms.

**Part III** of the program examines the regulatory professional frameworks that form the outer guardrails of our industry. The discussion identifies the laws, rules, and standards of compliance that capture our collective responsibilities. When we consider compliance, we do not intend to know these frameworks forward and backward; compliance, in this context, means that every regulation is a response to a loss. Each regulation was borne from failure. From that perspective, compliance is less about enforcing rules and more about responding to failures to continue; it represents an ongoing conversation about the history of failures and how we choose not to repeat them.

**Part IV** of the program focuses on people and adaptation. It includes learning about teams, how leadership mode culture works, and how communication nurtures trust. Construction management is not simply technical; it is so deeply human. The systems we build (and design) will succeed or collapse based on our ability to engage with those who build in them. This section will investigate the training, mentoring, and ongoing learning that will move safety and quality away from a program and into a mindset —a celebration of adaptation, the capacity of a team and organization to adapt to unexpected conditions without losing purpose or control.

**Part V** looks ahead to sustainability, technology, and risk evolving in a time when construction meets environmental change, digital transformation, and global complexity. Artificial Intelligence, automation, modular construction, climate resilience, and so on are complicating our plans to build. These advancements challenge us to adopt new tools as innovative while learning to ensure that innovation serves human welfare rather than undermining it. Progress that ignores the lessons of failure is repeating them by another name.

Across the previous parts, the book's layout parallels that of a successful project. It will be initiated with purpose and planning, progress through systems and leadership, and conclude with accountability and improvement. Each part builds upon the others, culminating in a complete cycle of understand, design, implement, and learn. In that cycle, you will find the same principle repeated; the strength of the system is not in its complexity, but rather its ability to communicate and adapt.

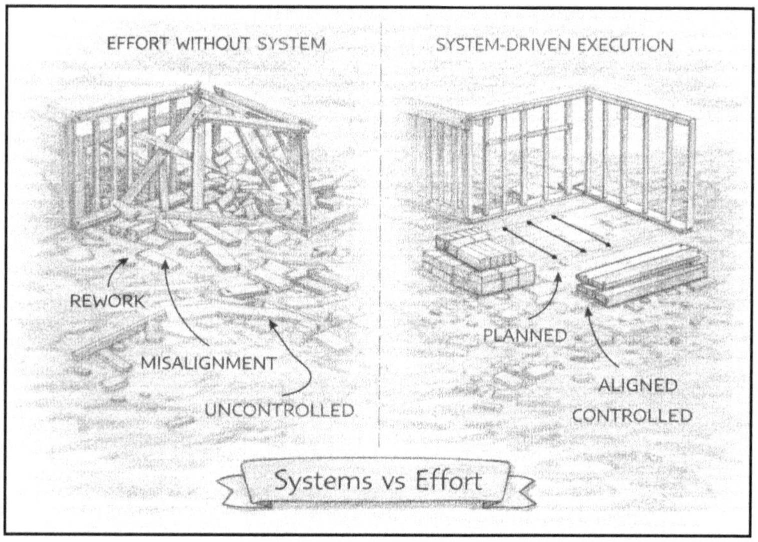


The roadmap of this textbook will not only outline the topics to be examined but also the sequence of transformation. It will begin by showing how projects fail, and conclude with organizations learning to prevent those same failures through discipline and foresight. In between, we will ensure a compelling analysis of historical case studies, a necessary breakdown of management frameworks, and an examination of the moral and operational aspects of leadership principles. The chapters will encourage you not only to memorize processes but also to think systemically—connecting to the safety of the human—from the drawing board to the final walk-through inspection of any project, which is vital to the organization's professional integrity.

If there is one message that synthesizes every page of this book, it is this – we cannot rely on luck or tradition to protect us. Resilience is not inherited - it is built intentionally, reflectively, and continuously. The systems that you, your team, and your project will thrive in, despite complexity, are those deliberately designed with the humility to expect failure in and through, and the discipline to prepare for it. The roadmap is both technical and moral. It asks not simply how to run a project, but how to lead a profession.

**Operational Takeaways for Today's Job:**

- Treat small issues as early warning signs of bigger problems. If you see a minor design error or safety lapse, address it immediately and share the concern – one overlooked mistake can snowball into a major failure.
- Think of your project as an interconnected system, not isolated tasks. During daily planning, consider how each crew's work affects others and ensure communication lines remain open so critical information isn't lost between teams.
- Build in "pause and check" moments. Rather than pushing nonstop for the schedule, take short breaks to verify that safety measures and quality controls are working – a few minutes of reflection can catch issues that momentum might hide.
- Encourage your team to speak up about near-misses or discomfort with a task. Create a no-blame environment where a foreman or field engineer who spots a potential problem is praised for preventing the next link in a chain of failures.
- After any incident or close call (even a minor one), do a quick root-cause huddle. Ask "What in our system allowed this to happen, and how do we prevent it next time?" – then implement one improvement immediately.

# Learning from Catastrophic Engineering Failures (2005 - 2025)

Over the past two decades, the United States has experienced a series of structural failures and disasters that offer important lessons. One of the most relatable incidents is the Champlain Towers South condominium collapse in Surfside, FL (2021), which took place in an ordinary residential building—a condo—in which 98 people died tragically because of the spontaneous collapse of a high-rise building. This incident was high-profile and highlighted how design flaws, compounded by long-ignored essential maintenance and regulatory blind spots, can lead to catastrophe, making it an ideal, relatable case study for an introductory chapter of a book on lessons learned.

While the Champlain Towers South condominium collapse was relatable, memorable, and certainly tragic, an important "underappreciated" case took place much earlier, with no fatalities, yet almost destroyed a community—the Taum Sauk Dam failure in Missouri (2005). The incident occurred, in part, because a forgotten sensor caused the reservoir behind the Taum Sauk Dam to overflow and breach the dam. The mass failure that followed nearly destroyed Lake Ozarks, Missouri, and the Taum Sauk Dam incident illustrate that failures and mistakes in landlocked, remote infrastructure can still cause mass damage.

Below, we provide a reflection on the Champlain Towers South condo collapse and other notable U.S. structural failures between 2005 and 2023, examining the failures, contributing causes, loss of life, financial costs, and regulatory failures. Each incident demonstrates that every one of these catastrophic incidents stemmed from a single mistake or act of negligence that compounded and led to the collapse of the infrastructure. As well, each incident could potentially have been avoided based on current ideas of proactive engineering and architecture.

# *Champlain Towers South Condo Collapse (Surfside, 2021) – Relatable Primary Case*

**What happened:** In June 2021, there was a partial collapse of an approximately 12-story condominium in the middle of the night. The condo was named Champlain Towers South. When this failure occurred, 98 residents died. Surveillance video footage and forensic examinations indicate that the probable mechanism of the structural failure was a punching shear failure at a column connection, which caused the grade-level slab to fail and led to the collapse of the building above. Most victims were in their beds and in a state of sleep when the building pancaked, which made this building collapse one of the deadliest in U.S. history.

## Contributing Factors:

The investigation has identified design flaws and construction defects, as well as discrepancies dating back to the building's original construction in 1981, and evidence of deferred maintenance. In the original flat-plate concrete design, there was insufficient rebar at the critical points (the pool deck-to-column connections) to accommodate the necessary tolerances if the concrete tub deteriorated. In fact, as late as 2018, the engineer issued warnings regarding water intrusion and concrete spalling to the pool deck and garage areas; however, repairs proceeded slowly. Up to the collapse, the community of residents voiced concerns about at least cracks in the planter wall, a shifted entrance gate, and significant water leaking into the garage—all located near the pool deck. All of the residents' concerns, which were tied to the pool deck slab area, were ignored as serious distress signals. On the evening of June 24th, the deck slab reached overstress and, due to the absence of redundancy, its collapse caused a substantial portion of the tower to suffer the same fate as the pool deck area.

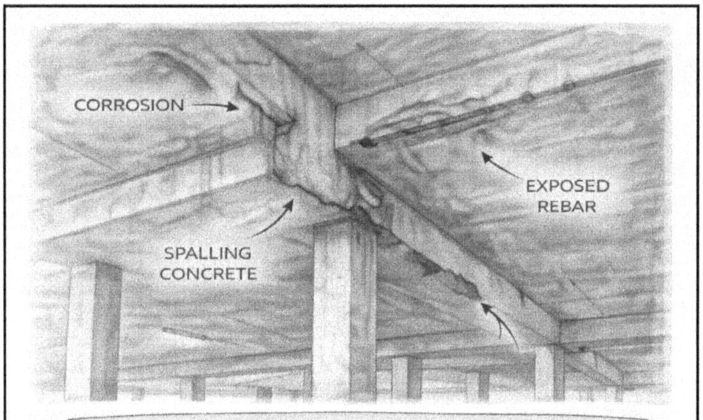

Champlain Towers Structural Deterioration (Pre-Collapse)

### Aftermath / Lessons Learned:

The Surfside collapse led to rapid reforms. Florida implemented rules requiring more rigorous condo inspections and reserve funding to prevent repairs from being postponed indefinitely. Analysts were quick to identify systemic issues, stating, "This was a tragic event that revealed flaws in our systems at all levels, and quality is at the center of it," a lead federal investigator commented. The main takeaway is not to overlook small warning signs—water and leaks, cracks, design deficiencies, etc.—that must be addressed before a complete chain-reaction failure occurs. Champlain Towers South is relatable for many reasons, but most importantly, it was an occupied residential building housing dozens of families, underscoring that building code compliance inspections and preventive maintenance in everyday buildings are literally life-safety strategies.

---

## *Hurricane Katrina Levee Failures (New Orleans, 2005)*

Hurricane Katrina Levee Failures (2005)

**What Happened:** When Hurricane Katrina struck New Orleans, Louisiana, in August 2005, the city's levee system failed catastrophically. Dozens of floodwalls and earthen levees either breached or were overtopped, flooding 80% of the city. Louisiana witnessed over 1,500 fatalities as a result of flooding, much of it as a direct result of inundation. This disaster was not primarily a natural one, and investigations into the disaster in New Orleans determined that it was an engineering failure of the flood protection system. The U.S. Army Corps of Engineers took the blame when Corps Chief, Lt. Gen. Carl Strock, said, "The levees failed because they were constructed in a piecemeal fashion and built with outdated information."

## Contributing Factors:

The levee system was a patchwork project that began in the 1960s but was not built to common standards, and many portions were poorly designed or constructed. The Corps' $19.7 million investigation found numerous engineering and design failures that allowed Katrina's storm surge to easily overtop the flood protection system. Outdated soil data: Designers did not fully account for soft soils and land subsidence in New Orleans, leading to problems with existing floodwalls. In some cases, floodwalls were constructed on short sheet piles that did not reach stable soils. Additionally, some components were built poorly and had sunk several feet over time. Inconsistent levee construction did not help either. Levees varied in quality and height, leading to weak links and gaps. Four canal floodwalls failed, with water levels below the design threshold, due to instability at the base that the design hadn't accounted for. Even breaches in the canals accounted for two-thirds of the flooding. There were also Corps mistakes (dating back to the 1980s) of using steel sheet pilings that were shorter than necessary, which put the floodwalls at risk of underseepage and failure.

## Aftermath and Lessons:

The loss of life and property was remarkable, with losses over $100 billion, marking the levee breaches from Hurricane Katrina as one of the worst engineering disasters in U.S. history. Investigations by IPET, ILIT, and others led to significant changes in the flood defenses in New Orleans and a review of Corps practices nationwide. The clear statement is that piecemeal, underfunded infrastructure is a recipe for disaster. Flood protection has to be continuous and built to the best estimates of acceptable flood risk. The Corps acknowledged the need to "build better flood defenses" and to gain public trust. In summary, Katrina was a lesson that engineers could never take soil and subsidence for granted and that adequate safety measures were necessary in the design of critical infrastructure, particularly given the growing risk of climate-based threats.

## *Boston "Big Dig", Tunnel Ceiling Collapse (Massachusetts, 2006)*

Boston "Big Dig", Tunnel Ceiling Collapse (Massachusetts, 2006)

**What Happened and Why It Matters:** On July 10, 2006, in the Ted Williams Tunnel in Boston, as part of the Big Dig highway project, a huge concrete ceiling panel broke free from above and fell, crushing the roadway. The panel, which weighed about 26 tons, fell and crushed a vehicle passing underneath, killing a passenger and injuring the driver. The tunnel collapse caused significant additional damage beyond the crushing of a victim's vehicle, as the tunnel had to be shut down indefinitely, subject to a criminal investigation. Immediately prior to this event was a dynamic, highly publicized infrastructure installation that was already receiving considerable discredit for no good reason, which compounded a public trust crisis. In an era of mega projects (like the Big Dig), the road under the roadway sparked additional skepticism after the project was completed. Although the Big Dig got more than its fair share of unnecessary and well-publicized respect as a significant "engineering" project that re-engineered American infrastructure, the emergence and video of the collapse were viewed with idle curiosity around the public as yet another example of failure in production, and one that did not inspire confidence in "mega-projects".

**Factors Involved:**

The National Transportation Safety Board (NTSB) determined the source of the ceiling failure to be the "creep type" failure of multiple anchor bolts that had been glued into the tunnel ceiling using epoxy adhesive. Over time, under the persistent loads of the heavy concrete panels, the epoxy adhesive slowly allowed the anchors to deform, or creep, until they all pulled out at once. With a small number of critical connections lost, the remaining bolts pulled out in a chain reaction, dropping the concrete slabs on vehicles in the tunnel area below. Since this mechanism of failure had been recommended against in a 1998 Inspector General report that noted the epoxy system was unsafe, the bolts were too short and not "up to standard" for the long-term overhead loading, it is concerning that these warnings did not result in sufficient changes. Further, the investigation showed that contractor and inspection officials had opportunities to identify the problem. There was no mention in inspection and construction records of the design of the ceiling in the area of the junction of the tunnel sections that had these heavy panels that had been added without proper support.

### Aftermath and Lessons:

The ceiling collapse at the Big Dig project led to the epoxy supplier being charged with manslaughter and to contractors being sued. Politically, it was a scandal. The Mass. Governor called the tunnel a "crime scene." Eventually, the thousands of ceiling-bolt locations in the tunnel system were reinspected, and many were either replaced or reinforced. The critical takeaways are the appropriateness of the materials and the inspection validation. Even when used under short-term loading, using an epoxy anchor to hold heavy concrete panels was a huge mistake. Engineers began to trust the red flags in the material tests and field reports rather than ignore them. This case emphasized the value of independent checks of complex designs. In the end, the tragedy prompted changes to tunnel inspection protocols and reminded builders that taking shortcuts and failing to meet specifications could lead to deadly outcomes. As NTSB determined objectively, the failure of the "epoxy anchoring system" and the decline in quality control caused an unsafe form of construction in the design to be in place.

---

## *I-35W Mississippi River Bridge Collapse (Minnesota, 2007)*

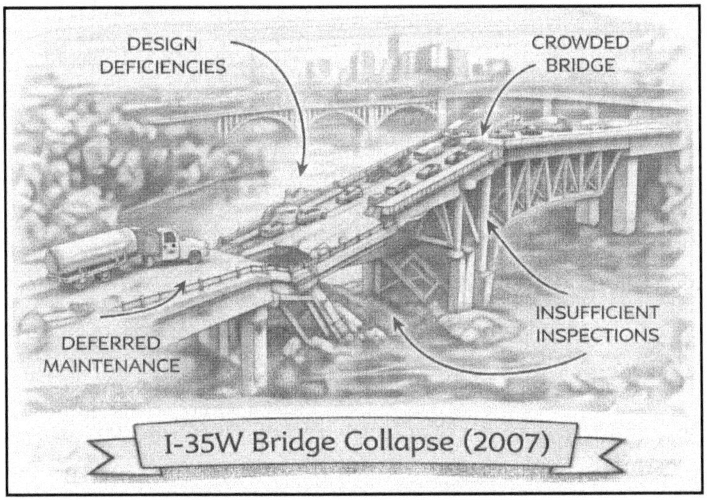


### What Happened and Why It Matters:

On August 1st, 2007, the busy I-35W highway bridge over the Mississippi River in Minneapolis suddenly collapsed during rush hour, sending dozens of vehicles into the Mississippi. The bridge collapse killed 13 people and injured 145. The steel truss bridge had been in place for over 40 years, and the sudden failure of the bridge horrified the nation. The entire central span of the bridge collapsed in seconds, which NTSB later called a "catastrophic" failure of the bridge structure.

## Factors Involved:

Based on the intensive investigation, the NTSB determined that a specific design flaw was likely the cause. The gusset plates were undersized. The inch-thick steel plates that connect the girders in the truss bridge were half as thick as they were supposed to be. In one critical node (U10), the original designers used 0.5-in plates when 1-in plates would have sufficed. Generations would pass before the mistake made by the original designers impacted the safety and design of the bridge system. To complicate the situation, the bridge was subject to increased loads over its lifespan: the addition of concrete pavement and heavy construction equipment on the bridge itself in 2007 placed a significant additional load on the already-stressed thin gusset plates. On the day of the collapse, there was also a stockpile of construction materials on the span, estimated at an additional 300 tons. The NTSB concluded that the untenable, but understudied, gusset plate at joint U10 tore along the line of rivets, causing a complete and immediate failure of the span that adjoined joint U10. In very basic terms, poor original design + additional loads = failure. Importantly, the state of Minnesota was also not checking the total thickness of gusset plates in the spans as part of inspection and monitoring (rather, they focused on cracks or corrosion). There was no redundancy. The visible signs of deformation of the gusset plates under load were the warning signs in this instance (later photographic evidence revealed small changes in configuration); this warning was not addressed. The I-35W collapse is a painful reminder that a single failure — here, a design mistake made 40 years prior — can evolve into a fatal defect if left undetected; therefore, periodic reevaluation of aging infrastructure is necessary.

## Aftermath and Lessons:

The collapse prompted an extensive emergency response and a complex recovery effort. Beyond the human toll, the economic impact included hundreds of millions of dollars for bridge replacement and related costs, along with months of regional traffic disruption. Federal and state authorities issued advisories advising the recalculation of gusset plate capacities on similar truss bridges. Many jurisdictions reinforced or retrofitted connections and moved toward risk-based inspection programs that account for design details, construction-staging loads, and the added dead load from overlays. Minnesota instituted independent peer review for major bridge designs and tightened procedures for construction loading on existing structures.

## *Dallas Cowboys Practice Facility Collapse (Texas, 2009)*

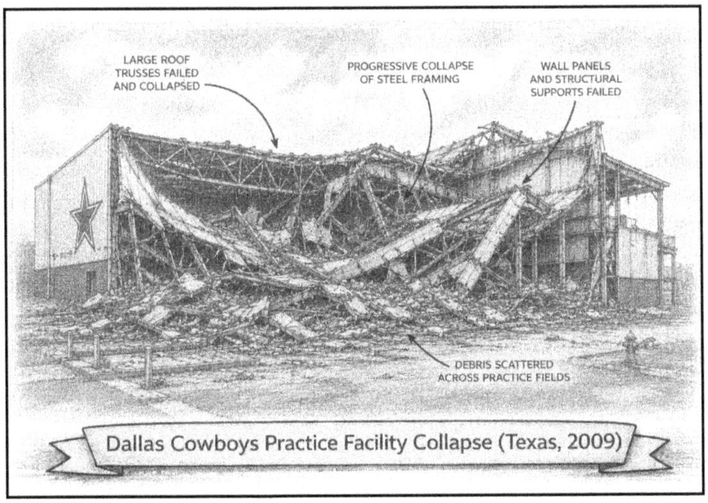


**What Happened:** On May 2, 2009, a large fabric-covered NFL practice dome for the Dallas Cowboys collapsed during a windstorm. The 80,000 sq. ft. covered facility in Irving, TX, collapsed during a thunderstorm with approximately 70 occupants (players, coaches, staff) in attendance. Miraculously, nobody died, but a scout assistant sustained permanent paralysis, and 11 others suffered injuries when the tent structure collapsed when its steel frame failed and fragmented.

### Contributing Factors:

Initially, the blame was placed on "severe weather." However, a federal investigation led by NIST uncovered serious design deficiencies in this structure, which was originally manufactured to withstand wind loads well below its required design wind speed. The original engineering was performed by Summit Structures (a corporation under the umbrella of Contemporary Cover-All Building Systems), and NIST found that the framed structure's capacity was deficient. Under the normal required wind pressure for a framed structure, the capacity for this structure was exceeded. In other words, the dome was not designed to meet wind code requirements. As there seems to be a (concerning) trend, it was also revealed that personnel were unqualified and warnings were ignored. As it turns out, the person who did the initial calculations and analyses of the dome and another dome structure that collapsed in Philadelphia was directed by an unlicensed, non-engineer trainee. In 2004, following a discovery of design shortcutting, Cover-All's own engineering manager warned, "We can't carry on this way, or we're going to kill somebody." In 2008, the Cowboys' facility added some supports after the team noticed deflection in the fabric roof; however, this work apparently proceeded without a permit and did not involve a full reanalysis of the load capacity. A re-setback effort had taken place; however, the eventual collapse confirmed this would be insufficient. Additionally, the connections at the frame joints ("knees" and ridge) were likely under-designed, causing the steel tubes to buckle apart when struck by 55-mph gusts (though this structure was likely designed against this wind load).

### Aftermath and Lessons:

The injured individuals were settled in subsequent lawsuits, and Summit Structures went out of business. Forensic reports, as well as the NFL's own review, eventually led to improvements in codes for these types of large fabric structures. The overarching lesson is a giant one: specialized structures always require proper credentials and an independent third-party review. The Cowboys simply trusted the vendor who had a history of problems (multiple collapses). Also, this case required oversight by an independent engineering review team. This case also shows the danger of dismissing near misses: the Philadelphia collapse of a sister structure and near-internal warnings were, in fact, almost massive red flags. The tragedy of a man being paralyzed from this eventual collapse should have never happened if a proper design review had occurred. NIST's report specifically claimed that the facility should withstand substantially larger wind speeds. In conclusion, the catastrophe instilled in organizations the idea to "trust but verify." That is, even when designers express confidence that a design works, it is sometimes prudent for owners to verify the designers' credentials or seek a second opinion. A rough fix (as with a 2008 patch) does not substitute for good engineering. The Cowboys' collapse is a cautionary tale indicating that cost-saving, hasty designs can literally fail when used as intended.

---

## *Indiana State Fair Stage Collapse (Indiana, 2011)*

Indiana State Fair Stage Collapse (Indiana, 2011)

## What happened:

Approximately fifteen minutes before a Sugarland concert on August 13, 2011, at the Indiana State Fair in Indianapolis, a gust of wind from a thunderstorm blew over the outdoor stage rigging. The entire canopy lighting and sound stage (essentially a temporary steel truss) came crashing down into a crowded area. Seven people died, and 58 were injured. It was widely televised because it happened in front of thousands of concertgoers, just after officials declared it unsafe and began evacuating the audience (but no one had yet been evacuated).

## Contributing Factors:

A state investigation determined there were numerous safety failures associated with both the construction and use of the stage, which consisted of a roof and lighting truss level by vertical masts secured by guy wires. On that evening, the approaching storm generated an hourly speed estimation of 59 mph, which is strong but not out of the ordinary. The stage should have been able to withstand this wind speed if it had been properly engineered and secured. The investigation established that the guy lines and the system failed against the wind load. The weight system (ballast, such as large concrete or water weights used to hold the bases of the towers) used for anchoring was insufficient to withstand the wind loads, and some guy ropes may not have been tightened. In effect, these were not engineered or inspected for the anticipated high winds prevailing. In addition, it appears that fair officials had not established clear evacuation criteria; the guest host for the concert was on stage when the gust hit, even though severe thunderstorm warnings were in effect. Videos depict the aluminum truss roof rising like a sail, then leaning to one side before collapsing. Additionally, Thornton Tomasetti's forensic report determined that improper beam-to-column connections failed. After one side's connections failed, the structure failed completely.

## Aftermath and Lessons:

The stage collapse in Indiana resulted in sweeping changes to safety practices and regulations surrounding events in the State of Indiana. Shortly after the fair, the Indiana Fair Commission accepted full responsibility and significantly changed its emergency action plans. Subsequently, the state government enacted stricter regulatory requirements for outdoor stage elements, including engineered reviews and wind-resistance standards for temporary structures. Within the industry, significant changes occurred nationally, as the entertainment industry created the Event Safety Guide ("ANSI E1.21"), which addressed outdoor stage design, weather monitoring, and best practices following this tragedy. The biggest lesson learned here is that temporary staging structures should be treated as nearly identical to permanent structures for the sake of life and developed accordingly. You cannot assume that the weather will remain calm; reliable anchoring solutions, an appropriate wind action plan, and live monitoring of weather events are necessary. Unfortunately, it took this tragic event, later deemed "preventable," to illustrate the essential role of construction safety. The collapse in Indiana showed that ambiguity about whether to evacuate (whether the individual in authority has already exited) and loose construction standards for event staging could lead to significant injury or worse. Because both mistakes were pointed out, fairgoers and performers are now safer in Indiana, despite a significant human toll.

---

## *Collapse of a Parking Garage at Miami Dade College (Florida, 2012)*

Collapse of a Parking Garage at Miami Dade College (Florida, 2012)

### The events:

In October of 2012, a nearly completed five-story parking garage at the West Campus of Miami Dade College (Doral, FL) collapsed while under construction, leading to four fatalities and seven injuries. The precast concrete parking garage was only a few months away from opening for public use when, in an instant, a large portion of the building pancaked to the ground, trapping workers under the rubble. First responders worked for more than a week to recover victims from the entangled concrete slabs.

### Contributing Factors:

OSHA, along with engineers, discovered numerous errors in construction practices and design. The building had a precast concrete system – prefabricated columns and tee beams made offsite and assembled at the garage. The evidence produced showed that major weld connections failed, which would then point to a structural failure. One reported theory (as mentioned in local news at the time) was that a crane may have bumped into a column earlier in the erection sequence, causing a hidden cracking or misalignment of the member. There was also evidence that some columns had inadequate grout and reinforcement at the joints, which weakened the effective capacity of the column. OSHA issued serious violations against the general contractor and engineer for the garage project and found that supports had been removed too early. In particular, temporary shoring and bracing had been removed before the structure had independently established stability, which may have caused progressive failure. Investigators traced the failure to the northwest corner of the garage: one column failed, causing the flooring slabs above to shear away from their connections, which led to the failure of an entire wing.

### Aftermath and Lessons:

The college, the contractor (Ajax Building Corporation), and the precast supplier engaged in litigation that resulted in settlements in the millions. The college tore down the garage and rebuilt it from the ground up. In the professional aspect, lessons learned stressed the importance of adhering rigidly to erection plans and careful post-erection inspection of connections. One major lesson learned was the need for independent peer review of the structural design for large precast. Moreover, contractors learned to always err on the side of caution with temporary supports: never take shoring off too soon just to reduce agenda time. Miami Dade College is now said to have structural inspectors on the premises for each of these projects, to emphasize urgent safety and quality and, above all, to protect lives. The tragedy reiterated a lesson that became axiomatic in practice: construction sites must focus on safety and quality, as they would for site work that could be opened to public occupancy in the future, even if no such occupancy existed at the time. The lesson was ultimately that construction is a critical phase in building toward a creative vision, as much as design is, and that if quality is not in place, quality can change and a failure can occur, with lives being lost.

---

## *The Collapse of Building Demolition in Philadelphia Downtown (Pennsylvania, 2013)*

## What Happened:

On June 5, 2013, a four-story masonry building under controlled demolition at 22nd and Market Streets in Philadelphia, Pennsylvania, collapsed onto an adjacent occupied thrift store. The incident killed 6 people shopping in the adjacent Salvation Army store, and, in addition, injured 13 survivors. In the absence of conditions related to urban collapse, during the day, and on a city-resourced street during the busy business workweek, there was a structural design failure, as much as it was a failure of oversight and demolition practice. It became a famous example of negligence: a contractor with relatively little experience was carrying out demolition using unsafe work practices on public property, and a massive, unbraced party wall had a domino effect, collapsing onto the smaller neighboring store.

## Contributing Factors:

A grand jury and several investigations subsequently discovered numerous egregious failings in what led to the unfortunate loss of life. The property owner had contracted a low-bid demolition contractor who apparently lacked a license or training to carry out the high-risk takedown. Furthermore, the contractor appeared to be using labor with no experience or digging rights to demolish a building. Rather than deconstructing the building in a considerate manner, such as removing its heavy structures first—its roof and floor—the laborers stripped the building, leaving a tall, unsupported brick wall standing. This wall loomed menacingly high over the one-story thrift store. In fact, witnesses had given warnings just a few days earlier that the wall was bulging and in trouble. Emails from workers at an adjacent construction site confirmed the imminent risk and alerted the city and the Salvation Army, but no timely action was taken. On June 5, as the excavation was underway below the wall, the unsupported wall collapsed onto the thrift store. During the trial, an expert summarized the deconstruction, calling it "gross negligence" in its planning and execution. In addition, the contractor had laborers using machinery (an excavator) that shook the structure, yet did not close the adjacent shop or cordon off the area, despite the probable risk.

## Aftermath and lessons:

The consequences were enormous. The contractor was charged with involuntary manslaughter and sentenced to prison time, along with the excavator operator, who was on drugs. The building owner and the Salvation Army (for failing to communicate the warning) ended up in a civil trial, and the damages reached a record-breaking $227 million for victims in the Pennsylvania case. The city of Philadelphia dramatically revised its demolition permitting process, raising standards for contractors, increasing monitoring of demolition sites, and now requiring a structural or geotechnical engineering survey and bracing plan prior to demolishing a building adjacent to a populated area. The takeaway: don't assume safety when demolishing a building. Building owners are responsible for vetting their contractors. In this case, the owner "did minimal due diligence" and hired based on price—with tragic results. The tragedy taught cities that they must be proactive and inspect buildings during demolition, and act without hesitation when faced with anything deemed unsafe. The collapse also showed that neighbors of an unsafe site have some responsibility. The Salvation Army was found at least partially liable for maintaining operations at its store next door, despite clear warnings about unsafe conditions. The absence or poor planning, training, and communication in demolition can be just as destructive as detonating a bomb. A heavy lesson that led to policy revamps in Philadelphia and other cities.

---

## *Berkeley Apartment Balcony Collapse (California, 2015)*

Berkeley Apartment Balcony Collapse (California, 2015)

### What Happened:

In the early hours of June 16, 2015, a group of young adults celebrating a birthday gathered on a fourth-floor balcony of a new apartment building in Berkeley, CA. Without warning, the small cantilevered balcony detached from the side of the building and dropped 13 feet or more than 40 feet to the street below. Six (five Irish students and one Irish-American) died immediately; the other seven sustained serious injuries. The event received international attention (especially in Ireland) and had intense media scrutiny regarding both construction techniques in California and the events leading up to the collapse.

### Contributing Factors:

The collapse was attributed to significant dry rot in the wooden support for the balcony. The balcony was a cantilever design, meaning wooden beams (joists) extend out from the floor system to support the area. In merely 8 years after construction, the joist ends had been penetrated by water intrusion and were rotten to a fraction of their original strength. Investigators found that the balcony's weatherproofing was improperly done—"improper construction caused the dry rot," not the weight of the people at the party. Essentially, the balcony had become a sponge; layers of waterproof coating were missing or failing, allowing rain and moisture to penetrate the joist bay. Pictures obtained after the collapse showed the joist stubs sheared off and crumbling from rot. To illustrate the extreme conditions on the balcony, tenants in the unit below had previously reported mushrooms growing there—a sign of internal decay, but it seems not to have been a "red flag" for the building managers involved. Furthermore, it was found that the construction included shortcuts, including the use of untreated wood for outdoor exposure and failure to seal in accordance with the plans, which allowed the fungus to thrive behind the drywall until the timber gave way. When 13 people stepped out (well within design live load limits if the wood was sound), the rotten beams failed suddenly.

## Aftermath-Related Regulations:

Following the Berkeley balcony collapse, California introduced strict new regulations regarding balconies and elevated exterior structures on buildings. In 2018, state law was amended to require periodic inspections (every 6 years) of apartment balconies and decks by qualified individuals. Builders and landlords also had to account for water intrusion. This incident also caused civil litigation and a lawsuit against the construction company responsible for the project. It has been shown that the owner/developer previously resolved water-damage issues in other properties, suggesting they were openly cutting corners, resulting in multiple contractors losing their licenses and incurring monetary settlements. The emphasis is on waterproofing and inspection. A decay issue can go undetected for months/years and culminate in sudden, catastrophic failure; thus, building codes now require architects to detail, contractors to construct, and inspectors to verify effective weather barriers on balconies and other areas, and to verify the details on-site during construction. The tragedy illustrated that tenant concerns of leaks or mold should always be taken seriously. In the case of Berkeley, had they made the small intervention when they saw the first signs of rot (mushrooms and sagging balconies), they might have mitigated more serious repairs that could have saved the balcony and perhaps even lives. Ultimately, a roof collapse inevitably results from neglected maintenance, and an entirely minor issue—missing a waterproof membrane—can become a structural issue in modern structures.

## *Sioux Falls Copper Lounge Collapse (South Dakota, 2016)*

### What Happened:

On the 2nd of Dec., in downtown Sioux Falls, SD, a century-old brick building collapsed during renovation. The building was being developed into what was called a Copper Lounge, and it happened with no warning. The building collapsed with no observable signs or warning that it would. Unfortunately, one of the construction workers was killed (trapped in the rubble). A second person was rescued after spending hours buried under the debris. The collapse happened during normal working hours when there were other people in the building, in a very populated area, and could have potentially resulted in many more injuries.

### Contributing Factors:

Investigators stated the collapse happened after a critical load-bearing wall was improperly removed. The contractor was completing the ground-floor space and knocked out a structural brick wall to create an open floor plan, but had not installed adequate temporary supports or a replacement beam. Essentially, the contractor had removed a wall that supported the upper floors. Photos and reports taken just before the incident show a large opening in the first-floor wall that was only minimally shored up. The overall building was an old brick masonry build (1916), where walls were continuous for structural support. By removing a 20-foot section of wall all at once, the crew compromised the building's structural integrity. There were also indications that an overloaded construction dumpster placed against the foundation may have contributed to stress, and that vibrations from demolition work may have loosened the mortar. However, the primary cause was determined to be the "gross negligence" of removing a bearing wall without shoring, as indicated by OSHA and the city. The reports also indicated that the renovation had not been properly reviewed by an engineer and that the contractor had not previously worked on alterations or remodels of structural portions of buildings.

### Aftermath and Lessons:

The contractor was penalized, and the incident prompted Sioux Falls to tighten its permit process for structural changes. In renovation work, the event further embodied another age-old lesson: you must know what is holding the building up before removing it. An engineer's assessment is vital when making alterations to old buildings. The other miscalculation was the "misidentification" of the wall—it was treated as a non-structural wall, simply because it was in the middle of the room and did not appear to support any framing or structural elements. The incident reinforced the importance of incremental wall removal with appropriate shoring (either needle beams or temporary columns) until a permanent support is provided. Even though the event was considered smaller-scale, it was consistent with the fallacies exhibited in larger failures; it would have been avoided had standard practice been followed. The post-event report found that the root cause was a "grossly inadequate" level of support for removing the wall. The bigger takeaway from the event is that, regardless of size, contractors in small towns also have to follow basic structural principles; ignorance or cutting corners can literally bring the house down. Following this, Sioux Falls implemented a system to provide proof of shoring plans (and, in some cases, engineering inspections) for any major interior wall removals. The hope was to prevent events like this from happening in the future and, if possible, to save a life from the unexpected hazard posed by renovation in Sioux Falls.

---

## *Oroville Dam Spillway Failure (California, 2017)*

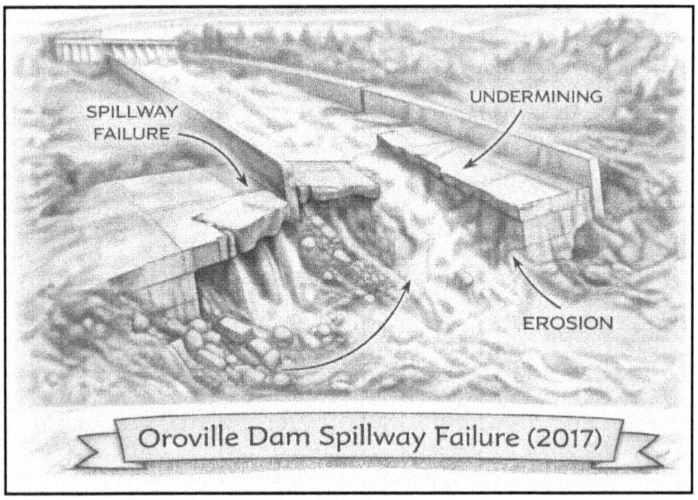

## What Happened:

In February 2017, after several days of heavy rain, California's Oroville Dam —the tallest dam in the U.S. at 770 feet— came close to disaster when its main concrete spillway cracked and collapsed, and the emergency spillway adjacent to it was also severely eroded. Though the dam itself did not fail, damage to the potential spillways prompted immediate evacuations for a population of nearly 188,000 due to fears of uncontrolled flooding. Rushing water carved a massive gash through the engineered hillside spillway, sending muddy rivers downstream and nearly overtopping it within hours.

## Contributing Factors:

Independent forensic teams found that the spillway failure was preceded by a long history of systemic issues. The 3,000-foot concrete chute had multiple design and construction flaws stemming from the original 1968 work: thin concrete sections, inadequate drainage beneath the slab, and poor repairs over the years. During the 2017 floods, water infiltrated a deteriorated section of the slab, causing soil erosion and ultimately a blowout of one of the concrete panels. After one panel failed, the high-velocity water scoured a massive gully, undermining more of the chute until most of it was destroyed. On the adjacent emergency spillway, essentially an unlined hillside, no water had ever flowed across that area since construction. Once water began flowing in 2017, the lack of appropriate armoring led to erosion that could have undercut the concrete weir. Another factor is deferred maintenance: vegetation and cracks on the main spillway, reported during inspections, were not adequately addressed. In addition, the dam operator (California DWR) underestimated the potential for failure modes; independent experts had warned in 2005 that the emergency spillway should be armored with concrete. That never occurred because of the cost and then complacency. The situation essentially came together of an aging, under-designed structure combined with years of inadequate maintenance.

## Aftermath and Lessons:

The Oroville crisis led to the reconstruction of both spillways at a cost of $1.1 billion and prompted a nationwide review of dam safety. The event presented some lessons related to dam safety: the main lesson is that for aging infrastructure, regular, rigorous inspections and proactive repairs are essential. Small crack conditions can, under stress, turn into massive washouts. The forensic report pointedly stated that "industry-wide complacency" existed because structures that had not failed in decades were assumed to be safe, leading to perfunctory inspections. Oroville definitely shattered the veneer that things do not fail. Dam operators are now beginning to implement improved inspection regimes, including advanced techniques like ground-penetrating radar to detect voids beneath the concrete. Another lesson is that systems of emergency systems should be truly robust. Relying on untested overflows (the emergency spillway) would have created a catastrophe; therefore, redundancy must be as robust as the primary systems. Oroville demonstrated that even if the main dam is in good condition, the failure of supporting components can also create an emergency. To summarize, the Oroville spillway collapse teaches us the need for addressing known deficiencies before an extreme event forces the issue, and it highlights that investing in maintenance and upgrades is really investing in public safety before a catastrophic failure results in greater losses.

---

## *Florida International University Pedestrian Bridge Collapse (Florida, 2018)*

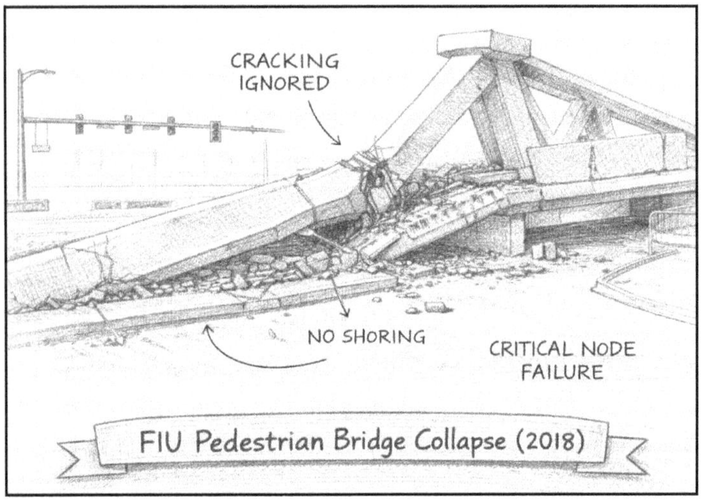


### What Happened:

On March 15, 2018, a pedestrian bridge that was newly opened to the public at Florida International University (FIU) in Miami collapsed onto the roadway below, crushing several cars. Six people were killed, and nine were injured as a result of the unexpected failure of the bridge. The bridge was 174 feet long, made of steel and concrete, with a truss design, and had been lowered into place only five days prior (when it was completed) using accelerated construction methods. At the time of the collapse, the bridge was still incomplete (the central support tower had not yet been installed), but it was open to workers, and traffic below was not closed. Surveillance video captured the event and shocked the world's engineering community, given that the construction had been completed just 5 days earlier.

### Contributing Factors:

The NTSB determined that the root cause of the collapse was a fatal design error. The connection at a node (Node 11/12) of the unique design by FIGG Engineers did not have sufficient capacity. The assumption of the capacity and load of a critical connection (diagonal-to-deck connection) to the floor deck of the bridge was flawed. The diagonal-to-deck connection failed due to cracking before it was put into service. Cracks had been observed at Nodes 11 and 12 just days before the collapse—extensive cracks. The cracks were 40 times the normal tolerance, indicating that the concrete was crushing. The design-build team did not recognize the cracks as a critical sign of failure. The peer-review firm (Louis Berger Group) did not catch the design error during its independent review of the plans. The day of the collapse, engineers were adjusting tension on the steel rods that were part of the bridge (which was going). The project manager's inaction to correct the error, coupled with the load calculations it performed, allowed the project to continue. As the workers continued tightening the rods, which placed additional stress on an already compromised node, the node failed completely, causing the concrete to crack and allowing the 950-ton span to fall. The National Transportation Safety Board (NTSB) found that FIGG made an error in its load calculations; therefore, elements 11 and 12 were grossly under-designed relative to what they should have been. The project-overseeing agency also failed, as the peer reviewer was neither qualified for nor given adequate time/budget to properly oversee this innovative design. FIU and Florida DOT were also criticized for failing to close the road after the first major crack was discovered.

## Aftermath and Lessons:

The FIU bridge collapse led to numerous lawsuits, for which settlements totaling more than $100 million were reached. Florida DOT revised its procedures and processes for approving innovative designs. There were so many lessons learned. First, designs must be checked and rechecked for correctness when they are not standardized. The NTSB noted that the FIGG design did not provide a backup (or redundant) load path, meaning that when a load-bearing member failed, the entire structure failed. This notion is fundamentally at odds with the "Fail-safe" philosophical consideration, which could have led to a different outcome if it had been considered. Second, peer review must be independent, and each peer review committee member must be independent. In this case, the peer review firm did not identify the calculation error, and the same firm was not adequately vetted by FDOT. Third, construction-stage safety: again, the member collapsed at the point just after post-tensioning adjustment to the span, which had been heavily compromised by loading. The current best practices in the country provide that access to the area be limited for safety, as this is the most critical stage of construction. In the aftermath of this risk, agencies are likely to be more conservative in design during active construction and when determining whether a road will be closed or opened, even with a less-compromised loading value. Ultimately, the FIU case serves as a powerful example of the unacceptable practice of simply disregarding evident signs of distress (e.g., large cracks). Several engineers associated with the project had observed the cracks but collectively decided they did not represent an immediate threat—a very serious and incorrect conclusion. As NTSB Chairman Robert Sumwalt stated, the "blame game" was warranted, "the oversight of the project collapsed much like the bridge itself." For that reason, the case has been utilized in engineering ethics courses ever since, as a reminder that we cannot simply ignore safety for schedules or budgets.

---

## *Hard Rock Hotel Construction Collapse (Louisiana, 2019)*

### What Happened:

On October 12, 2019, a portion of the Hard Rock Hotel under construction in downtown New Orleans collapsed without warning, sending the upper floors into the lower floors. Three workers were killed, and more than 18 were injured. The upper floors of the 18-story concrete and steel structure pancaked, with the streets below remaining partially upright and precarious, with horrific scenes of crumpled floors visible from the street. The collapse shut down streets for months in the busy French Quarter and complicated the recovery of the victims' remains, which would take more than 10 months.

### Contributing Factors:

OSHA's investigation determined that engineering design flaws in the building's support structure were the primary cause. Specifically, the structural engineer of record had failed to design and review adequate connections, namely the steel bolt connections for the floor spans. Reports indicated that several steel beams and girders either failed to meet minimum design criteria or were improperly built. Some witnesses (workers) reported seeing sagging and vibration in the weeks leading up to the collapse. One worker even recorded a video demonstrating the unpublished shoring condition, stressed the floors, and then raised an alarm on social media about the building's condition. The building's failure began on the 16th floor, where a transfer girder and deck failed, taking out the 17th and 18th floors with them.

Other factors that contributed were the poor sequence of the concrete; they had large pours in succession, which imposed very heavy loads on the concrete floor deck, which had not yet cured on the floors below. Essentially, portions of the building were on the verge of collapse even before substantial loads were applied, as they were not designed to carry the loads during construction.

The city's review also highlighted some lapses in inspections; it was found that a city building inspector falsely represented at least some inspection reports (representing that he checked the building when he did not). Three city inspectors were subsequently indicted (without ultimately being charged) for failing to ensure construction safety through proper verification. This also signals a lapse in the chain of enforcement.

## Aftermath and Lessons:

This partially collapsed hotel remained a haunting shell for over a year, ultimately being imploded in 2020 after lengthy court battles. It led the City of New Orleans to increase enforcement efforts, including the implementation of a whistleblower system. Workers can now anonymously report unsafe conditions in the building. It was documented that some workers had warned about unsafe conditions, but none of the warnings were acted on. The engineering firm was also cited by OSHA for serious violations and is in numerous lawsuits. Industry-wide, the Hard Rock collapse will be remembered as a lesson in shared responsibility: everyone involved has a role in safety (engineers, contractors, city officials), and when one person or party fails to fulfill their duties to uphold safety. Specifically, it stresses the need for peer evaluation of structural plans for unusual designs (such as long-span post-tension slabs in this example); a second pair of eyes may have identified the connection failure. The report also emphasizes the need for rigorous site inspections; had the owner promptly investigated sagging floors in the early stages of the construction, work may have been halted. New Orleans now requires more documentation of thorough inspections and is more cautious in approving high-rise construction. The Hard Rock collapse, in effect, taught that cutting corners in the design or inspections could literally bring a building down. Safety regulations and plan check-offs are written in blood, reminding us that a mistake in a calculation at each step can add up to tragedy.

# *Seattle Tower Crane Collapse (Seattle, Washington, 2019)*

Seattle Tower Crane Collapse (Seattle, Washington, 2019)

## What Happened:

A crane was dismantled on April 27, 2019, in Seattle's South Lake Union neighborhood in a crowded urban area. The tower crane's mast failed in high wind gusts, fell onto a new office tower for Google, and onto a crowded street, forcing the tied-back hinges to buckle. The crane fell, killing two ironworkers on the crane and two in cars below. Four others were injured. Seattle, a city nicknamed Crane City due to the number of construction cranes, was shocked by the incident.

## Contributing Factors:

A lengthy investigation by the Washington State Department of Labor & Industries demonstrated that the structural failure was caused by improper dismantling procedures. Specifically, the crew removed critical pins that the company failed to provide guidance on before the top mast could become unsafe at the prescribed wind speed. Typically, when deconstructing a tower crane, operatives will place a mobile crane to support each section of the tower crane before removing the steel pins that hold it. In this case, the three parties involved (the crane owner, the general contractor, and the disassembly subcontractor) did not follow the manufacturer's safety manual. The crew had removed almost all the pins securing the tower sections before successfully supporting the tower with a mobile crane. Once the personnel removed so many pins, the tall, free-standing tower and the relieved and unbraced section were dramatically weakened. Compounding the risks, it was a stormy day, with wind gusts over 45 mph. Upon removing the pins, a strong gust caused the pieces to buckle and descend catastrophically. The breaking and buckling of sections would not have been possible without the hasty actions of the personnel, who, at that moment, ignored the manufacturer's safety manual and protocols, thereby enabling speed in disassembly. The investigation found that the unsafe act of removing pins early was not an anomaly; it had been positively established that it had occurred in other locations and was cited as a shortcut that violated all safety norms. Other contributing factors should include the lack of a qualified and/or experienced supervisor proxy and failure to account for wind conditions during a teardown scenario.

### Aftermath and Lessons:

The state of Washington levied significant fines: the crane owner (Morrow) was fined for a "Willful Serious" violation for intent to approve removal of the pins; the general contractor (GLY) and sub-contractor were cited for several serious violations. In a broader context, the collapse has spurred changes to crane disassembly regulations; Seattle and other cities now require that the authority be called in when a crane is assembled or disassembled, and there has been a shift in focus to the manufacturer's method. Convince the reader that the paramount issue is that safety is more important than speed. The removal of as many as 50 pins at the outset was done to "likely aid in the disassembly," but one official exclaimed, "It is not OK to remove anything until the mobile crane is attached." The incident offered crane professionals and companies across the globe an immediate lesson: problems can occur when corners are cut with cranes. The situation would also reveal that weather must always be factored in; had the pins not been removed, wind alone would not have prevented the crane from failing. However, the best practice now is not to take a crane down in high wind, if at all possible. This incident put new emphasis not only on following the manufacturer's directions to the letter, but also on training and supervising all workers by an accredited individual. In total, after considering all related components, the Seattle crane collapse is a disturbing lesson and indication of the belief that one reckless transgression of policy related to construction can obliterate an entire project in a matter of seconds, and cause injury or death.

---

## *Merrimack Valley Gas Explosions and Fires (Massachusetts, 2018)*

## What Happened

On September 13, 2018, a series of gas explosions and fires erupted across three communities in the Merrimack Valley, Massachusetts. Within a short period, over 100 homes and structures were affected as natural gas systems became overpressurized.

Residents reported smelling gas, hearing explosions, and seeing fires ignite in basements, streets, and buildings almost simultaneously. One person was killed, dozens were injured, and thousands were evacuated from their homes.

The event was not localized. It spread rapidly across an entire distribution system, turning what should have been a controlled utility network into a widespread public safety emergency.

## Contributing Factors

The incident was triggered during a pipeline replacement project, where an old gas main was being taken out of service and replaced with a new one.

A critical failure occurred in the engineering and planning of the work. The system relied on pressure-sensing lines connected to the old pipeline to regulate gas pressure. When the old main was removed, those sensing lines were no longer properly connected, but this was not accounted for in the work plan.

As a result, regulators received incorrect pressure readings and responded by increasing gas pressure instead of controlling it. This allowed high-pressure gas to enter low-pressure residential systems, overwhelming appliances and piping inside homes.

The failure highlights a breakdown in management of change (MOC). The design did not fully account for how the system would behave once modified. There was no effective verification step to confirm that pressure controls would function correctly after the change.

Additionally, there was insufficient coordination between engineering, field crews, and system operators. A relatively routine infrastructure upgrade turned into a system-wide failure because no one validated the full system impact of the change.

## Aftermath and Lessons

The Merrimack Valley disaster led to major reforms in gas utility safety, engineering review processes, and project planning requirements. The utility involved faced significant penalties and long-term oversight.

The key lesson is that small system changes can create large system consequences.

Infrastructure systems must be treated as interconnected, not isolated. Removing or modifying one component without understanding its relationship to the entire system introduces hidden risk.

This case also reinforces the importance of engineering verification and field validation. Plans must be tested against real conditions, not just assumed to work based on design intent.

Ultimately, Merrimack Valley demonstrates that failures are not always caused by extreme events. Sometimes they are the result of routine work performed without a complete understanding of how the system will respond.

---

# *Husky Superior Refinery Explosion and Fire (Wisconsin, 2018)*

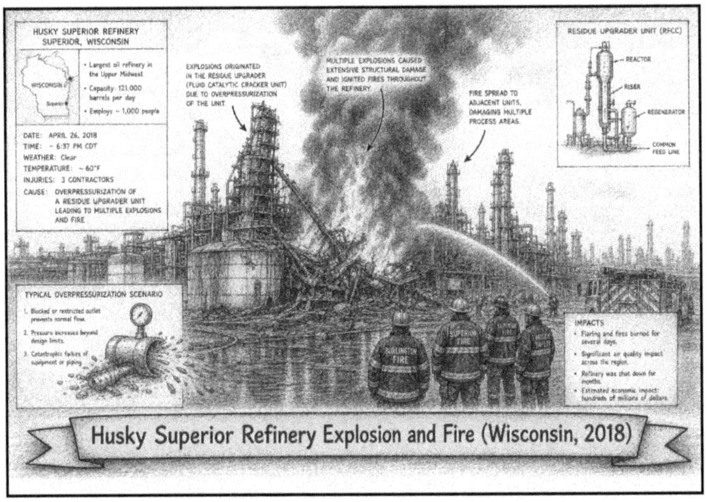


## What Happened

On April 26, 2018, an explosion and fire occurred at the Husky Energy refinery in Superior, Wisconsin, during a planned shutdown of the facility. The incident originated in the fluid catalytic cracking (FCC) unit, a critical process used to convert heavy hydrocarbons into lighter products.

As the unit was being shut down for maintenance, a sudden loss of containment occurred, releasing a large volume of flammable vapor. The vapor ignited, triggering an explosion that sent debris across the facility and ignited multiple fires.

Thirty-six workers were injured, and the incident prompted emergency concerns for the surrounding community due to the proximity of tanks containing hazardous materials, including hydrofluoric acid. The situation had the potential to escalate into a much larger public disaster.

## Contributing Factors

The investigation found that the explosion occurred during a non-routine operation, specifically a shutdown phase, where system conditions are less stable and more vulnerable to error.

Air entered the equipment that still contained flammable hydrocarbons, creating an explosive mixture. This condition was not adequately anticipated or controlled in the shutdown procedure. The safeguards in place were insufficient to prevent ignition once the mixture formed.

The event exposed weaknesses in process hazard analysis and planning for abnormal operations. While steady-state operations may have been well understood, the transition phase introduced risks that were not fully evaluated.

Additionally, the layout of the refinery contributed to the potential severity. Critical units and hazardous materials were located in proximity, meaning that a localized failure could quickly escalate into a broader event.

The case reflects a common pattern: systems are often designed and managed around normal conditions, while transitional states like startup and shutdown receive less attention, despite carrying equal or greater risk.

## Aftermath and Lessons Learned

The Husky incident led to increased focus on hazard analysis for shutdown and maintenance operations, as well as emergency planning for worst-case escalation scenarios.

The key lesson is that non-routine operations are high-risk operations.

Shutdowns, startups, and maintenance activities change the state of a system, removing layers of control and introducing new variables. These phases must be treated with heightened discipline, not reduced oversight.

This case also reinforces the importance of anticipating how systems behave when they are partially taken apart or reconfigured. Risk does not disappear when production stops; it often increases.

Ultimately, Husky demonstrates that many catastrophic events occur not during peak operation, but during the moments when systems are in transition and assumptions go unchallenged.

## *Fern Hollow Bridge Collapse (Pennsylvania, 2022)*

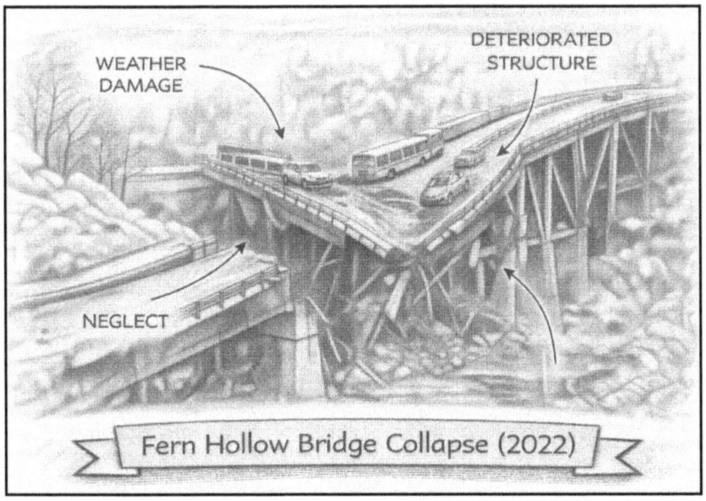


### What Happened:

The Fern Hollow Bridge collapsed suddenly in Pittsburgh in the early morning hours of January 28, 2022. The bridge was the main thoroughfare through Frick Park. The 447-foot-long bridge dropped over 100 feet into a ravine. A city bus, along with several cars, was on the structure at the time and was destroyed in the wreckage. Fortunately, no one was killed, since it occurred shortly before the start of rush hour, but approximately 10 injuries occurred. The collapse occurred just hours before President Biden was scheduled to visit Pittsburgh to discuss infrastructure, clearly highlighting a significant issue.

### Contributing Factors:

The NTSB's final report identified the cause as "long-term corrosion to a critical steel member" and deferred maintenance over the years. Fern Hollow was a K-frame, rigid-steel girder bridge constructed in 1970. The "transverse tie plate" on this structure was critical for tensioning the two frame legs. This tie plate, located on the southwest leg, was corroded and had lost much of its thickness over the years of rusting. The inspectors also reported severe drainage on the bridge, alleging that storm drains were clogged, allowing water and road salt to overflow onto the bridge legs for years and trapping debris and moisture at the base. The weathering steel, designed to form a rust patina, never had a chance to dry, and as a result, it just rusted through. Most concerningly, this tie plate was non-redundant and fracture-critical; once it cracked, the entire bridge would collapse, as occurred. The NTSB chastised the City of Pittsburgh for ignoring several prior inspections that had reported increasing corrosion and recommended repairs. The inspections noted rust and section loss over the years, and there had been no significant repairs. Furthermore, the NTSB documented that the inspections themselves were poor: some inspectors missed or downplayed corrosion, potentially because mud and debris obscured the bottoms of the connections, and the bridge's load rating calculations were outdated or inaccurate and did not account for corrosion that diminished the bridge's capacity. The Pennsylvania DOT (PennDOT), which supervised inspections of local bridges, also received notice of inadequate oversight.

### Aftermath and Lessons:

Fern Hollow Bridge's collapse prompted immediate actions to alert bridges across the nation to re-inspect for corrosion as quickly as possible, especially those with similar models. The bridge was promptly replaced within a year with federal infrastructure funds, but it served as a stark reminder of the essential maintenance lesson: small problems can fester silently until the bridge collapses. The small costs of periodically cleaning drains and painting or replacing rusted elements are minuscule compared to the costs of a collapse. The NTSB identified preferred changes, such as improved inspector training and not concealing corrosion from owners, who must promptly address deterioration to prevent fractures in critical elements. Also, it filled the need for oversight; PennDOT now audits municipal bridge inspection reports more closely. The collapse, which miraculously didn't result in death, proved that "state of good repair" was more than a catchphrase; being in a state of good repair is an essential safety requirement. The Fern Hollow Bridge collapse indicated that 50-year-old bridges in parks are prone to collapse if nothing is done. This incident will be remembered, likely in ways that support rebuilding and improving bridge asset management by enabling cities to address deferred maintenance and prioritize repairs before problems fail.

---

## *Davenport Apartment Building Collapse (Iowa, 2023)*

## What Happened:

On May 28, 2023, a portion of a six-story apartment building located at 324 Main Street, Davenport, Iowa, experienced a sudden partial collapse. The complete brick façade and structural component of a floor on the rear west side dislodged and pancaked downward. This collapse resulted in the deaths of three residents (whose bodies were subsequently retrieved from the debris some days later) and caused the displacement of dozens of low-income tenants. This catastrophe in a Midwestern downtown area was especially jarring because residents had previously expressed concerns and complaints about the building's condition, yet it was not cleared of residents until after the structure collapsed.

## Contributing Factors:

A thorough forensic examination showed that the collapse of this building was completely avoidable, and that the consequences of the owner's and his contractors' actions or inaction caused this incident. The building in question was a 1907 brick-masonry structure. In the weeks prior to the collapse, an engineering firm (hired by the owner) examined the building and recommended repairing the west exterior brick wall after finding it bulging. A masonry contractor was engaged in the replacement of what the contractor believed was a decorative brick "veneer." They were significantly wrong in assessing the wall's composition; it was neither cosmetic nor a veneer, but a load-bearing wythe intended to tie and support the floor joists above. The contractors removed portions of the wall without providing proper vertical shoring or temporary lateral support to carry the loads from the floors above. In essence, they were removing the wall that was supporting the building. The investigation report identified the "grossly insufficient" shoring as the cause of the building collapse. Had the wall been properly shored to support itself and the subsequent floor loads during repairs, it would not have fallen. In addition to the lack of safety supervision, the engineering oversight was extremely poor; not only did the engineer's work include numerous significant errors and incorrect assumptions about the building's framework, but they also overlooked warning signs of significant structural distress (i.e., progressive widening of cracks and sagging floors) and allowed work to continue as if no issues existed. Long-term water infiltration had compromised and deteriorated structural elements. The owner had neglected the building's exterior, resulting in rotting beams and crumbling mortar. The photos of the conditions inside the building, including evidence of substantial water damage (peeling paint, mold), indicated a significant compromise in integrity. The city also shared blame; while code inspectors had been on site (in response to bricks that fell from the front façade), they did not have repair plans ready for printing and review to determine whether the work complied with city codes or engineering best practices.

## Aftermath and Lessons:

No criminal charges have been brought against anyone, but lawsuits are pending against the owner and engineers. The City of Davenport commissioned an independent report to evaluate the incident and learn lessons from the failure. The first lesson is the extreme importance of having structural engineering expertise when doing rehab work; neither the owner's engineer nor any of the contractors had any idea what they were doing except for the owner's directions to remove bricks, and that ignorance would prove deadly. It is now a lesson learned that, with old buildings, an engineer must assume walls are load-bearing unless proven otherwise, and must convey that understanding to all parties involved. The next lesson learned might reinforce the idea that even a "small construction job" involves shoring and safety planning. Moving bricks always feels simple, but, as stated in the report, not even shoring was done to the minimal industry standard. The next lesson is regarding code enforcement on the city's part. Cities, for older buildings undergoing structural repairs, should require detailed engineering repair plans (which would help code inspectors avoid oversight gaps). Ultimately, the Davenport collapse underscores the human implications of ignoring chilling complaints, despite numerous reports from tenants that they were worried about the noise of creaking and their walls cracking; they were sadly right. Moving forward, in addition to several other cities, Davenport is discussing how to respond more obnoxiously to tenants' reports of structural concerns (such as requiring third-party structural inspections or, certainly, evacuation). In other words, the Davenport collapse illustrates very simply that primary errors, misunderstanding a wall's significance and failing to shore it, can take down a building. With adequate knowledge and caution, these failures can be easily avoided.

---

## *Surfside Champlain Towers North Partial Failure Warning Signs (Florida, 2018–2021) – Precursor System Failure Case*

## What Happened

Prior to the catastrophic collapse of Champlain Towers South in 2021, a nearly identical neighboring structure, Champlain Towers North, and the South tower itself exhibited years of visible structural distress and warning signs.

Beginning as early as 2018, engineering reports identified significant deterioration in the pool deck slab, parking garage, and structural elements. Observations included cracking, spalling concrete, exposed reinforcing steel, and ongoing water intrusion.

Residents reported worsening conditions over time: shifting surfaces, standing water, and visible degradation. Despite these warnings, repairs were delayed, debated, and only partially implemented. The structure remained occupied throughout this period.

While this case is tied to the ultimate Surfside collapse already included in your Chapter 2, this focuses specifically on the system failure before the event, where intervention was still possible.

## Contributing Factors

The primary issue was not a lack of information, but a failure to act on it.

Engineering assessments clearly identified structural deterioration and risk, particularly in the pool deck slab, which played a critical role in load transfer. However, the severity of these findings was not fully translated into urgent action.

A major contributing factor was governance and decision-making breakdown. The condominium association faced financial constraints, disagreements among residents, and delays in approving major repair work. The estimated cost of repairs led to hesitation, even as conditions worsened.

There was also a failure in risk communication and escalation. While engineers documented the issues, the urgency of the situation was not consistently conveyed or enforced in a way that triggered immediate structural intervention or evacuation.

Additionally, the system lacked regulatory enforcement mechanisms that could compel timely action. Inspections identified problems, but there was no structured requirement ensuring that critical repairs were completed within a defined timeframe.

Over time, these delays allowed deterioration to progress from a manageable condition into a critical structural vulnerability.

## Aftermath and Lessons Learned

This case became central to understanding that catastrophic failures are often preceded by visible, documented warning signs.

The key lesson is that information without action is failure.

Engineering reports, inspections, and resident complaints only have value if they lead to decisive intervention. When financial, organizational, or political factors override technical judgment, risk accumulates.

This case also highlights the need for clear accountability in ownership structures. When responsibility is distributed across boards, residents, engineers, and regulators, critical decisions can stall.

From a construction and management perspective, it reinforces the importance of treating deterioration as a time-sensitive condition, not a maintenance inconvenience.

Ultimately, this case shows that the most preventable failures are often the ones where the problem was already known, but action came too late.

---

# *Francis Scott Key Bridge Collapse (Maryland, 2024)*

## What Happened

In the early morning hours of March 26, 2024, the containership Dali lost electrical power while departing the Port of Baltimore. Without propulsion or steering control, the vessel drifted off course and struck a primary support pier of the Francis Scott Key Bridge.

Within seconds of impact, a large portion of the bridge collapsed into the Patapsco River. Several construction workers performing roadway work on the bridge at the time fell from the structure. Six workers were killed.

The collapse shut down a major transportation corridor and one of the busiest ports on the East Coast, creating immediate economic and logistical disruption far beyond the site itself.

## Contributing Factors

Preliminary investigations identified that the vessel experienced a total loss of power, including propulsion and steering, shortly before impact. A cascading electrical failure left the crew unable to control the ship's movement in a constrained navigation channel.

However, the event was not solely a vessel issue.

The bridge itself, while structurally sound under normal conditions, had limited protection against large vessel collision forces. The pier struck by the ship was not designed for the scale and mass of modern container vessels, which have increased significantly since many bridges were originally designed.

This highlights a critical gap in infrastructure risk modeling. The bridge met design expectations for its era, but evolving external conditions, such as vessel size and traffic volume, had outpaced those assumptions.

There was also a limited opportunity for mitigation once the failure sequence began. While the crew issued a distress call, the time between power loss and impact was extremely short, leaving minimal margin for intervention.

## Aftermath and Lessons Learned

The collapse prompted immediate national attention on bridge vulnerability to vessel strikes and the need for updated risk assessments across similar infrastructure.

The key lesson is that infrastructure must be evaluated against current realities, not original design assumptions.

As systems around infrastructure evolve, whether through increased traffic, larger equipment, or changing environmental conditions, the structures themselves must be reassessed for new risk exposures.

This case also reinforces the importance of redundancy and protective systems for critical infrastructure. Single-point impacts should not lead to disproportionate collapse where feasible.

From a construction and management perspective, it highlights the need to plan for low-frequency, high-consequence events, even when they fall outside normal operating conditions.

Ultimately, the Key Bridge collapse demonstrates that failure can occur not because a system was poorly built, but because the world around it changed and the system was never updated to match it.

---

# *San Francisco Millennium Tower Settlement Crisis (California, 2016–Present)*

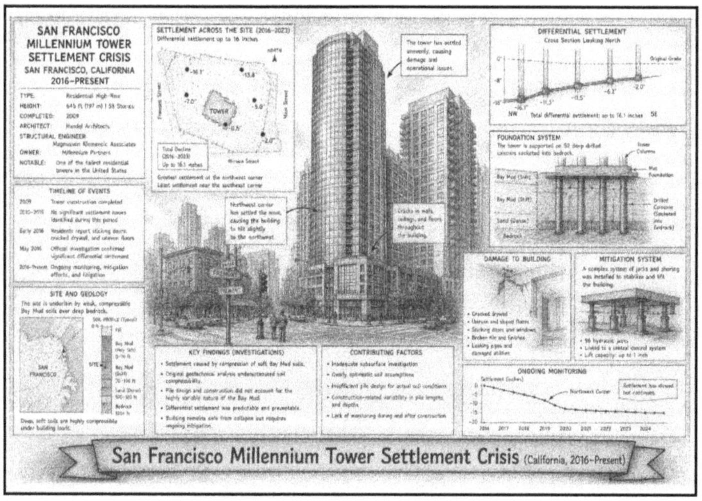


## What Happened

The Millennium Tower, a 58-story luxury residential skyscraper in downtown San Francisco, opened in 2009 and was initially considered a premier high-rise development. However, by 2016, it became widely known that the building had sunk over 16 inches and developed a noticeable tilt.

Residents began reporting issues such as cracked walls, sloping floors, and doors and windows that would not properly close. Engineering assessments confirmed that the tower was both settling and leaning, raising serious concerns about long-term structural performance and public confidence.

While the building did not collapse, the situation became one of the most high-profile structural performance failures in modern U.S. construction, with significant financial, legal, and reputational consequences.

## Contributing Factors

The primary issue stemmed from the building's foundation design. The tower was constructed on a deep concrete mat foundation supported by friction piles driven into dense sand layers, rather than bedrock.

At the time of design, this approach was considered acceptable. However, it relied on assumptions about soil behavior and long-term settlement that proved overly optimistic.

Over time, the compressible soil beneath the structure continued to consolidate, causing uneven settlement and tilt.

Compounding the issue was the interaction with nearby construction. Excavation and dewatering associated with adjacent high-rise projects may have altered groundwater conditions, accelerating settlement.

The case also exposed gaps in the design–construction interface and long-term performance evaluation. While the building met code at the time of construction, there was limited consideration of how surrounding development and environmental changes could impact performance over decades.

Additionally, early signs of settlement were not fully escalated as a systemic risk. The issue evolved gradually, making it easier to normalize until it became publicly visible and financially significant.

## Aftermath and Lessons Learned

The Millennium Tower led to extensive litigation, costly retrofit efforts, and ongoing monitoring programs. A major foundation retrofit involving new piles driven to bedrock was initiated to stabilize the structure.

The key lesson is that code compliance does not guarantee long-term performance.

Designs must account not only for initial conditions, but for how systems behave over time, especially in complex urban environments where adjacent construction can influence outcomes.

This case also highlights the importance of geotechnical uncertainty and conservative assumptions. Soil behavior is inherently variable, and small miscalculations can translate into large structural consequences.

From a management perspective, it reinforces the need for early recognition and escalation of performance issues. Gradual problems can be just as serious as sudden failures if left unaddressed.

Ultimately, the Millennium Tower demonstrates that failure is not always immediate or catastrophic. Sometimes it appears slowly, through movement, misalignment, and loss of confidence, until correction becomes far more difficult and expensive than prevention.

# *Lower Manhattan Parking Garage Collapse (New York, 2023)*

## What happened:

On April 18, 2023, a heavily used parking garage, 57 Ann Street in lower Manhattan, partially collapsed (pancaking another floor down on top of another), killing the garage manager and injuring five others who worked there. In this incident, many cars were damaged after concrete slabs and ramps were installed on top of this century-old building. This event raised alarm bells not only within the municipality of New York but also across other local governments with aging parking garages, allowing city inspections to proceed for every parking garage in the city.

## Contributing Factors:

The investigation was a joint relief effort over two years by NYC's Buildings Department and the District Attorney's office regarding the collapse and concluded, "It was the deadly combination of preventable errors, neglect of a load-bearing structure, and unpermitted work." In short, the collapse was linked to demolition activity that had begun on the heavily constructed column that was not supported by an architect. Private engineers who had visited the building had noted heavily cracked and spalling in one of the garage's load-bearing brick piers (columns). Rather than appropriately shoring and subsequently repairing or reconstructing this pier, the garage proprietor employed an unqualified handyman crew to unsupervised remove and replace bricks in that pier, without permits and without understanding the risks involved. The crew assumed the pier contained a steel column for support, but, as the expert confirmed, "there was no steel inside, the brick itself was load bearing." As the crew removed the brick, they inadvertently stripped the column of its load-bearing capacity. On the day of the collapse, the crew had removed so much brick that "very little was left to support that whole area of the floor." As the workers and others continued to remove the brick, a parked SUV was driven near the pier but never contacted it. The pier collapsed when the crew had removed the brick and continued to exert a dynamic load on the brick pier. The video evidence suggested that driving the SUV near the pier increased the already bowed brick pier's capacity. The investigation also revealed decades of deferred maintenance in the garage, as substantial concrete degradation and corrosion had never been repaired. Inspectors discovered concrete slabs with exposed, rusted rebar and significant evidence of structural overload (although the Commissioner acknowledged that the number of parked cars at the time of the eventual collapse was not beyond the building code). Ultimately, the garage collapsed due to the unsupervised removal of the brick pier (which constituted catastrophic negligence). The extent of liability for the structural failure is yet to be determined and is not confined to the negligent pier hire, as the garage had decades of deferred maintenance, adding another layer of liability.

## Aftermath and Lessons:

The City of New York immediately launched a "sweeping review" of similar garages. Building codes were fundamentally amended and modified so that many new and existing garages would, at a minimum, have periodic inspections (in the 80 years that had recently been planned to begin, but were now expedited). The main lessons are that structural repairs must be carried out with appropriate engineering review and permits. In the incident in question, unpermitted, unsafe work was performed, resulting in the collapse. Subsequent attempts to fix this critical element via a DIY process are a recipe for disaster. Another learning realization here is that assuming that hidden structural elements exist can lead to disastrous results. The work crew assumed that a steel column was within the pier; it turns out a column was missing, necessitating a probing investigation or, at a minimum, a review of structural drawings before altering the structural members. The collapse also confirmed the need for maintenance; the deterioration of the structural elements should never have gotten to the point of requiring this type of intervention. Moving forward, New York City's new inspection program (similar to the façade inspection program) will detect spalling concrete and overstressed structural framing members before failure. Unfortunately, the events leading to the loss of the manager's life are a reminder of how neglect and quick fixes can lead to death. The Department of Buildings Commissioner said it best, "These were human errors. And these were preventable." There may be no criminal charges, but liability is significant. In conclusion, the Manhattan garage collapse teaches building owners and contractors never to remove structural material without shoring, to always work from a signed-off repair plan, and to take corrosion and cracks seriously, as these conditions can bring an entire building down in seconds.

# *BP Texas City Refinery Explosion (Texas, 2005)*

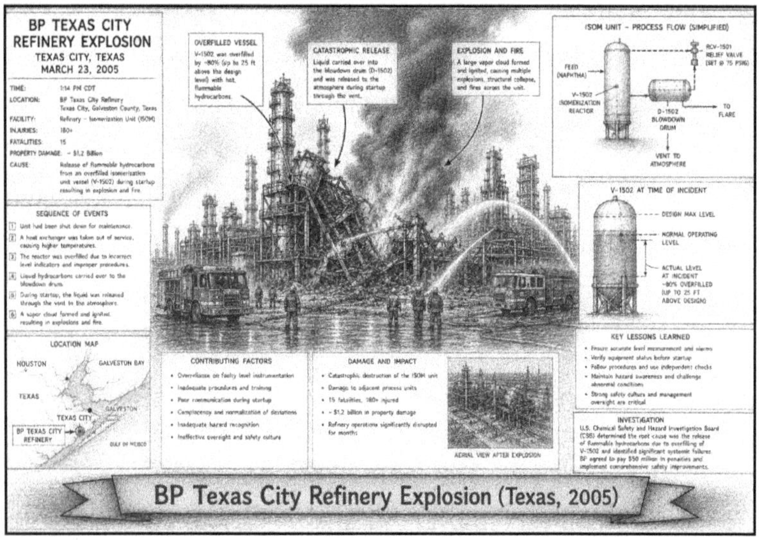


## What Happened

On March 23, 2005, the BP refinery in Texas City, one of the largest oil refineries in the United States, experienced a catastrophic explosion during the startup of an isomerization unit. The event occurred in the early afternoon as operators attempted to bring a key processing unit back online after maintenance.

During startup, hydrocarbons were introduced into a distillation tower that became severely overfilled. Liquid hydrocarbons overflowed into a blowdown drum and stack system, an outdated design that vented directly to the atmosphere rather than routing to a closed flare system. As the system flooded, a geyser-like release of flammable liquid and vapor discharged into the surrounding unit area.

Within minutes, a large vapor cloud formed at ground level and drifted through the congested refinery environment. Nearby ignition sources, including running vehicles and operating equipment, were present. The vapor cloud ignited, triggering a massive explosion followed by multiple secondary explosions and fires.

The blast was powerful enough to destroy nearby temporary office trailers, where many workers had gathered for meetings. Fifteen people were killed, and approximately 180 others were injured. The explosion caused widespread structural damage across the refinery and shut down significant portions of operations.

What made the incident particularly devastating was not just the scale of the explosion, but how many people were positioned directly in harm's way at the time of the release.

## Contributing Factors

The investigation revealed that the disaster was not the result of a single failure, but a layered breakdown across design, operations, and organizational systems.

At the core was the use of an outdated blowdown drum system that vented flammable hydrocarbons directly into the atmosphere. Modern refineries had long since transitioned to closed flare systems specifically to prevent this exact type of release. The continued use of this system represented a known hazard that had not been corrected.

Instrumentation and process control failures compounded the risk. The tower level indicator provided misleading information, causing operators to believe the system was within safe limits when it was actually overfilled. High-level alarms either failed or were ignored, and there was no effective independent verification of the actual liquid level inside the tower.

The startup procedure itself was poorly designed and not followed with sufficient rigor. Critical steps were either unclear or bypassed, and the operation lacked a structured, verified checklist process. Startup is one of the highest-risk phases in any industrial system, yet it was treated with routine execution rather than heightened control.

Another major factor was facility siting and workforce exposure. Temporary trailers had been placed dangerously close to the process unit. These trailers were occupied at the time of the explosion and absorbed much of the blast force, resulting in the majority of fatalities. This reflects a failure to evaluate risk zones and protect personnel from known hazards.

Underlying all of these issues was a deeper organizational problem. Investigations identified a culture that emphasized production and cost control over process safety. Previous near-misses and warning signs had occurred at the facility but were not treated as indicators of systemic risk. Maintenance was deferred, upgrades were delayed, and known hazards were tolerated.

In effect, the system had normalized unsafe conditions. What should have been seen as warning signals were instead treated as part of normal operations.

## Aftermath and Lessons Learned

The BP Texas City explosion became a defining case study in process safety and organizational failure. It led to multiple investigations, including a major report by the U.S. Chemical Safety Board, and resulted in significant financial penalties, lawsuits, and long-term reputational damage to BP.

More importantly, it shifted how the industry understands safety.

One of the most critical lessons was the distinction between personal safety and process safety. The refinery had strong metrics related to slips, trips, and minor injuries, which created a false sense of overall safety performance. Meanwhile, catastrophic process risks were poorly managed. This case demonstrated that low incident rates do not mean low risk.

The event also reinforced the importance of engineering controls over administrative controls. Reliance on procedures, alarms, and operator judgment cannot compensate for fundamentally unsafe system design. The presence of an atmospheric blowdown system in a hydrocarbon process environment was itself an unacceptable risk.

Another key takeaway is the importance of management of change and hazard recognition during non-routine operations. Startup, shutdown, and maintenance conditions introduce instability into otherwise controlled systems. These phases require increased oversight, not reduced discipline.

Finally, the explosion highlighted the consequences of organizational culture drift. When cost, schedule, and production begin to outweigh safety in decision-making, risk accumulates silently. The failure was not just technical, it was cultural.

The BP Texas City disaster stands as a clear reminder that catastrophic events are rarely surprises. They are the result of known hazards, accepted over time, until the system reaches a breaking point.

---

## *San Bruno Pipeline Explosion (California, 2010)*

San Bruno Pipeline Explosion (2010)

### What Happened

On September 9, 2010, a major natural gas transmission pipeline owned by Pacific Gas and Electric (PG&E) ruptured beneath a residential neighborhood in San Bruno, California. The rupture created a massive crater, releasing high-pressure natural gas that immediately ignited.

The resulting fireball engulfed homes and streets, producing flames over 1,000 feet high. Entire blocks were destroyed or severely damaged. Eight people were killed, dozens were injured, and hundreds of residents were displaced.

Emergency responders initially struggled to control the fire because gas continued feeding the blaze for over an hour before it could be isolated and shut off. What began as an underground infrastructure failure quickly became a large-scale urban disaster.

## Contributing Factors

The National Transportation Safety Board (NTSB) determined that the pipeline segment that failed contained defective seam welds dating back to its original installation. Over time, these defects weakened the pipe until it could no longer withstand normal operating pressure.

However, the physical defect alone was not the full story.

PG&E's asset management and recordkeeping systems were severely deficient. The company did not have accurate records of the pipeline's construction, materials, or weld type. As a result, the line was misclassified in risk assessments and not subjected to appropriate integrity testing.

Inspection and maintenance practices were also inadequate. The pipeline had not been properly evaluated for known risks such as seam failure, and pressure levels were not conservatively managed relative to the unknown condition of the pipe.

Control room response further compounded the event. Operators took too long to identify and isolate the rupture, allowing gas to continue fueling the fire well after the initial explosion.

At a systemic level, the failure reflected fragmented oversight and a false confidence in incomplete data. The organization believed it understood the condition of its system, but in reality, critical information was missing or incorrect.

## Aftermath and Lessons Learned

The San Bruno disaster led to sweeping regulatory changes in pipeline safety, recordkeeping, and integrity management. PG&E faced significant fines, legal action, and long-term oversight.

The central lesson is clear: you cannot manage what you do not truly understand.

Accurate records, verified data, and conservative assumptions are essential when dealing with aging infrastructure. When information gaps exist, risk increases, not decreases.

The case also highlights the importance of rapid emergency response and system isolation. In high-energy systems like gas transmission, the duration of release directly affects the severity of consequences.

Ultimately, San Bruno demonstrates that infrastructure failures are often not sudden. They are the result of long-term information gaps, deferred verification, and misplaced confidence in incomplete systems.

---

## *Silver Spring Apartment Explosion and Fire (Maryland, 2016)*

Silver Spring Explosion and Apartment Fire (Maryland, 2016)

### What Happened

On August 10, 2016, a massive explosion and fire occurred at the Flower Branch Apartments in Silver Spring, Maryland. The blast originated in a garden-style apartment building and quickly escalated into a multi-structure fire.

The explosion destroyed a large portion of the building and ignited a fast-moving fire that spread through adjacent units. Firefighters faced intense conditions as the structure partially collapsed during response efforts. Seven people were killed, dozens were injured, and hundreds were displaced. The event unfolded in a fully occupied residential setting, turning a utility failure into a large-scale life safety disaster within minutes.

### Contributing Factors

The National Transportation Safety Board (NTSB) determined that the explosion was caused by natural gas leaking into the building from a damaged service line. A critical issue was the failure of a gas regulator vent line, which had become disconnected and allowed gas to escape directly into the building structure.

Over time, gas accumulated in enclosed spaces until it reached an explosive concentration. Once ignited, the confined environment amplified the blast and resulting fire.

A major contributing factor was the lack of protection and inspection of gas infrastructure at the building interface. The vent line was vulnerable to damage and not adequately safeguarded from external forces such as ground movement or nearby work.

Additionally, there were gaps in maintenance, inspection practices, and system awareness. The condition of the gas system was not fully understood, and early warning signs were either absent or not recognized.

The incident also exposed how quickly conditions can escalate when gas enters an occupied structure. Once inside, even a small leak can become catastrophic if not immediately detected and controlled.

## Aftermath and Lessons Learned

The Silver Spring explosion led to changes in inspection requirements and safety practices for gas service regulators and vent lines, particularly in residential environments.

The key lesson is that system interfaces are high-risk zones. The transition point between utility infrastructure and occupied buildings often lacks the same level of scrutiny as major systems, yet it carries significant consequence.

This case also reinforces the importance of preventing gas intrusion into structures at all costs. Once gas enters a confined space, the margin for error disappears.

From a construction and management perspective, it highlights the need to treat utility coordination, protection, and inspection as critical components of life safety, not secondary considerations.

Ultimately, Silver Spring demonstrates that catastrophic events can originate from small, hidden failures at the edges of a system, where responsibility is often unclear and risks are underestimated.

# *Philadelphia Energy Solutions Refinery Fire and Explosions Pennsylvania, 2019*

## What Happened

On June 21, 2019, a major fire and series of explosions occurred at the Philadelphia Energy Solutions (PES) refinery, one of the largest refining complexes on the East Coast. The incident originated in the alkylation unit, a highly hazardous processing area that uses volatile chemicals under pressure.

In the early morning hours, a sudden release of flammable hydrocarbons ignited, triggering multiple explosions. One of the explosions was powerful enough to launch a large vessel fragment across the Schuylkill River. Fires spread rapidly through the unit, creating a large-scale industrial emergency.

While only a small number of workers were injured, the event caused extensive damage and ultimately led to the permanent shutdown of the refinery.

## Contributing Factors

Investigators determined that the incident was caused by the rupture of a corroded pipe elbow within the alkylation unit. Over time, the pipe had thinned significantly due to corrosion mechanisms that were not adequately detected or addressed.

The failure revealed critical weaknesses in mechanical integrity and inspection programs. Although the refinery had inspection processes in place, they did not effectively identify the severity of the corrosion or prioritize the component for replacement.

Additionally, the unit involved hydrofluoric acid, an extremely hazardous chemical. While safeguards existed, the presence of such materials increased the potential consequences of any failure. The incident demonstrated how a single point of material degradation, when combined with high-energy process conditions, can escalate rapidly.

At a broader level, the case highlighted issues with aging infrastructure, deferred maintenance, and risk prioritization. The degraded component was not an obvious or visible failure point, which made it easier to overlook until it reached a critical state.

## Aftermath and Lessons

The PES incident led to increased scrutiny of refinery inspection practices, particularly around corrosion management and aging equipment. The refinery itself was permanently closed, marking a significant economic and operational consequence.

The key lesson is that catastrophic failures often begin with small, unseen degradation.

Mechanical integrity programs must go beyond routine inspection and focus on identifying the most vulnerable components in a system. When dealing with high-energy processes, even minor material loss can become a critical failure point.

This case also reinforces the importance of prioritizing maintenance based on risk, not convenience or visibility. The most dangerous components are often the least obvious.

Ultimately, the PES refinery explosion demonstrates that when deterioration is allowed to progress unchecked, failure is not a possibility, it is a timeline.

## Conclusion

Out of all of the cases, nearly all of these failures had a precursor, some warning sign or mistake that was ignored until it was too late. From the advancing cracks in the Surfside condo and the insufficient gusset plate in the Minneapolis bridge, to the non-resorted demolition in Philadelphia and the rush to take down a crane in Seattle, tragedies are almost never random events. The failures are almost always attributable to compounded human behavioral errors. There were design errors, a lack of diligence, construction shortcuts, maintenance neglect, and a lack of oversight. It is easy to deduce the lessons learned. First, design matters; engineers should adhere to accepted practices, standards, and codes (whether for wind bracing, load calculations, or material selection) and build redundancy in, so that a single-point failure will not be disastrous. Second, construction vigilance and maintenance vigilance are paramount; remember, even minor defects or deviations in construction forms should be treated as issues that must be addressed, not blown off. Any defects or deviations should cause the inspector or contractor to stop. Third, the lapse in communication and accountability can never diminish; it should never be dismissed that someone on the line or in the field is raising, albeit prematurely or annoyingly, an issue that must be actioned, rather than being delayed or postponed because of a commitment to schedule or cost. The most regrettable part of so many events that did indeed have fatal results was that people or organizations simply did not act in light of the possibility of an action or condition (whether a crack, corrosion, or just an unsafe practice) that risked human lives.

Lastly, a positive trend can be identified: many of the accidents served as catalysts for new laws, codes, training, and cultures aimed at preventing or mitigating future accidents. As academics and journalists investigate the "whys" and "hows" of structural accidents and events, it is up to us to ensure that accidents are understood and that lessons are consistently conveyed to new minds entering the profession. The most relatable accident in the entire series, Surfside, is a reminder of the vulnerabilities of our homes and of our being; construction must not be ignored for these to remain safe. Taum Sauk is quite different since it is rarely acknowledged without a fatality or serious human injuries, but it reminds us all that a structural accident without a significant fatality still creates significant social and economic impacts worth noting. When looking at the reasoning of each accident, factual detail and unflinching avoidance to not only identify but also review mistakes made on construction or design, at the very least gives respectful honor to those dead and accepts responsibility moving forward to all to hold onto obtainable knowledge, intellect, and practices to make better, safer structures.

Every single case in this compilation, beginning at Champlain Towers South, has served as a faceless witness for time further educated in engineering as a commitment to learning more, a constant learning of the morals and practices we use for not only society, but there is another aspect to the profession of engineering ethics regarding safety. If we can all truly rest on the morals of learning not only from the series researched and written about in every accident, that at least all shown as temptations, will be important lessons and values in practice, causing the profession to change for the better examining the roots of complacency against repetitively leading to loss of life and ultimately us finding ourselves in the roofs of families as any engineered roof can hold thousands of tons and human lives, pedestrian or otherwise.

**Operational Takeaways for Today's Job:**

- Start your shift or safety meeting with a real lesson from past failures. For example, briefly share a case (like a crane collapse or structural failure) and the key mistake behind it – this keeps everyone alert that "it could happen here" if we're not careful.
- Don't ignore the "small" warning signs. If you notice water leaks, cracks, equipment alarms, or any anomaly on site, treat them seriously and get them checked out now – catastrophic failures often begin as minor issues that were brushed aside.
- Make inspections and maintenance a priority, not an afterthought. Ensure critical equipment (scaffolds, sensors, lifts) is inspected as required and fix known issues promptly – postponing repairs or skipping checks has led to disasters in past cases.
- Foster an environment of continuous learning. Encourage crew members to bring up lessons from incidents they've experienced or heard about, and discuss how to avoid those scenarios on your project – learning from others' mistakes is faster and less painful than learning from your own.
- before a high-risk activity, pause to ask, "How could this go wrong in a worst-case scenario?" Use that answer to put extra precautions in place (redundant supports, extra monitoring, emergency plans) – many disasters could have been averted by simply imagining them in advance.

Chapter 3

# Root Cause Categories of Failures (Across 24 Case Studies)

### *1. Design and Engineering Deficiencies*

- **Champlain Towers South (2021)** – insufficient slab reinforcement, long-term deterioration
- **FIU Pedestrian Bridge (2018)** – flawed node design, underestimated cracking
- **I-35W Bridge (2007)** – undersized gusset plates
- **Dallas Cowboys Facility (2009)** – under-designed for wind loads
- **Boston Big Dig Ceiling (2006)** – anchor system not designed for sustained load
- **Berkeley Balcony (2015)** – waterproofing failure led to structural decay
- **San Bruno Pipeline (2010)** – defective seam welds and flawed original fabrication
- **Millennium Tower (2016–Present)** – foundation design assumptions underestimated long-term settlement

### *2. Construction Method & Procedural Failures*

- **Miami Dade College Garage (2012)** – unsafe erection sequencing
- P**hiladelphia Demolition (2013)** – improper demolition sequencing
- **Sioux Falls Copper Lounge (2016)** – removal of load-bearing wall without shoring
- **Hard Rock Hotel (2019)** – improper sequencing and load management
- **Seattle Tower Crane (2019)** – improper dismantling procedure

- **Merrimack Valley Gas (2018)** – construction activity altered system behavior without proper procedure

### 3. Poor Communication and Coordination

- **Champlain Towers South (2021)** – resident concerns and engineering warnings not aligned
- **Oroville Spillway (2017)** – known risks not escalated across agencies
- **Katrina Levees (2005)** – fragmented communication across authorities
- **Hard Rock Hotel (2019)** – field warnings ignored
- **Merrimack Valley Gas (2018)** – disconnect between engineering, field crews, and operators
- **Silver Spring Explosion (2016)** – lack of coordination at utility-building interface

### 4. Insufficient Oversight and Supervision

- **Philadelphia Demolition (2013)** – lack of city enforcement
- **Davenport Apartment (2023)** – known unsafe conditions not acted upon
- **NYC Parking Garage (2023)** – deterioration observed but not enforced
- **San Bruno Pipeline (2010)** – weak regulatory and internal oversight
- **BP Texas City (2005)** – inadequate supervision of high-risk startup operations

### 5. Human Error in Execution

- **Seattle Tower Crane (2019)** – premature removal of critical pins
- **Indiana State Fair (2011)** – continued operation despite unsafe weather
- **BP Texas City (2005)** – operator misinterpretation of levels during startup
- **Husky Refinery (2018)** – failure to control conditions during shutdown

### 6. Violation of Safety Rules & Standards

- **Philadelphia Demolition (2013)** – unsafe demolition practices
- **Hard Rock Hotel (2019)** – code violations in structural work
- **I-35W Bridge (2007)** – inspection gaps allowed unsafe loading
- **BP Texas City (2005)** – use of outdated blowdown system instead of flare

- **Merrimack Valley Gas (2018)** – failure to follow safe engineering review practices

### *7. Organizational Culture & Safety Culture Deficiencies*

- **Champlain Towers South (2021)** – delayed action due to financial pressure
- **Hard Rock Hotel (2019)** – warnings not acted on
- **BP Texas City (2005)** – production prioritized over process safety
- **PES Refinery (2019)** – normalization of deterioration risk
- **Husky Refinery (2018)** – insufficient focus on non-routine risk

### *8. Fast-Track and Economic Pressure*

- **Hard Rock Hotel (2019)** – compressed schedule
- **Champlain Towers South (2021)** – delayed repairs due to cost concerns
- **BP Texas City (2005)** – cost-cutting impacted safety systems
- **PES Refinery (2019)** – aging infrastructure not replaced due to economics

### *9. Inadequate Training and Qualifications*

- **Seattle Tower Crane (2019)** – improper dismantling execution
- **Indiana State Fair (2011)** – lack of engineering oversight for wind risk
- **Merrimack Valley Gas (2018)** – inadequate understanding of system behavior during modification
- **BP Texas City (2005)** – insufficient training for startup risk conditions

### *10. Equipment and Component Failures (True Mechanical)*

- **Taum Sauk Dam (2005)** – faulty level instrumentation
- **Boston Big Dig Ceiling (2006)** – anchor creep failure
- **Oroville Spillway (2017)** – structural and material degradation
- **San Bruno Pipeline (2010)** – pipe rupture due to material defect
- **PES Refinery (2019)** – corroded pipe failure
- **Husky Refinery (2018)** – loss of containment during shutdown

### *11. Lack of Preventive Maintenance and Postponed Repair*

- **Champlain Towers South (2021)** – deferred structural repair
- **Fern Hollow Bridge (2022)** – corrosion ignored
- **NYC Parking Garage (2023)** – long-term deterioration
- **Oroville Spillway (2017)** – known issues unaddressed
- **PES Refinery (2019)** – corrosion not mitigated
- **San Bruno Pipeline (2010)** – poor asset maintenance and validation

### *12. Design–Construction Interface Gaps*

- **FIU Bridge (2018)** – cracking misinterpreted during construction
- **errimack Valley Gas (2018)** – design did not reflect field conditions
- **BP Texas City (2005)** – disconnect between design intent and operation

### *13. Unforeseen Loads and Environmental Factors*

- **Indiana State Fair (2011)** – wind event
- **Dallas Cowboys Facility (2009)** – wind loading
- **Katrina Levees (2005)** – storm surge and soil conditions
- **Oroville Spillway (2017)** – extreme water flow
- **Key Bridge (2024)** – vessel impact
- **Silver Spring (2016)** – confined gas accumulation leading to explosion

### *14. Lack of Redundancy and Fail-Safe Systems*

- **FIU Bridge (2018)** – no alternate load path
- **Champlain Towers (2021)** – progressive collapse
- **ig Dig Ceiling (2006)** – no backup restraint
- **Key Bridge (2024)** – lack of impact protection redundancy
- **San Bruno Pipeline (2010)** – no fail-safe containment once rupture occurred

## *15. Regulatory Gaps & Lag*

- **FIU Bridge (2018)** – fragmented oversight
- **Champlain Towers (2021)** – weak inspection enforcement
- **San Bruno Pipeline (2010)** – regulatory gaps in pipeline validation
- **BP Texas City (2005)** – insufficient process safety regulation at the time

## *16. Planning for Public Protection During Work*

- **FIU Bridge (2018)** – traffic left open under active risk
- **Indiana State Fair (2011)** – no evacuation trigger
- **Philadelphia Demolition (2013)** – adjacent occupancy not cleared
- **Big Dig (2006)** – live traffic under hazard
- **Merrimack Valley Gas (2018)** – public exposed during system overpressure
- **Silver Spring (2016)** – occupied structure exposed to gas hazard

## *17. Governance Fragmentation and Diffusion of Accountability*

- **Katrina Levees (2005)** – split authority
- **Oroville Dam (2017)** – multiple agencies
- **Hard Rock Hotel (2019)** – unclear responsibility
- **Merrimack Valley Gas (2018)** – unclear ownership of system risk
- **San Bruno Pipeline (2010)** – fragmented accountability between operator and regulators

## *18. Asset Management and Funding Shortages*

- **Fern Hollow Bridge (2022)** – maintenance deferred
- **Champlain Towers (2021)** – funding delays
- **Oroville (2017)** – deferred upgrades
- **PES Refinery (2019)** – aging system not reinvested
- **San Bruno Pipeline (2010)** – poor investment in system validation

## *19. Change Control and Management of Change (MOC)*

- **IU Bridge (2018)** – changes not properly evaluated
- **hamplain Towers (2021)** – warnings not formalized into action
- **Hard Rock Hotel (2019)** – undocumented changes
- **Merrimack Valley Gas (2018)** – failure to manage system modification
- **BP Texas City (2005)** – unsafe startup conditions not controlled

## Chapter 3

# Root Cause Categories of Failures (All 24 Case Studies)

In reviewing two decades of disastrous events and structural failures in the construction industry, one thing becomes apparent: no failure is in isolation. Each event is the result of a series of technical neglect, process failure, or poor management decisions that inevitably lead to failure. In every bridge, high-rise, dam, and temporary structure, the causal reasons behind are repeated themes, no matter what type of project it is or where it is constructed.

Recognizing these root causes is not about blame but a systems diagnosis. When a connection in a walkway fails or an earthen levee fails, the technical reasoning seems material in nature (i.e., a bolt, a gusset plate, a membrane), but below the surface is the basis for the failure: the input systems of design, review, communication, and decision-making were wrong. By utilizing these breakdowns of recurring categories, we can identify how risk arrangements navigated their way through the project and concealed the consequences until they became a catastrophe.

The following lists provide a living lesson for each category, allowing you to learn from them. Each root cause is examined in terms of both technical and organizational issues, providing real-world examples for each vehicle. Attached to each category are reviews of the regulations and procedures that were changed, the economic and liability consequences, and the relationship of human faults versus technological faults.

## Design and Engineering Deficiencies

In almost every catastrophic construction failure, there has always been a design or engineering flaw — either an engineered assumption, an untested assumption, or a mathematical error in the analysis. These deficiencies are seldom caused by uneducated practice; instead, they result from inadequate verification, untested assumptions, or broken communication between the design professionals and contractors. The drawings and calculations that specify how steel, concrete, and soil behave together, for all intents and purposes, are the DNA of every structure. If those drawings and calculations are incorrect, each subsequent decision will be compromised, whether it concerns procurement or construction.

The Hyatt Regency walkway collapse (1981) remains among the more disquieting examples of this type of failure. One innocuous revision to a shop drawing — to change the hanger rod configuration — resulted in a doubled load for a critical connection without a reanalysis by the engineer of record. The subsequent cascading failure killed 114 individuals and permanently altered the ethics of engineering.

The I-35W bridge failure (2007) demonstrated undersized gusset plates — a design error made as part of the original plans more than 40 years earlier. The plates functioned without issue until a repair project added construction loads that the bridge could not sustain. In the case of the FIU pedestrian bridge in Miami (2018), finite element models that indicated shear strength at a plate connection implied the failure model. The design team observed cracking days before failure but did not intervene because they were confident in their analysis. The Boston Big Dig ceiling collapse (2006) revealed that a series of adhesive anchors was inappropriate for continued tensile loading, which resulted from two separate design mistakes paired with the erroneous application of materials. Even in contemporaneous work, such as the Champlain Towers South failure (2021), we observe vulnerabilities in design and constructability — insufficient punching-shear margins at the slab-column connection and limited redundancy within the structure — that led to zero tolerance once necessary maintenance work advanced the deterioration. These scenarios expose a consistent truth: failure begins well before concrete is poured or steel is erected. It begins when assumptions are not tested. The technical culture that supports design sometimes relies on experience and precedent, but innovation, cost-cutting, and time-sensitive pressure can often outstrip the rigorous checking process needed to limit errors. If a design team does not return to significant assumptions — if models are not peer-reviewed or validated against independent calculations — failure begins quietly.

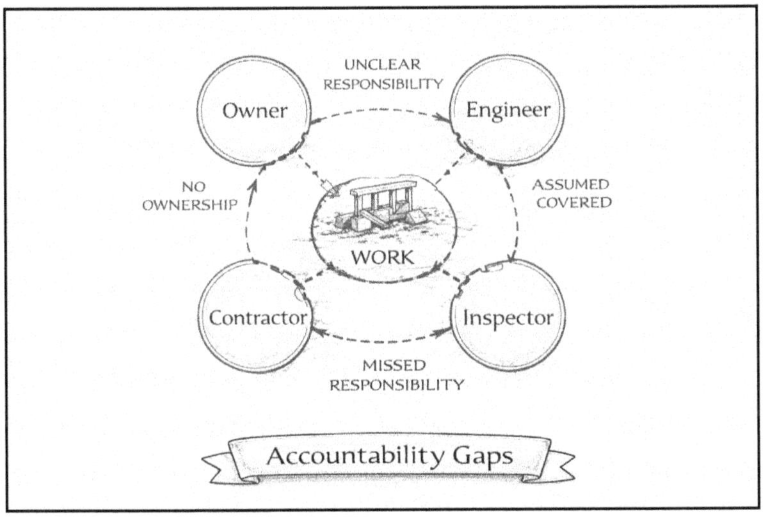


## Regulation and SOP Changes

As a result, the industry has established institutionalized practices to enhance the verification of design. In many jurisdictions, independent design review is now required for complex and/or unusual structures. Connection details, once considered delegated design items, are now treated as fully peer-reviewed and signed off by the engineer of record. Building codes, such as AISC 360 and ACI 318, require that design assumptions be documented and that load paths, redundancy, and other verification items be demonstrated. Firms have instituted "check sets" — sets of details and designs to be reviewed by a senior engineer before issuing them for construction. Manuals produced from software outputs need manual hand checks for verification, and deviations from standard design require justified documentation.

## Accountability and Economics

Project professionals, the professional liability insurance industry, and the wider profession will live with the consequences of design errors. The failures at Hyatt and FIU led to multimillion-dollar settlements, loss of licenses, and criminal investigations. The insurance industry now requires design quality assurance programs showing peer-review certification as part of professional liability programs. Delays often cost more than the project. The legal defense for a design issue or settlement may vastly exceed the project's original cost, often costing professionals their reputations and practice, with incremental pain. In major infrastructure, a design error or defect (even of the most minor scale) can lead to billions in remediation and replacement costs. Such situations reinforced that engineering review is not merely a formality of administration, but rather a mechanism for financial benefit and an ethical responsibility.

## Humans vs Technology

Technology enables precision in modeling, software, and parametric design, but it remains ultimately a human responsibility. Almost all significant design failures have arisen from human error — misreading the results, overconfidence in the software, or simply failing to ask questions. Technology is just a tool, not a measure of truth. The level of human oversight versus machine reliance is where our safety hangs in the balance. The best systems combine computers performing calculations with the professional's skepticism that always leads them to ask, "Does this all make sense?"

Design and engineering errors are the purest form of preventable failure. They remind us that design is not about documents; it is a practice of verification, challenge, and humility. The main takeaway is that each computation has an ethical implication, and not to challenge an assumption is to set up the potential for disaster.

## Construction Method and the Failures of Procedures

While design anticipates the theoretical limits of a structure, construction methods and procedures determine whether a design can survive the challenges of the field. Many of the worst structural failures were not due to a bad plan but to a failed process — unsafe sequencing, poor control of temporary states, or a decision to accelerate beyond what was reasonably safe or physically sustainable. Construction is a process of choreography between timing, curing, and load transfer. When those steps are omitted, ignored, or improvised, the edge between stability and failure evaporates.

The collapse of the Willow Island Cooling Tower (1978) is one of the most tragic instances of procedural failure in U.S. history. Workers were erecting a natural draft cooling tower through a continuous-lift method, tying scaffolding to newly poured concrete that had not cured. As the workers lifted the next pour, the concrete failed underneath, bringing down the entire scaffold and crew — killing fifty-one workers instantly. The L'Ambiance Plaza building collapse (1987) had similar procedural lapses in lift-slab construction. The intent behind the process was to save time and money by pouring entire slabs on the ground and then jacking them into place. When lifting began, the temporary seating and connections failed, and the building completely collapsed. Both events resulted from a similar cause — performing tasks or utilizing techniques that were either unproven or not monitored, while trusting that there was no need for redundant safety systems.

In modern examples, these practices persist. The Hard Rock Hotel collapse in New Orleans (2019) involved unsafe shoring and improper pour sequencing at upper levels, where incomplete formwork and inadequate bracing failed to hold back the fresh concrete. The Miami-Dade College garage collapse (2012) involved unsafe systems for erecting precast building elements; the panels did not have sufficient support during installation, ultimately leading to a progressive failure. The 2013 Philadelphia demolition collapse and the 2016 Sioux Falls renovation collapse each illustrated how non-engineered shoring of load-bearing walls can quickly turn a renovation into a deadly demolition. In all these situations, the failure mechanism was not complicated. It was procedural drift — the slow normalization of deviation from a proper sequence. As schedule pressure mounts, project teams take small risks that they might justify as "temporary" or "minor." In construction, those justifications culminate in the cooperation of physics. Something borrowed from safety in the name of time savings may result in an irreversible loss.

## Regulatory and SOP Changes

After each of these collapses, regulatory agencies and industry groups established changes to regulations and/or SOPs to instill procedural safeguards. OSHA and ACI have established specific provisions for verifying the strength of formwork and falsework before loading. Temporary works that previously were solely at the discretion of the general contractor must now be engineered, reviewed, and documented under seal. Many states now require a formal "Construction Engineering Plan (CEP)" to be in place before any critical lift, post-tensioning, or sequential pour takes place. The CEP requires a PE structural engineer to verify project stability in the temporary condition, not simply at the final configuration. Some states now require Special Inspections during erection and demolition to verify that design intent is maintained and that the structure is safe at each phase of construction.

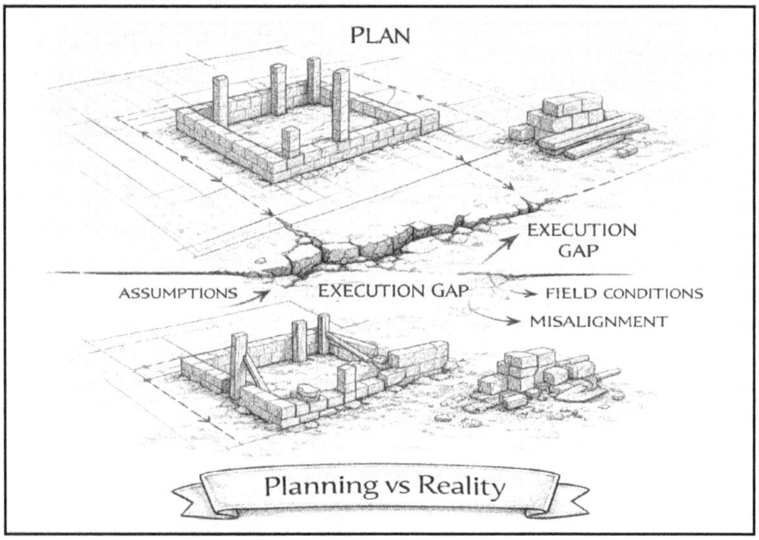


The industry is also embracing procedural "hold points" — portions of the process that include a formal pause point to verify that prior steps are solid in strength, alignment, and dimensional tolerances before proceeding. Hold points are also applied to curing concrete, stressing tendons, and removing temporary supports. The goal of the hold point is to discourage decision-making based on assumptions or convenience. The aim is to replace discussion and intuition with data.

## Accountability and Economics

The cost of procedural failure is astronomical. Each of these events results in not only loss of life but also complete loss of value. Rebuilding or demolishing compromised buildings requires far more resources, time, and materials than the savings gained from original shortcutting. Contractors and developers also endure litigation, contract terminations, and debarment from public projects. Insurers have become very sensitive to procedures that create risk and often require method statements, stability analyses, and same-day verification logs to monitor the performance of high-rise or complex structures. Economically, the lesson is crystal clear: safe sequence is time; unsafe sequence is loss of value.

## Human Versus Technology

Procedural failures occur at the intersection of human behavior and technological processes. Many modern construction tools — including sensors, load monitors, and loggers for real-time curing analysis — have the potential to provide precise data, rather than relying on anecdotal evidence. However, they are only effective if procedures require their use and personnel trust the results. In almost all historical situations, the technology existed but was not utilized. The human component — impatience, overconfidence, or communication failures — facilitates collapse.

Procedural stability is not bureaucracy; it is the unseen rebar in the foundation of the industry. Every sequence, checklist, and inspection point is a lesson learned, some costly, some fatal. Ultimately, a professional construction culture is not defined by how quickly a project is completed, but by how methodically it is done. Each sequence must begin with a review of structural plans to ensure sound construction before proceeding to the next phase.

## Poor Communication and Coordination

As is the case with every major construction failure, somewhere in the chain of events, a message was sent, a report filed, or a warning that went unheeded. While communication failures represent one of the least technical and most common root causes of construction failures, communication often plays a crucial role in determining whether a problem is fixable or has progressed to a catastrophic failure. Engineering awareness and field experience are worth very little (in fact, nothing) if they aren't communicated clearly throughout the organization. When communication fails to take place between the designer and the contractor, management and the workforce, or the agency and the operator, even the best design can be brought down by silence or misunderstanding.

Many of these lessons are articulated in the widely known and most infamous communication failure in engineering, the Hyatt Regency walkway collapse (1981). The engineer of record had a shop drawing revision to a fabricator's assembly approved by the contractor, but failed to understand how the approved fabrication adjustment changed the load path. One single note (which was miscommunicated between all parties) doubled the connection load and led to one of the deadliest structural collapses in United States history. The Champlain Towers South collapse (2021) represented a failure of communication across another dimension; the owners and governing boards were warned in multiple engineering reports about "major structural damage," yet couldn't come to an agreement about funding repairs for years. In the 2017 Oroville Dam spillway failures, the California Department of Water Resources had received engineering reports for years detailing the deterioration of the concrete. No comprehensive repairs were made to the dam until it partially failed. A similar situation occurred during Hurricane Katrina in 2005, where ineffective communication among federal, state, and local agencies resulted in varying levels of understanding regarding the capabilities of the levees, the necessary emergency actions, and the real-time flooding conditions.

These failures highlight an essential truth: communication breakdowns are rarely mere misunderstandings; they often reveal systemic weaknesses within project management and governance frameworks. They point to issues such as fragmented authority, inadequate change management protocols, and unclear roles in planning and governance. Construction projects are complex systems composed of a network of specialists—designers, engineers, subcontractors, inspectors, and clients—each using their own technical jargon. Without standardized communication or established pathways, messages can become diluted, delayed, or entirely lost.

## Regulatory and SOP Changes

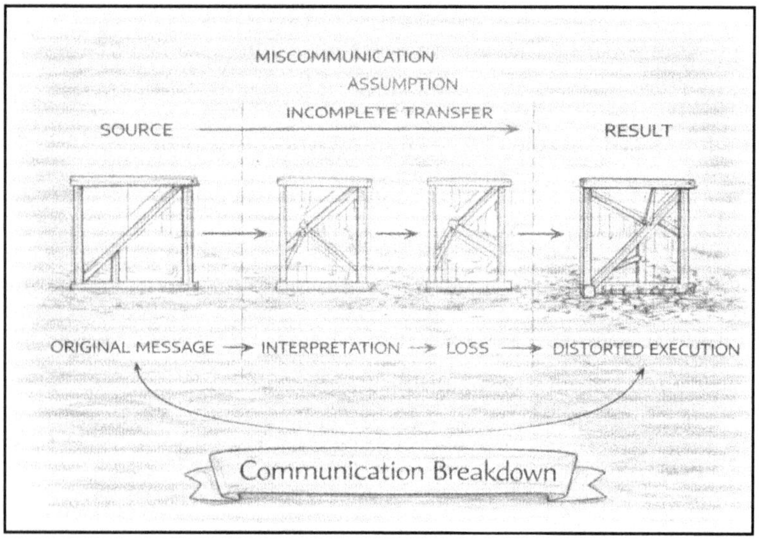


In the years after these events, formalized communication frameworks were created, underpinning construction management systems. The Management of Change (MOC) procedure is a relatively new, formalized method that documents the trail associated with construction and built environments, ensuring that any design changes or field modifications receive a documented, traceable assessment and approval. Similarly, Requests for Information (RFI) and submittal logs require the recipient to issue either a written confirmation or a written response indicating that the response or modification meets the requirements of the engineer of record. A significant advancement in the construction project management framework is the emergence of digital project management systems, including Procore, Aconex, and BIM 360. These systems centralize communications, document them, and track version changes.

In both design-build and public projects, contract language has evolved to mandate the tracking of revisions made for construction purposes. These revisions must be documented appropriately and redistributed, rather than communicated informally. Project communication plans have also developed into clearly outlined protocols that define pathways for escalation, for example, how issues are raised in the field (reported), who is notified, and how the decision is documented. The best organizations treat communication as a technical discipline no less critical than structural analysis and validity testing techniques.

## Accountability and Economics

The financial and legal ramifications of ineffective communication are severe. Structural failures often lead to litigation, where emails, reports, or meeting notes are invariably categorized as evidence that decision-makers ignored early concerns. Each communication gap indicates a missed opportunity for prevention. Even without considering mitigation, remediating issues during construction is the most cost-effective option compared to remediating after failure. Insurers and owners now scrutinize the documentation of communication during claims. In major settlements, claims have been supported in court by the "absence of documentation" of formal reporting processes.

Organizations that adopt transparent and traceable communication reduce risk and costs. Proper documentation enables faster decision-making, protection, and professional accountability. When the communication system is weak, costs compound — not only in terms of money, but also in terms of public trust and professional credibility.

## Human vs. Technology

Communication is human, but technology amplifies communication. Communication mediums have exploded; it may be easier to find information, but it can now be overwhelming. Technology can tag and timestamp every piece of correspondence, but it does not guarantee that everyone reads, interprets, and acts on that information. The failures at Oroville and Champlain Towers demonstrate that whether data is unknown or available to us, inaccurate data is just as dangerous as having no data at all. Therefore, communication is not simply about transmission; it involves understanding, verification, and accountability.

The construction industry's most significant issue is not ignorance but connecting the dots. The systems in place — MOC protocols, RFI workflows, digital tracking — exist to reduce the gap in protection that previously cost lives. They create an ongoing feedback loop linking design intent with field execution. When communication is effective, it can help prevent deviations from the contract documents from escalating into structural risks. When communication fails, risk can build cumulatively, even invisibly, until failure is certain. Every major failure reaffirms the same lesson: the cost of silence in construction is measured in lives.

## Insufficient Oversight and Supervision

No amount of sound design or planning can go forward without continuous, competent supervision ensuring that what is built reflects what was planned. Oversight is the industry's immune system, a final line of defense against small deviations, mistakes, or omissions that become catastrophic without supervision. When supervision fails, mistakes in construction that could have been corrected in minutes become failures that result in structural collapses and reputational damage.

The collapse of the cooling tower at Willow Island (1978) serves as one of the enduring examples of the results of absent supervision. Without verifying that the concrete had cured to strength, the fatal error of attaching climbing scaffolding to green concrete was made. Not only was there no engineer present to confirm if the structure was capable of carrying additional load, but there was also no inspector to verify. The same could be said of the lift-slab construction at L'Ambiance Plaza (1987), which also had little engineering oversight. The temporary jacking system was unmonitored and was not subject to an independent inspector verifying stability between lifts. After the slab connections failed,

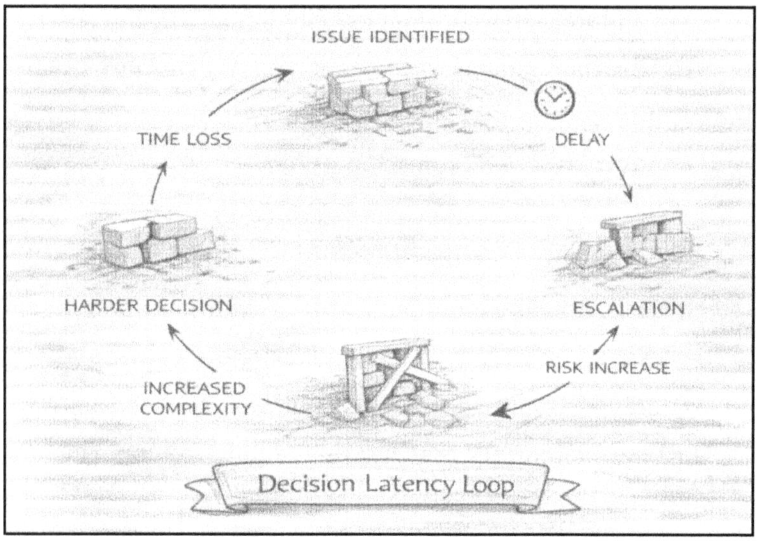


twenty-eight employees lost their lives. The collapse arising from demolition in Philadelphia (2013) was a recent example of the same trend. The demolition contractor worked without sufficient supervision or engineering guidance, removing important walls directly adjacent to a still-occupied store. Inspectors later observed that the demolition was conducted without a documented demolition sequence, and there was no competent person on-site, as mandated by OSHA regulations. The collapse of the Davenport apartment building in 2023 and the garage collapse at 57 Ann Street in NYC in 2023 highlight significant municipal failures in oversight. Both structures exhibited visible deterioration and had been flagged for issues, yet regulatory agencies took no action to enforce repairs or designate the buildings as unsafe beforehand.

By failures in supervision, we mean accountability has either become diluted or routine supervision has transitioned from investigative to procedural. Many organizations end up treating inspections as merely a paperwork requirement instead of an analytical process. When supervisors focus on "getting through the checklist" rather than observation and verification, the system becomes blind to deviations. Among construction professionals, there is often a mindset that "someone else has already checked the work," which creates a vacuum for hazards to go unnoticed.

## Regulation and SOP Shifts

In response to a series of tragedies resulting from supervision failures, the construction industry and regulatory agencies have increased the requirements for qualified supervision and inspection. OSHA clarified the legal distinction between a "competent person," someone who, with training, can identify and act on hazards, and a "qualified person," who has the training to design and verify corrective actions. In any project involving structural erection, demolition, or temporary works, both designations must now be defined and documented. The International Building Code (IBC) Chapter 17 now requires special inspections for critical components of the structure, including bolted and welded connections, concrete strength, post-tensioning, and fireproofing. Special Inspections must be carried out by certified independent inspectors, not the contractor's personnel.

Several jurisdictions now require the submission of an inspection plan during preconstruction, outlining the frequency of inspections, what is being inspected, and who will conduct them. For public works, third-party peer reviews and verification programs have become commonplace, meaning that no phase of construction can be advanced without some documented assurance of quality.

## Accountability and Economics

Without suitable oversight, there are real costs. In the instances of Willow Island, L'Ambiance, and Philadelphia, contractors and supervisors were sued, faced civil and criminal trials, and were assessed significant OSHA fines. There are costs associated with not having sufficient supervision, beyond the costs imposed by accident recovery and unsafe conditions, as your project has already been impeded by time and delays. This has consequences for cost and schedule, which ripple through financing, liability insurance, and the public's trust. Insurers now require risk audits and supervision logs, especially for projects classified as high risk for lifting, demolition, or deep excavations.

Economically, oversight has shifted from a type of overhead to a value-enhancing certainty. Projects with strong quality assurance and inspection records often experience fewer insurance claims, less rework, and lower lifetime maintenance costs. An owner or borrower should consider that the cost of supervision is cheaper than the costs of negligence.

## Human vs. Technology

Experience shows that oversight failures are human failures, but when supervised appropriately with the use of technology they can be significantly improved. Technologies such as drones for inspection, digital field reporting, and automated data collection platforms have transformed the way supervisors monitor work. While an image can be captured with a drone, the drone cannot interpret the pattern of a crack. Although a data logger can store curing temperatures, the logger cannot tell you whether to wait on a pour. Effective supervision relies on situational awareness, the trained eye for recognizing risk before it shows itself.

Where oversight fails, technology often fills the gap. Any system, including its sensors, is configured, calibrated, and interpreted by humans. Most modern collapses reveal this deeper truth: the absence of competent supervision is often the failure point that allows every other barrier to fail.

Oversight and supervision are the connective tissue between design, method, and execution. They are the last chance to stop a mistake from becoming a tragedy. The modern construction industry has evolved its perspective on quality and safety; they cannot be assumed through technology or documentation. Quality and safety must be demonstrated by individuals who know what to recognize, who can take action, and who understand that vigilance is not optional but is the foundation of safety itself.

## Human Error in Execution

Whatever the technology, the design, or the plan may be, the human element remains the most unpredictable variable in construction. Every structure relies on the judgment, attention, and discipline of people — operators, riggers, laborers, engineers, and supervisors — to turn plans into physical reality. When decisions are made in the moment without following the process, or without knowledge of consequence, the limits of control and chaos quickly fragment. Human error usually does not occur due to ignorance. It usually stems from a brief gap in concentration, overconfidence, or sleep deprivation interacting with systems that cannot account for those states.

One example is the collapse of the Big Blue crane in Milwaukee (1999). The crane, one of the largest in the world, attempted to lift a 450-ton roof panel to be placed on a stadium roof even though the wind conditions exceeded the safe operating limit of the crane. The crane operator was warned several times about the wind conditions, but the lift proceeded anyway. When the load started swinging, the crane tipped over, killing three workers and crushing the nearly completed roof of the stadium. The investigation determined there was no mechanical failure, only a lapse in judgment. In the 2019 Seattle South Lake Union crane collapse, construction crews failed to maintain panel alignment and removed the bolts prematurely, leading to the mast's failure; they lost track of timing and safe operating practices. When the crane collapsed, four people lost their lives as the boom tore across a nearby street. In both situations, procedures existed to prevent the events that unfolded, but they were not followed in real-time. In another case, the 2011 Indiana State Fair stage rigging collapse demonstrated a similar principle of risk outside the scope of traditional building construction. Event organizers and rigging contractors did not evacuate the temporary structure, despite forecasts predicting destructive winds. When the gust front arrived, the rigging collapsed onto the crowd, resulting in seven fatalities. The 1985 New York City crane accident serves as an example of how experience and certification can be overlooked for the sake of expediency. In that incident, an unlicensed crane operator failed to properly load a counterweight necessary for rotational control.

Human error during execution of work is not an incident; it is a process. When people are working under time pressure, not rested, and/or have incomplete information, human error is more likely to occur. It will escalate when supervision is weak, workers are complacent about safety culture, and procedures are viewed as guidelines rather than requirements. The real tragedy is not when people start making errors; it is when systems fail to anticipate those errors and to intervene before they are performed.

## Regulation and SOP Changes

Following these and similar events, regulations became significantly stricter regarding operator qualifications, procedure control, and safety systems. OSHA introduced, and later federally mandated, a national crane operator certification to standardize training and testing and required all crane operators to be recertified every five years at construction projects in the U.S. What was once advisory safety instruction in a manufacturer's manual became an enforceable document under OSHA 1926 Subpart CC. The consequences for not following or deviating from the procedures outlined constitute a regulatory violation.

Industry standards also changed. ANSI and ASME created a comprehensive code to follow for lifting and rigging activities. The codes covered limits of wind speed, planning for a lift, and worker communications. They now require that a lift plan be engineered for every critical lift, including load charts, wind data, personnel assigned to the lift, and signatures from all parties signing off on the plan before the lift. Procedural chains were established for dismantling and temporary works to ensure that no part of a lift can proceed without verification of safety and proper documentation. The goal is to shift from a reactive approach of assigning blame to a proactive approach that emphasizes control, thereby integrating human reliability into operational systems.

## Accountability and Economics

Human error resulting from poor decision-making on construction sites can have immediate and far-reaching consequences. OSHA considers willful neglect or disregard for safety procedures—such as those seen in various crane operation incidents—to be among the most serious violations. These violations can result in fines amounting to millions of dollars and may also lead to potential criminal liability. Big Blue led to extensive litigation and broad standards changes in crane assembly and lift planning, while the Seattle incident during dismantling led regulatory agencies to modify oversight for permitting and notifications in a multi-state area. Every consequence — criminal, civil, or regulatory — has a long-lasting impact on public trust. Stadiums, towers, or permanent installations, such as seawalls, are more than engineering feats; they are civic anchors. When human error causes a collapse, reputational damage affects the project, the industry, and all practitioners, leading to increased insurance costs and a decline in public confidence.

From an economic standpoint, human error is our most significant risk. Insurance data indicate that operator error is the leading cause of construction loss incidents, followed closely by deviations from procedures and miscommunication on-site. These factors account for more than half of all tracked events. Even in newer project delivery methods, human performance systems have integrated behavioral safety programs, fatigue management, and situational awareness training to address incremental risks and reduce lapses in performance. Engineering in human performance has consistently yielded benefits and returns over time, and one objective of human-centered training is to understand human limitations while also meeting the required level of compliance.

## Human vs. Technology

While human error is, by definition, a product of behavior, technology has a unique and multilayered role. It can prevent error or exacerbate it when relied upon as a crutch. New automated cranes and loading systems provide load sensors, stability monitors, and wind cutoffs to support safe operation. Yet even robust systems can be deactivated. In many instances, operators disregard alarms because they need to complete a task, or they disable alarms due to past positive experiences. The paradox in the relationship between humans and technology highlights the growing need for ergonomics and human-factors planning. This approach aims to ensure that technology provides a safe experience while aligning with the increasing trend of automation.

The goal should not be to replace humans with machines but to incorporate Human Factors Engineering. This means designing systems that account for human error, making it harder to make mistakes, and recognizing that neither perfection nor training alone can eliminate errors. Operational procedures must focus on the limits of human performance, and training should prioritize situational judgment over mere compliance with standard operating procedures (SOPs).

While we will never completely eliminate human error in execution, we can manage and reduce its impact. Evaluations should first identify errors as data and then analyze them as evidence of where systems failed to safeguard against common and predictable mistakes. The construction industry is embracing this narrative, viewing near misses and minor deviations as opportunities for learning rather than merely as a means for punishment. Progress will come when the industry stops relying on the rationale that accidents are the cost of productivity and instead views prevention as a demonstration of professionalism.

## Violation of Safety Regulations and Standards

When accidents occur in construction due to a disregard for established safety regulations, the tragedy is compounded by the fact that these incidents were preventable. These situations do not stem from a lack of knowledge or innovation; instead, they arise from a failure in discipline and ethics. The rules were in place, and the standards were clear. What is missing is the commitment to adhere to them. This group of disasters reflects not the unknown, but the ignored—events where the crew, foreman, or contractor knowingly overlooked rules designed to protect human life, property, and the public's trust in safety, security, and reliability.

The collapse of the Big Blue crane (1999) in Milwaukee demonstrates disregard for regulations. The crane's significant lift occurred when winds were well above the limits established by the manufacturer. Supervisors acknowledged the risks associated with meeting the schedule. When the load began to sway and the crane fell, three workers died, and a portion of a nearly completed stadium was destroyed. The demolition collapse at 22nd and Market Streets (2013) in Philadelphia depicts a similar pattern of willful negligence. A contractor not qualified for the work directed the crew to remove a load-bearing wall adjacent to an occupied retail store without conducting an engineering review or obtaining a permit. When the unsupported wall fell, six people were killed inside the store.

The collapse of the Hard Rock Hotel in New Orleans in 2019 highlighted a culture where meeting project deadlines took precedence over following building safety codes. Before the collapse, workers reported noticeable distortions and cracks in the framing components. Internal photos of these deficiencies were circulated, but the construction timeline remained a higher priority than compliance with safety standards.

Similarly, the I-35W bridge collapse in 2007, though primarily due to design flaws, also involved procedural violations during its rehabilitation. Crews placed concentrated loads of construction materials directly over critical first-order gusset plates, which were already pushing their tolerance limits. Each of these circumstances illustrates how applicable standards — whether lifting, demolition, shoring, or load management — were ignored in the name of expedience, cost, or convenience.

Violations occur when compliance becomes a barrier rather than a baseline. Teams in construction may misinterpret a safety code's language as optional practice rather than a legally enforceable minimum. Supervisors may observe workers cutting corners to save time, and previous involvement creates a false sense of control. This is the normalization of deviance, which obscures risk in the name of efficiency. Over time, violation becomes the norm — until it is not — and a structure fails or a life is lost.

## Regulatory and SOP Changes

Each of these violations resulted in regulatory tightening across the industry. Following Big Blue, OSHA implemented substantive changes to Subpart CC, 29 CFR 1926, which expanded safety and procedural standards for crane assembly, disassembly, and lift planning. Limits on wind speed and charts for load calculation were no longer advisory but enforceable. For large lifts, the presence of a wind meter (anemometer) was bound to a lift plan. Standards for demolition were rewritten after the Philadelphia collapse, requiring engineering stamps on demolition plans, a safe sequence of work, and operating exclusion zones as a safe refuge around occupied structures. Cities then established more stringent permitting with clear verification to ensure any work affecting adjacent properties was engineered, inspected, and communicated to the authority having jurisdiction. The incidents at Hard Rock and I-35W further demonstrated the need for engineered lift and shoring plans, as well as live oversight of temporary load situations. Regulations now emphasize that an engineer must approve critical lifts, include pre-task hazard analysis, and be subject to real-time monitoring. Violations once regarded as administrative mistakes have been reclassified as willful acts of wrongdoing, potentially leading to criminal consequences for supervisors and executives if negligence is proven.

## Accountability and Economics

When a safety violation leads to catastrophic circumstances, the financial and reputational costs are often terminal. In the case of the Philadelphia demolition contractor and property owner, both were criminally convicted of involuntary manslaughter and imprisoned. There were also civil litigations that resulted in multimillion-dollar settlements and broad reform to insurance related to demolition activities. Big Blue generated lawsuits through several tiers of contracting, including equipment suppliers and manufacturers. The implications of the Hard Rock Hotel collapse, even after all litigation is done, remain existential for the firm: loss of licenses, bankruptcy, and long-lasting public scrutiny.

This financial impact is not solely the outcome of a firm's regulatory violation; the collective impact of individual actions affects industry price. The cost of a regulatory violation manifests as higher insurance premiums, fewer investors, and fewer project approvals. A regulatory violation diminishes a company's reputation, and the collective effect damages the sector's reputation as a whole. A common announcement after every disaster is, "Compliance is not optional." Compliant actions ratify professional legitimacy.

## Human vs. Technology

The violation of safety standards is inherently a human failure. Rarely is it due to a lack of technology or inadequate tools. Each of the referenced cases involved established warning systems, design criteria, or manufacturer instructions. In each of these situations, technology provided relevant information — wind gauges, load charts, inspection reports — but humans decided to ignore it. That is the most sobering realization of all: technology cannot save you from knowable hazards that are deliberately ignored.

Current technology, however, is playing a greater role in enforcement and governance. Digital monitoring systems, timestamped logs, and automated safety alerts provide objective evidence for or against compliance. That is not to say they guarantee ethical behavior, but they make it more visible and traceable. In that sense, technology acts as a protective mechanism or witness.

Failing to comply with safety regulations and standards is by far the most preventable type of failure. It requires no advanced technology; it requires advanced integrity. Regulations protect lives when followed. No amount of engineering or technology will compensate for not following the rules. The simple lesson for every professional in this industry is this: compliance is not the ceiling of performance; compliance is the floor. Excellence starts where the law stops, when the decision is made to do what is right, even when no one is watching.

## Organizational Culture and Safety Culture Deficiencies

Every organization on a construction project — owner, general contractor, design firm, field crews — contributes to the culture that shapes decision-making. When that culture values production over protection, efficiency over ethics, and silence over accountability, it becomes the invisible force behind nearly every major accident in this industry. Cultural failure is not due to a single decision or act of negligence; it is a slow process of degrading standards through repetition, rationalization, and bad priorities. A poor safety culture allows risk to become normalized and responsibility to diffuse until violations and shortcuts become routine rather than dangerous.

The Hyatt Regency walkway collapse (1981) illustrated how fast-track construction jeopardized an engineer's duty of care. The project's cadence necessitated rapid design modifications and approvals, often leaving little time for a thorough review. That sense of urgency wore away professional vigilance and contributed to one of the deadliest collapses in United States history. The same production pressure contributed to the Willow Island Cooling Tower (1978) disaster; workers were instructed to proceed with the next lift even though the concrete below had not reached adequate strength, to meet a scheduling milestone. At L'Ambiance Plaza (1987), the decision to use the unproven lift-slab method reinforced a managerial culture focused on schedule and cost savings rather than a conservative engineering approach.

In recent events, the collapse of the Hard Rock Hotel (2019) in New Orleans brought to light a culture that ignored or suppressed field warnings. Workers documented deflection, deformation, and buckled support beams, yet those concerns never reached a decision-maker before the disaster. At Champlain Towers South (2021), the condominium board postponed contributing to reserves for essential structural repairs, and financial reluctance ultimately posed harm to residents. All these cultural failures shared one central theme: cultural drift. Over time, we become desensitized to risk, allowing the extraordinary to become ordinary. A strong safety culture does not mean there are "no accidents"; rather, it signifies accountability. It is a culture where concerns are acknowledged and rewarded, not suppressed; where the chain of command is clear; and where leaders exemplify transparency. When this culture deteriorates, an organization loses its ability to recognize danger in real time. People stop reporting concerns because they believe it won't make a difference or, worse, that it could jeopardize their jobs. When silence takes hold, failure becomes inevitable.

## Regulatory and SOP Shift

After several of these incidents, regulators and industry leaders accepted that compliance does not equate to safety. Safety culture became a priority, and programs such as Stop-Work Authority were implemented. Stop-Work Authority allows any worker, at any level, to stop an operation without retaliation if they believe a safety hazard exists. Today, safety management systems embrace leading indicators, which are proactive measures of success (such as hazard observations, near-miss reports, and participation in safety meetings) that provide early warnings of cultural inadequacy.

Industry certification programs, such as ISO 45001 (Occupational Health and Safety Management Systems), have established frameworks that assess not only operational procedures but also leadership commitment, worker engagement, and internal communication. There are indications of a shift from simply reacting to compliance requirements to actively building a proactive safety culture. Construction companies have started to implement behavioral safety training, field leadership programs, and hold executives accountable for safety performance. These initiatives recognize that an organization's culture is shaped by what leaders choose to tolerate, reward, and overlook.

## Accountability and Economics

The organizational and economic implications of a poor safety culture can be significant. A company will suffer reduced profitability in the long term, even if it does not immediately experience loss of life or property. When reputational risk develops around unsafe practices, long-term harm occurs through increased insurance premiums and reduced access to major public contracts and bidding opportunities. For example, after the Hard Rock collapse, contractors were pursued for debarment and civil litigation far exceeding any perceived "savings" from rushing the work. For the Hyatt Regency and Willow Island events, permanent changes in professional liability law were instituted, and insurance costs rose across the industry.

From management's perspective, a strong safety culture is also a strong economic culture. Organizations with stronger safety cultures tend to execute projects with greater efficiency, as they avoid rework, litigation, and downtime. Conversely, the cost of unsafe shortcuts always emerges in one form or another. Contractors are increasingly being evaluated by investors, developers, and insurers based on their safety culture. They recognize that a poor safety culture poses not only a moral risk but also a financial one.

## Human vs. Technology

While technology can support a strong safety culture, it cannot replace it. Tools such as wearable sensors, safety dashboards, and digital reporting systems provide valuable data; however, it is the culture that determines how this data is utilized. For instance, on the Hard Rock jobsite, the necessary technology to capture and share information was available, but the real barrier was not the lack of information—it was the attitude towards safety.

In essence, culture is human. It is articulated in daily actions, such as choosing to stop work because something appears wrong, a supervisor's willingness to listen, and whether leadership models accountability instead of excuses. A safety culture is developed through consistency — when management behaves in accordance with stated values, keeps promises, and every level of the organization believes human life is the ultimate measure of success.

Deficiencies in organizational and safety culture are the invisible scaffolding of failure. They will not appear on drawings or be measured with instruments, yet they are evidenced in every incident where a warning was not heeded or corners were cut. The takeaway for each leader is straightforward: culture is not a line item; it is the underpinning for every procedure, every system, and every life on the worksite.

## Fast-Track and Economic Pressure

Time is money in construction. Deadlines drive decisions, schedules drive sequencing, and budgets drive priorities. But when time and cost take precedence, safety, quality, and judgment are the first to pay. Fast-tracking — compressing design and construction schedules to accelerate delivery — is commonplace in competitive markets. When used appropriately, it involves careful planning, risk management, and open communication between disciplines. When misused, it becomes a breeding ground for oversights, shortcutting systems or procedures, and fatigue-driven mistakes that result in collapse.

The Hyatt Regency walkway accident in 1981 took place in a fast-paced environment where the design and construction processes overlapped, leading to a lack of proper coordination. Due to time constraints, engineering changes were made hastily through the submittal process without adequate review. This resulted in a critical load-path error, which can be directly attributed to the pressure of meeting a tight schedule.

Similarly, the Willow Island Cooling Tower accident in 1978 was driven by schedule pressures. The construction crew was instructed to continue raising the scaffold even though the underlying concrete had not fully cured. This decision violated established procedures in order to meet production deadlines.

In 1987, the L'Ambiance Plaza project employed lift-slab construction methods to reduce costs and time, but the approach left little margin for error. Unfortunately, temporary supports failed before the structure achieved self-stability, leading to a catastrophic collapse.

More recently, the Hard Rock Hotel collapse (2019) in New Orleans illustrated the dangers of compressed schedules and fragmented oversight. Multiple trades, subcontractors, design revisions, and overlapping work created a chaotic environment and a sense of urgency that replaced structure. Discussions of deformations and deflections went unheeded as haste replaced patience. The outcome was complete structural failure, killing three people and leaving a permanent scar on the skyline. In every example, the capacity to build quickly was viewed as a business advantage, until it was measured in human lives.

Fast-track systems increase risk by compressing the decision-making cycle. In the normal workflow, engineering designs are reviewed in sequence, peer-reviewed, and field-coordinated prior to construction. Under fast-track conditions, these steps are often skipped or overlapped. The feedback loops that typically flag design inconsistencies or misinterpretations in the field shut down. Errors that would be identified and corrected are instead moved forward into construction without remedial action. The larger systemic risk is not just one bad decision but the compounded occurrence of many small ones, all accelerated.

## Shifts in Regulation and SOP

In response to these repeated failures, both the construction sector and regulators have pushed for intentional pacing and mandated quality points in the workflow. Many states now have mandatory hold points on critical safety milestones, such as concrete curing, post-tension stressing, and critical lifts. Work can continue only after independent verification confirms that the readiness is met. Standard contract language is increasingly being revised to include buffer time in the schedule as a de-risking approach, allowing contractors and designers to avoid penalties for justified safety-based delays.

Safety culture initiatives have expanded beyond “on-time completion” to include performance indicators for safety. In high-consequence projects, Integrated Project Delivery (IPD) and progressive design-build models are increasingly utilized to align financial incentives with safety, quality, and schedule outcomes. These structures rely on collaboration among all stakeholders, sharing risk rather than facing adversarial time pressure. Similarly, modern project management standards, such as PMBOK and ISO 21500, recognize risk analysis and contingency scheduling as competencies essential to project outcomes, not optional extras.

## Accountability and Economics

The economic reality of fast-track construction is a contradiction. The goal may be to save time and money, but when shortcuts lead to rework or failure, the losses far outweigh any initial savings. The cases of Hyatt, Willow Island, and Hard Rock illustrate the exponential downstream costs of expediting without thorough controls: litigation, rebuilding, forfeitures, and insurance overruns. Sometimes, the price of haste is the survival of the corporation.

Today, financiers, insurers, and owners view schedule realism as a key risk factor. An unrealistic schedule can result in loss of coverage or added premiums. Regulatory agencies and sophisticated owners, especially public entities, expect contractors to show that safety protocols will not be sacrificed for accelerated delivery. Pursuing time savings through risk is not efficiency; it is exposure.

### Human vs. Technology

The collision between human effort and technological capability becomes evident under financial pressure. Scheduling tools, Building Information Modeling (BIM), and automated tracking systems can optimize workflow, but they do not replace sound human judgment. A Gantt chart may show a shorter duration, but it does not account for factors such as fatigue, stress, or the ethical boundary being crossed. When deadlines and metrics drive behavior, people often begin to justify unsafe actions by relying on technology or experience to compensate.

Technology provides speed, but it also amplifies errors when safety checks are skipped. The way forward is balance: use technology to inform conflict identification, not to justify acceleration. Fast-tracked construction should be viewed as a controlled risk trade-off between time and quality, rather than a business-as-usual model. Professionalism demands the ability to slow down when necessary and the discipline to prioritize quality over speed.

The universal message across decades of failures is simple: allowing schedule pressure to operate without boundaries poses a safety risk. Construction is not a race against time; it is a negotiation between time and precision. When time prevails, safety loses. This industry's progress will be measured by how wisely we choose to build, not how fast.

## Inadequate Training and Qualifications

Construction is an industry that relies on a combination of precision, experience, and judgment while working under high-risk conditions. Each position, from equipment operator to structural inspector, requires intricate knowledge to work safely and correctly. When personnel perform work beyond their training or without proper certification, the project is tainted by a hidden level of impairment that no design or supervision can salvage. Inadequate training and qualifications can be a silent threat: they exist in the short term in the shadows, but can be both crippling and deadly when systems are stressed.

The New York City crane collapse (1985) provides an early case in point for this root cause. The operator at the time of the incident had no license and had a limited understanding of the machine's capacity limitations. When the crane was overloaded and improperly configured, it failed catastrophically, leaving one person dead and several others injured. Notably, there was nothing wrong with the crane itself; the error was entirely operational. The South Lake Union crane collapse in Seattle (2019) demonstrated this lesson decades later. During the dismantling process of a crane, crews removed two critical mast pins out of sequence, effectively disregarding the manufacturer's requirements. The construction activity was not intentionally malicious; it was procedural, based on inadequate training and a lack of understanding of the engineered method for dismantling the structure.

At Willow Island (1978), many crew members were not familiar with the newly developed jump-form system used to raise the cooling tower. There was no formal training; the crews were following the verbal directions of supervisors with little technical background. When the scaffolding was attached to uncured concrete, this decision reflected not defiance but a complete lack of understanding of the load path for the structure. The lift-slab method was a relatively new technology in 1987 when L'Ambiance Plaza collapsed. There was little formalized training available. Site workers and supervisors improvised based on limited manufacturer guidance. The Indiana State Fair stage collapse (2011) also involved individuals who lacked the engineering training to make a wind load assessment on the temporary truss system.

These are examples where taking knowledge for granted leads to consequences. Qualifications are not just paperwork; they are the cradle of human reliability. In a technical setting, poor training creates a facade of safety until an abnormal event occurs. When this happens, untrained people are left without the means to react effectively.

## Converging Regulation and SOPs

In reaction to repeated incidents of unqualified personnel performing work, the industry has moved aggressively toward formal credentialing and documented competency. OSHA's Cranes and Derricks in Construction rule (29 CFR 1926 Subpart CC) now requires that all crane operators in the United States be certified through an accredited program and recertified periodically. States such as NY, CA, and WA have incorporated additional licensing and local registry requirements to ensure that operators and signal persons are trained.

Rigging, lifting, and specific assembly work require formal verification of competency and adherence to manufacturer instructions. Many companies utilize qualification matrices, a documented trail of each employee's training, certifications, and authorized levels for operational tasks, for each specific task. The intent is to ensure that no one operates or directs work outside their demonstrated capabilities. In high-risk activities such as scaffolding, confined space entry, and post-tensioning, safety plans must clearly identify trained and competent individuals.

Infrastructure and industrial owners are making training part of their prequalification requirements for contractors. Programs such as NCCCO (National Commission for the Certification of Crane Operators), AWS (American Welding Society), and ASNT (American Society for Nondestructive Testing) are now written into specifications. Contractors may invalidate their contracts and insurance coverage by failing to use trained personnel.

## Responsibility and Economics

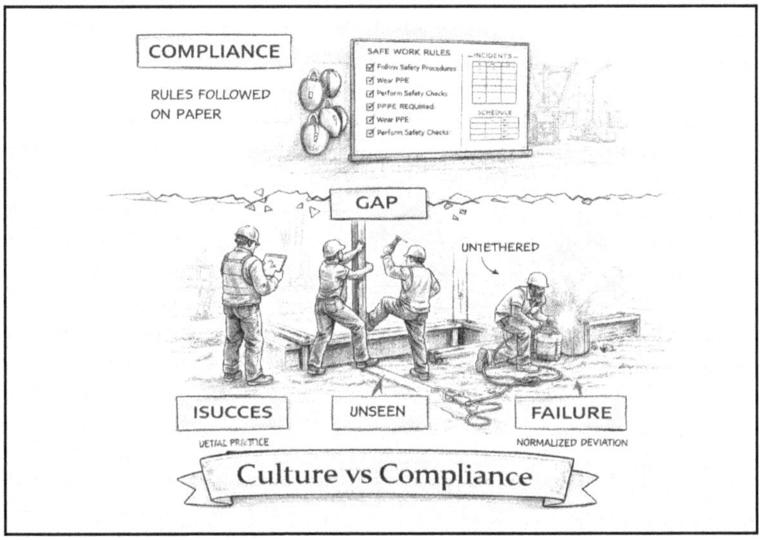


The ramifications of using unqualified individuals exceed the risk of failure. Economically, organizations will incur fines, work stoppages, and litigation if accidents occur under the supervision of noncredentialed individuals. Furthermore, reputational damage may be irreparable. Numerous lawsuits faced the dismantling subcontractor following the Seattle crane disaster, and the adverse effects extended to the entire industry, leading to altered training and supervision requirements for tower crane operations in North America.

The industry is beginning to realize that training is not an expense, but an investment in insurance. Credentialed and trained workers statistically have a lower chance of incidents, produce higher quality work, and identify risks in advance. In contrast, an absence of competence increases liability. New contracts and insurance policies require more proof of qualifications as a condition of coverage or payment. Owners are beginning to understand that saving money on training can increase total costs through claims, delays, and lost productivity.

## Human vs. Technology

Technology may provide new modes of delivering training, such as simulation, augmented reality, and digital modules, but it cannot supplant experience and mentoring. Training must evolve from compliance to understanding. Numerous contemporary systems fail not because workers choose not to follow procedures, but because they never fully understood them. Digital platforms for training are essential due to accessibility; however, hands-on, practical training cannot be substituted in construction environments, where decisions must be made in seconds under stress.

Similarly, technology can create an illusion of knowledge. Automated systems suggest ease of operation and lower the perceived need for training. When technology is pushed beyond normal conditions or fails, untrained personnel will not know how to respond. The key is balanced integration, where technology enhances human skill but does not replace it.

Inadequate training and qualifications, like other causes of failure described here, are preventable. Inadequate training often stems from a lack of organizational and personal perspective, as well as an undervaluation of professional development. Each time an untrained operator sits in a machine or an unqualified supervisor signs off on work, the entire project assumes an unseen risk that it will pay for later. Training is not an incidental cost of doing business; it is the business of ensuring construction is safe.

## Equipment and Component Failures (True Mechanical)

Although the majority of catastrophic events reported in workers' compensation cases start with human or procedural error, some events begin with the physical systems themselves. Equipment and component failures occur when machinery, materials, or assemblies break down due to stress. With respect to "mechanical" failure, even these failures are rarely strictly technical; there is often an additional layer attributable to a lack of maintenance, improper use, or inadequate system redundancy. Equipment is no more reliable than the systems that keep it maintained and monitored. The failure of the Taum Sauk Upper Reservoir Dam (2005) in Missouri is a vivid example of mechanical system failure compounded by human neglect. The sensors that indicated the reservoir level, which operated the automated pumps, were miscalibrated and provided faulty data to the plant's control logic. The pumps continued to run and overtopped the dam, leading to erosion of its rockfill structure and eventual failure of the wall. While the sensors malfunctioned, the problem was systemic: poor calibration, inadequate testing, and the removal of human oversight in favor of automation. The Boston Big Dig ceiling collapse (2006) shows a material and installation failure continuum. The anchors used to tether heavy concrete panels were never qualified for sustained tension loading. Over time, under constant tension, the epoxy creeped until the anchors pulled free, dropping the panel onto vehicles below.

In L'Ambiance Plaza (1987), it is conceivable that a jack or lifting seat malfunction contributed to the collapse following slab raising. While it was never definitively proven that the hardware was at fault, this incident illustrates the risk of using equipment not proven for the task or not sufficiently tested for heavy lifting operations. Mechanical reliability also contributed to other incidents, such as the Oroville Dam spillway failures (2017), where poor material performance in the concrete and underdrain systems caused erosion of the spillway structures, and the Big Blue crane collapse (1999), where excessive dynamic loading exceeded design capacity.

A common feature of these events is that the equipment did not fail in isolation. The failures transpired within contexts of human overreliance, process gaps, or neglected maintenance. In almost all cases, the failure could have been avoided through more rigorous testing, proper calibration, or monitoring redundancy. The reliability of machines is not constant; it tends to decline over time. Without disciplined intervention, decline turns into failure.

## Regulatory and SOP Changes

In response to failures associated with machines, both industry standards and regulations have shifted to emphasize equipment qualification, redundancy, and maintenance throughout the life cycle. After Taum Sauk, the Federal Energy Regulatory Commission (FERC) and the U.S. Army Corps of Engineers revised dam safety regulations to require multiple independent level sensors, physical overflow weirs, and human oversight of automated systems. Federal guidance on automating risk management was updated to indicate that any automatic sensor related to critical infrastructure must have manual verification and alarm redundancies.

In construction, the Big Dig collapse spurred reform in anchor and fastener design. The American Concrete Institute (ACI) developed ACI 355 standards for testing and qualification of post-installed anchors. The Federal Highway Administration (FHWA) required verification of adhesive anchor performance for long-term load applications. Suppliers must now provide test data showing that adhesive anchors meet sustained-load and environmental durability requirements.

Similarly, equipment inspection requirements have grown and must meet OSHA and ANSI standards. Cranes, jacks, and other lifting equipment must be visually inspected daily and periodically inspected according to a set schedule, with documentation to ensure traceability. Current codes emphasize the importance of lifecycle management, requiring components to be tracked, tested, and replaced at established intervals, regardless of their apparent condition.

## Accountability and Economics

When components or equipment fail, accountability often extends to multiple parties, including manufacturers, engineers, contractors, and owners. The Taum Sauk failure resulted in nearly $200 million in costs to AmerenUE in fines, settlements, and reconstruction. The arched roadway collapse in the Big Dig led to multimillion-dollar settlements and criminal charges and drove permanent changes in transportation oversight in Massachusetts. These events reshaped the economics of risk designation. Owners and insurers now demand delineation of responsibilities for calibration, maintenance, and testing. Large project specifications are increasingly including warranties and reliability clauses that assign responsibility for system and component performance. Preventive maintenance and monitoring programs, once considered operational expenses, are now viewed as prudent fiscal measures and essential protection. Equipment performance and component reliability have become integral to project risk assessment, alongside construction quality and design.

## Human vs. Technology

Mechanical failure may appear technical, but it is human in nature. Sensors fail because they are miscalibrated, components fail because they are misapplied, and systems fail because they are not checked. Technology is evolving to reduce human exposure to error. Real-time monitoring systems, load sensors, and predictive maintenance analytics allow engineers to assess degrading systems one step ahead of failure. Artificial intelligence and digital twins are emerging tools for forecasting mechanical vulnerabilities. Nonetheless, these tools require skilled interpretation and conscientious upkeep. The Taum Sauk disaster revealed the risks of overtrusting automation: a machine operates only within the boundaries established by people. Reliability rests not in the machine itself, but in the human systems that support it.

The lesson remains timeless: every piece of equipment and material has a lifespan, tolerance, and maintenance threshold. Safety relies not on unquestioning belief in reliability, but on repeatedly proving it. A single sensor, bolt, or anchor can make the difference between stability and failure. Mechanical systems do not fail society; society fails the systems when vigilance becomes complacency.

## Lack of Preventive Maintenance and Postponed Repair

No structure, regardless of design, material selection, or construction, is immune to deterioration once it is completed. Materials age, loads shift, and the environment continue to stress the structure. Preventive maintenance, the act of routinely inspecting, repairing, and renewing components, is the system that combats entropy. When maintenance is delayed until after an event or failure, the structure silently accumulates deterioration, leading to inevitable failure. Whether due to cost saving, oversight, or complacency, lack of repair and delayed intervention are chronic contributors to construction failures.

The collapse of Champlain Towers South (2021) remains the most devastating example of chronic repair practices. Engineering reports cited years of water intrusion, structural spalling, and corroded rebar as “major structural damage” requiring repairs. However, repairs were postponed for almost three years because the homeowners’ association could not agree on financial responsibility. The restoration cost, estimated at $15 million, remained under discussion until the structure reached an irreversible point of decay. This failure killed 98 individuals and changed how the nation understands accountability in maintenance.

The Fern Hollow Bridge collapse in Pittsburgh (2022) revealed substantial corrosion in critical load-bearing elements. Investigators determined that the drainage systems had been clogged for years, allowing water and de-icing salts to pool around the structural steel members. This caused significant section loss and fatigue cracking, which went unnoticed or uncorrected despite multiple inspection reports. The 57 Ann Street parking garage collapse in New York City (2023) was analogous. Decades of corrosion, concrete spalling, and exposed rebar were documented by engineers, but repairs were postponed indefinitely.

At the Oroville Spillway (2017), inspectors repeatedly noted surface cracking and ineffective underdrains, but no major repairs were completed. Erosion beneath the slab deepened until it failed during a flood event.

The common thread across these cases is not ignorance, but inaction. Engineers and inspectors were aware of the deterioration; however, the systems of funding, enforcement, and maintenance prioritization failed to convert this knowledge into action. When timely intervention is not made, repairable defects become structural failures. In construction, neglect is rarely an event; it is a timeline.

## Changes in Regulatory and SOP Requirements

In light of these events, jurisdictions across the United States strengthened maintenance and inspection requirements for aging structures. In Florida, condominium milestone inspection requirements mandate that each condominium building exceeding three stories be reviewed structurally every 30 years (or 25 years in coastal zones) and then every ten years thereafter. The requirement includes waterproofing and corrosion protection assessments, as well as structural assessments, by licensed engineers. Other municipalities, such as New York City, Pittsburgh, and Los Angeles, have instituted periodic inspections for garages, bridges, and balconies. In New York City, for instance, Department of Buildings rules and Local Law updates for facades and parking structures expanded required inspection intervals and obligate owners not only to inspect but also to report conditions and repair timelines. These regulations are reframing preventive maintenance as a requirement rather than an option.

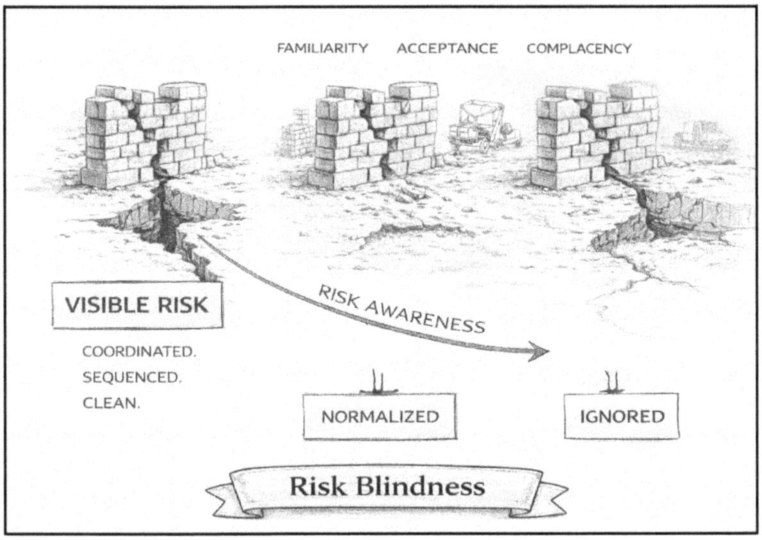


Industry standards have also evolved. The Federal Highway Administration (FHWA) updated bridge inspection standards in the National Bridge Inspection Standards (NBIS) to better address corrosion control and drainage maintenance. The American Society of Civil Engineers (ASCE) has updated asset management guidance to establish lifecycle maintenance programs, condition monitoring plans, and risk-based intervention plans. These standards recognize that maintenance is not a cost center but an engineering responsibility.

## Accountability and Economics

The economic consequences of deferred repair are among the most severe in the industry. At Champlain Towers South, the estimated $15 million restoration, delayed by budget disputes, was overshadowed by more than $1 billion in property loss, settlements, and demolition costs. The Fern Hollow Bridge collapse resulted in emergency reconstruction costs exceeding $25 million, far more than the projected cost of basic maintenance. These instances demonstrate false economy: saving money in the short run by expending lives and resources in the long run.

In response, insurance markets and lenders now require documented maintenance histories as part of risk assessment. Properties without documented inspections, or with evidence of deferred maintenance, face steep premiums or loss of insurability. For public infrastructure, federal funding is increasingly linked to asset management performance and quantifiable maintenance metrics. Deferred maintenance is no longer viewed as a budgeting exercise, but as a liability.

## Human vs. Technology

Technology can enhance preventive maintenance, but it cannot relieve human responsibility. Structural health monitoring (SHM) systems that employ embedded sensors, strain gauges, and corrosion detection provide real-time data on material condition and performance. Like any digital data, SHM information is of little use if no one in the organization is engaged enough to interpret and act on it. At Champlain Towers, engineers had more than enough visual evidence of distress affecting structural integrity; technology was not needed to confirm the necessity of repairs. The issue was not data, but decisions.

Human factors also influence maintenance, as budget constraints, bureaucratic delays, and the normalization of "it has always looked like that" can lead to deferred repairs. Technology provides early warning tools, but leadership and accountability determine whether warnings are heeded. Preventive maintenance requires scientific rigor and moral clarity: spending funds on preservation before the far higher costs of failure are incurred.

Every major structural failure caused by neglect reads as a warning in hindsight. Maintenance is seldom celebrated, yet it is central to resilience. Structures do not fail solely from age; they fail from neglect. Societies are responsible not only for who builds safely, but also for sustaining what has been built, with vigilance that extends beyond the ribbon-cutting ceremony.

## Design–Construction Interface Gaps

Theoretically, a seamless transition exists between design and construction. The engineer designs the structure, and the contractor constructs it according to the precise specifications. However, the design–construction interface is one of the most fragile and failure-prone transitions in the entire project lifecycle. Design–construction interface gaps occur when deliverables deviate from the designed specifications, or when the design overlooks factors related to construction staging, temporary loads, or field modifications. Design–construction interface gaps are not fundamentally “errors” when considered in isolation; they are failures of integration, communication, and shared responsibility.

The collapse of the Hyatt Regency walkways (1981) is the quintessential case study of an interface failure. While the final connection detail that resulted in the collapse of two suspended walkways cannot be classified as a design error per se, it was instead a revision to a shop drawing made by the steel fabricator that inadvertently exposed the hanger rod to double the load. This revision was passed along for approval, but it was never considered or evaluated by the engineer of record. The design–construction interface failed because the design team assumed the construction team would verify, and vice versa.

As was the case with the L’Ambiance Plaza (1987) model, the structural design was reasonable; however, the designer did not adequately consider the construction sequence of the lift-slab system. Although the final lift-slab configuration achieved the theoretical design requirements, the structure was deficient in stability during its intermediate lifting condition. The construction team built precisely to the design drawings; however, the drawings did not mitigate the construction team’s liability during execution.

The occurrence of the FIU pedestrian bridge collapse (2019) evidenced a contemporary iteration of the same issue. The design–build team did not coordinate the development of the design analysis with the construction process itself. When cracks were observed at a critical truss node during post-tensioning, the cracks were considered only cosmetic, not structural. The bridge was open to traffic below the structure when the node reached its failure point. Similarly, the Miami Dade College garage collapse (2012) illustrated a breakdown between engineering intent and field execution. The precast panels were erected without adequate bracing due to procedural errors resulting from poor detailing and a failure to verify with the designer and erector that the provided bracing would provide stability.

The design–construction interface is where responsibility tends to be blurred. In sophisticated projects, the design drawings will undergo multiple layers of review by engineers, fabricators, subcontractors, and field supervisors, each with their own interpretations and constraints. When it is unclear from the design intent what a buildable instruction is, assumptions will then be made in the field. Those assumptions, while practical, may misinterpret the design’s structural logic.

## Regulation and SOP changes

To close those gaps in the execution of the design, the construction industry has redefined how design and construction communicate. Most jurisdictions now require the review and approval of all shop drawings, field modifications, and delegated designs by the Engineer of Record (EOR) before execution. Any change that can affect load paths, connections, or stability must undergo analysis before being implemented. The once-perceived design–build delivery model of risk sharing now follows a duty-separation track of activities. Even when the builder and designer are under a single contract, the duty of the design remains the same.

Building regulations, such as the International Building Code (IBC), as well as codes and standards from AISC and PCI, now contain explicit requirements for construction stage analysis. Engineers must evaluate not only completed structures but also temporary conditions during erection, shoring, or lifting. In structural steel and precast construction, Special Inspections confirm that connections, welding, and assembly are performed in accordance with design intent prior to transferring loads. These precautions are taken with the understanding that safety is a function not only of the adequacy of design, but also of continuity of design. In other words, drawings, details, and decisions must be aligned from conception to completion.

## Accountability and Economics

Design–construction disconnects create a monumental economic burden. The two prominent case examples of Hyatt and FIU resulted in over $1.5 billion in combined settlements, excluding ancillary economic losses such as project cancellations and professional discipline. The cost of implementing outcome-based coordination from design, including additional reviews, pre-construction meetings, and shop drawing audits, is trivial in comparison to the consequences of failure. Owners and carriers now require traceability in the flow of design documentation, along with version control and sign-offs on design documentation at every phase of the process. In terms of accountability, these cases modified what is acceptably expected in professional liability. Engineers of record cannot typically pass responsibility for changes in shop drawings without a review, and contractors cannot claim ignorance of design limitations. The consensus position, legally and ethically, is that both sides have the responsibility to check and validate. Design and construction are two halves of a single system, and when either side neglects to check communication with the other, there is risk.

## Human vs. Technology

Technology has significantly improved the design–construction interface, but it has not eliminated human error. Building Information Modeling (BIM) platforms, digital twins, and 3D coordination tools now enable engineers, contractors, and fabricators to visualize conflicts and sequencing promptly. However, these tools require disciplined data management and ongoing involvement by all parties. Misalignment can still occur if updates are not made in sync or if humans interpret or render information in ways that diverge from the data.

The failures at FIU and L'Ambiance remind us that the best technology cannot replace professional judgment. Interface management must strike a balance between technical precision and human coordination, encompassing regular meetings, thorough documentation, and a culture of open communication. The interface is not a barrier, but a collaborative workspace where both disciplines must take equal ownership.

The design–construction interface is arguably one of the most fragile fault lines in modern construction. This is the juncture where creativity meets reality, where theory becomes real. When done well, it results in structures that are beautiful yet safe. When it fails, it blurs the line between design error and construction error, because the reality is that there is no clear distinction between the two. A gap in understanding is a gap in structure, and both can hinder a project's progress.

## Unforeseen Loads and Environmental Factors

There is no design, no matter how sophisticated, that can eliminate uncertainty. The physical world is dynamic. Wind, water, temperature, and time constantly test the limits of engineering prediction. Even if a structure is designed well, failure will occur when environmental forces exceed its limits, or when temporary or unplanned loads are introduced during construction. The failures described herein are not derived from nature alone, but rather exhibit human behavior in the failure to recognize nature's potential. In this regard, modern construction management has made the unknowable more knowable; however, failures appear when people deem unknown conditions irrelevant.

For example, during a wind event in Milwaukee on October 27, 1999, the field performance of the Big Blue crane was not adequately evaluated. The lift was attempted despite wind speeds being far above safe limits. The project team had access to weather data; however, the project schedule weighed into the decision-making process. The load being lifted was significantly large. Wind effectively added multiple times the load due to sail effects on both the load and crane, and the forces on the crane exceeded its capacity. When lateral forces overwhelmed the crane, nearly 50 feet above ground, it failed catastrophically. A similar sequence of events led to a stage collapse at the Indiana State Fair in August 2011.

High winds were forecasted, and the temporary rigging structure was neither evacuated nor reinforced. When conditions rapidly changed and a gust front took effect, the stage fell into the audience, killing seven and injuring many. In 2009, the Cowboys' practice facility roof in

Dallas collapsed due to wind loads that were within code but not properly accounted for in the flexible frame design. The structure was ultimately evaluated for local wind conditions and dynamic effects, taking into account the lightweight structure's intended use. The structural failures from Hurricane Katrina in 2005 vividly demonstrated the consequences of underestimating the environment and hydrology. With levees designed using outdated surge assumptions and unstable soils, levees failed in multiple locations, unable to withstand hurricane storm levels. The incidents involving the Oroville Dam spillway (2017) and the Willow Island cooling tower failure (1978) demonstrate that the environment can contribute to collapse and that extreme conditions, such as excessive water flow or cold weather with slow concrete setting, can initiate failure.

Environmental conditions coupled with unanticipated loads present a humbling challenge for the engineer balancing science and practice. Design codes are based on historical evidence and probability, but with climate change exacerbating dramatic weather events, and with advancing construction techniques displacing historical assumptions, those assumptions can quickly become outdated. A once-in-one-hundred-year event may now occur once per decade. Further, construction techniques that do not account for temporary or changing loads, such as wind gusts, partial dead loads, or shifting foundation loads, downplay the environment's ability to expose weak links in the plan.

## Regulateion and SOP Changes

Based on a history of disasters, engineering codes and standards have incorporated practices for more advanced modeling and monitoring. ASCE 7, in significant revisions to Minimum Design Loads for Buildings and Other Structures, has integrated site-specific wind, flood, and seismic criteria, with updated regional design maps reflecting current climatology and a focus on dynamic effects, including gust factors and turbulence.

Temporary works, including crane use and requirements, now follow safer practices with weather action plans that identify maximum allowable wind speeds, monitor wind in real-time, define actions when conditions approach or exceed limits, direct the cessation of work or evacuation in unsafe conditions, and reinforce adherence to established rules. For flood risk management and dam design, agencies such as FEMA, the U.S. Army Corps of Engineers, and FERC have reformed model criteria within hydrology and geotechnics. These reforms include probabilistic flood hazard analysis, soil–structure interaction analysis, and real-time monitoring of hydraulic pressures. Resilience design, the ability to absorb damage without catastrophic loss, is now required for projects located in coastal or seismic environments.

## Accountability and Economics

The economic implications of underestimating environmental loads are staggering. The costs from Hurricane Katrina's failures exceeded $125 billion in damages, not including long-term displacement and lost economic activity. The Oroville Dam spillway repairs alone exceeded $1.1 billion. More localized events, such as the Big Blue collapse or the Indiana State Fair incident, resulted in multimillion-dollar settlements and years of litigation. These events remind us that forces of nature are not simply environmental variables; they also pose financial risk when underestimated.

The modern insurance and financing industries have taken notice. Underwriters require climate and environmental risk assessments as part of project due diligence. Contractors are now required to demonstrate adaptive scheduling, weather monitoring, and protection plans prior to being insured. Owners are beginning to understand that environmental resilience is not optional; it is the cost of long-term economic viability.

## Human vs. Technology

Technology offers excellent tools for forecasting and mitigating environmental risk. Computational fluid dynamics, remote sensing, and real-time weather analytics provide greater accuracy than previous generations. Structural monitoring systems capture real-time data on vibration, load distribution, and environmental strains. However, these tools are only valuable if people use them and trust their warnings. In many historical failures, notifications from sensors and forecasts were often ignored in favor of intuition or prior experience. The decision to lift Big Blue or to continue the Indiana State Fair event was not due to a lack of technology; it was due to human overconfidence. Similarly, overreliance on models without a thorough understanding of their assumptions and limitations creates blind spots. Engineers must juxtapose data with professional skepticism and ask, "Do these models still reflect reality?"

There are unforeseen loads and environmental factors that remind the industry that nature is not static, nor can engineering be static. The challenge is not to predict every load or condition, but to design for uncertainty itself and plan to recover. The price of humility is small compared to the price of arrogance. Nature will continuously test our work. The question is whether we have given nature the respect it is due in our designs and decisions.

## Lack of Redundancy and Fail-Safe Systems

Redundancy is the unspoken ally of safety. In structural design, it is the concept that one failure does not equal total collapse. Redundancy, achieved through secondary load paths, backup supports, and connection continuity, provides multiple options. When one element fails, others redistribute the load, allowing time for detection and action. If redundancy is not present, the structure becomes brittle, or it remains strong only as long as nothing goes wrong. A lack of redundancy will turn localized failures into global catastrophes. The Hyatt Regency walkway collapse in 1981 is a clear representation of this principle. The hanger rod connection that failed had no alternative load path. When the first connection failed, the entire walkway dropped, and then the one below it dropped as well moments later. The failure of a single connection detail, a single point of failure, resulted in 114 fatalities. At L'Ambiance Plaza in 1987, the lift-slab system also had no redundancy during intermediate construction stages. When one lifting jack or seating insert failed, the load could not be redistributed, and the entire structure collapsed floor by floor. The 2018 FIU pedestrian bridge collapse illustrated a similar outcome. The truss-to-deck connection that cracked just days before the collapse was the only load path carrying forces through the central node. When that connection failed, the entire bridge failed almost instantaneously; there was no residual capacity for redistribution. The Champlain Towers South collapse in 2021 represented a progressive collapse with years of corrosion and deterioration preceding the event. Punching-shear failure at slab-to-column connections triggered a progressive collapse of the entire building from the localized failure of one connection. These examples highlight a fundamental truth: even when they perform for years, structures defined by a lack of redundancy provide no forgiving margin for error.

In temporary works, redundancy is often eliminated to gain efficiency. The Big Dig tunnel ceiling collapse in 2006 relied on a single adhesive anchor system with no alternate or backup restraint. When one set of anchors experienced creep and failed, the concrete panels fell, and a motorist was fatally injured. Such designs rely on perfect materials and perfect behavior, conditions that are rarely present in the real world. When redundancy does not exist, there is no variation, no error tolerance, and no time to react.

## Regulatory and SOP Changes

In response to these failures, codes and standards of practice have introduced changes to increase structural robustness and integrity. The American Society of Civil Engineers (ASCE) and the American Institute of Steel Construction (AISC) now emphasize integrity design, which provides alternative load paths to retain some capacity after damage and prevent total failure. The International Building Code (IBC) incorporated explicit integrity provisions requiring linkages for critical components to avoid differential failure.

For temporary works and specialty systems, the Occupational Safety and Health Administration (OSHA) and American Concrete Institute (ACI) now require catch systems or shores beneath lifted elements to limit total failure early in construction. Federal and state transportation agencies have mandated redundancy and multiple load-carrying members or continuous spans for bridges to prevent collapse if the first member fails. These changes codify a shift in design philosophy from strength alone to resilience through redundancy.

Modern structural health monitoring (SHM) systems now have an adjunct role. With embedded sensors detecting stress redistribution, engineers receive clear signs when one element is taking more than its share of system load and the system is trending toward failure. This approach allows for intervention before collapse.

## Accountability and Economics

The financial repercussions of insufficient redundancy are staggering because they create a multiplicative effect from a single failure. The consequences of litigation from the Hyatt and FIU collapses financially crippled or led to the dissolution of the companies involved. Rebuilding required more than physical reconstruction; it required rehabilitating reputations, a process that can take decades or may never fully succeed.

Owners and insurers now consider redundancy an economic resource. While it may be more expensive to design redundant systems, the additional cost is minuscule compared to the liability associated with the absence of redundancy. Increasingly, insurance underwriting evaluates redundancy as a factor in risk mitigation. Redundant systems can reduce exposure to catastrophic claims and enable faster recovery after localized damage, supporting long-term financial stability.

## Human vs. Technology

The absence of redundancy is more a matter of design ethos than technology. The tools necessary to create redundant systems, including advanced analysis programs, finite element modeling, and load simulations, are readily available and accessible. What is often missing is the human element that balances cost, efficiency, and safety. When time or budget constraints are present, redundancy can be mischaracterized as a luxury rather than a necessity.

Technology can provide monitoring and early warning, but it cannot replace redundancy. If a structure is designed without fail-safes, technology alone cannot ensure its safety. Redundancy must be part of the system from the outset.

The ethos of redundancy embodies humility in design. It acknowledges that failure, decay, and unknown conditions exist. Events like Hyatt, FIU, and L'Ambiance remind the industry that safety does not come from treating human imperfection as perfection, but from expecting imperfection and ensuring that failures cannot propagate. Every engineer and builder has a responsibility to design and build with forgiveness, so that a structure can withstand human imperfection.

## Regulatory Gaps and Lag

Regulations and standards are the bedrock of safety in construction, establishing the minimum level of protection that society is willing to tolerate in its building infrastructure. However, historically, regulation often follows tragedy; standards evolve reactively; and regulation lags. Consequently, considerable delays occur when construction processes progress faster than the regulatory codes that govern them, or when oversight agencies are slow to respond to an obvious safety issue. This results in an unlevel playing field, where innovative or high-risk practices can exist without regulation or guidance until the inevitable accident occurs, forcing us to reform.

The Willow Island Cooling Tower collapse (1978) is one of the earliest examples of regulatory lag in modern construction safety. At the time of the collapse, no OSHA standard required verification of cured concrete strength before loading formwork or scaffolding was placed. The crew's fatal decision to tie climbing scaffolding to green concrete was not, technically, a violation of the code, since no code prohibited it. The collapse, which killed 51 workers, led to immediate changes to OSHA rules related to cure strength verification and cure time documentation.

The L'Ambiance Plaza collapse (1987) demonstrated similar gaps. The lift-slab method is a relatively new construction technique that lacks specific regulations for ensuring temporary stability during the lifting process. Neither OSHA's safety regulations nor state agencies provide adequate technical criteria to govern these systems, leaving the decision-making largely up to the contractor's discretion. Following the collapse, the U.S.

Department of Labor developed emergency standards for lift-slab operations, which required engineered lifting sequences, independent inspections, and redundant supports. The FIU pedestrian bridge collapse (2018) brought a new manifestation of regulatory lag. The design-build project involved fragmented regulatory oversight among federal, state, and local entities, which meant that no single entity assumed responsibility for ensuring the project was safe or underwent sufficient design review. Most critically, there was no established process for stopping traffic below a construction span. There were indicators of structural distress, but no clarity of authority or accountability. FIU registered engineering compliance with existing regulations; however, the regulations did not anticipate the risks associated with accelerated bridge construction.

The collapse of the Champlain Towers South in 2021 highlighted a significant concern regarding regulatory lag, particularly in relation to aging infrastructure that lacks enforceable recertification and maintenance supervision. While Florida's building codes outline important regulations regarding the design and construction of buildings, they do not provide provisions for the long-term maintenance of buildings or consideration of structural integrity. Moreover, condominium associations have considerable freedom to self-direct, which may lead to ineffective oversight. Buildings degraded and became noncompliant within the regulatory framework of the code. Only after the loss of nearly one hundred lives was a statewide recertification law for older buildings passed in Florida.

Regulatory gaps are not merely technical issues; they reflect systemic problems. Regulatory lag occurs when new construction practices, materials, or ownership models outpace the bureaucracy designed to respond to them. It can also occur when a known safety risk is acknowledged but not codified, when everyone agrees on what should be done, but nobody writes it into law.

## Shifting Regulation and SOP

After each of these tragic failures, regulations were revised, often in a timely manner. OSHA Subpart Q was updated following the Willow Island incident, revising regulations regarding the verification of formwork strength, bracing design, and inspection of shoring before loading. The tragedy of L'Ambiance led to the establishment of 29 CFR 1926.705, which governs the construction of lift slabs and requires a temporary stability analysis as part of the construction design. The collapse at FIU led the Federal Highway Administration (FHWA) and the National Transportation Safety Board (NTSB) to release guidance on accelerated bridge construction. The guidance emphasized the need for clearer lines of authority, independent design review, and protocols for closing traffic during phases of construction when conditions could be hazardous. The collapse of Champlain Towers led to sweeping changes in Florida law regarding building maintenance, including milestone inspections for condominiums and the establishment of reserve funding for future repairs.

In addition to regulatory changes, there was also a change in regulatory philosophy. Agencies such as OSHA, ASCE, and ACI have begun to consider performance-based standards that concentrate on outcomes (for example, structural reliability or worker safety) rather than simply on prescriptive rules. There is some flexibility to innovate, but ultimately, agencies hold organizations accountable for outcomes. In parallel, enforcement mechanisms are evolving through the use of digital permitting, automated inspection tracking, and public reporting to increase transparency.

## Accountability and Economics

Regulatory gaps present enormous economic and reputational liability to responsible agencies or institutions. Public confidence is eroded when government oversight fails, a phenomenon similar to what occurs in the private sector. The FIU bridge and Champlain Towers failures led to congressional inquiries and insurance implications that crossed many sectors. Municipalities and agencies were liable to victims and faced lawsuits from insurers and property owners, who alleged regulatory negligence.

From an economic perspective, reactive regulation is expensive. The costs of repair after a failure or retrofitting other structures after a failure can be more expensive than the costs of proactive inspection and regulation. The "regulate after the fact" model is neither realistic nor ethical. Anticipatory maintenance and inspection programs are much more efficient than rebuilding after a disaster.

## Human vs. Technology

If positioned correctly, technology can close regulatory gaps more quickly than traditional bureaucratic cycles. Digital records, sensor networks, and information-sharing platforms enable regulators to monitor structural performance and compliance status in real-time. Machine learning models can reveal risk patterns across tens of thousands of assets, highlighting priorities for regulatory focus. Implementation still requires a strong political will and effective interagency coordination.

The primary cause of regulatory lag is human. Codes are written by committees and enforced by agencies constrained by budgets, politics, and tradition. Technology can facilitate the flow of information throughout the regulatory chain, but leadership is necessary to transform that information into actionable policy.

Every time a bridge collapses, a dam fails, or a building falls, regulations change, but that change comes at an unwritten price: human life. The challenge for today's engineers and policymakers is to reverse that equation. Human life cannot be the currency by which regulatory frameworks improve. We need regulation that is anticipatory, not reactive; preventive, not merely remedial.

## Planning for Emergencies in the Service of Public Protection During Work

Worksites are temporary, as they are typically temporary housing for the future. The risks associated with construction are not contained within site boundaries. When projects interface with the public — for instance, highways under bridge construction, occupied buildings adjacent to demolitions, or events under temporary structures — safety responsibilities extend beyond the workers. Emergency planning and public protection during active construction are among the most neglected aspects of construction management. Workers are protected through OSHA safety rules and project safety plans, but the public relies on the judgment and foresight of those managing safety. When judgment fails, the ramifications are sudden and unforgiving.

The FIU pedestrian bridge collapse (2018) is the most egregious modern example. Cracks were seen in a critical truss node days before the failure, yet the highway below (U.S. 41) remained open. Engineers met at the bridge the morning of the failure to discuss the cracks, but did not close the roadway or cordon off a hazard area. A few hours later, the bridge failed during post-tensioning, killing the occupants of six vehicles passing under the span. A significant disconnect occurred between engineering knowledge and emergency action — the risk was known, but no one acted to protect the public.

The Indiana State Fair stage collapse (2011) showed the same neglect in a different form. Organizers and engineers knew a storm was approaching with winds strong enough to bring down the temporary structure, yet the concert continued. When a gust struck the temporary truss system supporting the stage, the truss collapsed into the crowd, killing seven and injuring more than fifty. Investigators found no formal evacuation procedure, wind monitoring trigger, or established emergency chain of command. In the Philadelphia 22nd and Market demolition collapse (2013), a contractor removed an exterior load-bearing wall next to an occupied store without isolating the area or evacuating customers. The wall fell, killing six people inside. While part of the failure was structural, another part was moral.

Failures in public protection also manifest in infrastructure. In the Big Dig tunnel ceiling collapse (2006), anchor connectors failed, dropping overhead panels onto active interstate traffic, killing the motorist. Concerns about the performance of the anchors had been documented for a long time. In the Oroville Dam spillway failure (2017), vulnerability was underestimated, leading to an emergency evacuation of nearly 200,000 people. In both situations, insufficient emergency planning compounded the negative outcomes.

These examples show a systemic gap: construction and operations teams must plan for worker safety, but adjacent public safety is often underaddressed. Emergency planning must extend beyond the fence line to include motorists, nearby residents, and anyone within reach of the site's operational and physical impacts.

## Regulatory and SOP Changes

After these failures, federal and state agencies expanded public protection within overall safety planning. OSHA's Subpart T (Demolition) now includes exclusion zones and protective structures for adjacent occupancies. Bridge construction standards and DOT/FHWA procedures require traffic control plans that account for closures and detours when construction occurs above or adjacent to active roads.

Temporary structures and public assembly requirements have also expanded. The ICC, through the IBC and IFC, added language requiring engineered design and verification of wind load for temporary assemblies. NFPA 101 now requires crowd management and evacuation plans for temporary venues. In some states, such as Indiana and California, operators are required to provide stamped, engineered designs and emergency plans before an event is permitted.

The NTSB and FHWA issued joint recommendations for new public safety guidelines in construction zones following the FIU incident. These include addressing public exposure during construction phases, defining specific trigger points for closing roads, and delegating authority to an individual empowered to stop work if hazards arise. Emergency Action Plans (EAPs) are now expected to be living documents integrated with the project schedule, risk register, and communication protocols.

## Accountability and Economics

The financial and legal consequences of neglecting to keep the public safe can be severe, as evidenced by multimillion-dollar settlements and long-term consequences after incidents at FIU and Philadelphia. Both groups faced damage to their reputations and a loss of licenses to practice. The costs associated with any delays or alternate routes will be minor compared to the economic and ethical repercussions of public deaths.

Tort law has broadened the number of parties against whom an entity can be liable, as owners and contractors can now be liable for exposing the public to construction hazards. Insurers will not even underwrite a high-risk project unless they have evidence of an adequate emergency and public protection plan. Municipalities have updated permitting parameters to require the owner to include plans for exclusion zones, barricades, and public evacuation as part of the site safety plan. Culturally and procedurally, the public is now a stakeholder in the safety of construction operations in public spaces because they have multiple and shared interests in the safety of construction operations in public spaces.

## Human vs. Technology

The introduction of technology enhances emergency planning and response capabilities. Integrated site management systems can, for example, provide automated alerts about hazardous conditions based on wind speed, vibration limits, or distance to the equipment. Geofences and smart barricades can also limit access and communicate real-time exposure zones. Drones and IoT sensors can provide real-time situational awareness.

However, as seen in the FIU and Indiana cases, technology works only when paired with a clear human decision-making hierarchy. In both cases, data were available — wind forecasts, visible cracks, distress signals — but no one was positioned or empowered to act. Effective emergency planning and public protection are more about hierarchy, defined roles, established authority, and rehearsed procedures.

Construction takes place in a living environment. It cannot proceed without members of the public nearby, and it cannot be planned without acknowledging the public. Emergency preparedness and public protection should be treated as seriously as design and engineering. A construction site that places the community at risk is not progress; it is an expectation of failure. The actual test of professional integrity lies not only in how well we build, but in how we account for those who never signed up to be part of the project.

## Governance Fragmentation and Diffusion of Accountability

Modern construction projects are becoming increasingly complex, involving the interaction of multiple parties (owners, contractors, subcontractors, consultants, designers, regulators, and financiers) with diverse roles, contracts, and interests. Specialization can increase efficiency, but it may also contribute to a division of responsibility. Governance fragmentation occurs when oversight and responsibility for risk become so widely distributed that no one organization maintains holistic control. The diffusion of responsibility leads to everyone being confident that someone else is ensuring safety, compliance, or quality standards are maintained, creating an organizational vacuum that allows unknown hazards to persist until they fail.

The levee and floodwall failures during Hurricane Katrina (2005) are noteworthy instances. The levee network protecting New Orleans was designed, constructed, and maintained by a variety of federal, state, and local agencies, each operating under differing legally defined mandates and funding sources. While the U.S. Army Corps of Engineers was ultimately responsible for design and construction, local levee boards were supposed to inspect and maintain the levee system. Communication gaps and differences in responsibility complicated accountability for failures. When the storm surge eventually overtopped the system, no entity could legitimately claim full accountability for the series of failures of the levee network as a whole. This catastrophic event was understood as a failure of engineering, but it was also a failure of governance.

The Oroville Dam spillway failure (2017) exhibited similar characteristics. The dam, owned by the California Department of Water Resources and operated under a federal license regulated by the Federal Energy Regulatory Commission (FERC), had multiple consultants and reviewers, none of whom had significant authority, and no single entity could require design changes or major repairs. The spillway had long been documented as deficient, but many of the reviews would push responsibility to other agencies or to a future funding source. Eventually, the spillway eroded during a significant release, requiring the evacuation of nearly 200,000 people and resulting in over $1 billion in repairs.

The Hard Rock Hotel collapse (2019) in New Orleans displayed a comparable fragmentation of risk within the private sector. A developer, several subcontractors, an engineer of record, delegated designers, and city inspectors each held responsibility for controlling a portion of the risk. Workers observed deformation of the structure and issued warnings, but the alerts were not escalated to someone with authority to stop the work. The municipal observation was reactive, rather than proactive. Subsequently, investigators revealed that no one entity had a complete understanding of how design changes and field practices interacted with one another.

L'Ambiance Plaza (1987) likewise reflected diffused accountability within a single project team. More than forty contractors and subcontractors were on site. Each had defined scopes, but no one held complete technical oversight of the lift-slab operation. When the unexpected occurred, investigators did not assign blame to a single firm; risk had been so widely distributed that nobody was accountable.

## Regulatory and Procedural Changes

In response, both public and private sectors moved to clarify the chain of accountability. Large infrastructure projects began requiring a Single Point of Accountability (SPA) — an individual or entity legally responsible for ensuring construction safety and compliance with standards. Federal and state agencies have transitioned to integrated risk management frameworks that require cross-jurisdiction collaboration, risk sharing, and consolidated responsibility. After Katrina, the U.S. Army Corps of Engineers reorganized its levee efforts into the Levee Safety Program (33 CFR Part 208), publishing national standards for inspection and monitoring. The rationale was straightforward: a single program with defined standards to mitigate risk.

Contracts and delivery methods have also undergone significant evolution. Design-build and Integrated Project Delivery (IPD) employ shared risk but now rely on clearly identified RACI roles (Responsible, Accountable, Consulted, Informed). OSHA, state licensing boards, and insurers increasingly require project-specific safety management, competency designations, and identification of the ultimate responsible party for each critical process. Municipalities have intensified inspection coordination, requiring major decisions or variations to be approved by the engineer of record or a designated construction safety coordinator.

## Economics and Accountability

Fragmented governance is costly before and after failures. Projects without apparent authority suffer cost overruns, schedule delays, and scope disputes long before accidents occur. When failures occur, litigation becomes protracted and expensive as parties attempt to deflect blame. Legal fees and reputational damage often exceed the cost of proactive coordination.

Katrina-related litigation dragged on for years, costing agencies millions while victims awaited resolution. Oroville produced similarly complex cost-sharing disputes between state and federal agencies. The takeaway is clear: accountability may be expensive to implement, but confusion is always more costly.

From an owner's perspective, governance clarity now aligns with financial responsibility. Insurers, lenders, and investors assess risk using both technical indicators and governance structures. Projects that demonstrate oversight, authority, and interagency cooperation receive more favorable terms and lower premiums. In a world where trust is currency, governance transparency has tangible value in construction operating costs.

## Human vs. Technology

Even with technological advances supporting collaboration, technology cannot replace accountability. Digital project management systems, shared data logs, and BIM environments improve visibility, but they do not create responsibility. A shared model can reveal a problem; someone still must own it and execute a solution. Effective governance is not about generating more information; it is about establishing apparent authority to make decisions.

As seen at Oroville, Hard Rock, and other events, information availability does not guarantee success if no one is in charge of solving the problem. Accountability depends on people — leadership, communication, and ethical ownership — the foundations of good construction governance. Technology can facilitate coordination, but accountability must be embedded in people, contracts, and law.

Fragmented governance and diffused accountability remain pervasive root causes of modern construction failures. As projects grow in scale and scope, the issue becomes more acute. Viewed differently, clarity itself is a robust safety measure. Industry maturity will be measured not only by what we build, but by how clearly we designate who is accountable when things go wrong — because in construction, everyone's job is someone's life.

## Asset management and funding shortages

The management of a built asset does not begin and end with the design and construction; instead, a built asset's lifespan is a function of the management of the asset during and beyond its life. Management of built assets includes continuous monitoring, maintenance, and renewal to ensure reliable and safe asset performance. When funding shortages or poor management are in place, it can be challenging to resume asset management, leading to its deterioration at a faster rate than it can be repaired. The overall result can be disastrous. The result was not a disaster, however, because the assets were poorly built and designed, but rather the abandonment of the systems required to manage the assets adequately.

The collapse of the Fern Hollow Bridge in Pittsburgh (2022) provides a timely and symbolic example of this phenomenon. The bridge had been rated "poor" by engineering standards for over a decade. Inspection reports repeatedly indicated significant corrosion, clogged drainage systems, and section loss in critical structural members. Funding issues or commitments elsewhere allowed the owner to delay fixing the bridge year after year. When the bridge collapsed, injuring many and disrupting important access between two transportation corridors, it was not surprising to most people with knowledge of the bridge's history. This was a failure of asset management that assisted no one, as it was expected but not funded.

Similar to the consequences of asset management practice in the private sector, the collapse of the Champlain Towers South (2021) exemplifies the consequences. While condominium associations have long-term responsibility for managing the built assets, they often lack the financial means or clinical requirements to establish reserves. When engineers reported “major structural damage” to the building preceding the collapse, the ownership group disputed the $15 million estimated repair and simply declined to raise assessments. The overall ability of the building to withstand physical deterioration progressed until, finally, a localized interior slab failure propagated and extended throughout the structure. The Oroville Dam spillways failed in February of 2017, and so did levees from Hurricane Katrina in 2005; all follow the same principle at the asset infrastructure level. In all instances, significant issues existed in the location of deficiencies and vulnerabilities that were and are known to have been engineered into the designs of the structures. In all instances, underfunded maintenance budgets, disorganized repairs, and political paralysis hindered the timely management of mundane technical and physical conditions that are well known to pose risks. The outcome was an avoidable calamity, with recovery costs approaching billions of dollars, when an expense many times less could have addressed the asset management shock prior. In neither case is there a technical defect; rather, a choice, due to administrative inattention, resulted in asset failure. All underfunded maintenance budgets, maintenance repair policies, and long-term capital planning create a system of fragility. Structures can degrade in this manner as well, until an incident where a defect and failure to observe are combined constitutes a failure, leading to inevitable collapse.

## Regulatory and SOP Shifts

Once failing to function as expected or at risk of dysfunction, asset management has become considered a discipline within engineering and public policy. The Federal Highway Administration (FHWA) now mandates that each state develop and maintain a Transportation Asset Management Plan (TAMP) to inventory infrastructure, assess its condition, and prioritize funding based on risk. TAMPs must identify and prioritize large and medium-sized assets that are technically critical to life and the economy, and outline performance measures that the agency will maintain regarding asset maintenance and replacement.

The law has also undergone changes in the private sector, with the most notable developments occurring in condominium and facility management in Florida. Following the collapse of the Champlain Towers, the Florida legislature passed Senate Bill 4-D, which requires condominium associations to conduct milestone inspections of structural components and maintain a reserve for significant repairs. Although there are many other aspects of the law beyond those discussed in this point, it highlights how the law is being utilized to manage assets in civil infrastructure in both the public and private sectors.

All major cities, including New York, Los Angeles, and San Francisco, have local ordinances requiring periodic inspections of parking structures, balconies, and facades, as well as documented maintenance plans. At the global level, organizations such as the International Organization for Standardization (ISO) developed ISO 55000 – Asset Management Systems, which is intended to provide a framework for integrating maintenance, financial planning, and risk management. This helps owners and operators view maintenance not as a cost they can mitigate, but as an investment in value that extends the life of their asset while controlling life cycle costs. Taking responsibility and economic considerations is also essential: while it is assumed that there is no funding, it is a matter of priorities and policy. The economics of deterioration cost much more than the annual maintenance costs. A recent example would be the collapse of the Fern Hollow Bridge, which incurred an emergency reconstruction expense of approximately $25 million. The emergency reconstruction expense was not only multiple times more than the most recent maintenance work, but also captured only the most immediate concerns of the benefit-cost approach.

The Katrina and the Oroville disasters together produced costs of more than $150 billion in recovery and litigation. Deferred maintenance costs produce an economic debt that continues to accumulate in a compounding fashion—the debt, in this case, accumulates in the forms of corrosion, fatigue, and liability. From the standpoint of private ownership, the absence of maintenance funding decreases the value of the properties and the eligibility for insurance. Following the collapse of the Champlain Towers, many insurance companies required structural inspections and then established reserve funding for policy renewal. Financial institutions now factor this risk into the property valuations and lending terms. In the area of public infrastructure, performance-based budgeting (PBB) is a new trend that ties funding to objective safety measures and provides incentives to agencies that maintain asset condition within acceptable thresholds.

## Human vs. Technology

Although technology has disrupted the industry with powerful tools for managing assets, it does not supplant hard funding. Digital asset management systems provide a pathway to track deterioration and predict maintenance cycles, and drone inspections can gather data that informs detailed predictive analysis. Structural health monitoring (SHM) networks can have real-time data about physical changes, which may include strain, vibration, or corrosion. However, even if these tools can illuminate the problem, they are incapable of intervening to fix it. Therefore, when organizations lack the funding or authority to intervene to rectify even perfect data, an organization with imperfect information becomes a slow-motion record of failure.

The more profound challenge is cultural and political. Maintenance is not as politically appealing as the visibility of new construction. Politicians prefer ribbon cuttings to repair budgets, and owners prefer low dues to expensive assessments. The preference for short-term decision-making at the expense of long-term resilience of the built environment is a barrier to sustainability. A sustainable future means rebranding asset management as a moral and fiscal responsibility; a duty to maintain what we have constructed before constructing more.

Asset management and funding shortfalls are not abstract administrative problems. They are the quiet background noise of collapse that continues to erode, like corrosion and deferral, building year after year, until physics ultimately demands a reckoning. The question is when, and how do we pay that bill? In infrastructure, as in any system, neglect costs more than care, always.

## Change Control and Management of Change (MOC) Failure

Construction projects are systems in a state of constant change. No project moves ahead without some change, regardless of how accurately it was planned. Project conditions on-site change, the design changes, and unknown conflicts emerge that require adjustments to be made. The extent to which a project can be controlled, or moved toward chaos, is dependent on obtaining the ability to manage project Change (MC). The Management of Change (MOC) is the formal process of evaluating, documenting, reviewing, and certifying every change in design, material, or procedure prior to its use. When MOC fails, what may seem like an innocuous change can actually lead to catastrophic vulnerabilities.

The Hyatt Regency walkway collapse (1981) is a quintessential example that highlights MOC failure within the history of construction. To facilitate ease of fabrication, a shop-drawing revision modified the hanger rod layout supporting the suspended walkways, thereby reducing the loads.

The change increased the load to a connection detail by more than double without the design engineer performing or approving any reanalysis. The modification was informally reviewed through the submittal path and did not follow any of the practices required in MOC. The outcome resulted in tragedy, as two walkways collapsed onto a full atrium, killing 114 people and injuring over 200. This now becomes a case study of how a change that was not reviewed, or even documented, can destroy a project and potentially a career.

Despite the passage of time, there has been no exchange of lessons gained. The Hard Rock Hotel Fall (2019) in New Orleans applied several undocumented design and construction changes. Connection details were adjusted, shoring configurations were changed, and components were exchanged in the field without a complete engineering buy-off. Site workers observed positive deflections and structural deformation, but daily, the foreman continued to make decisions on site and felt that no changes were being made anymore.

The MOC became a daily activity on the project, indicating what work was to be performed rather than formal changes, which needed to be re-validated by an engineer. The building collapse in Davenport (2023) and the NYC parking garage collapse at 57 Ann Street (2023) were precipitated by unpermitted structural changes, which were classified as repairs. In the Davenport case, crews were removing load-bearing masonry and categorized it as veneer removal, while in New York, the "repair" of a concrete pier stripped away considerable capacity. Again, in both cases, these actions were taken without permits, without engineered drawings, and without inspections. Again, both cases signified a failure of change management, despite the inspections and oversight that were in place.

Failures in managing change have fatal consequences beyond obscure, anecdotal cases of infrastructure crises. The failure of the FIU pedestrian bridge (2018) led to design modifications and the development of a tensioning construction sequence, which were poorly communicated and involved a lack of review that spanned across design, construction, and inspection groups. The cracks that developed before failure were symptoms of changes that management implemented without proper coordination.

In each of the scenarios just outlined, failures of construction are seldom the result of the original design; they are issues related to what occurs following the first modification. In other words, the danger is not change itself, it is unmanaged change.

## Regulation and SOP Change

Due to significant failures arising from undocumented changes or insufficient documentation of changes, MOC principles derived from chemistry and nuclear energy were adopted in construction management.

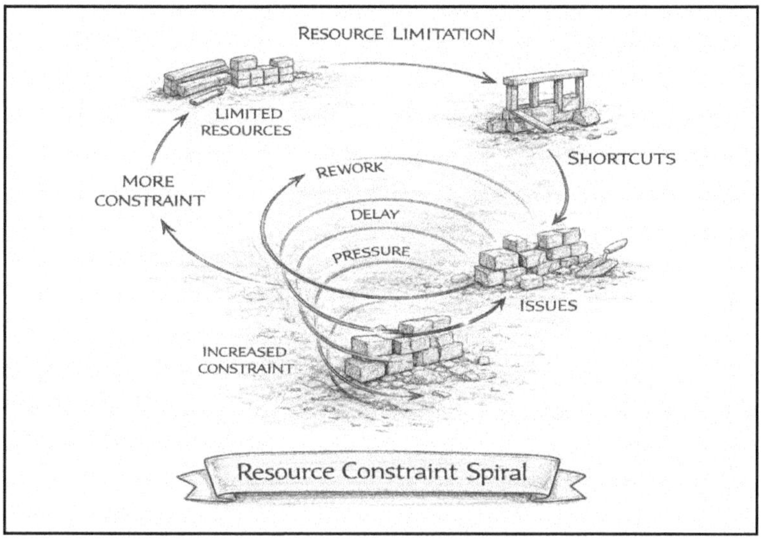


Formal Change Control Systems are integral to standard operating procedures in project management, as outlined in the PMBOK, ISO 9001, and ISO 45001. All changes that could affect safety, design intent, and risk to operational work must be documented, assessed, and approved for implementation by an identified and qualified engineer or authority employing formal protocols. Even OSHA and state regulators have integrated MOC principles into safety and quality plans where critical tasks require procedures for documentation and/or employer assessment and approval, such as lift plans, shoring plans, demolition plans, and confined space-entry plans requiring an updated assessment and validation of the original plan when work conditions, environments, or aspects of the plan itself change. For large construction projects, digital MOC workflows using construction management software (e.g., Procore, Aconex, and BIM 360) create audit trail documentation to compile what changes were made with details of who signed and the date of change for the trail to verify injury and liability responsibilities and accountabilities for maintaining safety and quality of work. Some regulators are now requiring Design Change Logs and Field Change Notices (FCNs) submitted in the record as part of compliance requirements. To perform structural work, any change to the design of load-bearing components must be documented with a stamped revision drawing to account for the changes.

Additionally, if the original plans or field analysis changes, each change, including temporary ones, must be accepted and approved by the engineer of record. As indicated in the written procedures of MOC, the system is transitioning from reactive problem-solving and reworking plans to pre-emptive risk management, documenting plans for those risks and proposed changes.

## Accountability and Economics

The costs of failing to manage change can be measured in both financial terms and human lives. For example, the Hyatt disaster resulted in over $140 million in settlements — which was a substantial figure in 1981 — and changed the liability model for engineers and contractors indefinitely. The Hard Rock collapse led to a financial disaster for several companies involved, resulting in lawsuits and fines totaling hundreds of millions. In instances of lesser consequence than Davenport and 57 Ann Street, property owners and contractors were similarly charged with criminal misdemeanors for work done without permits that jeopardized public safety.

Although MOC might seem like a protracted process from a financial viewpoint, it is a type of insurance. Every formal review serves as a safeguard against rework, lawsuits, and total loss. Increasingly, owners and insurers are requiring MOC compliance documentation for their coverage or to approve project funding. In today's contracts or documents, even a single change not approved could invalidate a warranty or expose the professional to liability for their coverage. The message is clear: any "change" that cannot be documented is considered an uninsurable risk.

### Human vs. Technology

Almost entirely, modes of change failures are about human behavior and a tendency to believe that a slight adjustment "won't matter." Technology may assist in managing change through automation, but it cannot be an alternative for professional judgment or accountability. Digital workflows and version-control systems may reduce the administrative burden and increase transparency, but they still rely on human honesty to accurately input information and faithfully follow directions.

Technology has created both opportunities and temptations. Modern modeling and communication tools now allow changes to be made and shared nearly instantaneously; however, absent an established process, this speed can outpace verification. As tools become more sophisticated, there is a corresponding increase in the need for rigorous processes.

The changing of discipline represents a fundamental cultural change in construction: from improvisation to intentionality. In years past, changes made in the field were indicative and valued signs of flexibility and ingenuity; in this era, these are seen as triggers to rise that now need to be governed. Treating change management does not mean project delays — it means protecting safety, quality, and responsiveness.

Change has always been a constant in construction and will continue to be so. The difference will be whether it is deliberately managed as a process or left to chance. The choice of how to manage change will dictate the stability of my given construction project. As the lessons from Hyatt, Hard Rock, FIU, and Davenport demonstrate, one change made without review can alter an entire industry.

**Operational Takeaways for Today's Job:**

- Double-check any unusual design or engineering detail before building. If something in the plans looks odd or overly complex, consult the engineer – catching a design deficiency or error early prevents costly fixes or failures down the line.
- Verify that field procedures match what's on paper. If crew members are deviating from the written safe work procedure or checklist because "that's how we always do it," take time to realign practice with the approved process or update the process – procedural drift is a common root cause of failure.
- Maintain strong oversight on critical tasks. Don't assume "someone else is checking" – assign a specific person (or yourself) to supervise high-risk activities, perform independent checks, and ensure quality standards are met. Lack of supervision often underlies major mistakes.
- Keep communication tight, especially at interfaces (e.g., between trades or shifts). Make a habit of daily cross-trade or cross-discipline check-ins when work overlaps – poor coordination is a frequent failure category, and a quick conversation can prevent costly miscommunications.
- Watch out for pressures that can lead to shortcuts. If schedule or budget stress is mounting, explicitly remind the team not to cut safety corners or skip steps. Many failures trace back to economic or deadline pressure – counteract it by reinforcing that doing it right is the only option.

## *Part II – Leadership, Planning, and Process*

**Roles:** *Field leaders* – project managers, superintendents, foremen, and safety officers.

**Focus:** Provides a day-to-day playbook for managing the job site through effective leadership and planning. Part II dives into communication, risk management, scheduling, standard operating procedures, and other tools that help run projects proactively rather than reactively.

**Operational Decisions Informed:** How to run a project systematically under real-world pressures. After Part II, a site leader will know how to structure a risk register, when to initiate a stop-work or corrective action, how to maintain disciplined communication loops, and how to implement standard procedures on site – all contributing to consistent, repeatable project execution.

Chapter 4

# Leadership and Communication

## The Leadership Burden in Construction

Leadership isn't just a title; it's a weighty responsibility that can mean the difference between a safe, successful project and a catastrophic failure. Every day on a jobsite, hundreds of decisions are made that affect worker safety, structural integrity, and the project's outcome. The burden on the construction leader is knowing that many of those decisions ultimately rest on their shoulders. One veteran superintendent described it as "having 500 people tied to your belt – if you slip, they all fall." In other words, a lapse in judgment or a breakdown in communication by the leader can set off a chain of events that puts everyone at risk. This chapter examines the burden and how effective leadership and communication can carry it wisely.

Even the best-laid plans and state-of-the-art engineering can be undermined by poor leadership. We have seen that many failures in construction trace back to human factors: unclear directives, ignored warnings, or a culture where team members were afraid to speak up. Technical skill alone cannot guarantee safety on a jobsite – it takes field-level leadership and disciplined communication to bind all the moving parts together. You can have a solid design and a sound schedule, but if your crew doesn't trust you, or if critical information isn't flowing both ways, you're building on sand. History provides painful reminders. The collapse of a pedestrian bridge during construction at FIU in 2018, the fatal power plant explosion in Connecticut, and the fall of the Hard Rock Hotel structure in 2019 – all involved situations where warnings were missed or leadership failed to intervene in time. In each case, the leaders on site were left to answer the toughest questions in the aftermath: What did you know? Who was in charge? Why didn't someone speak up or stop the work? These questions highlight the immense moral and practical weight on those in command. In one instance, after a major accident, a senior project director confided that the hardest part was facing the victim's family, knowing that leadership failures had contributed to their loss. That kind of reckoning is profound – and it illustrates why the burden of leadership is so heavy.

Great leadership in construction means preventing those questions from ever needing to be asked. It means creating an environment where problems are surfaced and addressed before they turn into crises. It means being the kind of leader who constantly thinks, “What could go wrong here, and what am I doing about it?” The burden includes sleepless nights worrying about the “unknown unknowns,” double-checking that safeguards are in place, and sometimes making unpopular calls like halting work for safety reasons or telling a client “no” when a shortcut is too risky. That is the price of stewardship. A project manager who treats leadership as merely giving orders from a trailer will sooner or later find themselves in damage-control mode. By contrast, a leader who actively engages with their team, listens to concerns, and leads by example can steer a project away from trouble long before a crisis ever materializes.

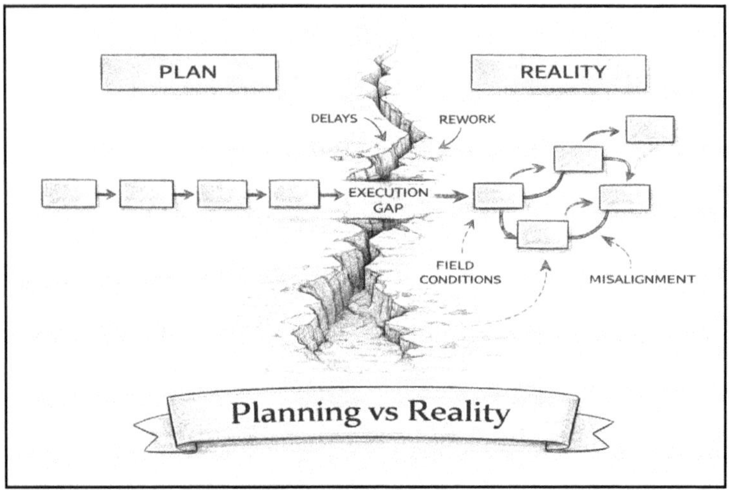


No one is immune to mistakes, of course – but effective leaders build layers of communication and trust so that when they stumble, someone else can catch the issue. In a way, carrying the burden of leadership is not about carrying it alone; it’s about distributing that responsibility through a strong team culture. As we delve into the following sections, we will explore how leaders can create active communication loops, provide clear direction (often through storytelling), uphold transparency, foster trust and accountability, and maintain control of the project climate. We’ll also confront how leadership is tested under stress and time pressure, and what happens when leadership falters or is absent. These lessons are drawn from hard-won experience – including cautionary tales from real projects. By the end of this chapter, it should be clear why leadership and communication are often the “make or break” factors in construction management, and how a reflective, professional leader can use these tools to shoulder the responsibility and keep their team safe.

## Active Communication Loops and Field-Level Feedback

Communication on a construction site must be a two-way street. Active communication loops mean information continually circulates upward from the field and back down from management. Effective leaders make it a habit to get out of the trailer and engage with crews on site, listening more than they speak. They set up regular forums – daily huddles, safety meetings, briefings – where workers can voice concerns and share observations. Great leaders keep information flowing and encourage even bad news to surface. When feedback from the field is taken seriously and acted upon, small issues can be fixed before they escalate into big problems.

In contrast, when these communication loops break down, the results can be deadly. During the construction of the Hard Rock Hotel in New Orleans in 2019, workers repeatedly warned of unsafe conditions, but their warnings were ignored. Because there was no effective avenue for field-level feedback to reach decision-makers, those warnings went unheeded until the structure tragically collapsed. This underscores that communication discipline is not just bureaucracy – it's a lifesaving practice. Frontline crews often spot the first signs of trouble; the project culture must empower them to speak up and ensure leadership actually hears and responds. A superintendent or project manager should treat a worker's cautionary comment as valuable data, not an annoyance. By creating an open communication loop – where every level of the team regularly shares updates, and leadership visibly listens and adjusts – construction leaders can catch weaknesses early.

Establishing an open feedback loop requires deliberate effort. It starts with accessibility – leaders who are regularly present in the field tend to pick up on concerns early. It also requires setting clear channels for reporting issues. Many successful projects institute "field feedback" practices such as anonymous drop-boxes, open-door policies at the site office, and scheduled walkabouts where supervisors ask open-ended questions ("Is there anything you've seen this week that worries you?"). The key is follow-through. If a crew member brings up a potential hazard – say, a formwork that seems shaky or confusion about a change in the drawings – management must visibly listen and act. Even if the concern turns out to be minor, the very act of responding sends a message that input is valued. Over time, this positive feedback loop snowballs: workers become more vigilant and communicative when they know their voices will be heard.

Active loops are maintained through continuous updates from leadership. Good communication isn't only reactive; it's proactive. Leaders should regularly push information out to the field as well – sharing updates on schedule changes, engineering fixes, new safety alerts, or lessons learned from other sites. When workers understand not just what they're being asked to do but why (the context and potential risks), they become partners in the plan rather than blind followers. For instance, if a certain scaffold tie needs inspection because a similar project had a scaffold collapse, a supervisor who explains that context will get far more diligent compliance from the crew. The field should also see that its feedback loops upward. If an apprentice reports a near-miss with a piece of machinery on Monday, by Friday, the crew might notice a new safety guard installed or a revised procedure issued because of that report. That tangible result "closes the loop" and reinforces the behavior of speaking up.

Consider the Florida International University (FIU) pedestrian bridge collapse in 2018 as another lesson in the importance of field-level feedback. In the days leading up to that collapse, significant cracks were observed in the bridge span. Engineers and workers noted the cracks, but the information did not trigger an immediate halt to work or a full safety stand-down. Communication remained trapped in silos – a voicemail was left for a state engineer, concerns were discussed in a meeting but ultimately downplayed, and no decisive action was taken in time. The bridge collapsed suddenly during live traffic, killing six people. In hindsight, the signs of failure were literally visible and noticed on the ground, but without a strong communication culture and clear feedback loop to the top, those warnings went unheeded. It's a stark reminder that for a leader, hearing about a problem is a gift – it offers a chance to fix something before it blows up. Active communication loops ensure that the gift is not lost.

In short, active communication loops connect the insight on the ground with the decision-making in the office, establishing a feedback-rich environment where information flows freely rather than getting stuck or filtered out. Field-level engagement is a hallmark of disciplined, proactive project leadership. When people at all levels regularly share updates and leadership visibly listens and adapts, the entire project team gains an early-warning system for safety and quality. Problems can be caught and corrected at the "whisper" stage, long before they turn into shouts or sirens.

## Storytelling and Directive Clarity

Clarity and narrative are two powerful tools in a leader's arsenal. On a busy job site, ambiguous instructions can be as dangerous as faulty materials. Crews need to know exactly what is expected of them and why. The best leaders, therefore speak plainly and leave no room for misinterpretation in their directions. But they don't stop at issuing orders – they also reinforce those directives with context and story. In practice, that means a supervisor might not just say, "Follow that bracing procedure," but add, "Let me tell you about a crew that didn't, and what it cost them." Leaders essentially become storytellers, turning past lessons into cautionary tales for the team. In fact, teaching through real failures is often the most effective way to instill discipline: the only way to truly drive home critical practices is to educate operators in the root philosophy and then demonstrate how failing to observe those practices has led to disaster in the past. These stories stick in the crew's minds, giving life to safety rules or engineering requirements that might otherwise seem abstract. When workers understand the real-world consequences behind an order, they're far more likely to execute it faithfully.

Equally important is making the mission and goals unambiguous. Construction veterans often compare this to coaching: you have to "recreate in the minds of the players the vision that you yourself have". In other words, a leader must communicate the project's mission, the team's potential, and the game plan clearly and unmistakably to everyone from the apprentice carpenter to the senior engineer. If the goal is zero incidents on a job, the crew should hear not only the slogan but also the concrete steps and standards to achieve it. Command clarity extends to the smallest instructions – there's no room for hints or half-measures when it comes to safety and quality. Directions should be specific: instead of "Be careful with that excavation," say "No one goes deeper than 4 feet without shoring because we've got a high water table here." Rather than assume workers will "use common sense," a leader spells out the exact expectations and the reasons behind them.

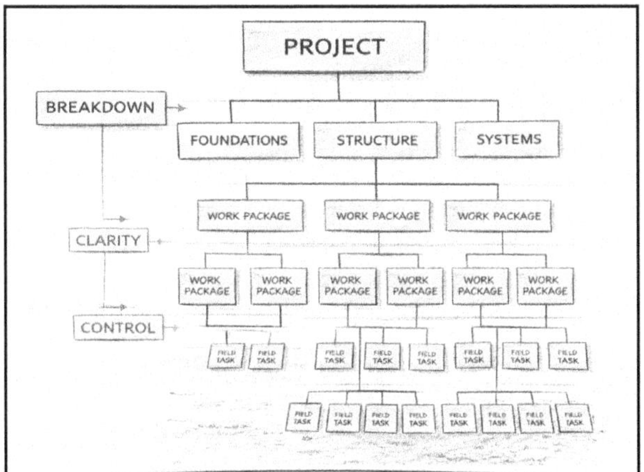


Planning Breakdown Structure

Why the emphasis on stories along with clear commands? Because stories create buy-in. Humans have passed down knowledge through storytelling for millennia; on a construction site, a five-minute story at the start of a shift can do more to shape behavior than a five-page memo. For instance, before a critical crane lift, a project manager might recount the story of "Big Blue" – the infamous crane that collapsed at a Milwaukee stadium project when operators pushed ahead in high winds. By reliving that scenario with the crew – describing how overconfidence and schedule pressure led to ignoring the wind limits and how the giant crane toppled, killing three ironworkers – the manager imprints a lesson: never ignore the limits, no matter the pressure. Suddenly, the wind gauge and the rules in the lift plan aren't just bureaucratic numbers; they represent real people's lives and a story everyone wants to avoid repeating. In the same way, an experienced foreman might tell young laborers about the 1978 Willow Island disaster, when a rush to stay on schedule led workers to climb onto uncured concrete and caused a collapse that took 51 lives. That tale is more chilling and memorable than any generic "work safely" admonishment. It shows in vivid detail what can happen when step-by-step procedures (like allowing concrete to reach strength) are skipped.

A sobering example of poor communication clarity was the Hyatt Regency walkway collapse in 1981. In that case, a critical design change – a slight modification to the hanger rod connections supporting elevated walkways – was approved without clear communication through the proper channels. The engineer of record never fully understood how the detail had been altered by the steel fabricator, and because the directive and its implications weren't clearly communicated, the change doubled the load on a connection that couldn't handle it. The result was a sudden structural failure that killed 114 people during a hotel event. That tragedy, now taught in engineering ethics courses, shows how a single misunderstood instruction or undocumented change can reverberate catastrophically. For today's leaders, the takeaway is to be crystal clear in every directive: say exactly what needs to happen, confirm that everyone understands, and explain the reasoning.

By pairing each "what" with a "why" – often through a real example or story – effective leaders ensure their team not only knows the plan, but believes in the necessity of following it. Over time, this discipline of clear, story-driven communication creates a workforce that doesn't just obey orders, but internalizes the lessons behind them. Crews start repeating those same stories among themselves and even to newcomers, which is the clearest sign that the knowledge has truly taken root. A laborer might warn a co-worker, "Don't even think about skipping that step – remember what happened at Willow Island," thereby perpetuating a culture of doing things the right way, every time.

## Transparency as a Leadership Tool

Transparency in leadership means aligning one's words with one's actions and being forthright about issues. A project team can sense immediately if a leader's behavior doesn't match their platitudes. Great leaders understand this and therefore don't just preach quality and safety – they personally follow the rules, show up prepared, and demonstrate the work ethic they expect from others. By modeling the standards they espouse, they send a powerful message to the crew: we're all held to the same rules here, from laborer to project executive. This consistency between word and deed earns respect. In contrast, a "Do as I say, not as I do" approach destroys credibility. Leading by example is the first step to creating a climate of trust and honesty.

Another aspect of transparency is open communication about project realities – especially the bad news. An effective manager does not hide delays, cover up inspection failures, or downplay safety hazards to avoid embarrassment or conflict. Instead, they bring issues to light early, framing them as problems the team will solve together. For instance, if an inspection reveals a flaw in a concrete pour, a transparent leader will promptly inform the crew and then explain the plan for remediation. This candor prevents small cracks from growing into big failures. When people know that management isn't sweeping things under the rug, they are more likely to speak up for themselves. By contrast, if leadership is known to suppress information or punish messengers, workers learn to stay silent – and that silence can be deadly.

Consider the case of the Champlain Towers South condominium collapse in Surfside, Florida (2021) as a cautionary tale in the importance of transparency. Years before the building's collapse, engineers had documented major structural damage and urged repairs. But the condo board – effectively the project's leadership in this context – became mired in disputes and delayed addressing the issues. In hindsight, the board's reluctance to fully confront and communicate the gravity of the problem proved disastrous. Had they been more transparent with themselves and the residents about the risk (and the costs of inaction), pressure to fix the defects might have mounted before it was too late. The lesson for construction leaders is clear: when you know of a hazard or a deficiency, face it and communicate it. Transparency can be uncomfortable in the short term – it may mean admitting a mistake or exposing a near-miss – but it pays off in long-term trust and risk reduction. When your team sees that you will tell them the truth, even when it's unwelcome, their confidence in you grows. They are more likely to bring you problems early, rather than hide them. In a very real sense, transparency is a tool that prevents disasters: it builds a culture where potential failures aren't swept under the carpet, but dealt with out in the open. And once an issue is out in the open, the whole team can rally to resolve it.

In sum, a leader who communicates honestly and acts with integrity creates an environment where accountability isn't feared, it's shared – everyone knows the status of the work, and everyone can take part in keeping the project on track. Problems get acknowledged, owned, and fixed – not hidden until they explode. This isn't about airing every minor concern to everyone; it's about being candid when it counts. A transparent leader tells the client about a delay as soon as it's evident, explains it to the crew, and works on a recovery plan, instead of hoping to paper it over. They admit "We had a safety incident because we missed something," instead of blaming a worker or pretending it didn't happen. Over time, that approach earns loyalty. The crew learns that management won't cover up issues or leave them in the dark, and in turn, they reciprocate with honesty and commitment. The site becomes a place where truth is the norm – even when it's uncomfortable – and that is the bedrock of a resilient, responsive team culture.

## Trust, Accountability, and Psychological Safety

Trust and accountability are the glue that hold a team together, while psychological safety is the atmosphere that allows that glue to set. Trust is not instantaneous – it's earned over time through consistency and honesty. As one leadership principle puts it, trust is predicated upon honesty. When team members see their leaders doing what they say they will do, admitting when they don't know something, and owning up to mistakes, it creates a bedrock of credibility. In that kind of environment, workers trust that if they speak up about a hazard or confess to an error, management will respond constructively rather than with punishment. This is the essence of psychological safety: the shared belief that the workplace is safe for interpersonal risk-taking. In concrete terms, psychological safety means a laborer can tell a foreman, "I'm not comfortable with this lift plan," or an engineer can admit "I miscalculated that load," and instead of ridicule or reprisal, they receive support and problem-solving. Such an environment doesn't materialize by accident; it starts with leadership. Leaders cultivate it by how they react to bad news. If the first time a crew member raises a concern the boss dismisses or scolds them, you can bet it will also be the last time anyone speaks up.

Achieving psychological safety requires leaders to actively encourage input and normalize speaking up. Many construction failures have occurred in organizations where people noticed warning signs but stayed silent out of fear. In a trusting team, the opposite happens: crew members will halt work if something looks wrong, confident that management wants them to do so. For example, forward-thinking contractors often tell everyone on site that they are a "safety officer" for the day – empowering each worker, regardless of title, to call out risks and stop work if necessary. Crucially, when someone does so, leadership backs them up. Implementing this kind of bottom-up accountability takes conscious effort. It might involve setting up an anonymous hazard reporting box or starting every shift meeting by asking, "Any safety concerns or observations today?" and rewarding those who raise their hand. Over time, these practices send a clear signal: looking out for the project's welfare is everybody's job, and we've got each other's backs. The result is a team that is far more aware of developing issues – because people are actually talking about them – and far more likely to catch and correct errors before they cause harm.

Accountability goes hand-in-hand with trust, but in a different sense. While trust is about believing your teammates or leaders will do the right thing, accountability is about defining what the right thing is and who is responsible for it. On a complex project, it must be crystal clear who is responsible for each critical task, each safety check, and each decision. Ambiguity in roles can breed a dangerous diffusion of responsibility. If "everyone" is supposedly accountable for something, often no one truly is. A tragic illustration of diffused accountability comes from the levee failures during Hurricane Katrina in 2005. The flood protection system around New Orleans was managed by a patchwork of agencies, and no single entity had clear ownership of the system's integrity. After the disaster, investigators found that unclear accountability between the U.S. Army Corps of Engineers and local levee boards allowed maintenance and safety issues to fall through the cracks. Each party assumed someone else was addressing certain risks, and as a result, known vulnerabilities went unfixed. Construction projects can suffer a similar fate if oversight is fragmented or if responsibilities are not clearly assigned. A project executive might think the site superintendent is enforcing a certain safety rule, while the superintendent thinks the subcontractor's safety officer has it covered – and in the end, nobody actually checks that the rule is followed.

The remedy is straightforward: assign clear responsibility for every significant risk, task, and deliverable, and make sure everyone knows who that person is. A good leader doesn't shy away from being the one ultimately accountable. In fact, openly accepting accountability ("The buck stops with me on safety") is a strong trust-building move. It shows the crew that their leader will support them and take responsibility if something goes wrong, rather than just point fingers. When crew members see management holding themselves accountable – for instance, a project manager candidly telling the team, "I didn't coordinate that crane pick properly yesterday, that's on me and here's how we'll fix it" – it inspires them to take ownership of their own areas as well. Accountability, in this sense, is contagious: a culture of accountability arises when people at the top model it and expect it at every level.

Ultimately, all these aspects – trust, clear accountability, and safety – converge into a unified culture of reliability. It falls to leadership to set this tone. By relentlessly focusing on core principles day in and day out, the team minimizes risk and avoids complacency. They understand that true success isn't the absence of all risk (an impossible goal in construction), but the minimization of risk through disciplined habits, keen awareness of changing conditions, and continuous improvement. When the crew sees that their leaders prioritize safety and open communication over short-term gains or saving face, it cements a culture where everyone from management to the newest apprentice feels ownership of the project's outcomes. In such a culture, team members don't just look out for themselves; they look out for one another. An ironworker will double-check the carpenter's harness if something seems amiss, and the carpenter won't hesitate to shout a warning if a load looks unstable – not because it's "his job" specifically, but because it's everyone's job when you're part of a true team. This is the high level of team awareness that world-class construction crews achieve. They operate with a collective mindfulness, confident that any member can speak up and be heard. In a high-trust, high-accountability environment, no one is afraid to say, "Hold on, something isn't right here." And that simple phrase, empowered by a culture of open communication, has probably prevented more accidents than any piece of technology or engineering ever could.

## Command Climate and Role Modeling

Every construction crew takes its cues from the leadership's tone and behavior – this is often referred to as the command climate. It's the overall atmosphere that leaders create through their actions, attitudes, and priorities. A positive command climate is one where the team feels motivated, respected, and clear on expectations; a toxic or weak command climate breeds fear, confusion, or apathy. Leaders establish this climate whether they intend to or not. How they handle daily challenges, how they treat people, and what they emphasize (safety, quality, or just speed at all costs) all trickle down to shape how the team operates. In practical terms, the leader is the thermostat for the jobsite: their behavior sets the temperature for everyone else.

A critical part of command climate is role modeling. Construction crews are quick to observe what the boss actually does, not just what they say. If a project manager wears all required PPE, double-checks critical calculations, and approaches problems methodically, the team gets a clear signal about what's important. If, on the other hand, the boss skips safety meetings, dismisses concerns, or cuts corners to save time, the crew gets an entirely different message. People mimic the tone set by those at the top – often subconsciously. For example, on a site where the superintendent routinely bypasses lockout-tagout procedures to get machinery running faster, it won't be long before technicians and laborers start taking similar risks, figuring it must be okay since the boss does it. By contrast, if that superintendent visibly follows every safety step, no matter how inconvenient, it builds a norm that safety is sacrosanct.

Command climate also encompasses how leaders handle stress and conflict, which in turn influences team morale. If a leader stays calm under pressure and treats mistakes as learning opportunities, the crew is more likely to remain level-headed and transparent when issues arise. But if the leader panics, screams, or starts blaming whenever something goes wrong, the team's morale and openness will plummet. Consider a scenario where a crane lift is going awry: perhaps a load starts swinging unexpectedly. If the supervisor in charge yells incoherently or – worse – freezes in indecision, that chaos will permeate everyone's actions and confidence. Conversely, if the supervisor quickly radios "All stop, all stop!" (using a clear, predetermined signal) and then calmly directs the crew to secure the load and regroup, he not only prevents an accident but shows the crew how to respond to crises. In the follow-up, if he frankly discusses what went wrong and focuses on solutions rather than scapegoating, the team stays cohesive and learns from the scare. In short, the leader's conduct in tough moments directly teaches everyone else how to act.

Negative command climates have been factors in many construction disasters. In some high-profile failures, investigations later describe an "atmosphere of intimidation" or a "culture of silence" on the site. This often tracks back to project leadership that did not welcome input or showed open disdain for regulations. One notorious example was the Willow Island cooling tower collapse (1978). The site leadership's overriding priority was meeting an aggressive schedule, to the point that workers felt they couldn't object to patently unsafe practices. When crews were ordered to climb formwork and continue concrete placement on a tower section that hadn't fully cured – a practice they knew was questionable – they complied because the climate was one where schedule trumped safety. The result was catastrophic. Post-accident reports indicated that the "we must not fall behind" message had been so strongly driven from management that crews simply didn't feel empowered to push back or demand proper curing time. Essentially, leadership's pressure created a climate of silent acquiescence to risk.

Another example of how leadership's tone dictates outcomes is the Big Blue crane accident mentioned earlier. The site managers decided to proceed with the lift despite dangerously high winds, likely because the project climate emphasized meeting the deadline above all else. Some workers later said they had misgivings as the gusts picked up, but since the bosses were signaling to move ahead, no one pressed the issue. Here, the command climate lacked an avenue for dissent; the role modeling (of ignoring a fundamental safety limit) overpowered any checklist or rule on paper. In both Willow Island and Big Blue, the leaders perhaps never explicitly said "ignore safety if we're behind schedule," but their actions and urgency created that impression – and the crews followed suit.

On the flip side, a strong, positive command climate can be the unsung hero in averting disasters. There are cases – less famous than the failures – where a potential catastrophe was avoided because someone spoke up or a leader hit the pause button in time. These "non-events" don't make headlines, but insiders often share them as success stories. For instance, veteran contractors talk about near-misses where a crew member shouted "Stop!" upon sensing something wrong, and the team listened immediately, only to discover a critical rigging issue or an unseen danger. In such stories, the common thread is a leadership that had consistently encouraged vigilance and empowered the team to act. The crew knew that the project manager genuinely valued safety over schedule, because he had demonstrated it repeatedly – canceling a concrete pour when a formwork tie looked suspect, or taking a subcontractor to task for skimping on scaffold bracing. So when a worker yelled out, nobody hesitated or worried about repercussions; they all trusted that halting work was the right thing because that is the climate they lived in. It's impossible to quantify how many accidents have been prevented by these kinds of strong leadership climates, but seasoned professionals will attest that it's likely many.

Leaders also shape the climate by how they treat their team on a personal level. Respect and care go a long way. A project where management belittles workers or constantly threatens firing will breed either resentment or fear – neither of which leads to productive, safe work. But a leader who shows that they care – for example, by taking an injured worker to the clinic personally and then visiting them later at home to check on recovery – sends a powerful message of solidarity. That worker, and their co-workers, will remember that gesture; it builds loyalty. A loyal team is more likely to go the extra mile to keep each other safe and quality high, partly out of respect for a boss who clearly values them as people. Role modeling isn't just about demonstrating technical tasks; it's also about demonstrating values like empathy, fairness, and integrity. If the crew sees the boss owning up to his own mistakes ("I misread the plan, that's my fault") and treating people right, they'll emulate those behaviors with each other. You'll see more cooperation and less cover-your-back behavior among the ranks, because the tone set at the top is one of trust and mutual respect.

The command climate is the invisible hand steering daily behavior on site. A leader's example and attitude radiate outward and set the norms – "how we do things here." If the climate is one of discipline, care, and openness, that's how the job will run. If the climate is one of corner-cutting, fear, or indifference, the project is sailing with a dark cloud over it, just waiting for lightning to strike. Leaders should periodically pause and read the climate they've created: Are people freely bringing issues to me? Do they look motivated in the morning or defeated? Is safety truly treated as paramount or just given lip service? These reflections help a leader adjust their own approach. Command climate isn't static; it's something a leader must actively cultivate every day through countless small actions. The payoff for getting it right is huge – not just in preventing accidents, but in building a team that takes pride in doing things the right way even when no one's watching.

## Communication Under Stress and High-Energy Conditions

A construction site can go from calm to chaotic in an instant. It might be a sudden change in weather, an accident, a critical equipment failure, or simply the pressure of an impending deadline that cranks everyone's stress level to max. In those high-energy, high-stress moments, communication is often the first casualty – and that is dangerous. Under stress, people tend to talk faster (or not at all), listen poorly, and revert to raw instinct. That's why one hallmark of elite construction teams (and indeed any high-risk team, like firefighters or pilots) is drilled communication protocols for emergencies and fast-moving scenarios. When adrenaline floods the system, having pre-established communication habits can make the difference between a controlled response and total chaos.

Stressed communication can manifest in several ways. One common issue is information overload or confusion – multiple people shouting instructions, or conversely, nobody communicating because each assumes someone else will. Another is tunnel vision, where leaders or workers become so fixated on one aspect of the problem that they forget to share critical updates with others. To counter these tendencies, good teams use simple, clear communication structures under duress. For example, many sites adopt an incident command system (ICS) when things go wrong: one person takes charge as the incident commander and all information flows through them to avoid overlap and contradiction. Everyone should know who that point person is (it could be the site safety officer or the superintendent by default) and defer to their direction once an emergency is declared. This isn't about ego – it's about efficiency in chaos. When a formwork collapses or a fire breaks out on site, someone needs to quickly say "I've got command," and others respond with concise status updates or await instructions, rather than everyone acting on impulse.

Another technique is using specific emergency phrases or signals that cut through noise. Just as ship crews use "Mayday" or the military uses "Abort" as unambiguous terms, construction teams can establish their own. One high-rise crew decided that the word "Freeze" shouted three times would be their all-hands stop-work signal after a near-miss where multiple people yelled conflicting things during a scaffold failure. They practiced it: if anyone yells "Freeze, freeze, freeze!", every piece of equipment is shut down, and every worker stays put until a leader assesses the situation. They rarely had to use it, but knowing that protocol was in everyone's back pocket meant that if a crisis came, they wouldn't be tripping over each other in confusion. Drills and simulations are incredibly useful for this. Just as office buildings hold fire drills, a construction site might run a "man down" drill or a "crane failure" drill to rehearse who does what and how information flows. The goal isn't to scare people – it's to build muscle memory so that when a real emergency hits, everyone instinctively communicates and acts in sync.

Staying calm is easier said than done, but leaders must set the tone in a high-energy moment. The moment the supervisor's voice on the radio betrays panic, the whole crew will feed off that fear. Experienced leaders often speak more slowly and evenly under extreme pressure – almost artificially so – as a way to steady their own nerves and reassure others. In a crisis, the leader's voice becomes a lifeline. Consider a scenario of a partial trench collapse with a worker trapped. A panicked reaction might have everyone shouting the worker's name, random people jumping in to dig (potentially making it worse), and confusion about who called 911 or whether heavy equipment should be used. A trained, calm reaction would sound like this on the radio: "Emergency, trench collapse at Sector C. John is trapped up to his waist. All crews not involved, clear the area. Excavator operator, shut down and hold position. Safety officer, call 911 now. I need two volunteers with shovels at the top, everyone else stay back." That kind of composed, step-by-step direction can save a life because it focuses the team on immediate, coordinated action and prevents well-meaning but dangerous chaos.

Maintaining communication under stress also means keeping information flowing after the initial emergency stabilization. In the heat of an urgent situation, time dilates – seconds feel like minutes. People not directly involved can grow anxious and might start speculating or spreading rumors ("I heard the whole scaffold came down!"). A good leader makes a point to update everyone frequently, even if it's brief: "Listen up – the fire department is on the way. The injured worker is being cared for. The situation is under control." These periodic broadcasts calm the wider team and keep the focus on productive efforts. It's when people feel in the dark that panic or frustration spreads.

We can learn from events outside construction too. Think of an air traffic controller calmly guiding a pilot through an emergency landing, or the composed voice of a 911 operator talking a caller through CPR. Those are extreme examples of communication under duress that we can emulate on site. During the 2011 Indiana State Fair stage collapse incident, one critique was that there was no unified communication or clear chain of command as the storm hit the fairgrounds. Multiple officials were unsure who should give the evacuation order, and as a result, precious minutes were lost and the message to evacuate wasn't delivered effectively before the stage structure gave way. That tragedy shows how stress can paralyze communication when leadership roles are unclear. By contrast, on projects that have faced sudden emergencies – say a structural collapse or a serious injury – those with a well-drilled emergency comms plan often reflect that "everyone knew what to do and who was doing what, even though we were all scared." Preparation and clarity are the antidotes to panic.

Not every high-energy scenario is an accident; sometimes it's just an intense phase of work. For example, a concrete pour racing against an approaching storm, or a complicated lift that has to beat nightfall. In those cases, energy is high because the team is trying to beat the clock or manage complexity, not because of a crisis. Here too, disciplined communication is key. The best crews use short, standardized calls and responses to coordinate under pressure. During a big concrete placement, you might hear the pour boss say: "Pump A, start on gridline 5 now," and Pump A operator replies, "Copy, starting gridline 5." Others stay off the radio unless they have something critical. If an issue arises – a form starts bulging or a truck is delayed – the communication is clear and directed: "Hold pour – formwork issue at grid 5, need carpentry lead here ASAP." Because the team isn't cluttering the channel with chatter, that crucial message gets through and everyone pauses safely.

In practice, maintaining good communication under stress comes down to training, culture, and personal discipline. Leaders should train their teams on what to do when adrenaline spikes: speak slowly, use the agreed emergency keywords, confirm messages by repeating back, and stick to facts ("We have a fire in the storage area, evacuating now") rather than emotions or assumptions ("Oh God, it's a huge fire!"). Culturally, having a climate (as discussed earlier) where people are used to listening to each other and not afraid to pipe up is crucial – because in a crisis you need all eyes and ears working, not just the boss's. And individually, it helps if each person has a mental game plan for emergencies. Even something as simple as "If something bad happens, I will first take a deep breath, then clearly announce what I see," can prime a person to communicate rather than freeze.

High-stress conditions are inevitable in construction – what's not inevitable is chaos. Leaders can enforce calm and clarity by preparing their teams for those moments. Think of these protocols and habits as the fire extinguisher for communication breakdowns: you hope you never need it, but if a blaze of stress erupts, you'll be glad it's there and everyone knows how to use it. With practiced communication and steady leadership, a team can walk through the fire of an emergency and come out the other side intact. Without them, even a minor crisis can needlessly spiral into a disaster.

## Decision-Making Under Pressure (Bias, Drift, and Urgency)

Decision-making in construction is rarely done in a calm, quiet office with unlimited time. More often, critical calls are made in the field under pressure – with the schedule ticking, money burning, concrete setting up, or a client demanding answers. These pressure-cooker conditions can lead even good leaders to make bad decisions if they're not aware of the pitfalls of **bias, drift, and urgency**. Let's break those down and see how they creep into judgment.

**Bias** under pressure usually takes the form of cognitive biases – mental shortcuts or tendencies that skew our thinking. One common bias is **confirmation bias**, the tendency to favor information that confirms what we want to believe and discount the rest. For example, a project manager under the gun to meet a deadline might downplay or reinterpret signs of trouble to fit the narrative that "everything is on track." We saw a version of this in the lead-up to the FIU pedestrian bridge collapse: cracks in the structure were plainly visible and concerning, but the team convinced themselves (and each other) that these cracks did not indicate a full structural failure in progress. They had a theory (that the design was sound and the cracks were not critical) and they latched onto that belief, disregarding how ominous the evidence really was. Another bias is **optimism bias** – the "it'll be fine" mindset. Teams might skip contingency plans or ignore warning signs because they simply can't imagine the worst-case scenario happening. We're all susceptible to assuming that because something has never failed before, it won't fail this time – even when conditions have changed.

**Drift** refers to the gradual departure from safe practices or standards, often because of repeated small concessions that don't immediately cause harm. This is often called normalization of deviance. Over time, what was once recognized as a risky shortcut can become "how we always do it," because nothing bad happened last time. Construction history is replete with examples of this. Prior to the 1981 Hyatt Regency collapse, plenty of projects had approved field design changes on the fly without thorough vetting – and usually got away with it. That likely bred a false sense of security that a small, unreviewed change was no big deal. In the case of the Hyatt walkway, that complacency met a deadly tipping point. In the Willow Island case, as discussed, workers had successfully gotten away with climbing on green concrete and rushing the next lift a few times before, which likely gave them and supervisors the fatal confidence to keep doing it, until the day everything fell apart. Drift is insidious because no single decision feels like a blatant breach; it's the cumulative effect of "just this once" or "we'll be okay" repeated until the safety margin is gone. Leaders have to actively fight drift by insisting on fundamentals even (or especially) when nothing bad has happened in a while. A savvy project manager might say, "We've poured dozens of walls without bracing that one side, and it's been fine – but that doesn't make it right or safe. We're not going to roll the dice again." It takes discipline to pull a team back from drift, because to the team, it can feel like the cautious leader is overreacting. But more often than not, that leader is proven right eventually – it's just a question of catching it before an accident rather than after.

**Urgency** is the intense pressure of time or performance that can hijack decision-making. Urgency creates a sense that "we must do something now," which can short-circuit our normal checks and balances. It often goes hand-in-hand with bias and drift. When the schedule or budget is on the line, people tend to dismiss inconvenient information (confirmation bias: "I'm sure that warning can be ignored, we have to finish today") and cut corners (drift: skipping steps to save time). Leaders under urgent pressure sometimes fall prey to what pilots call "get-there-itis" – an obsession with achieving the immediate goal (pour the concrete before the storm hits, open the road by morning) that overrides caution.

Another classic example outside construction is the Space Shuttle Challenger launch decision in 1986. NASA managers, under political and schedule pressure to launch, overrode engineers' concerns about cold temperatures affecting the O-ring seals. They convinced themselves the risk was acceptable – a mix of confirmation bias ("the data isn't conclusive, probably fine") and urgency ("we've delayed too many times, we need to launch"). The result was catastrophic. That disaster is often cited in project management circles as a warning: beware of letting schedule or external pressure push you into discounting technical risks. The parallels to construction are clear – whether it's a concrete pour or a crane lift or any critical operation, a leader must be vigilant that "we have to get it done tomorrow" doesn't become an excuse to turn a blind eye to safety or quality concerns.

So how do leaders make good decisions under pressure? First, by **acknowledging these mental traps**. Simply pausing to think, "Am I seeing this situation objectively, or am I filtering it through wishful thinking? Have we gradually relaxed our standards without realizing it? Would I still make this decision if we weren't in a rush?" can prevent a lot of mistakes. Many seasoned leaders develop a habit of playing devil's advocate against their own decisions. For instance, if leaning toward a risky go-ahead, they'll ask a colleague, "Tell me why this is a bad idea." Encouraging a quick dissenting opinion can surface biases or overlooked issues. Even within your own mind, it helps to articulate the opposing case: "What's the worst that could happen if we push on? Do I have any solid evidence that it won't happen? Or do I just really want to believe we'll be fine?"

Another strategy is having **predefined no-go criteria** – essentially, rules that force a deliberate decision process under pressure. For example, you might institute: "We do not pour concrete if the temperature is forecast below X, without written engineering approval." Or "Any deviation from approved design must be signed by the engineer of record – no exceptions, no matter schedule." By making some decisions effectively in advance (through such policies), you shield yourself from on-the-spot urgency bias. It's easier to say "We have to stop, it's policy" than to summon the will to stop when everyone's eyes are on you and the clock is ticking. Predefined thresholds act as circuit breakers that slow the momentum when it might be carrying you over a cliff.

Learning from past near-misses or failures is also a huge help. Leaders who have personally experienced the consequences of a rushed decision gone wrong tend to carry those lessons forward. If you've ever had a form blow out because you skipped a few tie rods to save time, you become far less likely to cut that corner again, even under pressure – the memory of cleaning up that mess (or worse, someone getting hurt) will rein you in. It's even better if you can learn vicariously by studying case histories like those in this book. They provide mental "scripts" that hopefully play in your head at the right time. For instance, if you're considering lifting in high winds, perhaps you'll recall Big Blue and realize you're re-living that cautionary tale. If you're contemplating ignoring an engineer's red flag, maybe Challenger pops to mind and gives you pause.

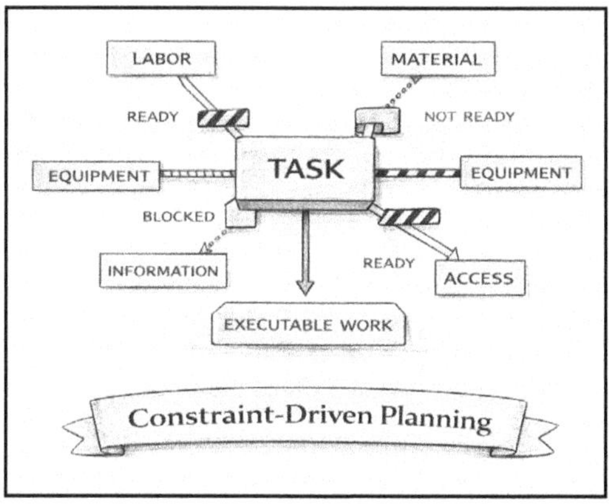


Being a leader under pressure also means managing groupthink. When the whole team is anxious to get something done, they can collectively reinforce each other's biases. A crew might all agree to proceed, each one thinking, "The others seem fine with it, so it must be okay." To counter this, actively invite a contrarian view: "Does anyone see a reason we shouldn't do this right now? Speak freely." You might be surprised – often someone had a doubt but didn't voice it until asked. Even a brief discussion of "what could go wrong" can reveal that the supposed consensus was masking individual concerns. Many companies implement formal last-minute risk assessments (LMRAs) or "Stop Work Authority" for exactly this reason – to break the spell of groupthink by empowering anyone to yell "Stop!" if something doesn't feel right, no matter how eager the group is to forge ahead.

Making sound decisions under pressure demands a combination of self-awareness, process discipline, and cultural support. Pressure will always be part of construction – whether from time, cost, or unexpected events – but it doesn't have to sabotage good judgment. The best leaders cultivate what might be called vigilant patience: they maintain urgency in action but patience in decision. That means they act quickly when needed, but not hastily. They reserve a mental fraction of a second to double-check their instincts ("Are we rushing into a mistake?") even when everyone else is screaming to go. They balance the drive to finish the job with the wisdom to sometimes say, "Not this way, not right now." In doing so, they steer their teams through the hair-raising moments without falling victim to the very dangers we're racing against.

## Diffusion of Responsibility and Leadership Vacuum

On a construction project, nothing is more dangerous than when everyone assumes that "someone else must be in charge of that." **Diffusion of responsibility** happens when roles and accountability are so blurry that each person or organization thinks another party will catch the problem or make the call – and in the end, no one does. This is closely related to unclear accountability, but here we focus on those moments when leadership is absent or fragmented – a **leadership vacuum.**

A leadership vacuum can occur in various ways. Sometimes it's literal: the key leader is off site or indisposed at a critical moment, and nobody has clear authority to act in their place. Other times it's structural: there are multiple entities with overlapping duties (for example, a general contractor and multiple subcontractors on safety), each assuming the other is handling a particular risk. It can also be a cultural vacuum: a leader is technically present but not truly leading – decisions are avoided, issues drift, and the team is left rudderless. In all these cases, the vacuum tends to be filled by confusion, delay, or unchecked hazards.

We already touched on one stark example of diffused responsibility: the Hurricane Katrina levee system. With multiple agencies involved and no single accountable leader, known problems languished until disaster struck. One of the harsh lessons was that when everyone's in charge, often **no one's in charge**.

On construction sites, diffusion of responsibility often shows up in day-to-day safety and quality oversight. Picture a scenario with multiple trades working in one area – say, electricians, plumbers, and carpenters in a cramped mechanical room. Everyone notices that a certain scaffold setup looks a bit shaky. The electricians assume the scaffold belongs to the carpenters, so the carpenters will fix it. The carpenters figure the general contractor's safety inspector must have OK'd it, so it must be fine. The safety inspector hasn't even been in that room today, thinking the trade foremen are monitoring their own setups. In the end, nobody actually takes action on the shaky scaffold, and it remains a looming hazard. This is a micro-level leadership vacuum: lots of people around, yet no one is clearly responsible for a glaring risk. If the scaffold collapses, injuring someone, you can bet an investigation will hear "Well, I thought someone else was handling that" from several folks. That is diffusion of responsibility in action.

How do we avoid leadership vacuums? Clarity and assertiveness. Clarity means explicitly assigning who is "captain of the ship" at all times and for all key aspects of the work. On a multi-employer site, there should be no doubt who has overriding authority on safety – perhaps the general contractor's superintendent by default, or a jointly designated safety lead. Within a team, if a supervisor steps away, they delegate their authority clearly ("While I'm at the meeting, Joe is in charge here"). It's also vital to clarify interfaces: if multiple contractors are working adjacent, they should agree on how to coordinate and who calls the shots if a shared risk (like a crane swing radius or a confined space) becomes an issue.

Assertiveness means encouraging people to fill a void the instant they perceive it. If something goes wrong and the usual leader isn't available, someone needs to step up and say, "I'm handling this until further notice." For example, if a serious accident happens and the superintendent is injured or absent, the next person in line – maybe the foreman with the most experience – should immediately act as incident commander. Crew members should be trained that it's okay to take initiative when leadership is missing. The often-cited saying in emergency management is: *When in charge, take charge; when not in charge, communicate*. That is, if it's your role to lead, don't hesitate – people are counting on you. If it's not your role, then make sure those who are supposed to lead are aware of the situation and that you're ready to assist. What we want to avoid is everyone standing around assuming someone else will handle it.

A good project culture has overlapping "safety nets" in leadership. For instance, crew members should know they all have stop-work authority if they see something life-threatening, which temporarily gives them a leadership role until the issue is resolved. That way, even if the designated leader misses something, another person can fill the gap by halting work. After all, it's better to briefly have "too many leaders" (like several people shouting stop) than none at all when it's needed. The key is then to sort out the situation and assign a single coordinator as soon as possible so that once stopped, the recovery is organized.

We should also be mindful that diffusion of responsibility can allow not just accidents, but **ethical lapses.** If everyone thinks someone else is checking the quality, errors can slip through. If everyone thinks "management must have approved this," unsafe shortcuts might persist. Thus, creating a culture of personal accountability at all levels is crucial. Each person should feel a degree of ownership: If I see it, I own it (at least to report or fix it). Leaders can cultivate this by praising those who step up. If a junior engineer flags a problem outside her usual scope, a good manager doesn't say "that's not your job" – they thank her for catching it and then ensure it gets addressed. Over time, people learn that seeing a problem = taking responsibility, not "mind your own business."

One more illustrative story: On a certain large project, it was noticed that a critical lifting lug on a heavy beam was badly welded. It was a subtle defect and easy to miss. The ironworkers thought the weld inspectors had caught it; the inspectors hadn't seen it and assumed the ironworkers would report if anything looked off. For a couple of days, that beam sat there with a compromised lug, and nobody acted because of those assumptions. Finally, during a walkthrough, a young field engineer pointed it out and directly called for a hold on using that beam. At first, a supervisor was annoyed – "We have sign-offs on these, it's fine" – but the engineer persisted, and they re-inspected and found the weld was indeed insufficient. The lug was repaired before any lifting happened. In the post-mortem, it was clear this could have been a disaster if the beam had been hoisted with that weak connection. The field engineer basically prevented an accident by breaking the diffusion loop – she didn't assume someone else would voice it, she took leadership in that moment. Afterward, the project manager made it a teaching point: "Don't assume someone else has it. If you see it, speak up or take action. We're all leaders of something here."

**Leadership vacuums** are avoidable through clear roles and an empowered team. Make sure at any given time, everyone knows who the go-to decision-maker is – and have a backup plan if that person is unavailable. Beyond formal assignments, foster a sense of shared vigilance: anyone can and should light a beacon if they sense no one is steering at a critical juncture. Far better to have a moment of overlap or to cry wolf than to have a gap where a wolf gets through. Construction is a team endeavor, but every team needs direction at critical moments. Don't let leadership be an assumed, amorphous concept – pin it down, person by person, task by task. And if the pin falls out, be the one who grabs it before everything unravels.

## Leadership as Stewardship

At the end of the day, all the talk about leadership and communication can be distilled to a simple principle: leadership is stewardship. A steward is a caretaker – someone entrusted to manage and protect something of value. In construction, a leader is entrusted with many valuable things: the lives and safety of the crew, the client's money and vision, the integrity of the structure being built, and the reputation of everyone involved. Embracing leadership as stewardship means recognizing that being in charge is not about ego or power, but about service and responsibility. It's about taking care of your people and your project as if they were your own – because in a very real sense, they are.

All the concepts we've discussed – active communication, storytelling, transparency, trust, accountability, command climate, poise under stress, sound decision-making, and stepping up to lead – are ultimately expressions of caring for the team and the mission. When a superintendent insists on double-checking a critical lift plan and halts the job to do so, it might frustrate some in the moment, but it's an act of stewardship: protecting lives and the project. When a project executive openly admits a mistake to the client and crew, it's not weakness – it's stewardship of trust and integrity. When a foreman mentors an apprentice, drilling into them the lessons of past failures so they won't repeat them, that's stewardship of the craft and the next generation.

Seeing yourself as a steward can also sustain you through the toughest days. Construction is high-stress, and things will go wrong. There are times when the burden of leadership feels overwhelming – when you're standing in front of an injured worker or a furious client or a pile of rubble that used to be your structure. In those moments, remembering why you lead is crucial. It's not for the paycheck or the glory (those evaporate pretty quickly in a crisis); it's because you have the duty and privilege to safeguard others and learn from mistakes. I recall a site manager who, after a fatal accident, met with the victim's family and vowed that he would never let something similar happen again on his watch. He spent the rest of his career pouring that pain into better training, stricter protocols, and a more open culture on every project he touched. That is leadership as stewardship – turning personal burden into collective betterment.

Stewardship implies continuity as well. We inherit the lessons (often written in blood) from those before us, and we carry the torch forward. The stories of Hyatt, Willow Island, Big Blue, Katrina, Surfside, and so on – these become part of our moral and professional compass. A steward leader makes sure these lessons aren't forgotten. They encourage reporting and recordkeeping of near-misses; they contribute to company libraries of lessons learned;

they talk about safety history in orientations not to scare, but to educate. In doing so, they ensure that each new batch of builders is a little wiser than the last. Think of it this way: every time you prevent an accident or solve a problem by applying a lesson from the past, you're honoring the stewardship of those who passed that lesson down. And each time you develop a protégé or share a hard-earned insight, you're fulfilling your duty to the future.

As we conclude this chapter, imagine two veteran construction managers – perhaps much like those who guided us through these pages – speaking with a young supervisor. What final advice would they give? Likely something like this: Don't strive to be a hero-leader who knows everything and never errs. Strive to be a guardian-leader who listens, learns, and lifts others up. Recognize that leadership means your actions set the course for everyone around you. When in doubt, err on the side of caring too much rather than too little. Care enough to question a risky plan. Care enough to spend an extra hour training your crew. Care enough to admit when you need help. And care enough to uphold standards even when no one is looking.

Our duty is to ensure that what we hand over – to the next crew, to the public, to posterity – is sound. Sound in construction, and sound in the precedents we set. If you lead with that mindset, you may not be perfect (nobody is), but you will earn the respect of your team and the peace of mind that you did right by those under your charge.

Leadership in construction is often the "make or break" factor, we said at the outset. It can make a project an environment of pride and safety, or break it into chaos and tragedy. By viewing leadership as stewardship, you choose the higher path – one of responsibility, empathy, and continuous learning. It's not the easier path, but it is the path of the true professional. Carry that mantle with humility and determination, and you'll not only build projects – you'll build people and a legacy that endures.

**Operational Takeaways for Today's Job:**

- Begin each day with a focused briefing that isn't just top-down – ask questions and invite feedback. This two-way communication ensures crews understand the plan and feel comfortable voicing concerns or clarifications before work starts.
- Be a visible leader on site. Take regular walks through the work areas, check in with individuals, and actively listen to what they report. Showing you're present and attentive builds trust and encourages your team to raise issues early.
- Model the behavior you expect. Follow every safety rule, wear all required PPE, and adhere to procedures scrupulously – your crew will take cues from your actions more than your words.
- When things get hectic or go wrong, keep your composure and give clear direction. Even under pressure, avoid yelling or giving vague orders – steady, calm communication prevents panic and mistakes.
- Promote a culture of accountability and learning. If a mistake happens, focus on fixing and understanding it rather than blaming. Openly acknowledge if you as a leader made an error too, showing that the goal is improvement and team safety, not finger-pointing.

# Chapter 5

# Risk Management Framework

## Identifying Project Risks (Safety, Technical, Regulatory, Financial)

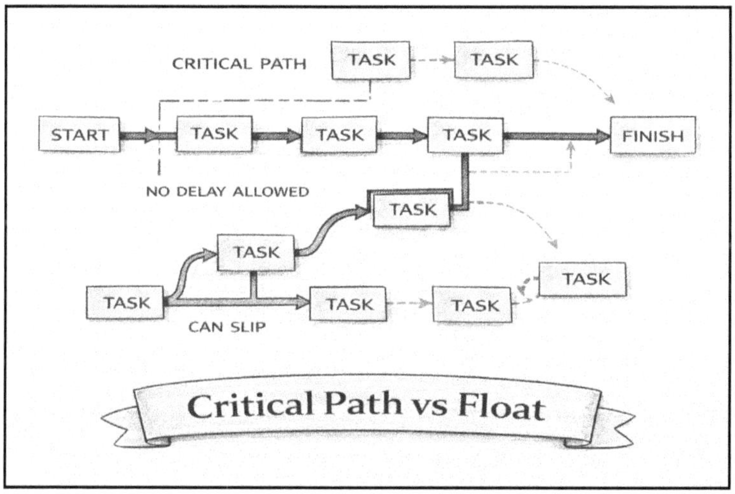


It is impossible to manage what you haven't first taken the time to see clearly. Construction risk management does not begin with a spreadsheet or software platform – it begins with the discipline of seeing. Before a risk can be ranked, mitigated, or transferred, it must be recognized as real, specific, and connected to the way the work is actually done in the field. On most projects, the early conversation about risk is short and optimistic. "We've done this before." "The design is straightforward." "The contractor has a good reputation." These statements feel reassuring, but they are not analysis – they are sentiment. Identifying project risks requires something different: a deliberate effort to look past familiarity and habit and ask the uncomfortable question, "**Where, exactly, could this go wrong?**"

## The Four Lenses of Risk

For practical purposes, project risk in construction can be viewed through four primary "lenses": Safety risk (the potential for harm to workers, the public, or occupants), Technical risk (the potential for failure in design, methods, sequencing, or temporary states), Regulatory risk (the potential for violations of laws, codes, or permitting conditions), and Financial risk (the potential for instability in cost, schedule, cash flow, or contractual exposure). These categories overlap constantly in the real world. A single decision can activate all four. For example, changing a shoring scheme without engineering review is a technical risk; if it results in a partial collapse, it becomes a safety risk; if it violates local code, it becomes a regulatory risk; and the ensuing claims, delays, and reputational damage manifest as financial risk. The categories are not separate boxes – they are different ways of looking at the same system.

Effective risk identification uses these lenses deliberately. For each phase and each major activity, the team should ask questions from each perspective: What could hurt someone here? What could fail technically, even if everyone has good intentions? What permits or requirements govern this work, and how might we drift out of compliance? What happens to cost and schedule if this assumption proves wrong? When such questions are asked early – and answered honestly – risk stops being an abstraction and becomes something visible and concrete.

## From Hazard to Scenario

Risk begins with a *hazard*, but it is understood through a *scenario*. A hazard is a source of potential harm: an unshored excavation, a live electrical system, a congested intersection, a deteriorated façade, a compressed concrete pour sequence. These are the **what** of danger. A scenario links that hazard to people, time, and context. It asks and answers: **Who** is exposed? **Under what conditions? How often? What else is happening at the same time?**

A scaffold tied into a sound structure is a normal temporary condition. But a scaffold tied into a deteriorated or partially demolished wall – with no understanding of the load path and no engineer involved – becomes a scenario for failure. Sadly, scenarios like that have played out in real projects with real consequences. One such incident involved a century-old brick building where long-term water infiltration had weakened structural members and widened cracks. On paper, those conditions were dismissed as "maintenance issues" – in reality, they were unacknowledged hazards. When a contractor was hired to replace what was thought to be a **cosmetic veneer,** portions of an exterior wythe that actually carried floor loads were removed without shoring. The hazard was not just old brick; it was a misunderstood structural system. The scenario only became clear in retrospect: a load-bearing wall misclassified as decorative, years of hidden deterioration, and active work that removed support from beneath occupied apartments. The result was a sudden, lethal collapse.

Had the risks been identified systematically beforehand, the crucial questions would have been simple and direct: *What is this wall actually doing? Where are its loads going? What happens to the residents above while this work is underway?* Instead, those questions were never asked, and gravity answered them with tragedy. The lesson is not limited to aging masonry. Every project has its own version of that wall – something everyone assumes is simple or benign until it fails. **Risk identification is the practice of turning assumptions into questions before reality turns them into failures.**

## Where Risk Hides in a Construction Project

On most projects, risk hides in three places: **interfaces, temporary conditions, and exceptions**. These often represent blind spots where normal planning might not immediately focus.

### Interfaces

Many failure case studies share a common pattern: each party understood their own piece of the work, but no one owned the interface. For instance, a bridge girder that "belongs" to the steel contractor and a concrete diaphragm that "belongs" to the concrete contractor meet in a detail that no one fully questions. A mechanical unit supported by structural steel and tied into a roof deck is designed by one team and installed by another, with a third party responsible for waterproofing. Each assumes the others have considered the load path, the penetrations, the access, the sequencing – but the gap between their scopes is where danger breeds.

In risk-identification sessions, these interfaces should be explicitly called out – the points "*where systems meet*" and "*where responsibilities cross*." These are often the spots where risk quietly accumulates, because everyone is busy with their own work and no one wants to be seen as slowing the job by raising a concern at the boundaries.

### Temporary Conditions

Most catastrophic construction failures occur during **temporary states:** shoring in place, forms loaded but not yet stripped, partial demolition in progress, staged lifting, traffic switches, initial energizing of systems. These conditions are inherently unstable, yet they are often treated informally as "just for today" problems rather than as critical design situations.

A sound risk identification process treats temporary conditions as their own design problems, worthy of the same scrutiny as the final permanent structure. The team should ask questions such as: What temporary loads are we introducing (from construction equipment, stored materials, wind on incomplete frames)? What are the acceptance criteria for this temporary state? Who actually engineered this condition, and has someone independent reviewed it? How will the load path change as we move from this state to the next? If these questions feel excessive, the history of construction collapses suggests otherwise.

### Exceptions

Almost every project has a sentence that begins, "We don't normally do it this way, but...." That phrase is a red flag. Each **exception** represents a deviation from the organization's tested habits and established procedures. Exceptions are not inherently wrong – many are necessary to adapt to unique constraints. What makes them dangerous is when they are adopted casually, without explicit risk identification.

When an exception is proposed – using a new material, altering a sequencing step, accepting a substitution, working outside normal weather limits – it should trigger a focused examination. The team should ask: *What hazard are we introducing or increasing by making this exception? What controls will keep that hazard within a safe envelope? How will we know if this exception starts to drift beyond what we intended?* If an exception cannot be clearly described in these terms, then it is not yet understood well enough to be safe.

## Tools for Identifying Risk – Beyond the Checklist

Construction teams too often treat risk identification as a formality – a box to check in preconstruction meetings, a few lines on a JHA form, or a generic list copied from a past project. That approach confuses record-keeping with thinking. The tools themselves (hazard checklists, "what-if" analyses, HAZOP studies, constructability reviews) are only as good as the questions and discussions they provoke.

In practice, an effective risk-identification toolkit for construction includes a mix of techniques. For example, teams should hold structured brainstorming sessions with the core project team (owner, designer, CM/GC, key subcontractors) focused specifically on how the project could fail – not just on schedule delays. They should conduct phase-by-phase hazard scans, walking through the project timeline to spot what is unique or high-energy in each stage (e.g. a deep excavation, a tower crane erection, a massive concrete pour, a tie-in to an existing structure, an occupancy transition). They should perform early field walks, where planners and engineers physically stand in the future work area to examine access, adjacent activities, utilities, and environmental constraints firsthand. Teams should review historical analogues – case studies and internal "lessons learned" – to ask, *Where did similar projects get hurt? What surprised those teams?* And they should do stakeholder mapping to identify where external risk might come from (neighbors, utilities, regulators, nearby structures, ongoing facility operations).

The goal of using these tools is not to generate a "perfect" list of hazards – no such list exists. The goal is to create a living catalog of credible scenarios that can be revisited and refined as the work evolves.

## Making Risk Visible and Shared

A risk that lives only in a manager's notebook or the back of one person's mind has not truly been identified in a useful way. In construction, a risk must be made concrete and widely understood. That means it should be named in plain, specific language; located in time and space (tied to particular activities or phases); owned by a specific person or role; and communicated to the people closest to the work.

Translating abstract risk concepts into concrete statements tied to operations is key. Instead of saying, "fall hazard," we might specify: "*During steel erection on grid lines 3–7, workers will be exposed to open edges at 30 feet with no permanent guardrails in place.*" Instead of a vague "regulatory risk," we might note: "*If we begin interior demolition before the asbestos variance is issued, we will be in violation of state law and can be shut down.*" Instead of "cost risk," we might warn: *"If the waterproofing detail at the plaza is wrong, we won't discover it until after turnover, and the owner will lose confidence in every other hidden system."*

When a risk is described with that level of specificity, people can act on it. Foremen can plan their crews around it. Superintendents can adjust the schedule to accommodate it. Engineers can double-check assumptions. Owners can better understand the trade-offs being made. Consider the brick wall collapse scenario again: the true risk was never articulated in concrete terms to anyone. There was no plain statement saying, *"If we remove this wythe without shoring, we are in effect removing a load-bearing wall under occupied units, and a floor collapse is a very real possibility."* Instead, the work proceeded under the softer (and misleading) narrative of "repairing veneer." That seemingly small difference in language is the difference between vigilance and complacency.

## Dynamic Risk – Treating the Register as a Living System

Risk identification is not a one-time ritual at the beginning of a project. Conditions on a construction job change daily – weather shifts, crews and subcontractors come and go, designs evolve through RFIs and revisions, materials get substituted, and regulatory interpretations emerge. A risk that was properly identified at the start can become obsolete, or far more acute, within weeks.

For that reason, the project's risk register (discussed in detail later in this chapter) must be treated as a living system, not a static document. New information and changes in conditions should continually trigger fresh risk identification. For example, a drawing revision that alters load paths or tolerances should send the team back to look for new hazards. A change order that compresses the schedule or adds work in tight quarters should prompt a review of risks introduced by the change. A near-miss incident that exposes a previously unrecognized danger should lead to that hazard being added to the register. An inspection finding that reveals a pattern (say, repeated missing rebar or recurring design questions on a certain detail) should spark a targeted risk assessment. A supplier substitution that changes material properties, lead times, or installation methods should trigger the question: What new risk does this pose?

Each of these changes loops back to the fundamental questions: *What new hazards have appeared? How do they interact with the existing ones? Does our safe operating envelope still hold, or have we quietly narrowed it?* Projects that treat risk identification as a continuous, ongoing activity – not just an early-project deliverable – are the ones that catch slow drift before it turns into a sudden crisis.

## Asking Better Questions

At its heart, identifying risk in construction is less about having the right form or template and more about asking better questions. Technical knowledge matters, and so do codes and standards – but the quality of risk identification ultimately tracks with the courage and curiosity of the people doing it.

The most useful questions are often pointed and imaginative. For example: *"What would have to go wrong here for someone to get seriously hurt?" Or, "What are we assuming will stay true for this plan to work safely?"* Or, *"Where are we relying on one person, one sensor, or one procedure with no backup?"* Another good question: *"What is the worst 'bad day' this system or activity could have, and how would we know it was coming?"* Perhaps the most uncomfortable question of all: *"If this goes wrong, could we honestly say we did not see it coming? Or would we have to admit that we chose not to look?"*

When a project's culture actively encourages questions like those, risk is no longer hidden. It becomes something that can be named, mapped, and managed. That is the purpose of this entire section: to move risk identification from a reluctant, cursory exercise into a core leadership habit – one that ties together safety, technical judgment, regulatory obligation, and financial stewardship.

In the sections that follow, we will translate identified risks into structured registers, contingency plans, and crisis-response frameworks. But none of those systems will matter if the first step – seeing risk clearly is skipped or rushed. **Identifying project risks is not decoration around the work; it is the work that makes everything else possible.**

## Risk Registers & Classification Systems

In the previous section, we focused on the discipline of seeing risk clearly – turning vague worries into specific scenarios tied to real work. A risk register is the tool that captures that vision so the project can act on it consistently over time. It is not meant to be a bureaucratic formality or an exercise in "checking the box." Properly used, the risk register becomes the living memory of the project's worries and the record of what the team plans to do about them.

Too often on jobs, the register is created only because a contract requires it. Someone copies a template from the last project, changes a few names, fills in some generic phrases, and files it away on a server where no one ever looks at it again. The project can technically say, "We have a risk register," but in reality, it has nothing at all – just a document with a title.

A useful risk register is something very different. It is an operational tool. It tells the evolving story of what might go wrong on this project, at this site, with this design and these people – and it connects what we fear to what we are actually going to do about it.

## Purpose of the Risk Register

At its core, a risk register serves several purposes. It collects all the risks in one place so that nothing important is forgotten. It clarifies each risk by forcing the team to describe it in specific, operational terms. It prioritizes risks based on their likelihood and consequence. And it assigns ownership, ensuring that each significant risk has a responsible party and an action plan attached.

Those four functions sound straightforward, but achieving them requires a different kind of thinking than everyday project chatter. A risk register that only lists vague categories like "safety risk," "design risk," or "schedule risk" has not done the real work; those phrases are headings, not hazards. By contrast, a well-built risk entry might read something like:

"During the staged demolition of the existing mezzanine on Level 2, there is a risk that the remaining beams will be overloaded because the original load path is not fully understood. **Consequence**: partial collapse, worker injury, significant delay."

Now the risk is something the team can visualize. It has a specific context, a location, an activity, and a plausible failure mode. In short, it's no longer invisible.

## Anatomy of a Risk Entry

While register formats vary, most effective risk logs share similar elements. At a minimum, each significant risk entry should answer a series of key questions about the scenario:

- **What could happen?** (A brief description of the risk event or condition.)
- **Why would it happen?** (The underlying causes or assumptions that could lead to it.)
- **So what?** (The potential consequences in terms of safety, quality, cost, schedule, or reputation.)
- **How likely is it?** (A considered judgment of probability – an informed estimate, not a blind guess.)
- **How bad would it be?** (An assessment of severity if it occurs – often a ranked category of consequence.)
- **What are we going to do about it?** (The mitigation steps or contingency plans in place to reduce the likelihood or impact.)
- **Who owns this?** (The person or role responsible for monitoring and managing this risk.)

When these elements are captured for each major risk – typically in a simple table or database – you end up with more than a random list. You have the beginnings of a risk story for the project. Over time, that story will evolve: new "characters" (risks) will appear, some will be resolved or retired, others will grow in importance. The register becomes a living logbook of that evolution.

## Turning Qualitative Judgments into a Shared Language

Construction projects rarely have the time or data to perform full quantitative risk analyses on every hazard. Most of the time, risk assessments rely on the experienced judgment of the team. That doesn't make them weak if the judgment is structured and shared.

To structure subjective judgments, teams often use simple rating scales. A common approach is to rate each risk on two axes: Probability (how likely is this event to occur during the life of the project?) and Consequence (if it does occur, how severe could the outcome be?). Each axis is divided into a few categories. For example, probability might be categorized as Rare, Unlikely, Possible, Likely, or Almost Certain. Consequences might be categorized as Minor, Moderate, Major, Severe, or Catastrophic. Crucially, these words are explicitly defined in terms of the project. "Rare" might be defined as "less than a 1% chance during this project," while "Almost Certain" might mean "expected to happen multiple times." "Minor" might be defined as "no injury, negligible cost or schedule impact," whereas "Catastrophic" might mean "a fatality, a structural failure, or a project-threatening cost/schedule impact."

By agreeing on such definitions upfront, the team transforms individual opinions into a shared language. A superintendent may not calculate an exact numeric probability, but he can say, "Based on similar sites we've worked on, this scenario is more than 'possible' but less than 'likely.'" An engineer might add, "If this assumption is wrong, the consequence is at least 'Major' – because we'd have to rip out finished work and redo it." Those qualitative statements can then be translated into the agreed-upon ratings on the register.

## The Risk Matrix

Once each risk is given a probability rating and a consequence rating, it can be plotted on a simple matrix – one axis for likelihood, the other for severity. The cell where a given risk falls is often labeled with an overall risk level (for instance, low, medium, or high risk).

The risk matrix is not a precise instrument; it's more like a map. Its value lies in how it focuses the team's attention. The high-likelihood, high-consequence risks congregate in one corner of the matrix (demanding immediate attention), while low-likelihood, low-consequence risks fall in the opposite corner (requiring minimal oversight). In between, there's usually a band of "medium" risk where more judgment is needed.

- In practice, teams use the matrix in a few key ways. First, to **prioritize discussion:** the highest risks are reviewed and discussed first in coordination meetings or risk workshops. Second, to **justify action:** if a risk is rated high (or even medium-high), the team is expected to develop concrete mitigation plans, not just note the issue and move on. Third, to explain decisions: the matrix provides a visual rationale for why some risks received more resources or schedule contingency and why others were accepted as-is. Without a classification system, debates about risk can easily devolve into subjective opinions or loud personalities. With a structured matrix and clear categories, the team can remain calm and systematic: *"We all agree this scenario is unlikely, but if it did happen, the consequences would be catastrophic. That combination still warrants serious preventative measures."*

## Risk Classification: Naming the Pattern, Not Just the Event

- In addition to likelihood and consequence, it helps to tag each risk by type or category. The specific event - say, "crane may become unstable during erection in high winds" - belongs to a broader class of risk (perhaps *"temporary works"* or *"lifting operations in weather"*). Classifying risks in this way provides several benefits.

First, it allows the team to see **clusters of risk**. If, for example, half of the high-risk items on the register fall into the category of "temporary conditions," that is telling you something important: the project's biggest vulnerabilities may lie in the states between permanent, engineered configurations.

Second, it supports **lessons learned** after the project. When the job is complete, the team (or the broader organization) can look back and ask, "Where did we end up spending most of our risk management effort? What patterns kept recurring?" If many issues fell into, say, "underground utilities" or "public interface," that knowledge can shape planning on future projects.

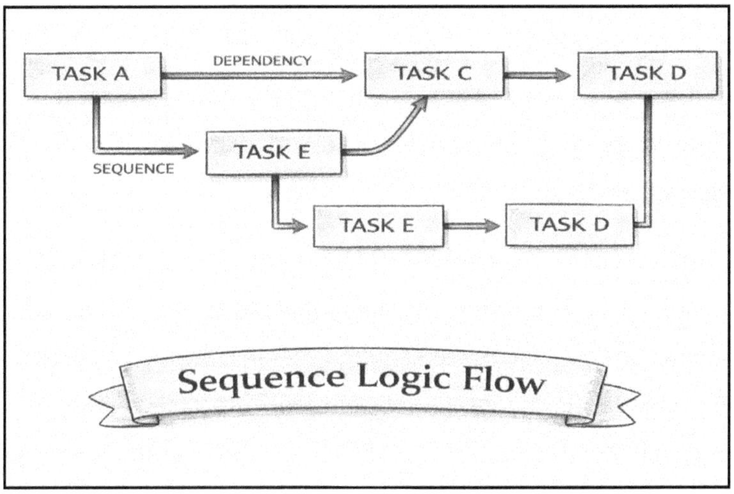


Third, classification can expose **blind spots**. If certain classes of risk - perhaps "adjacent structure" or "public safety" - never appear in your register, you should pause and consider whether this means the project truly has none of those risks, or if the team simply overlooked them due to familiarity or assumptions.

There is no single "correct" taxonomy of risk types; what matters is that the categories reflect the real contours of risk on your project. That said, typical classes in construction might include: Safety (both worker and public safety), Design (structural, geotechnical, MEP, and temporary works issues), Regulatory (permitting, code compliance, environmental conditions), Commercial (contract interpretations, change orders, potential claims), Schedule (long-lead items, critical path delays, seasonal/weather windows), and External (impacts from neighbors, owners, agencies, or utilities). Within each broad class, you can always add more detail later, but even a simple set of tags will dramatically increase everyone's understanding of where risk tends to concentrate.

## Ownership and Accountability

Every significant risk in the register should have a clearly identified owner. Not a committee, not "the project team" in general – but a specific person or role accountable for that risk. Ownership in this context does not imply blame if the risk materializes; it implies responsibility for shepherding that risk through its lifespan on the project.

The risk owner is expected to ensure the description of the risk is accurate and kept up to date, to monitor the conditions that influence its likelihood and consequence, to drive the agreed mitigation actions to completion, and to propose reclassification or closure of the risk if circumstances change substantially. For example, on a well-run project, a field superintendent might own the day-to-day safety risks for work happening on a particular elevation or area. The project manager might own commercial and scheduling risks. The design manager might own design coordination risks spanning multiple disciplines. Senior executives might retain ownership of rare but high-consequence risks that could threaten the entire enterprise.

Without clear ownership, risks tend to drift. Items languish on the register month after month with the same notes, the same ratings, and no movement. Over time, people stop believing that the register has any real connection to what's happening in the field.

## Keeping the Register Alive

A risk register that is updated only when the client requests a copy is not a management tool at all – it's a prop. To be useful, the register has to be woven into the normal rhythm of the job. This doesn't mean adding more long meetings or thick binders; it simply means incorporating risk review into existing processes.

At project kickoff and during preconstruction, the team should populate the register with the major known risks (design complexities, site challenges, logistical constraints). During look-ahead meetings, upcoming tasks should be checked against the register: *"Do we have any high-risk activities in the next 3–6 weeks, and are we ready for them?"* As part of change management, any time there is a significant design change, scope change, or re-sequencing, the register should be updated (new risks added or existing ones modified to reflect the change). After any incident or near-miss, the investigation should include adding or refining register entries to capture the newly revealed hazards. And in monthly reports or owner updates, it's wise to include a summary of the top risks, any changes in their ratings, and the status of mitigation actions.

Integrating the register in this way doesn't require a special standalone "risk meeting." It can be as simple as pulling up the risk log on a screen during a coordination meeting and asking: *"Has anything changed? What are we doing about these top risks? Are we seeing any early warning signs?"* Some teams find it useful to visually highlight the current "top ten" risks and post them in the site office, updating that list as the project evolves. The specific risks may change, but the message is constant: **these are the things that demand our attention and discipline right now.**

## Common Failure Modes in Risk Registers

Like any process, maintaining a risk register can fail – and it often fails in predictable ways. One failure mode is **the cosmetic register**, which is filled with generic statements ("safety risk," "weather risk") that are true of every project and specific to none. Another is **the historical copy**, which has been lifted from a previous job with only minimal editing, thus ignoring the unique hazards of the current project. A third is **the one-time dump** – a register heavily populated at the start of the job and then never seriously revisited or updated. There is also **the wishful register**, full of mitigation actions that sound good on paper but that no one on the team actually has the authority, budget, or intent to carry out. Finally, **the secret register** is the risk log kept quietly among a few managers and never shared with the people actually performing the work.

Each of those dysfunctional approaches leads to the same place: a disconnect between the risk document and the real operation of the project. To gauge whether your risk process is truly working, ask a few blunt questions: *Does our current register describe the work our crews are actually doing this month? Can each superintendent on the job name the top three risks for their portion of the work? When a risk's rating changes in the register, does anything out in the field change because of it?* If the honest answer to any of these is "no," then the register is not yet serving its purpose.

## From Register to Action

Ultimately, the test of a risk register's value is simple: Does it change how we build? If a certain lift or operation is classified as high risk due to wind exposure and proximity to the public, do we actually see changes in the rigging plan, larger exclusion zones established, or adjustments to the schedule to avoid high winds? If a temporary shoring condition is flagged as high risk, do we see that translated into a requirement for a stamped engineered design and perhaps an independent third-party review before the work proceeds? If a regulatory compliance risk is rated high, do we engage with inspectors earlier or adopt a more conservative interpretation of an ambiguous code requirement?

A good risk register becomes a bridge between analysis and behavior. It takes what the team worries about in the abstract and translates those concerns into decisions about manpower, equipment, time, and money. In essence, it ensures that the invisible architecture of risk management is as carefully planned and executed as the visible structure rising from the ground.

In the sections that follow, we'll look at how contingency planning, crisis management, and emergency response all build on the foundation laid by a strong risk register. None of that later work will succeed if the register itself is weak or neglected. The register is the ledger of what we know can hurt us – and it's the first place we commit (in writing) what we plan to do about those dangers.

## Contingency Planning for Critical Path Items

When a project is young, and the bar-chart schedule still looks neat, contingency planning is often treated as an optional "nice-to-have" – something to consider "*if we have time.*" Once the job is underway, it becomes painfully obvious that contingency planning should have been core work from the start. Critical path items are, by definition, the activities that determine whether the project finishes on time. If those same activities also carry high safety risk or high technical risk, then a failure in one of them can cost not just schedule but lives and structural integrity.

Contingency planning is the disciplined act of asking **"What will we do if this doesn't go as planned?**" – and answering that question before the work begins. It's a commitment to maintaining control of the work even when conditions deviate from the ideal scenario. In systematic operations terms, contingency planning is how we protect the safe operating envelope when the project is pushed toward its limits.

On many projects, the default "contingency plan" is nothing more than hopeful improvisation:

- *"If the crane isn't ready, we'll figure something out."*
- *"If the concrete supplier is late, we'll spin the crews onto something else."*
- *If the inspection fails, we'll just work around it somehow."*

These are not plans – they are hopes. Contingency planning is different. It is deliberate, written down, resourced, and tied to specific trigger conditions.

## Why Critical Path Items Deserve Special Contingency

Not every activity warrants the same level of backup planning. The effort spent on contingencies should be proportional to the risk involved and to how close that risk is to the project's critical path. Three key questions can help pinpoint where rigorous contingency planning is required:

**If this activity fails, how bad is the outcome?** Consider the dimensions of failure. Could someone be seriously injured or killed (a safety impact)? Could permanent works or critical temporary works be damaged (a structural integrity issue)? Would the project miss an immovable milestone like a scheduled rail outage, a regulatory shutdown window, or a seasonal weather opportunity (a major schedule/commercial impact)?

**If this activity slips or is delayed, how quickly do we run into downstream conflicts?** Will other trades back up behind it, causing stacking, inefficiency, or idle time? Will permits, road closure permissions, or outage windows expire if we miss the planned dates? Will an entire weather season be lost if the window closes? In short, how much real float (if any) exists for this activity before other commitments unravel?

**How much slack or margin do we truly have?** If the answer is "none" or "almost none," then we are not just managing risk – we are managing inevitability. In such cases, something going off-plan isn't just possible, it's probable, and the only thing preventing a cascade of delays or failures will be the quality of our contingency planning. Projects with zero slack on critical, high-risk tasks are essentially betting on perfection; contingency planning is how we prepare for anything less than perfection.

History shows why these questions matter. In many high-consequence failures documented over the years, the sequence leading to disaster had at least one common element: once the primary plan started to unravel, **there was no viable Plan B ready to go**. Work continued on a failing trajectory under the hopeful assumption that "we'll get through it somehow," even as warning signs piled up. That is exactly what contingency planning is meant to prevent.

## Contingency vs. Padding

It's important to distinguish real contingency planning from mere padding of a schedule or budget. Padding is just throwing generic extra time or money at an activity. Adding an arbitrary week "just in case," or tacking on a flat 10% "contingency" line item in the budget with no specific trigger or plan behind it. Padding might make people feel better, but because it isn't tied to particular scenarios, it often leads to complacency and waste (and the extra time or money gets absorbed without actually improving safety or certainty).

Contingency planning, by contrast, means identifying specific alternate paths and actions for specific risks. For instance: establishing an alternative method of hoisting in case the primary crane goes down; lining up a backup supplier for a critical material in case the first supplier fails to deliver; developing a re-sequencing plan if a key area of the site suddenly becomes inaccessible; or having pre-engineered temporary measures ready to deploy if a vital inspection fails. In each case, the contingency is purposeful; it addresses a known risk with a pre-planned action, rather than just adding a fudge factor.

Padding without thinking invites a false sense of security, whereas contingency planning demands thought, documentation, and discipline. It aligns with the same systematic, conservative operating philosophy that says we should anticipate abnormalities and prepare responses before emergencies arise.

**Step 1:** **Tie Contingency Directly to the Risk Register**. The starting point for contingency planning is the risk register itself (from the previous section). For each high or medium-high risk item that touches the critical path, the team should explicitly ask: *"If this risk materializes, what is our plan – one that doesn't involve panic – to deal with it?"*

Suppose the risk register includes an entry that says: "Temporary shoring for the main lobby slab may not meet expected stiffness, leading to excessive deflection at pour." If that concrete pour sits on the project's critical path, then contingency planning needs to answer several questions. *What if the pre-pour inspection fails and we can't pour on schedule? What if we begin pouring and observe deflection beyond acceptable limits? What if the shoring supplier is delayed or delivers components that don't meet spec?* The contingency plan for that item should be explicitly linked to the risk register entry (by ID or description) so that the register and the schedule are connected. In this way, identifying a risk on paper is not the end of the process – it's anchored to a specific future activity on the schedule, with an alternative ready.

**Step 2:** **Define Triggers – When Does the Contingency Activate?** A contingency plan that is never triggered does no good; one that is triggered too late is just theater. The key is to define clear triggers – observable conditions that tell the team, "It's time to switch from Plan A to Plan B." Triggers can be set in various ways:

- **Time-based triggers:** e.g. *"If procurement of the structural steel for Grid Line 5 is not 100% confirmed by March 1, we will activate the alternate vendor plan."*
- **Condition-based triggers:** *e.g. "If sustained wind speeds exceed X mph on the morning of the critical lift, we will postpone the lift and activate the backup sequencing plan."*
- **Performance-based triggers:** *e.g. "If the 7-day concrete break tests come in below Y psi, we will defer stripping the forms and adjust the pour sequence for the remaining slabs."*

By establishing these trigger points in advance, the team removes ambiguity and hesitation in the heat of the moment. There's less room for "Maybe we can get away with it just this once," because the decision threshold was defined calmly, ahead of time. This approach mirrors disciplined operational thinking in other domains (aviation, nuclear, etc.): the rules are decided before the pressure is on, not invented on the fly when everyone's stressed.

**Step 3:** **Build Realistic Alternatives, Not Fantasy Plans.** An effective contingency plan must be something the project can actually execute with the resources and constraints it has. A "plan" that relies on wishful thinking – like "*we'll just find another 500-ton crane on short notice*" or "*we'll add an extra crew (even though none are available)*" – is not a real plan.

For critical path items, realistic contingencies often fall into a few categories:

- **Alternate methods or means:** Perhaps a different erection or construction sequence that relies on smaller, more readily available equipment instead of one huge piece. Or a backup formwork or shoring system that could be used if the primary proprietary system is delayed. Or a temporary bracing scheme that would allow partial progress to continue while waiting for a permanent element to arrive or cure.

- **Alternate schedule paths:** Maybe re-sequencing other work so that if Task A is halted, crews can switch to Task B or C in the meantime and stay productive. Pull some off-critical tasks forward (if they can be done out-of-sequence safely) to buy time. In some cases, plan for night or weekend shifts (within reasonable fatigue limits) to recover lost time after the issue is resolved.

- **Alternate suppliers and logistics:** Identify pre-qualified backup vendors for essential materials or components (post-tension cables, structural steel, specialty items) in case the primary supplier falls through. Line up alternative delivery routes or laydown areas if the primary access becomes blocked or restricted.

- **Temporary performance degradation:** In operational settings, sometimes the contingency is to run a system at reduced capacity rather than full capacity until it can be fixed properly. In construction terms, that might translate to accepting a temporary state of partial completion or reduced loading while a fix is designed – rather than pushing ahead and risking total failure. (For example, operating a facility at half load for a period instead of forcing a quick full startup through an incomplete system.)

A sobering real-world example of failing to have realistic alternatives comes from infrastructure: an aging city bridge had been deteriorating and showing signs of distress for years. But there was no robust contingency or mitigation plan. No one had set thresholds for when to reduce loads or close the bridge; inspections noted increasing corrosion, but traffic continued as normal. In effect, a critical path asset had zero meaningful contingency – only hope that it would last until repairs. We know how that story ended: a sudden, catastrophic collapse. The lesson is that for vital elements, "hope" is not a strategy. If an element is crucial and failure-prone, you need a backup plan that doesn't rely on luck.

**Step 4:** **Pre-Commit Resources.** A contingency plan that requires resources the project hasn't reserved in advance is essentially a broken promise. To be credible, each major contingency tied to a critical path risk should come with pre-committed resources. That means the team has set aside:

- **Budget**: Allocated funds earmarked to cover things like rework, alternate methods, or acceleration if the contingency must be executed. (This is different from generic padding; it's budget tied to a specific scenario.)
- **Time:** Some "contingency float" around key milestones – explicitly shown in the schedule – that can be used if Plan B is activated. This float should be protected and not casually eaten up by minor changes.
- **People:** Specific individuals (and crews) identified who will lead and perform the contingency actions, including a clear decision-maker who will give the go/no-go call when trigger conditions are met.
- **Technical support:** Designers or engineers on standby who have agreed in advance to be available for rapid analysis or modifications if the contingency needs engineering input.

In disciplined operations (think of emergency drills in industrial plants or on ships), this kind of pre-commitment is akin to having practiced the fire drill rather than just hanging an evacuation map on the wall. When trouble hits, the reaction isn't chaos or confusion, because everyone knows their role and the plan is familiar.

**Step 5: Embed Contingency into Planning Rituals**. Even the best contingency plan is useless if it's hidden in a binder that no one reads. For critical path items, contingencies should become part of the everyday planning conversations.

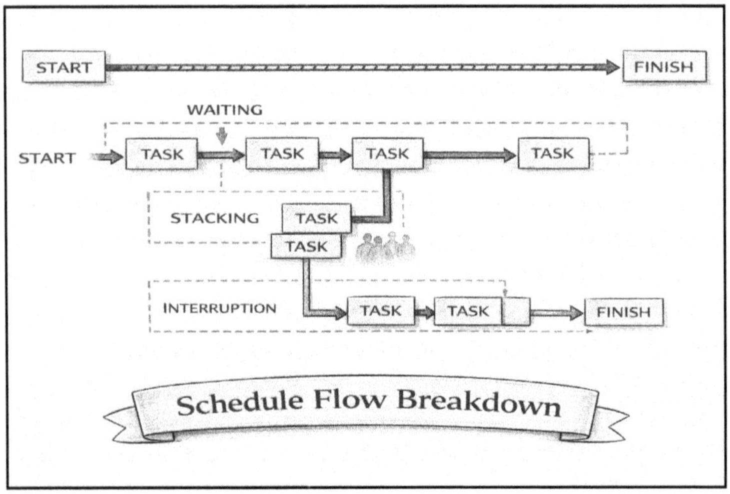


For example, at daily prep meetings or crew huddles, supervisors might routinely say: "*Here's what we plan to do today, and here's what we'll do if X happens.*" At weekly look-ahead meetings, the project manager might point out: *"Next week's critical concrete pour has three known risks; these are the backup plans we've prepared for each one." In client or stakeholder meetings discussing an upcoming high-risk operation, the team can demonstrate professionalism by explaining: "This*

*is our primary plan for the power shutdown, but we also have contingencies ready if the test results or the weather don't cooperate."*

When contingencies are spoken about openly and matter-of-factly – when they are rehearsed and understood by everyone from managers to the crews on the ground – they inspire confidence. The message is clear:
**if conditions change, we will not be scrambling in an improvisational panic that pits safety, quality, and schedule against each other. Instead, we will execute a Plan B that we prepared earlier with just as much care as Plan A.**

**A Simple Example:** Critical Lift with Weather Exposure. Consider a scenario involving a major steel truss lift over a busy roadway. This is a classic critical path item: if the lift cannot occur as scheduled, the entire project's completion date will slip. The primary plan calls for using a 500-ton crane, closing several lanes of traffic on a weekend, and executing a carefully choreographed sequence of picks to set the truss.

A weak approach to this lift would treat it as an all-or-nothing event that must happen on the chosen weekend, assuming ideal weather. The schedule might simply bank on a sunny, calm day, with no serious thought given to alternatives if that window doesn't hold.

A stronger approach includes a robust contingency plan. For instance, a clear trigger could be set: "*If the weather forecast 24 hours before the lift shows sustained winds above X mph – or if actual winds exceed Y mph during crane setup – we will stand down and postpone the lift." The team would have alternate methods ready, such as breaking the truss into two smaller sections that could be lifted in higher winds on separate weekends if the single-piece lift is canceled.* They would plan alternate schedule paths – for example, using that "lost" weekend to perform other structural work or bolt-up and decking in adjacent bays, so the time isn't wasted while waiting for the next closure opportunity. They would also pre-commit resources to make postponement feasible – for example, having contract terms with the crane company that define the cost and logistics of a rescheduled lift, so the decision to abort isn't financially ruinous.

When the stressful moment arrives – say clouds roll in and wind speeds start creeping toward the limit – the site discussion is no longer a desperate argument about whether to "push through" and risk it. It's a straightforward comparison against the predefined criteria: *"We agreed that if these conditions existed, we'd activate Plan B. That time is now – let's do it."* In this way, contingency planning turns anxiety into prepared decisions. The project may take a short-term hit (e.g., a week's delay), but it avoids a potential catastrophe and keeps control of its destiny.

## Common Failure Patterns in Contingency Planning

It's also useful to recognize how contingency planning efforts themselves can fail, so we can avoid those pitfalls. Some common failure patterns include:

- **The invisible contingency:** A plan exists on paper but is never communicated to the people who would actually have to implement it. In an emergency, those on the front line don't even know a Plan B exists.
- **The unfunded contingency:** The plan depends on resources (money, equipment, staff) that the project doesn't have readily available. When the time comes, the backup can't be executed because no one set aside what was needed.
- **The over-complicated contingency:** The plan is so complex or onerous that, in the moment, no one believes it's feasible. An over-complicated Plan B can be as useless as no plan at all if the team lacks confidence in it.
- **The "hero" plan:** The plan assumes extraordinary effort or perfection – for example, expecting people to work continuous double shifts or assuming zero mistakes will be made – rather than building in realistic margins. It relies on heroics instead of sound strategy.
- **The stale contingency:** The plan is created early but never updated as conditions change. By the time it's needed, it no longer fits the situation.

Contingency planning for critical path items is *not* about being pessimistic or expecting failure at every turn. It's about respect – respect for the complexity of the work, for the unforgiving nature of physics, for the limits of materials and human performance, and for the fact that reality seldom unfolds exactly as our baseline schedule predicts. When done well, contingency planning reinforces the same disciplined, conservative approach that underlies safe operations throughout this book: we acknowledge that failure modes exist, and we prepare in advance so that if one threatens to occur, we already know how to keep it from cascading into disaster.

## Crisis Management – From Prevention to Response

No matter how carefully we identify risks and build contingencies, there will be days when the project is truly tested. A piece of equipment fails at the worst possible moment. A hidden condition emerges at the exact point when the structure is at its weakest. A sequence that looked sound on paper starts to spiral out of control in the field. When that happens, the job is no longer in the realm of "normal operations." It has crossed into the territory of incidents, emergencies, and crises. How the team responds in those moments will determine not only the outcome of that event but also the project's culture going forward. People will remember: did leadership lose control and panic, or did they bring order to the chaos?

It's important to emphasize that crisis management is not a separate discipline from risk management; it is simply the extension of the same mindset under high-pressure conditions. In a systematic operation, the goal remains the same in both normal and abnormal phases: maintain control of the work and keep the operation inside its safe boundaries, even when conditions are far from ideal.

## Incident, Emergency, Crisis – Knowing What You Are In

In the heat of a bad day, words matter. If every problem is labeled a "crisis," people become desensitized and may either overreact or stop reacting appropriately. Conversely, if the team refuses to use the word "crisis" until it's far too late, they might not escalate their response fast enough. It helps to distinguish three levels of escalation and be clear-eyed about which is which:

- **Incident:** an unplanned event that disrupts normal operations but can still be controlled with routine responses. It might be a small equipment fire that's quickly extinguished, a minor injury where first aid is sufficient, or a piece of formwork that shifts slightly but doesn't collapse. Incidents are hiccups – they require attention but not a full emergency mobilization.

- **Emergency:** a situation where people or critical systems are in immediate danger, and rapid, coordinated action is needed to prevent serious harm. Think of a worker dangling from a failed harness anchor, a trench wall beginning to cave in, or a crane that has lost stability and is at risk of toppling. In an emergency, lives or critical assets are on the line right now, and the response must be fast and organized.

- **Crisis:** essentially a sustained emergency that fundamentally threatens the project, the organization, or the public. This could be a major structural collapse, an accident resulting in multiple severe injuries or fatalities, or a failure that triggers regulatory shutdowns, lawsuits, and intense public scrutiny. A crisis isn't just a bad incident – it's a disaster that can have long-term organizational consequences.

On a real site, these categories won't always be crystal clear. An incident can escalate into an emergency if it's mishandled or if circumstances worsen. An emergency can deteriorate into a full-blown crisis if the team is unprepared or if the initial response fails. The labels are less important than the clarity: everyone on the team should pause and ask, "*What do we have here, and what level of response does it require?*" Being able to say "This is an emergency" vs. "This is a crisis" helps marshal the right resources and urgency.

Projects that handle crises well rarely start with last-minute, heroic feats of crisis management. They start by being honest early. They call an incident what it is and respond with appropriate urgency, instead of pretending everything is fine. They step up the level of response as soon as the situation demands it, rather than waiting for things to "settle down" on their own when all indicators are pointing toward escalation. In short, they manage escalations in stages instead of letting a slow-boiling problem explode.

## Prevention Does Not End When the Alarm Sounds

Mentally dividing the world into "prevention" (what we do to stop bad things before they happen) and "response" (what we do after something bad has happened) is not an exclusively bookended assessment. In reality, even after an accident or failure has occurred, there are still opportunities to prevent the situation from getting worse.

When something goes wrong, we can think of three phases: the **lead-up** (a period of drift or deviation where conditions deteriorate and early warning signs may appear), the break (the moment when a component, process, or system crosses a boundary and actually fails), and the aftermath (the period when the team is trying to stabilize the situation, rescue or care for people, protect what remains, and eventually recover).

Our risk management and contingency planning efforts are aimed primarily at the lead-up – recognizing signs of drift, tightening controls, adjusting plans before the break happens. But even after the break, good crisis management is essentially late-stage prevention. We can no longer prevent the initial failure, but we have enormous influence over what happens next.

We can prevent a bad structural event from becoming a multi-fatality disaster by how quickly and effectively we secure the area and account for all personnel (so no one wanders into a collapse zone, and anyone injured is found and helped immediately). We can prevent a serious injury from becoming a fatality by how competent and rehearsed our rescue and medical response is. The difference between life and death might be a matter of minutes and the availability of proper first aid or emergency extraction. We can prevent a technical failure from collapsing a company by how we manage communication, documentation, and cooperation with investigators and authorities. Handling those poorly can lead to legal and reputational fallout far worse than the immediate physical damages.

In this sense, crisis management is really "**late-stage prevention**." It means recognizing that even though a bad thing has happened, our job is still to control the damage and avoid compounding the failure. The mindset is proactive: **What can we still save or protect? What secondary accidents or impacts can we head off? What longer-term consequences can we mitigate by acting decisively now?**

## Command Structure Under Stress

On an ordinary day, a construction project's org chart and decision-making can afford to be pretty flexible. Foremen handle issues on the fly. Engineers and superintendents work out sequencing or design details in real time. Decisions are decentralized, which is healthy for efficiency during normal operations.

In a crisis, however, that loose, everybody-jump-in structure can become a liability. If multiple people simultaneously try to take charge – or if everyone assumes someone else must be making the critical calls – the result is confusion, wasted time, and even conflicting orders being given. Seconds and minutes count during an emergency, so clarity of command is paramount.

A basic crisis command structure answers three questions unequivocally: **Who is in command of the site during the emergency? Who has the authority to communicate externally (to media, regulators, etc.)? How does information flow from the field to the command post and back to everyone on site?**

On many projects, the incident commander is designated as the senior superintendent or the project manager who is present on site. That person, once an emergency is declared, is responsible for a short list of critical tasks: declaring that an emergency exists (i.e. saying out loud that normal operations are suspended and emergency procedures are in effect), activating the emergency plan (initiating evacuations, rescue operations, system shutdowns as needed), assigning a few key roles (for example, appointing one person to focus on site safety/evacuation, one to liaise with incoming emergency services, one to manage operations or engineering decisions, one to handle logistics like headcounts and equipment), and ultimately demobilizing the emergency once the situation is controlled and it's safe to return to normal work.

Having one clearly identified person in charge does not mean everyone else stands around waiting for orders. It simply means that in the fog and noise of an unfolding event, there is one central node for decision-making and information integration. The incident commander's job is to listen to inputs (from field personnel, foremen, engineers, emergency responders), make the big decisions, and coordinate the overall response.

Equally important is clarifying **who is not in charge**. For instance, a well-meaning senior manager or executive might rush to the site upon hearing bad news. If that person starts issuing directives without understanding the situation, they can distract or even endanger the response. The crisis command structure must be strong enough and clear enough that even high-ranking visitors know: "*This is the person running the site right now – everyone else (no matter their title) supports that structure*."

For teams accustomed to a more informal or egalitarian culture, this can feel rigid. But on a day when seconds matter, formality saves lives. A practiced crisis team knows that **clear communication** is as critical as clear command.

On-site, the incident commander and key responders should communicate precisely. For example, a report like, "*We have a partial trench wall collapse, one worker potentially buried, air supply unknown, we have heavy equipment standing by to assist as directed,*" provides clarity that guides responders effectively. Everyone uses common terminology and sticks to facts.

With the outside world – which includes the client, regulators, neighbors, media – the challenge is to control the message without covering up the truth. There is often a temptation to either say too little (to avoid liability or bad press) or to say too much too soon (to appear in control of the narrative). The safest approach is a middle ground built on facts and empathy. Designate a single spokesperson for the company (usually a senior manager or communications officer). Release only confirmed, factual information – what is known to have occurred, what actions are being taken – and avoid any speculation, blaming, or premature promises. Express concern and commitment to those affected (e.g. "Our focus right now is on the safety of our workers and the public, and we are cooperating fully with responders."). Meanwhile, within the project team, enforce a strict rule that no one else should be texting pictures to friends, posting on social media, or giving off-the-cuff statements. Mixed messages or rumors can cause confusion, panic, or legal issues. A disciplined communication plan is as much a safety tool as it is a PR strategy – it keeps everyone focused on managing the situation rather than fighting misinformation fires on the side.

## From Chaos Back to Control – Recovery and Learning

Even the best-managed emergency will leave behind a mess. There may be damaged work, disrupted schedules, upset clients, injured workers, and shaken team members. Crisis management doesn't end when the sirens fade and the immediate danger is gone. It continues through the recovery phase and into the learning phase.

In terms of recovery, think of at least four dimensions that need attention:

- **Physical recovery of the site:** Make the area safe again for investigation and eventually for resumption of work. This could involve shoring or removing unstable elements, isolating utilities or energy sources, and establishing safe access for inspectors or recovery crews. The site might need to be re-secured (fencing, barricades) to protect both workers and the public during this stage.

- **Operational recovery:** Determine which parts of the project can safely resume work and which parts must remain shut down pending repairs or clearance. The schedule will have to be revised to account for lost time and possibly altered methods. The risk register should be revisited and updated based on everything the incident has revealed – often, new risks become apparent, and previous assumptions must be rethought.

- **Human recovery:** Support the people involved. This includes caring for any injured workers and supporting their families, but it also means addressing the psychological impact on those who witnessed or were close to the event. Construction crews take pride in getting the job done, but they need space and support (counseling, debriefings, time off if needed) to process what happened. A serious incident can shake confidence; ignoring that and just saying "back to work" can compound the harm.

- **Reputational and regulatory recovery:** In the aftermath, the project and company must work cooperatively with investigators (OSHA or other agencies, insurance investigators, possibly police if there were fatalities). That means being transparent with information and honestly identifying causes. It also means communicating with the client and other stakeholders about what happened and how the project will move forward safely. Reputational recovery is about demonstrating – through actions, not just words – that the team is learning from the event and is committed to preventing a recurrence.

One of the most dangerous impulses after a crisis is the urge to "get back to normal" as fast as possible and treat the event like an unfortunate scheduling slip. A serious incident is not just a schedule disruption; it's a diagnostic for the project and the organization. It has exposed weaknesses – maybe in the design, maybe in the plan, maybe in supervision, training, or culture – that were there long before the failure occurred.

A disciplined project (or organization) will resist minimization. Instead, it will use structured post-event reviews – similar to the root-cause investigation techniques discussed elsewhere in this book – to ask the hard questions: *What did we miss in our risk identification? Were there early warning signs that we saw but discounted or normalized? Did our contingency plans exist only on paper, or were they actually usable when we needed them? How did our command structure and communication hold up under stress? What will we do differently tomorrow as a result of this experience, so that we do not replay this event?*

Those questions aren't about pointing fingers at individuals. They are about strengthening the system. And they close the loop back to the very beginning of this chapter: *crisis management is part of risk management, not separate from it.* In a mature organization, the story of a crisis doesn't end with "...and we survived." It ends with specific changes – perhaps a design modification, a new procedure, improved training, tighter operating boundaries – that reduce the chance of that scenario happening again. The failure, in effect, gets folded into the company's operating philosophy. It becomes one more story that future leaders can recount in kickoff meetings years later: "*We know what happens when these basics are ignored, because we've lived it. We're not going to repeat that mistake here.*"

When you reach that point, you've done more than just manage a bad day. You've turned an uncontrolled event into a controlled source of learning. That is the silver lining of a crisis, and it reinforces a principle that underpins this entire book: **either you control the work, or the work will control you.**

## Case Applications of Successful Risk Mitigation

Up to this point, many of the stories we've discussed have started with something going wrong – a warning sign missed, a formwork stripped too early, a critical assumption overlooked – and ended in failure. Those stories are necessary because they show the high cost of neglecting risk. But they are not the whole story of construction.

There is another category of project event that we rarely see in the news: **the non-disaster.** The day when somebody says "Stop" and everyone actually listens. The night when a team willingly sacrifices a week of schedule to redesign a detail rather than gamble on it. The weekend when a critical lift is postponed in front of an anxious client, and nothing dramatic happens except a tense meeting and then a safe outcome later.

These quiet saves are just as real as the failures, and they are the true proof of effective risk management. They show what it looks like when the tools from this chapter – risk identification, risk registers, contingency planning, and crisis discipline – are actually used and pay off.

What follows are a few representative examples. They're drawn from real project patterns, though told here in composite form, since successful prevention doesn't always have a single headline event. The point is not the specific project details, but the habits and decisions they illustrate.

## Design and Planning – Catching the Problem on Paper

Consider a large civic plaza being built over a below-grade parking garage. The architect's concept calls for deeply planted gardens, reflecting pools, and heavy stone pavers across the plaza. On the initial drawings, the plaza is modeled with a relatively light live load, and the weight of planters and water is only roughly estimated. The structure beneath – thin post-tensioned slabs spanning long distances – is efficient, with not a lot of excess capacity.

During an early design-phase risk workshop, someone on the team asks a question rooted in the risk-identification mindset from the first part of this chapter: "*Where is the worst combination of weight*, water, and long-term load on this deck?" In other words, where could saturated soil, accumulated water, and heavy paving converge on a span in a way that challenges the design assumptions?

The team flags this as a technical risk: possible overloading of localized slab areas due to landscaping and water ponding. They log it in the risk register as a high-consequence item. From that moment, two things happen that would likely not have happened on a business-as-usual project:

First, the structural engineer goes back and re-runs the analysis with more realistic assumptions – the actual weight of fully saturated soil in the deepest planter, full water loads in the fountain, heavier paver materials, etc. The updated calculations reveal that several bays of the slab are much closer to their capacity limits than anyone realized. In a couple of spots, if you added just a bit more weight or some future tenant equipment on the plaza, you'd be flirting with structural overstress.

Second, the project manager refuses to let this issue drop or be value-engineered away. Because it's now documented as a high-consequence risk, it stays on the design coordination agenda week after week until it is resolved. The team collaborates on mitigation measures: they add reinforcing steel in certain areas of the slab, adjust some planter depths and detail them so their loads bear directly over beams or columns, and they establish strict limits on any future changes that could add weight to the plaza design. None of these actions is flashy or outwardly visible – in fact, the client might only see them as extra costs or minor delays with no obvious benefit. No ribbon will ever be cut for "the plaza that didn't collapse."

Yet, twenty years later, that plaza is still in service, not on the front page of the newspaper as the site of a failure. The "case history" of that project is simply a quiet structure doing its job year after year. The decisive moment happened not during an emergency, but in a mundane meeting when someone chose to treat a potential risk as real – to follow it through, run the numbers, and enforce conservative changes to the design. That is what successful risk management looks like at the design stage: a concern is raised and captured, that entry in the register drives further analysis, and that analysis drives a change, all before anything is built or anyone is harmed.

## Operations and Procedures – Staying Inside the Envelope

On another project, a contractor is tasked with erecting a series of long-span steel trusses over an active rail corridor. The erection plan calls for a large crawler crane positioned on one side of the tracks, making multiple big lifts during a tightly scheduled weekend rail shutdown. It's a classic critical path activity: if the lifts fail or if the work can't be completed in the window, the entire project schedule (and major stakeholder commitments) will slip.

In the months leading up to this operation, the team does three important things. They classify truss erection as a high-risk construction activity in the risk register, explicitly citing concerns such as wind sensitivity, ground-bearing pressure on the crane, and rigging complexity for tandem picks. They tie those risks to clear operating boundaries: they establish a maximum allowable wind speed for the lift, a required crane configuration and boom length (with no deviations), defined exclusion zones around the pick radius, and no-go criteria for ground conditions (for example, if there's been heavy rain and soil bearing pressure is in doubt, the lift won't proceed). And crucially, they build contingency plans ahead of time: an alternate plan to erect the trusses in smaller sections if needed, a backup weekend identified on the schedule with the rail operator's tentative approval in case the first window is missed, and even pre-negotiated contract terms with the railroad about costs/penalties in case of a postponement (so everyone knows what to expect financially if a reschedule happens).

Come the morning of the big lift, the weather forecast that looked acceptable a few days earlier has deteriorated. Wind gusts are edging toward the predetermined cutoff. In many historical failure stories, this is the point where schedule pressure overrides caution – someone says, "We're all set up, we've got to get this done," and they roll the dice in marginal conditions.

Here, the outcome is different. The superintendent has the anemometer readings and the agreed wind limit written on a whiteboard in the site trailer for all to see. Everyone on the team, including the client and the railroad representatives, has been briefed that the safety limit = X mph. When wind gusts actually hit that threshold, he calls it: "All stop. We're not lifting." The crews secure the partially lifted truss back down on the ground, the crane is shut down and locked, and demobilization begins. The client is frustrated; the rail operator is annoyed because an expensive shutdown window was "wasted" with nothing completed.

But no one is injured, and nothing is damaged. The risk control measures did their job – the decision was essentially made weeks prior, when the team agreed on the boundary conditions. Two weeks later, in calmer weather during the backup window, the lift is executed safely and successfully.

There was nothing heroic about the superintendent's decision to stop that day. It was the opposite of heroics; it was routine, almost boring, execution of a rule everyone had agreed to. But that's precisely the point: successful risk mitigation often looks like **nothing special is happening**. Headlines will never read "Crane Lift Postponed, Nothing Bad Occurred" – yet that is a victory for the project. By following the operating limits they set, the team avoided a scenario where they might have become a case study in failure. The risk register entries, the safe operating envelope, and the contingency plans all came together to take a potentially emotional, high-pressure decision out of the realm of improvisation and put it firmly into the realm of procedure.

## Asset Management – Choosing Maintenance Over Luck

Risk management isn't only about new construction projects. Some of the most dangerous situations arise in existing aging structures where deterioration and modifications over time create hidden risks. Imagine the case of a 40-year-old parking garage in a coastal city. Decades of exposure to salt have caused chloride-induced corrosion in the concrete's reinforcing steel. Occasionally, chunks of spalled concrete fall, and the owner patches them, only for new spalls to appear the next year. It has been like this for ages, and everyone's gotten used to it. Tenants grumble about the leaks and falling concrete, but the attitude is, "That garage has always been that ugly but basically fine."

A new facilities director comes on board, someone aware of high-profile parking garage collapses that have made the news in recent years. She's not content with "it's fine" as a conclusion. She commissions a proper structural condition assessment: cores, half-cell corrosion potential testing, and a detailed analysis by a structural engineer. The findings are sobering. In certain areas, the reinforcing bars have lost significant cross-section due to corrosion. Some critical connections and slab spans have much thinner margins of safety than assumed. In plain terms, the garage has spots where, under normal service loads, a localized failure or even a progressive collapse is plausible if things get just a bit worse.

From a risk perspective, the situation gets reframed. The hazard is no longer thought of as "some minor concrete cracking" – it is recognized as the potential for a localized punching-shear failure or even a progressive slab collapse under everyday use. The consequence is no longer "unsightly appearance or inconvenience" – it is understood to be "possible sudden structural collapse with multiple fatalities." And the timeframe is not "maybe someday in the distant future" – it becomes "this could realistically happen within the remaining service life of the structure if nothing is done."

They document that risk in the owner's enterprise risk register. Suddenly, upper management and the board are looking at a high-consequence safety risk tied to a specific asset, rather than a vague maintenance issue that could be deferred. That clarity forces decisions that had been avoided for years: it immediately restricts loads on the most deteriorated top level and closes a couple of bays entirely to reduce stress. They allocated funding for a phased repair and strengthening program, far beyond the cosmetic patching they had been doing – essentially investing now to drastically reduce risk. They also institute a formal inspection and re-certification interval for all similar parking structures in their portfolio, making sure this same problem doesn't creep up elsewhere unexpectedly.

No one is going to congratulate them in the newspaper for these actions. In fact, from the outside, users of the garage might complain about the inconvenience of closed bays and ongoing repair work. But five or ten years later, people are still parking in that structure, blissfully unaware of how close it once came to being "the next big collapse" on the news. The success story here is an owner willing to reclassify a familiar nuisance (leaks and cracks) into a serious risk – and then actually treating it with the seriousness it deserved. It's a case of overcoming normalization of deviance: refusing to accept a degrade-and-patch cycle and instead making bold changes to reset the structure's safety margin.

## Team and Culture – The Day Someone Spoke Up

Not every success story starts in a boardroom or design office. Many of the most important ones begin with a single individual in the field deciding that something is not right and trusting that the system will back them up if they speak out.

Picture a high-rise concrete construction project in a dense urban downtown. A crew is preparing for a large elevated deck pour, one of the biggest concrete placements on the job. The schedule is tight, as usual; the pumps are set up and ready; concrete trucks are queued around the block. The formwork and shoring for the deck were inspected and signed off early that morning. Everything is a go.

As the pour begins, one of the veteran carpenters notices that a certain area of the formwork is deflecting (sagging) a bit more than he's seen on similar decks. It's subtle, but he has that gut feeling that doesn't look quite right. In a project culture where production is king and questioning the plan is seen as weakness or whining, that carpenter probably stays quiet. He might even rationalize it: "It's probably fine, I'm not an engineer." The crew would keep pouring. And in some real historical cases, that's exactly how disaster has struck: the early signs were there, but no one acted until it was literally collapsing.

On this project, however, the carpenter has heard a very different message from his leaders for months. In safety meetings and on the site signage, they've empowered everyone with stop-work authority. More importantly, he has seen them practice what they preach: on a couple of occasions, work was paused by a foreman or even a laborer over a safety concern, and management supported the decision, even though it cost time. So in this moment, though he feels a knot in his stomach, the carpenter follows his training and intuition. He yells out, "Hold up! Stop the pour!"

The immediate reaction on site isn't ridicule or anger. The foreman doesn't roll his eyes; he comes over to see what's up. He, too, notices the unusual deflection. He calls the pump operator on the radio: "Shut it down for now." Within minutes, the superintendent and the site engineer are at that section of the deck, measuring deflection and checking the shoring against the drawings. Sure enough, they discovered that one bay was missing some post shores; a miscommunication between crews on a shift change led to less shoring under a critical area than the design required. The structure as-is likely would not have failed catastrophically, but it could have sagged badly and required significant rework, or in a worst case, it might have collapsed locally once the full load was on. Thanks to the stop-work, they are able to stabilize it: additional shoring is quickly installed, calculations are double-checked, and only after that do they resume pouring (with continuous monitoring in the flagged area).

The result? The project "loses" maybe half a day. No one is hurt. The structure doesn't crack or collapse; it's built as intended. The only evidence is a line item in a safety or quality report: "Near miss, excessive deflection noticed; pour stopped and re-shored; no damage." But that dry description masks the true significance: this was a full application of the risk management principles in this chapter:

- **Risk identified in real time** by an alert crew member who wasn't afraid to trust his eyes and gut.
- **A live contingency decision** – stop the work, inspect, reinforce – rather than plowing ahead and hoping for the best.
- **A culture that made it possible** for a single worker to protect the entire project by speaking up, because he knew he would be supported.

Every major accident investigation report praises workers for noticing something was wrong, after the fact. The real goal is to create conditions where those workers speak up before the accident, and are praised and supported in the moment. This example shows that the ideal in practice. It's not that the project was perfect, a mistake happened with the shoring, but the recovery mechanism (someone willing to shout "stop!" and a team willing to listen) was in place, so an accident was averted.

## Pulling It Together – What Success Has in Common

The success stories above are diverse; they involve design decisions, on-site operations, maintenance of older structures, and day-to-day safety culture. Yet they share a common architecture:

First, **someone clearly named the risk**. They didn't hide behind polite phrasing like "this might be an issue" or assume it was someone else's problem. Whether it was an engineer in a meeting or a carpenter on a deck, they pointed at a specific potential failure and described how it could happen and what the consequences could be.

Second, **the risk lived in a system, not just in one person's head**. It showed up in a tangible way, entered into a risk register or a meeting agenda, written on the site whiteboard as a wind limit, captured in an inspection report, etc. It wasn't a private worry; it became a team concern that multiple people could see, discuss, and act on.

Third, **there were pre-planned responses**. When conditions crossed a predetermined line, be it weather conditions, measured deflection, test results, or anything the team knew what to do next. Nobody had to invent a solution under duress because the "if-then" steps had already been thought through calmly in advance.

Fourth, **leaders were willing to take some pain now to avoid catastrophe later**. They were willing to postpone a major lift, add unplanned rebar and cost to a design, absorb schedule delays to fix a shoring issue, or have tough conversations with clients and executives about why the plan needed to change. In each case, they chose a controlled, smaller loss over the roll of the dice that could lead to a much bigger loss.

Finally, the **culture supported conservative judgment**. From upper management to field staff, people believed that stopping work for safety or quality was not a career-ending move but a core responsibility. That empowerment is what allowed the engineer to raise the plaza overload question and the carpenter to yell "hold up" without fear.

If you trace those elements backward, you can see they correspond to the same foundations we've built throughout this chapter: structured risk identification, meaningful risk registers, real contingency planning, and disciplined crisis response. Successful risk mitigation is not magic or luck. It is the repeated, sometimes boring, act of honoring those foundations even when nothing seems to be going wrong.

The construction industry will, unfortunately, always have its share of failure stories, and we will continue to learn from them. But perhaps the more important measure of our progress is how many failures didn't happen precisely because someone, somewhere, applied the practices from this chapter quietly on an ordinary Tuesday. Those "non-events" won't make newspaper headlines or TV documentaries. They will instead live on in the long, uneventful service of buildings, bridges, and structures that do exactly what they were meant to do: stand, serve, and never surprise anyone.

**Operational Takeaways for Today's Job:**

- Treat risk management as a daily habit, not a one-time paperwork exercise. Each morning or before a major task, take a moment to scan for new risks (e.g. changed weather, site conditions, staff changes) and adjust your plans or controls accordingly.
- Keep a simple "risk log" on site. Jot down any hazard or near-miss that comes up during the week on a whiteboard or notebook and review it at the next safety meeting – this makes risk visible and keeps the team alert to evolving issues.
- Assign clear ownership for critical risks. For example, if operating a crane is today's high risk, designate a competent signaler and a safety spotter explicitly. Make sure everyone knows who is responsible for watching each major risk area so nothing falls through the cracks.
- Always have a Plan B (contingency) for key operations. If you're doing a critical pour or lift, discuss with the team what you'll do if equipment fails or conditions change suddenly. Thinking ahead about "what if" scenarios means you can respond quickly rather than scramble in chaos.
- Use the risk matrix mindset on the fly. If a decision comes up (like whether to push on in bad weather), quickly weigh the likelihood and severity of something going wrong. Empower the team to speak up if the risk feels too high – sometimes the best action is to pause work until it's safer.

# Chapter 6

# Why Procedural Discipline Matters

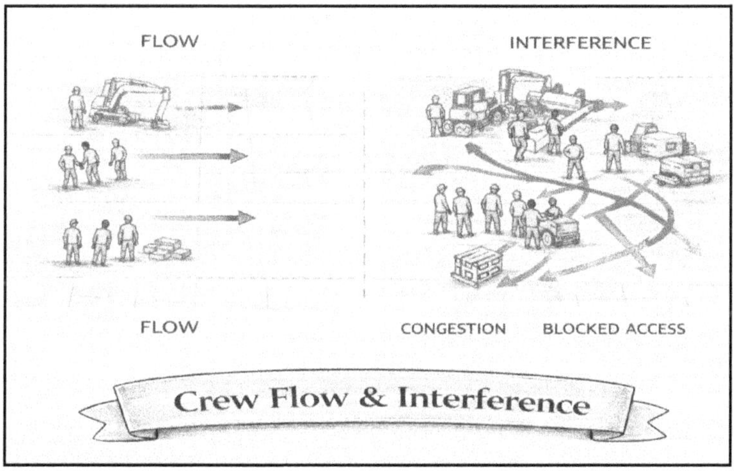


Construction is one of the few industries where the work itself carries inherent, unavoidable risk. Loads are heavy, structures are incomplete, conditions are constantly changing, and the safety of the job depends on a thousand small choices made by different people across all hours of the day. In such an environment, it's not enough for workers and leaders to **know** the right thing – they must do the right thing the same way every time, especially when conditions are stressful, schedules are compressed, or an unforeseen problem disrupts the plan.

Procedural discipline is what turns knowledge into consistent behavior. Think of it as the operating system that keeps a project within its safe boundaries even when individual judgment varies from person to person or day to day. Without disciplined procedures, success depends on personality, experience, or luck. With them, success becomes predictable. This isn't about paperwork or mindless compliance; it's about creating reliability in human performance – the kind of reliability that makes complex, high-risk work behave in a stable and controlled manner. When we talk about "disciplined operations," we are not calling for rigidity or blind rule-following. We are describing a system that protects teams from the very human tendencies that often precede failure: shortcuts, assumptions, improvisation under pressure, and the gradual drift away from the original engineering intent.

Construction environments do not forgive drift. Small variations that seem harmless can quietly accumulate until they push a system past its limits. A slightly different rigging method, a slightly different formwork layout, a slightly shorter curing time, a slightly altered sequence—none catastrophic on their own, but together they erode the margin between safe operation and failure. This erosion is often invisible until the moment of collapse; by then, the system has no buffer remaining. Procedural discipline matters because it is the countermeasure to this subtle degradation of safety.

At its core, procedural discipline is rooted in a simple operating philosophy: **control the work, or the work will control you**. That philosophy rests on three pillars: recognizing hazards and designing around them; building systems that stay well within known safe boundaries; and ensuring that every action preserves the safety analysis on which the project depends. Procedures are not "extras" or paperwork formalities – they are the mechanism that ensures the design intent survives contact with the real world. If the purpose of engineering is to make the future predictable, the purpose of procedures is to make people's actions predictable. Together, engineering and procedures form a system that resists chaos, surprises, and dangerous improvisation. In this sense, procedural discipline is not just a management preference – it is a technical requirement for any operation carrying significant physical risk.

## Procedural Discipline as Protection Against Drift

One of the most valuable insights from high-reliability industries is the concept of drift – the natural tendency for operations to slide gradually toward higher-risk states unless something actively resists that movement. Drift isn't usually malicious; it's the product of convenience, habit, fatigue, and optimism bias. Over time, a shortcut becomes "the way we always do it," and the original method (and its safety margin) is forgotten.

Consider an example of how drift can quietly set the stage for disaster: A contractor erects a network of aluminum post shores beneath a series of elevated concrete slabs, following an engineered shoring plan with exact post spacing and requirements that each post be fully threaded and locked. For the first few pours, the crew follows the plan closely; inspections find everything in order. But as the project progresses, subtle pressures chip away at discipline. One day a shipment of props arrives damaged, and the crew substitutes a different model "just for today." Another time, the pour is running late, so they widen the post spacing slightly to speed up installation. A month later, some posts are left only partially threaded because "they never move anyway." None of these deviations causes immediate failure – the slabs don't crack, the shores don't buckle, and the schedule even benefits – so these behaviors become normalized.

Weeks later, during a particularly heavy pour, an area of the slab deflects far more than expected. An investigation finds a cluster of mixed shore types, inconsistent spacing, and several improperly locked posts. The engineered capacity had been eroded piece by piece. Nothing in this story involved obvious negligence – every small step felt logical in the moment. The danger came from drift: a series of modest deviations that collectively pushed the system silently into unsafe territory.

Procedural discipline is the counterforce to drift. It's the mechanism that says, "We do it this way every time," not because people can't think for themselves, but because the system depends on consistency. By adhering to established procedure, the crew in the story would have prevented the gradual degradation of the safety margin. In short, disciplined procedures keep those tiny, compounding variations from accumulating into a serious problem.

## The Gap Between Knowing and Doing

One consistent finding in failure investigations is that workers and supervisors often knew the right step but failed to follow it in the moment. This isn't ignorance – it's human nature under pressure. When a rule is repeatedly bent with no immediate consequence, it gradually becomes mentally recategorized as "optional." That quiet drift from compliance can, over time, turn a safe operation into a precarious one.

Procedural discipline is designed to close that gap between knowing and doing. It recognizes that people do not always operate at their best: they get tired, distracted, optimistic, or rushed. They interpret ambiguous situations differently; they sometimes assume someone else has checked the critical detail. Good procedures aren't written for perfect days – they're written for the days when judgment is strained, and margins are thin. A crew following disciplined procedures should be safer on their worst day than an undisciplined crew is on their best day.

## The Limits of Personal Expertise and Heroics

Construction culture has traditionally glorified problem-solving and last-minute heroics. There's pride in the foreman who can "figure it out" when something unexpected happens, or the superintendent who can "make it work" when equipment doesn't arrive or drawings conflict. But improvisation is a double-edged sword. Many near misses – and many tragic accidents – begin with a bit of winging it under pressure.

Procedural discipline shifts the definition of professionalism. Instead of celebrating last-minute saves, it celebrates consistency, preparation, and adherence to proven methods. The best-performing teams are not those who can fix mistakes in a flash, but those who avoid getting into situations that require heroics in the first place. This cultural transformation reduces error at the source by dispelling the myth that "exceptional skill" can replace systematic controls. In a disciplined operation, consistency is valued over improvisation because consistency is safer.

In complex construction work, no individual – no matter how skilled or experienced – can perceive all the interactions at play on a project. Today's decision about sequencing might affect tomorrow's structural stability. A minor field change in a rebar layout could compromise shear capacity elsewhere. A material substitution might impact curing time, which affects when you can remove shoring and, in turn, safety.

Expertise alone is not enough. In fact, some construction failures occur under the leadership of smart, seasoned professionals who rely too much on personal judgment and too little on structured procedures. Procedural discipline acknowledges the limits of even the brightest individuals. It replaces the unreliable variability of human decision-making with a repeatable system, allowing the project to perform at a higher level than any one person could guarantee on their own.

## Procedural Discipline as the Foundation for Quality, Safety, and Reliability

A large construction project involves many layers of responsibility – engineers, inspectors, superintendents, foremen, subcontractors, craft workers, and more. If procedures are vague, outdated, or applied inconsistently, each layer ends up operating on its own assumptions. Misalignment becomes inevitable.

Disciplined procedures create a **shared operational language** across all these players. They define how the work is to be done, what prerequisites must be in place, what to check before proceeding, when to stop, and who is responsible at each step. In this way, clear procedures transform individual efforts into a coordinated, predictable system. When everyone follows the same playbook, the project operates with the coherence of a single team rather than a collection of siloed groups. Accountability also becomes clearer – deviations are obvious and traceable, which means problems can be caught and corrected rather than passed along.

Procedural discipline is the backbone supporting many other critical elements of project success. Without it, even the best plans and practices will fall short. In a project lacking discipline, **Standard Operating Procedures** become symbolic paperwork that people ignore. **Change control** becomes optional or gets bypassed. **Lessons learned** are quickly forgotten. **Quality assurance/quality control (QA/QC)** turns into a reactive policing activity rather than a preventive process. **Equipment reliability** becomes wishful thinking, and even clearly defined **operating boundaries** start to feel negotiable under pressure.

Conversely, with strong procedural discipline, each of those elements becomes a true safeguard. The purpose of procedures is not to restrict intelligence, creativity, or craftsmanship – it is to protect them. Only when the routine work is performed consistently and correctly can skilled workers and leaders devote their creativity to the real challenges: complex technical details, logistics and coordination, and continuous improvement. In other words, procedural discipline frees people to excel by ensuring that the foundation of operations is solid, stable, and predictable.

With these principles in mind, the remainder of this chapter examines how procedural discipline underpins specific operational domains: developing and using Standard

Operating Procedures, controlling changes and version documentation, building institutional memory, executing effective QA/QC processes, maintaining reliable equipment, and enforcing operating boundaries. Each of these domains provides practical structure to keep a project safe and on track – and each relies on procedural discipline to function as intended.

## Standard Operating Procedures (SOPs)

Standard Operating Procedures are the backbone of procedural discipline. They are the written form of the project's operating philosophy – the translation of engineering intent, organizational values, and safety boundaries into step-by-step methods that crews can execute consistently. SOPs do not exist to fill binders or check a box for auditors. They exist because a construction project has too many moving parts, too much uncertainty, and too thin a safety margin to rely on memory or improvisation alone.

A well-designed SOP captures the correct way to perform a task before stress, fatigue, or time pressure can distort judgment. It pre-decides the steps that reduce risk, distribute responsibility, and protect the operation from the drift we discussed earlier. When SOPs are followed, the project behaves in a stable and predictable way, even on difficult days. When SOPs are ignored or poorly designed, construction becomes dependent on luck and individual experience – both of which have limits.

Writing an SOP is the easy part; writing a valid SOP is much harder. A procedure is only "valid" if it meets several criteria at once. It must be based on a clear understanding of the work and its hazards, and it should reflect the actual safety analysis that the engineers or designers used. It has to be technically correct – consistent with codes, manufacturer requirements, and engineering review. It also needs to be practical for crews to execute under real job-site conditions. A good procedure not only defines how to perform a task correctly, but also defines how not to perform it (what behaviors or shortcuts are off-limits). And critically, it must be kept current: as designs change, equipment is updated, or new lessons are learned in the field, the procedure should be revised so it never lags behind reality.

In short, a technically sound procedure that nobody follows in practice is not a valid procedure; a very easy-to-follow procedure that is technically wrong is not valid; and a once-great procedure that no longer matches the latest drawings or specs is not valid either. Validity lies at the intersection of engineering accuracy, human usability, and operational discipline. If any one of those pillars fails, the procedure collapses.

**SOPs as Defense Mechanisms.** Construction work presents certain recurring hazards that simply cannot be managed casually or "on the fly." For example, consider temporary states of a structure with unstable load paths, high-energy operations like heavy crane lifts or demolition, activities with irreversible outcomes (once you pour concrete or tension a cable, you can't undo it), conditions that evolve hour by hour (weather changes, concrete curing, excavation stability), or work at the interfaces of multiple trades where each assumes something about the other. Failures in these situations tend to occur not because people lacked skill or knowledge, but because they lacked a consistent method. SOPs define that method. They establish a safe path through complex tasks and, by repeating it each time, ensure that safety doesn't rely on any one individual's memory or judgment. Good SOPs also reduce the cognitive load on individual workers by embedding the critical thinking upstream – into a written, reviewed, and verified plan – so that moment-to-moment decisions in the field are simpler, controlled, and predictable.

**Field-Realistic Procedures**. An SOP that looks perfect on paper is useless if it doesn't work in real life. Many SOPs fail not because they are technically wrong, but because they were written in a vacuum without regard for actual field conditions. A "field-realistic" procedure accounts for the messy realities of construction. It assumes a crew size and skill level that will actually be available for the task. It considers the site's real geometry and how difficult it might be to access certain areas. It specifies the actual tools and equipment crews have on site, not some idealized tools nobody actually owns. It fits within the time window the crew will realistically have to do the work. It factors in environmental conditions like noise, wind, dust, heat, or poor lighting that might be present. And it acknowledges that other activities will be happening simultaneously, so coordination and sequencing are addressed.

To keep procedures realistic, the people who use them need to be involved in creating and updating them. A procedure written in an office can drift out of alignment with the field in surprising ways. The surest way to prevent that drift is to include foremen, superintendents, and experienced craft workers in the procedure's design and review. A field-realistic SOP doesn't just work on paper; it works at 5:30 AM when the crew is actually trying to set forms in the cold and dark, and at 2:00 PM when the concrete pump is on site, and everyone's under the gun to get the pour done. It's grounded in the reality of the jobsite.

**When Procedures Become Mandatory.** Not every activity on a jobsite requires a formal written SOP – you wouldn't write one for every single everyday task – but certain categories of work absolutely demand them because improvisation in those areas is simply too dangerous. These include, for example, **critical lifts and complex crane operations; formwork and reshoring installation or removal; deep trenching or excavation beyond a specified safe depth; and installation of major temporary works such as scaffolding, falsework, or heavy bracing**. In such high-risk tasks, a documented procedure isn't a bureaucratic hoop; it's a mandatory safety measure. The consequences of "winging it" in those scenarios are unacceptable. A good rule of thumb is: if an operation has a very thin safety margin or a high potential energy (lots of stored energy or weight that could release), plan it out and write it down.

In summary, SOPs capture organizational knowledge and intent so that every crew, on every shift, approaches critical tasks the right way. They make safe and quality execution repeatable. The disciplined use of SOPs turns what could be a chaotic, variable process into a controlled one. And in the pressure cooker of construction, that consistency is what keeps people safe and projects on track.

## Change Control, Version Tracking, and Documentation

Construction is an industry defined by change. Drawings evolve, site conditions shift, materials arrive with slight differences, and external pressures (weather, supply chain issues, regulatory clarifications) alter the work in ways no initial plan can fully predict. Change itself isn't the enemy – **unmanaged** change is.

Indeed, many construction failures, both catastrophic and subtle, can be traced to one of two basic scenarios: (1) a change occurred that no one recognized as significant, or (2) a change was recognized but never formally controlled. Change control is the discipline that prevents those failure modes. It ensures that when something deviates from the plan – whether intentionally or unintentionally – the project does not drift outside the bounds of the engineering assumptions, safety margins, or operating limits that everyone believes are in place.

Put simply, procedural discipline tells crews **what** to do; change control ensures that **what they do** stays aligned with the design intent. You can have extremely robust SOPs, but without change control, those procedures can become disconnected from reality. The work might gradually turn into a hybrid of old and new intentions – none of which were properly engineered to work together. Change control keeps the map and the territory in sync.

**The Nature of Change in Construction**. Construction is fluid by nature. Foundations might uncover unknown utilities. Structural steel could arrive with minor differences in camber or hole alignment. A shoring tower might need to shift over a few feet to accommodate an excavator. A concrete delivery delay could force the team to resequence a pour. A vendor substitution might present a slightly different connection detail. The design team will issue clarifications and answer RFIs (Requests for Information) as questions arise. A superintendent in the field might tweak a pour sequence or joint location for practicality.

Each of these changes might seem harmless in isolation. And indeed, nine times out of ten, minor adjustments are harmless. The danger is in the accumulation. History is full of major failures caused by a buildup of incremental changes that, at a certain threshold, fundamentally altered the safety picture. In high-reliability terms, this is the erosion of the safety envelope – a slow bleeding away of the design's built-in margin, caused by decisions that seemed logical at the time but were never re-evaluated against the original engineering assumptions. Change control is the mechanism that halts that erosion by insisting that changes are noticed, evaluated, and approved (or disapproved) deliberately.

**Treat Change as a Technical Event, Not a Convenience**. In many construction cultures, changes in the field are treated informally – a matter of convenience or expedience, decided on the fly. But **every change has technical consequences,** even if the immediate motive is logistical or financial. For example, a foreman might think nothing of shifting a temporary support a foot to the left to make room for a pump hose: "Sure, move it over, it's only 12 inches." To the field crew, it feels like a harmless adjustment. But to the structural engineer who designed the shoring, that 12-inch shift could place the post under a thinner part of the slab or at a point that wasn't intended to carry the load. It might reduce the tributary area the post supports, or eliminate redundancy in a way the engineer accounted for. Field logic and engineering logic are not opposites – they're just looking at different dimensions of the problem. **Change control is the process that reconciles the two**. It mandates that any change for convenience is also treated as a technical event. And **technical events** require technical review.

**Formal Change Control (Management of Change)**. High-risk industries like chemical processing and nuclear power long ago developed rigorous Management of Change (MOC) processes to manage operational changes safely. Construction can benefit from the same principle. An effective construction change-control (or MOC) process typically includes a few key steps:

- **Description of the proposed change:** Clearly and explicitly describe what will be different. This should be in writing – not just a quick verbal "yeah, do it" in the field – so there's a record and clarity.

- **Reason for the change:** Explain why the change is being requested. Is it driven by a field condition, a schedule issue, a safety concern, or an unforeseen site condition? Understanding motivation helps evaluate necessity and risk.

- **Technical review:** This is the heart of MOC. An appropriately qualified person (often an engineer) must evaluate the structural, safety, and operational impacts of the change. If this step is skipped, the change process collapses into mere paperwork. The technical review asks, "If we do this, are we still within safe limits? Do we introduce any new hazards or invalidate any assumptions?"

- **Risk assessment:** Beyond pure engineering, consider whether the change introduces new hazards or reduces existing safety margins. Sometimes a change might technically "work" but leaves you with a lot less cushion than you had.

- **Approvals:** Define who has the authority to approve the change. For example, a superintendent might approve a change in pouring sequence, but they cannot approve a change that affects structural design without an engineer's consent. Make sure the right people sign off.

- **Communication plan:** Outline how the approved change will be communicated to everyone who needs to know – the crews doing the work, other subcontractors affected, inspectors, QA/QC personnel, and even the next shift coming on. Everyone operating under the plan must be aware of the change.

- **Documentation and version tracking:** Update relevant drawings, procedures, or specifications to reflect the change. Ensure the new version is available and that outdated versions are removed so no one accidentally works off the old information.

When a Management of Change system is functioning, nothing "just happens" as a casual field tweak. Every adjustment is evaluated, recorded, communicated, and verified. That may sound onerous, but it's the essence of disciplined change management – it's how you ensure that the project's current reality stays aligned with its engineered plan.

**A Cautionary Tale:** For example, imagine on a concrete frame project, a crew encounters reinforcement congestion at a beam-column joint. The rebar, as detailed on the drawings doesn't fit nicely with a formwork penetration, so the crew leader asks the site engineer, "Can we shift these bars over slightly to get them in easier?" The engineer, busy with other issues and seeing no immediate red flags, gives a quick **okay verbally**, thinking a small adjustment won't hurt. No formal change paperwork is filed out. No drawing is marked up to reflect this tweak. No structural calculation is performed to check it. The pour proceeds. A month later, a different engineer reviewing as-built conditions spots something off: those bars were shifted out of the zone where they engage properly with the concrete and other reinforcement. The "slight" shift moved critical reinforcement away from where the design assumed it to be, reducing the shear capacity of that joint. In the end, the team had to retrofit carbon fiber wrap and add extra stirrups to reinforce the area – an expensive fix after the fact. The slab didn't collapse, and the problem was caught in time, but the key is **why** it happened: what the crew saw as a minor constructability tweak was in reality a violation of a load path assumption in the design. A proper change control process would have forced that realization up front by requiring a review. It would have caught the issue early (or prevented the change altogether) and saved the team from flirting with a potential structural failure.

**Version Control – Fighting "Multiple Truths."** Even with good procedures for evaluating changes, construction projects often struggle with **version control:** making sure everyone is working off the same, most current information. It's all too common to find multiple conflicting "sources of truth" in circulation. You might see an old revision of a drawing still taped inside a gang box, while someone else has a screenshot of a newer plan on their phone, another person saved a PDF of the drawing on their laptop that isn't the final issue, and a foreman has some redlined sketches in a notebook that differ from all of the above. A subcontractor might be referencing their shop drawing that hasn't been updated to match the latest engineering change. This scenario is a recipe for errors. A crew cannot execute the correct procedure if they don't have the correct information.

In fact, mismatched plans and documents have caused real errors: post-tensioning tendons installed to the wrong profile because an earlier drawing was used, holes drilled in the wrong locations because someone worked off an outdated layout, concrete poured to the wrong elevation, or an excavation dug to a larger extent than intended – all simply because different team members were using different versions of reality.

Discipline in **document control** is the cure. A disciplined project implements robust version control by having one central, authoritative source for all current drawings and documents (and making sure everyone knows what that source is). When plans or specs get updated, there's an immediate distribution – notifications go out, and the new info is pushed to every relevant supervisor and crew, rather than hoping they find it buried in an inbox. It also means actively **removing or marking as void** the outdated versions – take those old drawings down off the walls and replace them, or lock older digital files so they can't be mistakenly used. Additionally, every document in use (SOPs, checklists, permits, etc.) should clearly indicate the version or date of the information it's based on (for example, an SOP might state it's for Drawing Rev. 5 dated such-and-such). That way, if someone is holding an instruction, they can quickly verify that it matches the latest plan. Inconsistent documents and drawings are like mismatched maps; if the project "terrain" has changed, but some of your team are still following an old map, nobody can navigate safely.

**Field Documentation and Hold Points**. Procedural discipline extends to how we document work in progress, especially around changes. Documentation is not bureaucracy for its own sake – it's memory, accountability, and verification all rolled into one. Every significant change or deviation – particularly those that affect structural behavior or safety limits – needs to tie back to **clear documentation** that can be understood months or years later. Some key tools and practices include things like pre-installation checklists (to verify conditions before starting a task), in-process inspection forms (records of checks done at various stages), marked-up field sketches that show modifications or clarifications with proper revision notes, and photographic documentation linked to specific locations or milestones.

A crucial element of disciplined workflow is the use of **hold points** – predetermined points in the sequence of work where progress stops until a certain verification is done and documented. Hold points act like brakes to prevent the project from racing ahead into unsafe territory. For example, you might have a hold point that **no rebar placement**

**can proceed beyond a certain level** until an engineer has inspected and signed off that everything so far is as per plan. Or no post-tensioning of tendons can commence until a QC inspector signs off on the tendon placement and stressing setup. Or a rule that formwork or shoring cannot be stripped without confirming (via testing) that the concrete has reached its required strength and minimum curing time. These are deliberate pause-and-check moments. A hold point is essentially the system telling everyone: "Stop here until we confirm it's safe to continue." It's the opposite of the "just keep going" mentality – and it requires a culture that values safety over speed in those critical moments.

To illustrate the importance, imagine a utility trench being excavated in phases. The geotechnical plan says that beyond 8 feet deep, the trench walls must be benched or shored for safety. The first crew follows the plan and benches the trench sides properly at 8 feet, but suppose they don't clearly document exactly where they reached 8 feet depth, nor take good photos tied to specific points. A week later, a new crew picks up the excavation further along. They see a trench that looks about the same and assume the previous sections were fine; they continue digging straight down without benching, not realizing they have passed the 8-foot mark in some spots because there's no clear record. On the fourth day, part of the trench wall sloughs off and injures a laborer. Investigation shows the second crew inadvertently went beyond the safe depth – the first crew had stopped at 8 feet but didn't leave behind clear documentation or physical markers, and no hold point enforced a check-in. Here, the failure wasn't due to lack of knowledge of the requirement; it was due to a lack of **transfer of information**. Good documentation (like daily logs noting depths reached, and tagged benchmarks in the trench) and a hold point requiring a supervisor's sign-off at each 8-foot interval could have prevented that incident. The lesson: documentation is not merely a record of what happened – it's a tool to prevent the repetition of what almost happened or could happen next time.

**Common Change-Control Pitfalls.** There are some familiar patterns that projects experience when change control isn't strong. One is the "**verbal approval trap**," where a superintendent or manager gives a quick verbal "go-ahead" to a change to avoid delaying work (like in our rebar example) – and later, nobody clearly remembers what was decided or why, and there's no paper trail. Another is the "**partial field fix**" – a crew encounters a conflict and implements a hasty workaround, intending to come back later to do it properly; but that temporary fix quietly becomes permanent, and no formal review ever happens. There's the **last-minute substitution**: materials or parts arrive wrong or late, so the team swaps in something "similar" without fully checking if it meets spec or is compatible – often not realizing the subtle differences that could matter. There are **unreviewed sequencing changes** where, say, the team flips the order of assembly or removal for convenience (strip the formwork earlier than planned or change the stressing sequence of the tendons) without analysis of the structural effects. We also see **hidden changes across shifts** – the night shift might adjust something and by morning, it's forgotten or not communicated, so the day shift proceeds unaware until a problem surfaces. And of course, **changes under intense pressure:** if concrete trucks are lined up or a crane is costing thousands per hour, people are tempted to bypass formalities "just this once" to keep things moving.

All these situations have one thing in common: a deviation happens outside the control system. Procedural discipline reframes these moments. Instead of "Just fix it and go, we'll sort it out later," the mindset has to be: **Pause. Document the issue. Analyze it. Get the right approvals. Only proceed when we're back in alignment with a safe plan**. It takes leadership resolve to enforce that in the heat of the moment, but that's exactly when it matters most.

**Unmanaged Change Erodes Safety.** Why such emphasis on controlling changes? Because every project's design and plan establishes a **safety envelope** – a range of conditions within which the work is considered safe and controllable. Engineering analyses assume certain loads, sequences, materials, and methods. When changes occur without proper evaluation, the project can stray outside that certified operating envelope without anyone realizing it. Collapses or serious failures often "slowly" incubate through a series of seemingly minor changes. Inch by inch, decision by decision, the work diverges from the conditions it was engineered for, until one day there's no safety margin left to absorb an unexpected event or extra load.

If we view safety as layers of defense, uncontrolled change bypasses several layers at once. Change control is the counterforce to that threat. It's the disciplined act of preserving the assumptions on which safety and design depend. By treating changes as technical events that must be reviewed and recorded, the team keeps the project within the boundaries of its safety analysis even as it adapts to field conditions. In essence, change control lets you remain flexible and responsive without losing control of risk. It ensures adaptability doesn't turn into dangerous drift.

## Institutional Memory: Recording Lessons Learned

Construction is also an industry with a notoriously short memory. Projects finish, teams disperse, foremen and supers move to new jobs, and the hard-won lessons of one project often fade before the next project starts. Companies end up repeating mistakes not because they lack intelligence or skill, but because they lack **institutional memory** – the ability to convert individual and team experience into lasting organizational knowledge.

If procedural discipline is the backbone of consistent execution, institutional memory is the nervous system that carries information from one part of the organization to another. Together, they keep history from repeating itself. Institutional memory is the antidote to organizational amnesia. It's the process by which near misses, failures, and successful innovations are captured, analyzed, preserved, and then fed back into future work. Without a strong memory system, each project operates like an isolated experiment, and errors recur with a tragic regularity.

**Why Memory Matters in Construction**. Every project generates insights, some of them paid for dearly. Perhaps a crane mat unexpectedly punched into soft soil that everyone thought was firm. Or a particular type of rebar coupler consistently failed its torque tests on a job. Or a certain sequencing choice created an unanticipated load path that almost caused an issue. Maybe a subcontractor came up with a brilliant workaround that reduced risk significantly. These insights are gold – but only if they're retained and shared.

What usually happens instead is more like this: a problem occurs → the team fights the fire and fixes it → everyone moves on to the next task. Later, on another project, a different team encounters the **same problem** and makes the **same mistake**, because none of the knowledge from the first incident was documented in any way that the new team could access. It's not that people are incapable of learning; it's that the **organization** failed to remember.

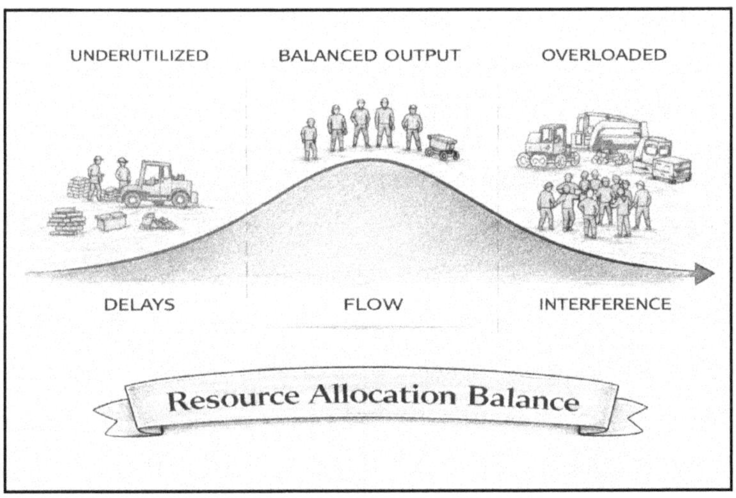


Institutional memory isn't about nostalgia or writing lengthy reports that nobody reads; it's an operational defense mechanism. When an organization's memory is strong and active, you see real effects: SOPs get better and more refined over time because they're updated with past lessons. Change reviews (MOC processes) become sharper and more proactive, because past issues taught what to look out for. Risk assessments become more predictive, because they're informed by patterns seen in previous projects (not just theoretical hazards). Crews inherit wisdom instead of constantly reinventing their workflows from scratch – maybe they've heard about a pitfall from another job and avoid it. And leadership decisions reflect the full history of what has or hasn't worked before, not just the limited experience of whoever is in the room at the time. Without that memory, a construction company is doomed to replay avoidable incidents as if it's the first time every time.

**Capturing Lessons – Incidents, Near Misses, and Successes.** Capturing lessons learned requires more than writing up an incident report and filing it away. It requires interpretation and action. A useful lesson doesn't just record what happened, but digs into why it happened, how it was detected or realized, and how it was (or wasn't) corrected. More importantly, a lesson has to lead to some change in the system – a procedure update, a training point, a design modification, something tangible – not just live as a story in a binder.

Traditionally, companies focus on lessons after *negative events* – accidents, errors, costly rework. Those are critical, of course. But mature organizations also capture positive lessons: the times someone's foresight prevented a disaster, or an innovation significantly improved safety or efficiency. These "near misses" or proactive success stories are treasure troves of know-how. They highlight what went right and can be repeated elsewhere, not just what went wrong.

In general, we can think of three categories of lessons worth capturing: - **Failure lessons:** Mistakes, oversights, design flaws, or any errors that led to incidents, defects, or significant rework. These teach us what to avoid or change. - **Near-miss lessons**: Situations where something bad almost happened but was narrowly avoided by luck or timely intervention – essentially free warnings of a latent problem. - **Success lessons:** Cases where a particular behavior or method notably reduced risk or improved reliability, even if it wasn't prompted by a prior failure (perhaps a creative planning idea that worked exceptionally well).

Each type of lesson strengthens a different part of the organization's "immune system." You need all of them: knowing what not to do (from failures), recognizing danger signals (from near misses), and knowing what to do more of (from successes).

Consider a temporary falsework (support scaffolding for a concrete pour). On Project A, carpenters notice one of the falsework towers settling a few millimeters during a deck pour – nothing collapses, but it's odd. They shim the base, take a couple of photos, and move on, thinking they've addressed it, and since nothing failed, it's not worth paperwork. They don't formally report it because, in their eyes, it wasn't an "incident." Months later, on Project B, a different crew (maybe even a different contractor division) erects a similar falsework tower on a slightly sloped grade with some uncompacted fill beneath. Under load, that tower settles a few inches – much more severe – causing a misalignment that requires expensive jacking and rework to fix, and it nearly causes a serious injury when a formwork panel shifts unexpectedly. During the investigation, someone recalls that something like this happened before. They dig around and find out that on Project A, there was indeed a minor settlement event, but it was never analyzed or widely communicated. What appeared to be a trivial anomaly in Project A was actually an early warning of a weakness – one that Project B didn't heed because the lesson wasn't captured.

Had there been a robust lessons-learned system, Project A's team would have documented the falsework settlement: how much it settled, what the soil conditions were, how the tower was configured, what they did (shim it), and importantly, they would have flagged it for future jobs (e.g., why did it settle? Is our standard falsework base detail adequate on fill?). That information could have prompted a precaution for Project B (such as verifying soil compaction or using a wider base). Institutional memory, in effect, could have broken the chain between those two events. Without it, the company paid the price twice for the same underlying issue.

**Mechanisms for Institutional Memory.** Institutional memory isn't a single database or document; it's a system with multiple components reinforcing each other. At the project level, teams need tools to preserve lessons within the job so they don't forget what they figured out last week. This includes things like well-kept redlined drawings and as-builts that record what was actually built versus what was planned (and why), detailed daily reports or logs that note deviations or unusual occurrences and how they were handled, "read-before-build" briefings for crews that highlight critical lessons or precautions before starting a new phase, regular debriefs among field leadership (foremen, superintendents) where they discuss what's going right or wrong, and even photo logs or journals tied to specific locations or milestones. These all create a project memory that can be handed off to new crew members or used in the project's later stages, so mistakes aren't repeated within the same job.

At the **organizational level**, the goal is to ensure that each new project benefits from the previous project's experience. Mechanisms here include a centralized lessons-learned repository – some kind of database or library where anyone can search for issues or topics and find past experiences. It might also involve maintaining a library of standard details or methods annotated with notes like "this detail failed in X case due to Y, so now we always add Z for safety." Some companies issue periodic bulletins or memos to all sites summarizing incidents that happened (either within the company or notable ones in the industry) and what should be learned from them. Others hold recurring seminars, safety stand-downs, or roundtable meetings where project teams convene to share recent challenges and insights. And importantly, there should be a process to update **corporate** SOPs, standards, and checklists based on field feedback – not just project-specific ones. If one project learns a better way to do something, that should become the new standard for all, not remain tribal knowledge.

Then there are the **human memory channels**, which might be the most powerful of all. People remember vivid stories and personal mentorship far more than they remember reading a report. Savvy organizations create structured ways for this knowledge to transfer person-to-person. For instance, rotational training sessions or workshops in which seasoned veterans present real case studies and "war stories" to less-experienced staff. Formal mentorship programs pair up less experienced supervisors with grizzled superintendents who have seen it all – and those mentors will inevitably pass on lessons in the course of daily work and conversation. Pre-construction or pre-mobilization kick-off meetings can include a review of relevant lessons from similar jobs ("Before we start this bridge, let's talk about what we learned from the last bridge project"). Even storytelling in regular safety meetings – taking five minutes for someone to recount, in first person, a mistake or a save they experienced and what everyone should take away from it – can imprint lessons in ways a written memo might not.

The bottom line is that **human memory is how wisdom travels** in an organization. Procedures and databases carry information, but stories and personal experiences carry wisdom. A mature construction culture will leverage both: the written systems to store and spread information and the human systems to internalize it and give it life.

**From Lessons to Action**. Capturing knowledge isn't enough; it has to loop back into doing things better. Institutional memory is useless unless it's reintegrated into the organization's operational systems. There are several feedback paths for this. One is updating SOPs and policies – every time a lesson highlights a gap or a superior method, the relevant procedure should be revised, formally re-issued (with version control so everyone knows it changed), and crews trained on the update. Procedures should be living documents that evolve as the organization learns, not static paperwork.

Another path is improving the risk **registers and planning checklists** for projects. If a previous job discovered a certain risk or issue, the next project's planning phase should include that item in its risk assessment. Maybe you could add a line in the pre-task plan checklist that says "Check soil compaction under falsework" because of that earlier incident, or add a known issue to a risk register so it gets monitored. Patterns identified from past issues can thus become prompts that alert future teams, so they're not blindsided by the same problem.

Training programs are another feedback avenue. Lessons learned (good and bad) from real projects should make their way into training materials for both field crews and management. It's one thing to teach generic fall protection or equipment training; it's another to say, "Here is a scenario that happened on one of our sites and how it was handled – what can we learn from it?" People perk up because it's real and it happened to people like them. Incorporating those specifics can prevent new employees from falling into the same traps.

Design and engineering practices can benefit too. If field experience shows, for example, that a particular design detail was very error-prone or confusing in execution, that feedback should reach the engineers so they can adjust their standards or at least be aware and watch out for it. Perhaps after multiple incidents of a certain type of connection failing to be assembled correctly, the engineering department might alter that detail in future designs for more clarity or robustness. Or they might simply add a note in design guidelines to contractors: "This detail is sensitive – double-check field installation."

To put it plainly: a lesson that doesn't result in some kind of change or action is a lesson wasted. Setbacks and mistakes are costly; the return on that investment is what we learn and do differently next time. Organizations that capture those lessons, disseminate them, and most importantly, use them to shape future projects transform their failures (and near failures) into fortifications. That is what institutional memory looks like in a mature, disciplined construction firm – it turns every project's experiences, good or bad, into smarter practices going forward.

## QA/QC Processes as Safeguards

Where procedures establish the intended method of work, Quality Assurance/Quality Control (QA/QC) verifies that the method was followed and that the result meets the requirements. In essence, QA/QC is the eyes and ears of procedural discipline – it catches the mistakes, deviations, or omissions that inevitably slip through human performance, and it does so early enough to prevent harm. A strong QA/QC program is not a nagging police force or an afterthought inspection at the end of a job; it's a continuous loop of verification and feedback woven into the construction process.

Think of it this way: procedures tell us what we should do and how to do it correctly. Operating boundaries (which we'll cover soon) tell us the limits we must not exceed. QA/QC is the mechanism that checks in real time that we are, in fact, doing what we said we'd do and staying within those limits. It ensures the design intent – all those assumptions and specifications the engineers and planners established – are actually being honored in practice. If procedural discipline is about doing things right consistently, QA/QC is about confirming that they were done right and catching them quickly when they're not.

It's important to clarify that QA/QC is not about creating a culture of blame or catching people out for the sake of it. It's about creating a culture of transparency and prevention. In a disciplined operation, quality inspectors or engineers are not the enemy of progress; they're part of how you ensure progress doesn't come at the cost of safety or rework. In fact, high-reliability industries treat quality control as an integral part of operations – another set of controls to prevent drift. Without active quality checks, any process left to humans will gradually drift off course (we are all prone to cut corners or making small errors). QA/QC provides the timely corrections that keep things on course.

**QA vs. QC – Two Sides of the Coin.** Often, we use the terms QA and QC together, but they actually refer to different levels of the quality process. Quality Assurance (QA) usually means the overall system or program in place to ensure quality – this includes the planning, procedures, training, and standards that are set before work begins, as well as audits and process checks during the work. It's proactive and systemic. Quality Control (QC), on the other hand, usually refers to the specific inspections and tests carried out on the work itself – verifying that a weld is good, that concrete strength is achieved, that bolts are torqued, etc. It's more reactive and product-focused.

In practical terms, QA is building the framework that should produce quality (such as qualified welders, approved welding procedures, an inspection plan, etc.), while QC is the act of inspecting the welds produced to ensure they meet the criteria. They interlock: QA sets it up so things should go right; QC checks that they did go right.

**Quality as Risk Control, Not Box-Checking.** Sometimes teams fall into the trap of seeing QA/QC as just a compliance requirement – something you do to satisfy an ISO standard, the client, or the spec, but not something that meaningfully affects the project. This mindset is dangerous. In reality, QA and QC are among the active countermeasures that prevent drift and failure. They are as essential as having brakes on a car. Construction projects are dynamic and full of variability: materials have natural variation, conditions change, multiple crews interface, and even the best procedures might be misinterpreted or misapplied. QA/QC steps in to detect those variations and errors early – ideally while they're still minor and correctable, long before they cascade into a serious problem or collapse.

In high-reliability organizations, an inspector or quality engineer is not seen as an obstacle to production; they're seen as a guardian of the project's integrity. The goal is not to have them come in at the end and point out flaws when it's too late – the goal is to embed them throughout, so that issues are identified and fixed when they're still small. Many construction teams have learned the hard way that skipping or glossing over quality checks doesn't actually save time or money; it just accumulates hidden defects that later result in big surprises (and usually much more time and money to fix).

**Building QA/QC into the Workflow.** A well-designed QA/QC system strategically places checks at the **natural "choke points"** or critical moments of construction – points in the process where, if something were wrong, it would have the biggest impact or would soon become very hard to fix. Essentially, these are hold points we talked about, but from a quality perspective.

**Before a concrete placement:** This is the last moment to verify that the rebar, embeds, sleeves, post-tension ducts, and formwork are all as per the design. Once you pour concrete, any mistake in rebar placement or a missing component is literally buried. So a QA/QC inspection before pouring (often called a pre-pour inspection) is crucial. - **Before stressing or grouting post-tensioning tendons:** Once you stress tendons, you lock in certain conditions (you can't rearrange tendons or fix anchor placements easily afterward). So QC must check tendon placement, anchor alignment, concrete strength around anchors, etc., before giving the green light to stress. Similarly, once tendons are grouted, you can't inspect their condition easily again. - **Before backfilling an excavation that contains utilities or structural components:** After backfill, you lose visibility. So you check that pipes are joined correctly, that coating is intact, that required sand bedding or protective measures are in place, etc., before covering them up. - **Before erecting steel on new concrete**

**foundations:** Once a steel column is erected, you might not easily test the anchor bolts or see the condition of the concrete footing. It's prudent to verify anchor bolt size, placement, and torque on any nut if applicable, and that the concrete reached its strength, before putting the big load of a steel structure on it. - **Before heavy lifts or setting large pre-fabricated elements:** Ensure rigging is correct, the piece is built to spec, and the receiving supports are ready. If something is wrong with the rigging or the connection points, you want to catch it before that precast panel is mid-air. - Before stripping formwork or removing shores: Verify concrete compressive strength via test results, check that the curing time has met the requirement, and confirm that any required re-shores are in place. Removing supports too early is a classic cause of collapse. - **Before energizing a new electrical or mechanical system:** Ensure all tests (like pressure tests for pipes, insulation resistance for cables, rotation checks for motors, etc.) have been done, and that the system is ready for safe startup.

Each of these QA/QC checkpoints acts as a gate that the project must pass through. If you blow through the gate without stopping (i.e., skip the inspection), you're essentially entering an unknown territory with no safety net. Passing through with verification, on the other hand, means you've contained the risk up to that point – you aren't carrying forward an unknown defect. Skipping a critical checkpoint is like crossing a bridge without knowing if it's fully built on the other side.

To illustrate how QA/QC can save the day (or how a lapse can create trouble): imagine a structural steel erection where the contractor is using tension-control bolts (the ones with the twist-off spline) to connect steel beams. The QC inspector notes that the splines have sheared off, indicating the bolts were tensioned, but because the steel work is behind schedule, the crew postpones doing a thorough torque check with a calibrated torque wrench – they plan to do it "at the end of the week" after more steel is up. In the meantime, they proceed to lay down the metal deck and start other work, which ends up covering some of those connections. By the time QC comes back to check, a few critical bolts are no longer accessible without undoing finished work. Sure enough, a few weeks later, after some loads and vibrations, one of the main girder connections shifts slightly. It turns out some critical bolts in that connection were never fully tensioned – maybe the tension-control indicator didn't work as expected, or a few bolts were missed – and now that connection is moving under load. No one was hurt, but it required an expensive retrofit of adding bolts and supports, and caused a project delay.

The point of this story: the QA/QC process existed (they knew they needed to torque check), but it wasn't **enforced at the right time**. Had the QC process demanded, "No decking installation or further work until we verify all those bolts," the crew might have grumbled about the delay for an afternoon, but they would have found the loose bolts and fixed them easily, avoiding the later issue entirely. This is how most quality problems manifest – not as dramatic, blatant negligence, but as small compromises ("we'll do that later" or "it's probably fine") that chip away at reliability. A disciplined QA/QC program forces those small compromises back into the open and addresses them when they're still small.

**Designing QA/QC for High-Risk Activities.** Some activities in construction are so complex or safety-critical that the QA/QC around them must be especially robust. Think of things like a major bridge tendon tensioning operation, a large-scale concrete pour for a foundation mat, or a critical lift of a heavy piece of equipment over live facilities. In such cases, a generic approach to quality isn't enough; you need a tailored plan. Key elements of a strong QA/QC process for these critical operations include:

- **Pre-defined inspection points:** Don't leave it up to memory or chance when inspections happen. They should be built into the work plan from the beginning. For example, the plan might say: "After the first 50 cubic yards of concrete, do a slump/temperature/air test and check formwork; at 100 cubic yards, check again," or "Inspection of all rebar splices to occur after rebar installation, before any concrete is placed." Everyone should know in advance that "At this step, we pause and inspect."

- **Hard hold points:** This is related, but emphasizes authority – points where work *will not continue until* QC gives a sign-off. If concrete arrives and something's not right, the hold point means you genuinely stop the pour – even if the pump is ready and people are yelling about the schedule. A hold point that can be overridden due to pressure isn't a real hold point. So for critical steps, it must be clear that the inspector or responsible engineer has the authority to halt work.

- **Clear acceptance criteria:** Inspectors and crews both need to know what "pass/fail" means for a given check. Vague statements like "ensure good workmanship" are not enough. It should be like, "Rebar cover must be 2" ± 1/4"," or "Concrete must reach at least 3000 psi before stripping forms," or "Bolt torque must be 200 N·m – 100% of connections tested with calibrated wrench." Having this clarity prevents arguments and guesswork in the field. It also speeds up inspections because everyone knows exactly what they're looking for.

- **Independent verification:** This means involving someone in the quality process who is not the person who did the work. People checking their own work can sometimes be blind to their own mistakes or have a subconscious bias to approve it ("I installed those bolts, I'm sure I did it right..."). An independent QC person, or a peer from a different crew, or a third-party inspector, brings fresh eyes and no pressure of "I'll make myself look bad if I find a mistake." Independent checks often catch things that might otherwise slip through.

- **Integration with engineering:** When an inspector finds a deviation or something that doesn't look right, there needs to be a clear path to involve an engineer or higher authority quickly to evaluate and decide on corrective actions. QC should never be left in a position of "I guess it's probably okay" if they're not sure. For example, if concrete strength is a little lower than specified on break tests, the QA/QC procedure might be: notify the structural engineer immediately for assessment. This integration means that when the unexpected occurs (and it will), the team can get expert input and not just make ad hoc judgments in the field.

- **Traceability**: For critical work, tracking exactly what was inspected, when, and by whom is vital. If an issue surfaces later, you want to trace back and see, for example, which batch of concrete was used in that pour and who tested it, or which welds were done by the welder whose qualification later came into question. Good QA/QC keeps records that can answer these questions. That often means tagging elements (label welds or structural members with IDs), logging locations of tests (e.g., "moisture test conducted at gridline X/Y"), and naming individuals involved (so you know who might need additional training or follow-up).

When QA/QC is fully integrated into the project planning, it doesn't feel like an interruption – it feels like part of the work. Crews come to expect that, say, after they tie rebar, they wait for the inspection because that's just how the job flows. Quality checks get scheduled just like any other activity in the look-ahead plan. This proactivity prevents the all-too-common friction of QC "stopping production," because quality is part of production.

**Documentation, Traceability, and Nonconformance Handling**. We touched on traceability above, and it's worth expanding. If a problem arises – say some concrete in a certain area isn't reaching strength, or a batch of bolts is discovered to be counterfeit – the project's ability to respond appropriately hinges on how well things were documented. Can we pinpoint exactly which structural elements were used that batch of concrete or those bolts? Or do we have to treat everything as suspect because we have no idea? Strong QA/QC systems include measures like: - Tagging or labeling materials and components (e.g., marking which lot of high-strength bolts went into which connections; labeling tendon ducts or rebar bundles by batch). - Keeping material certifications and test reports linked to installation locations (e.g., concrete batch tickets filed by pour location and date). - Unique identification for critical work (like numbering welds and recording which welder and which inspector signed off each one). - Meticulous logbooks for tests like torque checks, pressure tests, etc., noting exactly which part of the structure they correspond to.

These practices allow the team to answer precise questions when needed: "If weld #27 failed testing, which other welds did that same welder do around that time?" or "If this pile's concrete is under-strength, which other piles were from the same concrete batch?" Without traceability, the discovery of one problem often means you're forced into extreme, costly measures: either you undertake a widespread tear-out or investigation to find all instances of the problem (if you even can find them), or you assume the worst-case scenario and reinforce or redo large areas unnecessarily "just in case." Both outcomes waste time and money and can be avoided with upfront traceability.

Equally important is how you handle **nonconformances** – those moments when QA/QC finds that something doesn't meet requirements. A disciplined approach treats a nonconformance as something to be managed, not something to be hidden out of embarrassment or blasted with blame. The goal is to contain any risk from it and correct it properly. Typically, a good nonconformance process will: 1. **Identify and document the issue clearly:** What exactly is out of spec? Write it down, mark it on drawings, take photos – whatever is needed so it's clear. 2. **Assess the impact:** Does this matter? Sometimes it's minor (a bolt hole is slightly misaligned but still okay), other times it could be critical (the concrete is way under-strength, which could compromise the structure). If the impact isn't obvious, that's a red flag to involve engineers or others. 3. **Escalate if needed:** If the nonconformance could affect structural integrity, safety, or anything significant, it should be brought to the attention of someone qualified to judge – often an engineer of record or a senior manager. Frontline QA/QC should never feel empowered to just shrug off a serious deviation on their own. 4. **Decide on corrective action:** This could range from "accept as-is with justification" to "repair it like this" to "remove and replace that work." The corrective action often needs approval from engineering and perhaps the owner or authority having jurisdiction, depending on severity. 5. **Implement and document the fix:** Once a fix is done, QA/QC should verify that it indeed brings the work into compliance. And beyond fixing it, they should consider whether the process needs changing or if people need re-training to prevent it from happening again.

One key rule: **never sweep nonconformances under the rug**. If something is out of spec and the crew just quietly fixes it without telling anyone, the immediate piece of work might end up okay, but the organization misses the chance to learn that it happened. Maybe ten different crews at ten sites are all encountering the same issue and "fixing" it quietly – that indicates a systemic problem, perhaps a drawing error or a procedural gap. But if none of those instances are reported, management remains blissfully unaware until one day the issue doesn't get fixed and a failure occurs seemingly out of nowhere. A culture that encourages reporting of nonconformances (instead of punishing it) is essential. People should be able to say, "Hey, I messed up these welds, we need to repair them," without fear – because that honesty allows the system to respond and improve.

**Embedded vs. Independent QC.** Many construction contractors have QA/QC personnel embedded within their own project teams. For instance, the general contractor might employ the quality control inspectors who check the work, or each subcontractor has their own QC doing the inspections for that trade. This model has advantages: the inspectors are on site full-time, they know the work intimately, and they can often catch issues quickly in real time. However, it also has potential downsides. An inspector employed by the same organization that stands to gain from moving fast could feel pressure (explicit or implicit) to not slow things down. They might also become too familiar with the work and develop blind spots – a form of confirmation bias where they expect things to be fine and so they miss signs that they are not.

To counter this, many projects or clients incorporate some form of **independent verification** as well.

## Reliable Equipment and Facilities

Equipment reliability is often treated as an afterthought in construction—something to worry about only when a machine breaks down and causes a delay. In reality, equipment reliability is a core element of procedural discipline and overall project safety. The stability and predictability of cranes, hoists, lifts, pumps, excavators, scaffolds, formwork systems, and all the other mechanical and temporary devices we use directly determine whether crews can maintain control of the work. If those systems are unpredictable, it undermines every plan and procedure, no matter how well thought-out.

In high-reliability operations theory, having reliable equipment is considered just as important as having valid procedures or well-defined operating limits. Construction is no different. You cannot operate safely if the tools and machinery you depend on behave erratically. An unreliable crane, for instance, might suddenly drop a load or swing uncontrollably; an unreliable scaffold might shift or collapse under a routine load. These events can instantly defeat all other safety measures. Conversely, reliable equipment *prevents* a lot of chaos. It reduces surprise events that force people into a reactive mode. And as we've said, **surprise is the enemy of control.**

Ensuring equipment reliability starts long before the equipment shows up on site. It begins with **conservative design and procurement choices**. The reliability of a given piece of equipment is largely "engineered in" by the manufacturer and by the specs you decide to use. If you consistently choose the cheapest option with minimal safety features, or rent older equipment without thoroughly vetting it, you're essentially building latent risk into your project. Many equipment-related failures trace back to decisions made in the planning and procurement phase. For example, if a project selects a tower crane model that meets the bare minimum capacity but lacks certain redundant braking or safety systems (maybe because it was more affordable), that crane will obviously be less forgiving if something goes wrong. Or consider renting an aging hydraulic crane without checking its maintenance records – you might inherit someone else's maintenance neglect. Accepting scaffolding or shoring components from multiple suppliers might introduce compatibility issues (one piece doesn't quite lock properly with another, or ratings differ). Using generic, "one-size-fits-all" lifting beams or rigging hardware, instead of ones specifically engineered for your loads and configuration, also compromises reliability. Procuring pumps or generators with no backup capacity or no redundancy for critical operations means you're running on a razor's edge – any hiccup, and you have an outage or failure.

The general rule is: **reliability is rarely an accident; it's a result of conscious choices.** A crane that has an additional safety factor built into its load chart is far less likely to be overstressed by a slight misjudgment than one that's always operating near its limit. A shoring system that uses, say, redundant locking pins or bracing provides tolerance if one component slips or isn't tightened fully. A pump outfitted with an automatic shut-off if pressure or flow goes outside safe bounds can protect itself from catastrophic damage (imagine a pump that stops itself before running dry and destroying its seals). A formwork panel that is designed extra stiff won't warp or deflect as much under repeated use or humidity, reducing the chance of formwork failure or misalignment. These kinds of features often cost more or require more upfront scrutiny to procure, which is why they're often cut when budgets are tight or when decision-makers underestimate their importance.

In construction, unfortunately, focus on schedule and budget can overshadow equipment reliability considerations. The fastest or cheapest vendor might get picked, or a piece of gear that "mostly fits the need" might be forced to serve in a role it wasn't ideal for. To combat that, leadership has to explicitly prioritize reliability in specs and procurement. That means, for example, setting minimum requirements for critical equipment beyond just what's "lowest bid." It could mean requiring that any crane on site be within a certain age or service hour limit, or specifying that backup systems must be present. It could mean thoroughly vetting a rental company's maintenance program before signing the contract. A disciplined organization will have reliability requirements (often informed by past lessons) baked into its procurement process, not as an afterthought. They chose the crane model not just on capacity but also on safety track record. They might pay a premium to get all the scaffolds from one compatible system rather than mixing and matching just because someone had some pieces lying around. They might decide not to push a piece of equipment to its limit usage just to save renting an additional one, because they know the risk isn't worth it.

Even the best-designed equipment can fail if it's **assembled or installed improperly.** Reliability is born at assembly. Construction often involves putting together complex equipment on site – erecting a tower crane section by section, assembling scaffolds, installing formwork and falsework towers, joining heavy segments of a tunnel boring machine, etc. The quality and precision of this assembly work directly impact performance. However, it's an area sometimes overlooked because everyone assumes the equipment is "designed" to work, but if you put it together wrong, all bets are off.

Consider some common scenarios: If a tower crane is erected and the crew doesn't torque the mast bolts to the exact spec (maybe they're rushed or their wrench wasn't calibrated), that crane is now inherently less reliable and safe, even though on paper it's the same crane. If a shoring tower is plumb and perfect, but you place it on uneven ground or soil that wasn't properly compacted, the whole system can lean or settle regardless of the tower's rated capacity. If workers use makeshift connectors or slightly wrong pins to join scaffolding components (or mix parts from different systems, as mentioned), those joints may not behave as expected. Or if they improvise a connection - say, tying things together with wire or using a random bolt that "kind of fits" - they've undermined the scaffold's integrity. If someone extends a boom lift or a crane beyond its recommended configuration, or attaches something like a personnel basket that it wasn't designed for, you introduce new failure modes.

These are not rare one-off issues; they happen all the time where there isn't a culture of strict adherence during setup. Many incidents have roots in assembly errors: a supporting truss missing some bolts, a temporary bridge support not level, a formwork system not braced per instructions, etc. The takeaway is that **superior construction and installation practices translate directly to reliability**. It's not enough that the equipment has a good design - it must be erected, installed, and used in accordance with that design and the manufacturer's guidelines to actually deliver the reliability it promises.

A contractor was building a multi-level shoring tower to support a high-rise concrete pour. The tower components themselves were of good quality and engineered for the loads. However, the base of the tower was sitting on what appeared to be compacted fill (ground that had been backfilled and compacted). The crew didn't perform a separate bearing capacity test or otherwise confirm the soil strength; they just adjusted the screw jacks at the base until the tower was level and firm and proceeded. During the concrete pour, the fill beneath one corner of the tower compressed more than expected (perhaps it wasn't as well compacted in that spot), causing the tower to tilt slightly. That introduced some sideways force and uneven loading in the tower that it wasn't designed for. It didn't collapse, but it was definitely a dangerous situation - the kind of thing that could lead to failure if the pour continued or if it went unnoticed. Luckily, a QC inspector noticed the lean as it started, and the pour was halted, and the area was shored up further. The incident drove home a critical insight: **the tower's reliability was not just a function of the tower design; it completely depended on the conditions of its installation (the ground in this case)**. A shoring tower is only as reliable as what it's standing on. A crane is only as safe as its foundation and tie-ins to the building. A scaffold is only as stable as the connections and anchoring that the erectors actually implement in the field.

To generalize: reliability isn't an intrinsic property of a piece of equipment alone - it emerges from the combination of good design, proper installation, and correct usage. Construction teams need to be just as disciplined in setting up and caring for equipment as they are in following procedures.

Another major factor is **maintenance** – keeping equipment reliable through the wear and tear of the job. Maintenance is often seen as downtime or an overhead cost that eats into production, so there's a constant temptation to defer it or do the bare minimum. But consider the environment construction equipment operates in: dirt, dust, heavy loads, shock vibrations, heat, sometimes corrosive conditions, sometimes around-the-clock operation. Without regular and precise maintenance, even a high-quality machine will degrade and eventually fail in unpredictable ways.

Maintenance needs to be **scheduled, not reactive**. That means don't just wait for something to break. Follow the recommended service intervals (or even more frequent if conditions warrant) – whether that's hours of operation or calendar time – for lubrication, parts replacement, calibrations, etc. It also needs to be documented thoroughly – record what was done, when, by whom, and what was found. Don't rely on "I think we changed that filter last month?" Good records allow you to track trends (e.g., this motor has needed adjustment twice in a short time – maybe something else is wrong) and prove that maintenance was done if an issue arises.

Maintenance checks should be **measured** and based on data, not just eyeballed. For example, instead of just visually inspecting a hoist cable and saying "looks ok," one should measure its diameter and compare it to replacement criteria, or test oil samples for metal particles to see if a gearbox is wearing, or use a thermometer to see if a bearing is running hot. Subjective checks like "feels fine" can miss early warning signs that objective measurements would catch.

Also, maintenance must align with the **manufacturer's recommendations and the original safety analysis**. That means using the correct spare parts (not some similar part that "might work"), using the right fluids, and following the specified procedures (tightening sequences, calibration steps, etc.). Ad-hoc fixes – like "we lost a pin, so we just welded something there" – are basically un-engineering the equipment. You're changing what the designer intended. There's a principle in high-risk industries: maintenance should preserve the original design conditions under which the equipment was judged safe. If you let maintenance lapse or you improvise, you're effectively eroding the assumptions that made the equipment reliable.

If you skip or stretch out lubrication intervals on a machine, friction increases, which might cause components to heat up or wear faster, potentially leading to forces and stresses beyond what was accounted for. If critical pins, bolts, or connections aren't inspected and tightened or replaced as needed, you can't trust the equipment's structural capacity anymore – that pin might be half sheared off from fatigue, and you wouldn't know. If hydraulic seals deteriorate and you don't replace them, one day a cylinder might not hold pressure or might jerk unexpectedly, and the whole hydraulic system becomes unpredictable (imagine a crane's outrigger slowly bleeding down without anyone noticing, eventually making it unstable). If you don't fully torque structural bolts on a piece of equipment or formwork, that joint is basically not doing its job and could slip or fail under load. Each of these maintenance misses introduces uncertainty.

A properly maintained machine, by contrast, behaves consistently and as expected, which means your procedures that rely on it (like load charts, speed settings, etc.) remain valid. A poorly maintained machine is a wild card – it might work fine today and break tomorrow with no warning, or it might subtly underperform (like not actually achieving the pressure it says it is, etc.), leading to hidden deficits.

Beyond normal maintenance, one of the most dangerous habits in construction is casual **field modification** of equipment or temporary structures. We touched on this with procedures, but to reiterate in equipment terms: making an undocumented change to a piece of gear can completely undermine its reliability. This includes things like drilling holes in a manufactured beam or bracket to attach something new, using a different pin or bolt than the one specified, cutting or grinding parts to make them fit, welding on attachments or extensions, bypassing safety features (taping down a bypass switch, etc.), or "Frankensteining" parts from different systems together. Each of these seems like a handy solution to a problem ("we need this now and we don't have the exact right piece, but we can modify this one and make do"). However, once you do that, **all bets are off** with regard to the equipment's rated performance.

The reason is straightforward: no engineer calculated the effects of that modification. No tests were done to see if it still holds the same safety factor. The manufacturer's whole design and warranty basically go out the window. It might still work for a while (sometimes these modifications appear to succeed at first, which only emboldens people), but you've potentially introduced a new failure mode that no one understands until it's too late.

Imagine a crew needs a lifting beam for a crane that's a bit longer than what they have. They decide to weld on some new padeyes (lifting lugs) at the ends of an existing shorter beam to extend the pick points. The beam is heavy steel, the welder does a solid-looking job, and they even test lift something light, and it holds. Everything seems fine. But when they go to lift the actual load, the beam bends sharply at those new welded points and nearly fails. Why? Because the weld added heat and residual stresses, it altered the beam's behavior, and essentially created a weak point that the original beam (designed without those padeyes) didn't have. Without a proper engineering analysis (calculating bending moments, checking welds for those loads, etc.), the crew had no way of knowing the true capacity after their modification – they assumed it was still as per the original spec, which was wrong. They were lucky it only bent and didn't break mid-lift completely.

The lesson here is that what feels like a minor tweak can completely invalidate the equipment's known performance. **Disciplined operations require that no equipment modification be made without formal review and approval**, period. It doesn't matter how trivial it seems. If you need to modify something, treat it like any design change: get an engineer involved to analyze it, consult the manufacturer if possible (they might say "no, don't do that" or "here's how it could be done safely"), re-calc the capacity or get a new rating, and document it so everyone knows this is now a "one-off" piece with perhaps custom limitations. Often, the best answer is: don't modify – get the right equipment. But if the schedule absolutely forces a modification, elevate it to a formal process with real engineering input. Under no circumstances should a crew in the field be hacking equipment together, thinking it's fine – that's a recipe for a catastrophic surprise.

Another key aspect of reliability is regular **inspection processes** for equipment in use. This overlaps with maintenance but is slightly different – maintenance might be done by specialists on a schedule (e.g., mechanics servicing a machine every 200 hours), whereas inspections are something operators or supervisors do frequently, even daily, to catch issues between maintenance or anticipate needs.

For equipment to remain reliable day in and day out, you need systematic inspections that are frequent (for critical gear, that could mean a quick check every day or every shift, and more thorough checks weekly or monthly), structured (following a checklist so nothing important is missed – humans are forgetful, especially for boring routine tasks, so a checklist ensures you actually look at the tire pressure, the oil level, the safety latches, etc., every time), **documented** (if no one records the inspection, it's easy to start pencil-whipping or skipping them; plus documentation helps see trends – e.g., "we've noted low hydraulic fluid three times this week on this machine, there must be a leak"), competency-based (the person inspecting should know what they're looking at – they need training on what good vs. bad looks like for that equipment), and tied to specific criteria (for example, saying "no cracks in the welds" is okay, but better is "no visible cracks – even hairline – in welds; if found, do not use and report"; or instead of "check oil," specify "oil level must be between min and max marks on dipstick and no milky appearance, which could indicate water ingress," etc.).

Strong inspection protocols in construction might include things like: a daily walk-around of all heavy equipment at the start of the shift (looking for leaks, checking fluid levels, verifying tires or tracks are in good condition, lights and alarms work, etc.); function tests for safety devices (make sure backup alarms, emergency stops, limit switches, etc., actually function before you start work); focused inspections on critical components such as crane cables, hooks, shackles, lifting slings (any fraying, kinking, or damage – take them out of service early, not after they snap); weekly inspections of all scaffolding and temporary supports (which is actually required by OSHA and good practice – conditions can change daily on these); checking gauges on pumps or compressors to ensure pressures are within normal range; measuring wear items (like the teeth of an excavator bucket, or a drill bit, or the cutting edge of a dozer blade – if they wear too thin they can break unexpectedly); and visually/tactilely checking bolts or connections on equipment (is anything starting to loosen? any unusual vibrations?).

The point of disciplined inspections is to catch the **early signals** of something going wrong – signals that an operator might ignore because the machine "still works okay." A tiny hydraulic fluid drip might seem fine today, but tomorrow it could turn into a burst hose if pressure spikes. A slightly bent support member might still hold today, but under the next heavy load, it could buckle. A faint hairline crack in a weld can propagate with each load cycle until it fails catastrophically. Bearings running a bit hot or a strange noise in a gearbox are often precursors to a breakdown. A frayed sling or worn cable might lift fine a few more times... until the day it snaps unexpectedly under a load it used to handle.

By having formal checks, you force these small issues to be noticed and addressed while they're still small. And then you **act** on them – replace that hose now, don't wait for it to blow; schedule a controlled repair for that crack now, not after it breaks completely; swap out that sling for a new one rather than "let's just finish these next two picks with it." This mindset of proactive equipment care is very much part of a disciplined culture.

Temporary works (like formwork, falsework, temporary bridges, etc.) deserve special mention regarding reliability. These structures are often built with reusable equipment (like modular shoring frames, steel beams, jacks, etc.), and they are by nature not permanent, so people sometimes treat them with less formality than the permanent works. But they carry enormous risk because they support loads at intermediate stages when the permanent structure isn't there yet to provide redundancy or final strength. A failure of temporary works (like a formwork collapse) can be just as deadly and costly as a failure of the final structure, yet we often skimp on analysis, inspection, or maintenance for them.

Equipment and components used for temporary works must be managed more tightly than permanent ones. Why? They're often older (used over and over across projects), maybe dinged up or corroded; they get assembled and disassembled frequently, which is an opportunity for damage or incorrect assembly; and people might mix pieces from different sets. Also, conditions for temporary works can vary widely each time (different ground conditions, different spans, etc.). So what to do? Inspect them **more frequently** than you would a one-time installation. Perhaps inspect key scaffold or formwork connections daily or before each major load (since things can loosen). **Track compatibility:** ensure you aren't mixing parts that don't belong together – if you do, you need an engineer's explicit okay. Maintain an inventory where you know what component goes with which system and don't interchange without analysis. **Protect them from misuse**: for instance, don't take a formwork beam rated for a 6-foot span and decide to use it for an 8-foot span just because "it looks solid enough" – stick to rated limits or get engineering approval to deviate. **Controlled installation:** temporary structures should be erected by qualified crews following an engineered layout and plan – it's not a free-for-all because "it's just temporary." And have **documented load ratings** and instructions readily available – ideally posted right on the equipment or in the task plan that the crew has in hand. If workers know, for example, "this jack can carry X kips max and must be straight up (not angled), and you need one every Y feet," they are less likely to overload or misuse it than if they're guessing.

On a project, workers built a multi-level scaffold using components from two different manufacturers (perhaps because pieces were left over from previous jobs or they didn't have enough of one set). Everything fit together physically, so it seemed fine. But unknown to them, the load capacity of one system's ledgers was lower, and the locking mechanisms between the two systems weren't fully secure in combination. Under load (workers, materials, plus some wind force on the plastic sheeting around the scaffold), one of the mixed components failed, and part of the scaffold collapsed. This wasn't because any single piece of equipment was defective; it was a compatibility issue and an oversight in assembly. It underscores that reliability in temporary works demands uniformity (don't mix stuff arbitrarily), traceability (know what pieces you're using and their source), and disciplined assembly (no "just make it fit" solutions).

Finally, a truly disciplined construction organization will integrate equipment reliability considerations into every phase of the equipment's life on the project. To recap in a lifecycle sense:

- **Pre-mobilization (before equipment arrives):** They'll ensure the chosen equipment is the right tool for the job with a margin of safety. That includes an engineering review of its specifications against the project needs, verifying that its maintenance history is acceptable (if renting, maybe even rejecting a piece that doesn't have proof of recent maintenance), ensuring compatibility with site requirements (does it reach where we need it, does it run on fuel we have, is it too heavy for our temporary decking, etc.), and possibly doing a pre-use inspection either at the rental yard or upon arrival to catch any issues (basically not assuming "new" or "just serviced" means perfect).

- **Installation and commissioning:** They'll assemble and set up the equipment following a checklist. For a tower crane, for instance, every bolt torque might be logged, limits tested, etc., before declaring it ready. They might conduct a proof test – like a load test for a crane or a test run for a hoist under full load – to ensure it behaves correctly. If an outside inspector or engineer should verify the installation, they should be involved. And they document the initial state: baseline measurements, levels, etc., so they can detect if something shifts later.

- **Daily use and monitoring:** They train every operator on the safe operating boundaries of the equipment – never assume a veteran knows the exact wind limit or load chart by memory, make sure they do. They enforce routine pre-shift inspections (and logs for them). They create a culture where if an operator hears a strange noise or feels an unusual vibration, they report it immediately, and someone investigates – rather than ignoring it to get the work done. Small symptoms are treated as signals to heed, not nuisances to ignore.

- **Maintenance cycles:** They adhere strictly to maintenance schedules recommended (or even more frequently in harsh conditions). If a crane needs a 100-hour inspection, it gets it at 100 hours, not 150. They use manufacturer-approved parts and fluids because a cheap knock-off filter or incorrect hydraulic oil can degrade performance or even cause damage. They maintain logs of maintenance and monitor wear items – say they measure the thickness of brake pads or the tension of a cable periodically and trend it, so they can replace these proactively before failure. Basically, fix it before it breaks (planned downtime) instead of after (unplanned downtime, often accompanied by safety incidents).

- **Decommissioning and post-use review:** When the equipment is taken off the project or at the end, they inspect it as it's dismantled. Check for any abnormal wear or damage that occurred during the project and document it. They might have a brief after-action review: Did the equipment meet expectations? Were there any close calls or maintenance issues? This information should loop back to those planning the next project – if, for example, the crane was frequently near capacity, next time maybe specify a larger crane; if a certain machine kept breaking down, maybe it's time to not use that model or to adjust the maintenance strategy; if an operator error almost caused an issue, feed that into training. Essentially, don't just send the machine away and forget about it – capture any lessons to inform future procurement, planning, or training.

Following a life-cycle approach like this makes equipment reliability **predictable rather than reactive**. Instead of wondering when something will fail, you operate with high confidence that it *won't* fail unexpectedly because you've managed every phase of its use. And importantly, any lessons at the end (like "this type of pump struggled with the sediment in the water on site, next time we need a filter or a different pump type") are captured to improve future jobs.

Lastly, equipment reliability is as much a cultural issue as it is a technical one. You can often tell a lot about a construction crew or company by how they treat their gear. If you see a site where machines are battered, oil leaks are ignored, and busted equipment is cannibalized rather than fixed, it likely reflects a broader mindset of accepting risk and disorder. On the other hand, if you see a crew meticulously inspecting their equipment each day, keeping it clean, addressing issues promptly, it usually indicates a disciplined and safety-conscious culture.

You can pose some litmus-test questions to gauge this culture: Do operators truly do their daily inspections, or are those checklists pencil-whipped at best? When a minor issue is noticed (like a funny noise or slight leak), do crews report it and address it, or do they hide it to avoid downtime (maybe adding a bit of oil quietly each day to keep running)? Do leaders enforce the manufacturer's limits and recommendations (like not exceeding load charts, not skipping maintenance), or do they look the other way when someone overloads a machine to get a job done faster? Is scheduled maintenance baked into the project plan (cranes have to stop for an hour every week for checks, etc.), or is maintenance viewed as optional and always pushed off to tomorrow? Are unauthorized modifications and "creative" field fixes strictly forbidden unless approved, or are they quietly encouraged as ingenuity to save time? Is equipment seen as a **critical safety system** that deserves care, or just a commodity tool to be used up and replaced when it breaks?

Reliable operations require a culture where maintenance is respected as part of the work (not something to shirk), inspections happen consistently and people take them seriously (not checking the box mindlessly), operators are well-trained and empowered to shut down an operation if a piece of equipment isn't right (rather than fearing they'll get in trouble for causing a delay), any deviations in how equipment runs are treated as warnings (not shrugged off as "probably fine"), and boundaries like load limits and service intervals are obeyed rigorously. When a team truly treats equipment with that kind of discipline, not only do machines last longer, but the whole project becomes more predictable and safer. People aren't constantly reacting to breakdowns or near-misses caused by equipment issues, because those issues rarely get the chance to happen.

In summary, equipment reliability is a **pillar of the safe operating envelope** in construction. Without reliable equipment, all the boundaries and procedures we establish can be blown apart. Think of what happens if a crane's brakes slip – suddenly your well-planned lift goes out of control (boundary exceeded). If a generator dies during a critical concrete pour, your planned procedure for curing might be compromised. If a formwork jack fails prematurely, the load path changes beyond what was intended. Unreliable gear introduces chaos and uncertainty: boundaries drift because, say, a wobbly crane makes an operator feel they must rush before winds get worse, or an unreliable pump forces people to push it harder. Procedures assume that tools and machines do what they're supposed to; if they don't, the procedure can't be followed, and people improvise. Engineering designs assume equipment will apply loads in a certain way; faulty equipment might apply unintended forces. Unreliable equipment can fail without warning, giving no chance to implement contingency plans.

Reliable equipment, on the other hand, is a stabilizer against those uncertainties. It acts just like any other safety measure – it keeps conditions predictable and within known limits. In fact, it is as important to safe operations as having good procedures, effective training, and solid QA/QC. When machines behave predictably and do what we expect, the work stays within the planned safe envelope, and surprises are minimized. When machines start acting unpredictably, the whole system becomes vulnerable, no matter how good your plans are.

## Well-Defined Operating Boundaries

Every construction activity operates within certain **operating boundaries** – limits beyond which the activity becomes unsafe or uncontrolled. These boundaries might be things like load limits (how much weight a crane can lift or a floor can support), speed limits (how fast a hoist can go or how quickly you can stress tendons), pressure limits (in formwork or hydraulic systems), environmental thresholds (maximum wind speed for lifts, temperature or weather conditions for concrete pours), curing requirements (minimum time or strength before applying loads), spacing tolerances (how far apart supports can be), sequencing constraints (the order of operations that must be followed), and so on. None of these numbers is arbitrary; they come from engineering analysis, manufacturer specs, physics of materials, and safety factors. They define the **safe envelope** in which work can occur.

When these operating boundaries are respected, construction tends to be stable and predictable. When they are ignored or treated as mere suggestions, even a routine task can escalate into an accident. Enforcing boundaries isn't about being overly conservative for its own sake; it's about preventing chaos. The highest-performing construction teams are not the ones that brag about pushing the limits "and getting away with it" – they're the ones that manage their work so well that they rarely even approach the danger limits. They plan and execute with enough buffer that they're always in a safe zone, not right at the edge of failure.

Before you can hold or enforce a boundary, you have to establish it based on a real understanding of hazards and risk. A boundary is essentially a numerical representation of risk tolerance – it says, "beyond this point, we move from controlled work into potentially dangerous work." In construction, boundaries typically come from several sources: the **structural design** gives limits like allowable loads on members, how much deflection is okay, how long concrete must cure to reach design strength; **equipment specifications** provide max capacities (like the crane's load chart at certain radii, the maximum pressure a jack can exert, the max incline for an earthmoving vehicle); **material science and testing** inform limits like concrete temperature during cure, maximum pour rates to avoid cold joints or formwork pressure spikes, safe soil bearing capacities; **geotechnical analysis** yields things like maximum slope steepness or trench depth without shoring; **environmental conditions** add boundaries (you might decide no lifting above X wind speed, or no welding if it's raining heavily, or temperature limits for applying coatings); and **human factors** might set some boundaries too (like no more than N hours of continuous work for crane operators, or at least a certain crew size for a heavy task).

The point is, a boundary is not made up by someone on a whim – if it's a proper boundary, it's anchored in analysis and experience. It marks where **controlled** ends and **uncontrolled** begins. When a crew doesn't even know what the boundaries are, they're basically running an experiment every time – testing how far they can go by trial and error, which in construction often means errors come with injuries or damage.

A useful concept is the **operating envelope** which can be visualized in three zones: the **Normal Operating Zone** (well within all limits, low-risk, lots of safety buffer), the **Safety Margin Zone** (near the limits – you're using up the built-in safety factors, and there's less room for error), and the **Failure Zone** (outside the limits – the probability of an accident or failure jumps way up because you've exceeded what the system was designed or tested to handle). Strong, disciplined operations aim to stay firmly within the normal range. They might occasionally use some of the safety margin (that's what it's there for – contingency), but they do so knowingly and briefly, and with caution. Poor operations, often unknowingly, operate at the edge of the normal zone or in the margin all the time – basically flirting with the limits as part of normal work – and therefore they are shocked when one day that thin margin isn't enough and they slip into failure.

Why do crews drift toward boundaries? The realities of the construction environment tend to push people that way. Schedules are tight, so someone thinks, "Well, maybe we can strip that form a little early to gain a day." Weather windows are closing, so "We'll go ahead with the lift even though the wind is a bit above what we planned." Equipment and crews cost money every hour, so there's pressure to "make the most" of them, which might mean using a crane in conditions near its threshold rather than waiting, or cramming in one extra pick at the end of the day. Coordination issues might tempt a team to bend a rule – e.g., another trade is held up unless we hurry and pour faster than the spec'd rate. Limited access or site conditions might force heavier loads or longer reaches than ideal – and instead of saying "we can't do it that way," sometimes teams will just try anyway. Plus, if the company or workers have incentives for productivity (bonuses for finishing early, or just pride in being fast), there's a natural push to cut corners. And then there's complacency: if people have exceeded a limit before and nothing bad happened, they start to think the limit is overly conservative or not real. "We went a foot deeper in that trench last time with no shoring, and it was fine; probably can do it again."

All these pressures – time, money, convenience, overconfidence – encourage boundary pushing. And often it happens gradually. Under pressure, teams might start to "flex" the rules just a little: lift in winds 5 mph higher than the official limit, strip forms after 5 days instead of the required 7 because last time 5 was okay, pour concrete at a faster rate than recommended, dig the trench 9 feet deep even though the rule was shoring required beyond 8 feet, use one less shoring post here or there because everything's holding so far, skip a scheduled maintenance on equipment to avoid downtime, etc. Each instance feels like a practical choice in context.

Each of those decisions might not immediately cause an accident. In fact, usually they don't – that's why people become emboldened. But boundaries are set where they are for reasons. When you treat them as rubber bands, you're essentially gambling with the safety margin. The problem with drift is that nothing seems wrong until the moment when everything is wrong. People get used to operating closer and closer to the edge, thinking it's fine because nothing bad happened last time or the time before – then one day, inevitably, those safety factors run out. The wind gust that you didn't account for hits, or the concrete hasn't actually reached strength, and the formwork collapses, or the trench wall that held at 9 feet deep yesterday caves in today at 9.5 feet.

The truth is, boundaries are often discovered through failure. If an organization is smart, it doesn't wait for the failure – it respects the boundaries as drawn from analysis and doesn't test them in the field. And if they have to approach a boundary, they do it deliberately (like a planned overload test with all precautions) rather than by accident or neglect.

So part of the solution is to be aware of these common **pressures** and plan around them rather than falling victim to them. For the schedule: build in contingency so you're not forced to cheat to meet a date. For weather: have backup plans or just accept delays rather than push into unsafe conditions. For costs: remind everyone that one accident will wipe out any money saved by cutting a corner. For coordination: communicate clearly so one team's rush doesn't force another into unsafe practice. For incentives: make safety and quality part of performance metrics, not just speed. And for complacency: continually educate that near-misses and times "nothing happened" are false reassurance – luck is not a strategy.

Now, **operating limits** are the hard lines within those boundaries that must not be crossed. These include very clear, quantifiable thresholds like: the maximum wind speed allowed for crane operations (if it's 30 mph per the crane's chart or company policy, that is a hard stop – above that, you shut it down), the maximum load a lifting lug or temporary beam is rated for (you do not exceed it without redesign/reevaluation), the minimum curing time or concrete strength before you can strip forms or apply load (you must meet or exceed that, non-negotiable), the maximum excavation depth allowed without shoring or a different protection method (if it's 5 feet by rule, you don't go to 6 hoping it'll be fine), the required configuration for something like shoring posts (if spec says every post must be locked and every level braced at X spacing, you do exactly that), and administrative safety rules like lock-out/tag-out on live systems (zero energy state before work – you don't make exceptions because it's quicker not to). These limits are essentially the embodiment of "if you cross this line, you are in the danger zone." They are often derived from calculations and regulations that have very little wiggle room.

Ignoring an operating limit isn't being efficient or aggressive – it's conducting an unplanned experiment on your jobsite. It might go okay a few times, but eventually, nature wins (the structure fails, the anchor pulls out, the un-shored trench collapses, etc.). That's why these are usually encoded in safety rules and standards. A disciplined project treats these like ironclad laws of physics.

It's also worth noting that some boundaries are not just numeric but **administrative prerequisites** – conditions that must be true before work starts. If those aren't met, you're already outside the safe envelope from the get-go. For example, ensuring all workers are properly trained and certified for what they're about to do is a boundary. If you have someone operate a forklift without being trained or certified, you have crossed a safety boundary (regulatory and practical) before they even turn the key. Making sure all equipment inspections are up to date and everything is tagged as fit for use (like the daily crane inspection, scaffold tagged "safe to use" each day, etc.) – if you skip those, you're starting work under unknown conditions (maybe the crane has a defect that day, you didn't check). Ensuring the latest drawings and any engineering changes (MOC approvals) are available and being used – if the crew is accidentally using an outdated plan, they are effectively operating out of bounds because the conditions they think are safe might not be. Setting up barricades and exclusion zones is another prerequisite – if you don't establish them in, say, a crane lifting zone or around a floor opening, you've left a hazard uncontrolled. Having required permits for certain high-risk tasks (confined space, hot work, excavation, etc.) is also crucial – the permit process ensures certain conditions and checks are in place. If you bypass it, you bypass those safety checks.

All these administrative items might seem like paperwork, but they create a baseline safe condition. Without them, you are essentially outside of procedure and outside of a controlled state. So part of maintaining boundaries is verifying these prerequisites each day or each task: Are the people qualified? Is the gear inspected and good? Do we have the correct, current plan? Have we isolated hazards (lockout, barricades)? If not, we don't commence work.

**Managing boundaries** properly requires commitment from leadership and vigilance from everyone. Boundaries should be communicated clearly – everyone on site should know the key limits that affect their work ("This scaffold is rated for 2000 lbs per bay – do not overload," "No one goes beyond this point without fall protection," "We cannot excavate past 10 feet here without engineering a new shoring plan," etc.). They should be measurable – it's one thing to say "don't overload," a better approach is "we measure the wind with an anemometer and stop at 30 mph" or "we have a scale on the lifting device so we know the weight and will not exceed X." In other words, give people the tools to know if they're near a boundary. They should be visible – if something has a limit, display it. A tag on a sling that says "Capacity 1 Ton," or a notice on the formwork "Max pour height = 4 feet at a time," or a red line on a hopper "Do not fill above this line" can all help. Boundaries should be integrated into planning documents – lift plans, SOPs, permits should explicitly include the limits to be monitored, so it's part of the conversation from the start. And absolutely, boundaries must be enforced as truly **non-negotiable**. If a reading or condition says stop, **you stop**. There's no "well, maybe just a little more." Finally, boundaries aren't static forever – they should be reviewed if conditions change or new info comes. Maybe initial calcs were conservative, and later an engineer says it's safe to adjust a limit a bit – then do it through formal review and communicate the new limit. Or if you learned from a near-miss that the current boundary wasn't enough (maybe the wind limit should be lower in a particular configuration), update it.

A construction team that treats boundaries as nearly sacred will operate with much more stability. They don't allow the cumulative drift to nibble away their safety margin. And that takes leadership constantly reinforcing, "We do not break the rules for convenience. We will re-plan or wait or get help, but we won't intentionally violate a limit." Leaders also should explain the why behind boundaries. If workers only hear "don't do that," they might do it when no one's looking. But if they understand, for example, "We require a 28-day concrete strength test not because we like waiting, but because the engineering shows the slab could fail under load if we remove forms earlier, since early concrete can be unpredictable," then they're more likely to respect it. Explaining the invisible reasons – like "this is to prevent internal stresses you can't see," or "the wind can gust higher up there even if it's calm down here," or "this soil can look stable until it suddenly isn't, which nobody can predict by eye" – helps people appreciate that boundaries aren't arbitrary bureaucracy. They are acts of **stewardship:** protecting everyone's safety and the integrity of the work.

It's natural that workers sometimes chafe at boundaries if they don't understand them. They might see it as someone needlessly slowing them down or not trusting their judgment. That's why leadership's role is to communicate both the letter and the spirit of each critical boundary. And to model that behavior themselves – if a leader says "safety first" but then pressures a team to finish a task even when a limit is being reached, the crew will follow the pressure, not the slogan. Leaders must never set the precedent of asking for forgiveness after breaking a rule "just this once." Instead, they should plan the work such that those rules can be observed every time without putting progress in jeopardy.

One can think of boundaries as the final layer of defense when all else is starting to slip. People get tired (attention to detail slips – boundaries catch that by forcing stops). People get overconfident ("it was fine last time" – boundaries don't care, they stop you anyway).

Pressure builds to go faster (boundaries force you to pause or get extra help instead of cutting safety corners). In systematic safety thinking, we often say defenses are layered, and boundaries are like the outer perimeter fence. If even those are breached, you're definitely in the danger zone.

In summary, tying it all together: **Procedures** tell us what to do and how to do it correctly. Change control ensures that, as reality deviates, we update what we're doing so it remains correct. QA/QC verifies at each step that what we did was indeed what we were supposed to do and catches errors. Equipment reliability provides a stable platform so that all our plans and checks aren't undermined by a mechanical wild card. Operating boundaries set the safe limits to all our actions, providing clear lines that we do not cross.

These elements interlock to form a resilient, disciplined, high-reliability operating system for construction. If an organization can internalize procedural discipline, rigorously manage changes, maintain strong QA/QC feedback loops, keep its equipment reliable, and enforce its safety boundaries, it will systematically prevent failures that might blindside less disciplined projects. Each component reinforces the others: procedures won't be effective if people constantly improvise changes – that's why you need change control; change control can't catch everything if people quietly deviate – that's where QA/QC and a culture of reporting issues comes in; none of it works if the gear you rely on is breaking or behaving erratically – equipment maintenance and reliability are fundamental; and even if one layer slips, boundaries provide a backstop to prevent disaster. In such an organization, accidents and major failures don't just become low-probability – they are actively and continuously prevented by the way work is planned and executed every single day. The result is not only safer, but it's also more productive and predictable, because a project that avoids accidents, rework, and catastrophic surprises is invariably going to meet its goals more efficiently as well.

**Operational Takeaways for Today's Job:**

- Reinforce the basics every day. Even if the team has done a task many times, start by reviewing the proper procedure or checklist. This prevents "routine" jobs from slipping into unsafe shortcuts and keeps everyone's mind on doing it right.
- Don't let anyone "shortcut" the process, even under time pressure. If a step like locking out equipment, checking a measurement, or getting a permit seems like it's slowing things down, insist it be done anyway. Skipping procedural steps to save a few minutes can lead to hours of rework or an injury.
- Treat changes carefully. If a field condition forces you to deviate from the standard procedure or plan, stop and get the change documented/approved through the proper channel (e.g., a quick call to the project engineer or a signed change order). Uncontrolled changes are a recipe for errors.
- Use checklists and hold points to your advantage. For critical tasks, implement hold points where work stops until a supervisor signs off that everything up to that stage meets the standard. This kind of discipline catches mistakes early – for example, verifying rebar placement before pouring concrete.
- Capture lessons learned today, not later. If you discover a better way to do a task or encounter a procedural gap, make a note of it and inform the team or document it. Good procedures evolve – field input ensures the written process stays relevant and effective.

Chapter 7

# Systematic Project Planning

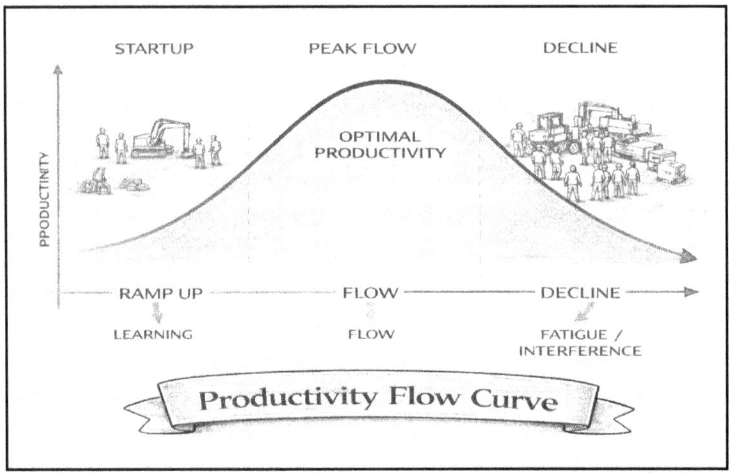


From a field leader's perspective, systematic project planning in construction means breaking a massive job into manageable pieces, aligning everyone's efforts, and keeping a firm grip on time, cost, and scope even when pressure mounts. A project plan isn't just a document – it's the playbook that turns chaotic undertakings into coordinated success. This chapter explores four core aspects of planning that are essential: structuring the work (Work Breakdown Structures), mapping out who's involved (Stakeholder Mapping & Accountability), balancing the big three constraints (the Time-Cost-Scope "Iron Triangle"), and planning for those temporary states in construction when things are most fragile. It provides practical steps, candid lessons, and real examples from the field. The goal is to show how a disciplined, systematic approach to planning can keep a project predictable and safe even under tough conditions. This isn't abstract theory – it's the hard-won planning discipline that seasoned construction teams apply on every multimillion-dollar job.

## Work Breakdown Structures (WBS)

On any new project, the first requirement is a solid Work Breakdown Structure (WBS). A WBS is a foundational planning tool that visually and hierarchically breaks a project into smaller components, mapping out all the work to be done and ensuring nothing is overlooked. In construction, a WBS organizes the total scope of work into logical chunks – often by phase, system, area, or deliverable – to ensure nothing is omitted. Seasoned managers often describe it as the project's family tree: at the top is the completed project, and beneath it branches every element of work down to the smallest task. If something isn't in the WBS, then effectively it isn't in the plan. That's why a good WBS inspires confidence – it captures 100% of the scope so the team isn't hit with surprises later.

For a commercial building job, a top-level WBS might include major phases like Design, Procurement, Site Preparation, Foundation, Structure, Envelope, and Finishes. Each of those then breaks down further. For instance, Structure could split into Substructure (foundations) and Superstructure (frame), and those break into concrete work, steel erection, etc. On an infrastructure project like a highway, it might make sense to organize the WBS by geography or segment (e.g. Segment A – Roadway, Segment B – Bridge, Segment C – Interchanges). For a power plant, a systems-based WBS can be used: Site Work, Civil/Structural, Boiler System, Turbine/Generator, Electrical Systems, Commissioning, and so on. The key is that everything planned to be built falls somewhere in the WBS – it's exhaustive by design. When reviewing a draft WBS, an effective question is, "What's missing?" If any scope item is absent from the WBS, it will bite the project later.

## Step-by-Step WBS Development

A structured process can be followed to develop a WBS. Experienced planners often use steps like these:

- **Identify Major Deliverables:** Start at the top with the end goal – the completed facility – and then list the big chunks of work or outcomes needed to get there. These become the Level 1 WBS elements. In building construction, typical Level 1 elements include Contract Milestones, Design, Permitting, Procurement, Construction, and Commissioning. Think of these as the major chapters of the project.

- **Break into Subcomponents:** Take each Level 1 element and break it down into its parts (Level 2). For example, under Construction, you might have Sitework, Structure, MEP (Mechanical/Electrical/Plumbing), Interiors, etc. Then go further: each of those can be broken into Level 3, Level 4, and so on, until you reach work-package level – the smallest tasks you'd assign to a crew or subcontractor. In practice, break down tasks until they are manageable and assignable. If a task is still too broad or long in duration, break it down further.

- **Maintain Logical Groupings:** Be consistent in how you categorize at each level. If at Level 2 the breakdown is by phase (e.g., Phase 1 – Foundations, Phase 2 – Superstructure), then every Level 2 entry should be a phase – do not mix a phase and an area in the same list. Mixing approaches at the same level causes confusion. (It's acceptable to use different breakdown approaches in different branches – for example, a project might break into geographic sections at Level 1 and within each section break down by phase – but each list should be internally consistent.)

- **Code and Describe Each Element:** Assign every WBS element a unique code (like an outline number) and a title, and often include a brief description in a WBS dictionary. This may feel bureaucratic, but it prevents misunderstandings. For instance, if one WBS element is called "Site Utilities," does that include both electrical and water lines? A one-sentence description can clarify scope boundaries so team members aren't guessing. Coding also allows data to roll up easily – for example, all costs coded under "3000" might roll up to total Structural costs.

- **Review for Completeness and Uniqueness:** Always review the WBS with the project team (designers, superintendents, etc.) and ask: Is everything included? Every deliverable and major task should have a place in the structure. If something doesn't fit, add a branch for it. Just as important, nothing should appear twice – no duplicate scopes. Every piece of work belongs in one and only one spot in the WBS to avoid double-counting or overlapping responsibilities.

- **Use the WBS for Schedule and Budget Planning:** Once the WBS is solid, it becomes the backbone for the schedule and budget. Link every schedule activity and every cost item to a WBS code. Modern scheduling software makes this easy – tasks can be indented under WBS headers or assigned codes, allowing the schedule to be collapsed or expanded by WBS. For example, the schedule can be collapsed to just see totals for Sitework or Electrical, which is great for high-level tracking. A well-structured WBS means that when the team later looks at schedule or cost reports, they roll up neatly and nothing slips through the cracks.

Building a WBS takes effort and foresight, especially on complex projects, but the payoff is huge. With a good WBS, the project has a clear map. It's far easier to spot if something is missing or if a plan is unrealistic. Many teams that jump straight into a schedule and forgo a proper WBS pay for that mistake later when they discover major scope items were forgotten, causing late scrambles and change orders. In practice, a robust WBS brings clarity and control. On a tunneling project, the WBS was structured in a linear sequence of tunnel drives (Shaft S2, then S1, then S3) to mirror the actual construction sequence. This made the schedule intuitive – anyone could look at the WBS and instantly see the order of work. In another project, a high-rise office tower – the planners included top-level WBS branches not just for the usual construction trades, but also for Submittals and Inspections (tasks often buried in schedules and easily overlooked). That way, every required permit, approval, or inspection had a place in the plan and got the attention it needed. The result was fewer surprises and a team that was proactive about paperwork and quality control, not just bricks and mortar.

Today, software makes WBS creation and use much easier. Programs like Primavera P6 or Microsoft Project are built around a hierarchical task structure (the indented outline of tasks is effectively a WBS). On large infrastructure jobs, Primavera P6 allows defining WBS levels and rolling up activity durations and costs at each level. Microsoft Project similarly uses indented task lists. Newer cloud-based tools (e.g. Smartsheet and other project management platforms) also offer WBS views or mind-map style WBS builders. The benefit of using these tools is that once the WBS is in place, everything stays connected – if a task is delayed, it's clear which branch of the WBS is affected, and vice versa. Some tools come with templates for different project types (for example, residential construction vs. industrial plant projects), which can be a helpful starting point, but it's important to customize the WBS to the specific job conditions.

## WBS Best Practices

Industry experience has led to a few best practices that every project team should follow:

- **Include Pre-Construction and Support Tasks:** Don't break down only the physical building work. Include sections for Design/Engineering, Permitting, Mobilization, and other preliminaries. Experienced planners list things like Contract Milestones and Submittals as their own categories – these are items a generic textbook WBS might overlook, but seasoned teams know they're crucial.

- **Milestones and Deliverables:** Consider dedicating a WBS element to major milestones. Whether it's "Turnover to Owner" or interim handovers, putting milestones in one branch makes them easy to track. They can be reported on directly – this keeps those critical dates from getting lost among hundreds of tasks.

- **Keep Each Level Consistent:** As noted, choose a categorization and stick to it at a given level. For example, if Level 2 is broken down by area, then every Level 2 entry should be an area; don't mix an area with a phase on the same level. If a WBS at Level 2 had "Phase 1" alongside "Building A," it would confuse the team. Different branches can use different breakdowns if needed (perhaps the project naturally breaks into geographic sections, and within each section by phase), but never mix categories side-by-side at the same hierarchy level.

- **Align with How You Will Manage and Report:** Think ahead to how progress will be measured and reported. If the owner expects updates by building wings, ensure the WBS includes those wings so you can easily roll up progress for each. For example, if the client cares a lot about the building envelope, you might create a WBS branch for the Exterior Façade across the whole project. That way, all façade-related work can be tracked as one package (even if it spans multiple areas and trades), making it easy to report on as a focused item.

- **Depth vs. Clarity – Find the Balance:** More detail isn't always better. A WBS with too many levels (say 7 or 8 levels of microscopic detail) becomes unwieldy and understood only by the planner. Too few levels (e.g. only 2 levels for a huge job) means each element covers too much scope. Most projects end up with around 4–5 levels of WBS – enough to break things into clear chunks, but not so much that the team drowns in complexity. There's a balance between detail and usability.

- **Treat the WBS as a Living Document:** Projects change – scope gets added or altered – and the WBS must be updated accordingly. If a major piece of work is added (say the owner introduces a new feature or alternative), insert the new scope in the appropriate branch of the WBS. Cost accounts should be tied to the WBS as well, so any budget changes align with it. This way, cost reports, schedules, and change orders all speak the same language (nothing is worse than a budget code that doesn't tie to any WBS item – people lose track of what it's for).

In short, the WBS is essentially the "skeleton" of the project plan. If effort is invested up front to build it right, everything else – schedule, budgets, roles – attaches to it and benefits. A strong WBS creates a common reference point for the entire team, from the crane operator to the executive steering committee. Everyone can locate the piece of the project they're working on and see how it fits into the whole. It's often said that if you feel lost in the details, zoom out to the WBS level – it will show you where you are. And when a change or surprise comes, the WBS helps figure out where to slot it in without losing control of the overall plan. Ultimately, a well-crafted WBS turns a daunting, complex project into a series of bite-sized, trackable pieces. That's the essence of systematic planning: it makes complexity manageable.

## Stakeholder Mapping & Accountability

Even a perfect schedule and budget can unravel if stakeholders aren't aligned. Construction projects live or die by people's commitments and expectations. Stakeholder mapping & accountability means identifying all the players who have a stake in the project – not just the client or the contractor, but everyone influential – and making sure they're engaged and on the same page. On a large job, the list of stakeholders can be long: the owner (or multiple owners), architects and engineers, the general contractor, and a host of subcontractors, regulatory agencies, utility companies, nearby neighbors or community groups, investors or lenders, insurers, future tenants, and more. Each comes with different interests and success criteria. If these stakeholders are ignored or mismanaged, it will almost certainly lead to miscommunication, delays, or conflicts down the road.

Let's break this down systematically. **First, identify and categorize the stakeholders**. This often starts with the core project team brainstorming every entity or person who can influence or is affected by the project. They are usually categorized to ensure none are missed:

- **Primary stakeholders (internal):** These are the people directly involved in executing or decision-making on the project. Obvious ones include the Owner/Client, the Project Management team, the Design Team (architects, engineers), the General Contractor (and perhaps key subcontractors), and major suppliers. These folks have skin in the game on a daily basis.

- **Secondary stakeholders (external influencers):** These might be nearby businesses or residents, local community leaders, or special interest groups (imagine a neighborhood association concerned about construction noise, or an environmental group watching the project's impact). They're not running the project, but their support or opposition can sway things (permits, public opinion, etc.).

- **Regulatory stakeholders:** Always list the agencies and authorities: city building department inspectors, OSHA (safety regulators), environmental regulators, the fire marshal, and so on. They have the power to stop work or enforce changes, so they must be managed closely.

- **Financial stakeholders:** If the project is financed by a bank or there's a bonding company or insurer heavily involved, they are stakeholders too. In some projects, the lender even sends their own inspector monthly to monitor progress – these parties care about budget and risk.

- **End-users / Operations stakeholders:** Often overlooked, these are the people who will eventually use or operate the facility once it's built – for example, the owner's operations and maintenance team, or the tenants who will move in. Design or construction decisions could affect how they run the place later. Getting their input early helps prevent delivering a building that doesn't meet the end-users' needs.

Once the list is compiled, the next step is a **stakeholder mapping exercise** to prioritize engagement. A popular method is the Power-Interest Grid – basically a 2x2 matrix mapping each stakeholder's level of power/influence versus their level of interest in the project. This tool quickly indicates how to tailor the stakeholder management effort:

- **High Power, High Interest:** These are the critical stakeholders – typically the client/owner, and perhaps a lead investor or a key government official if it's a public project. These stakeholders must be managed closely. The project manager (or a senior leader) will have frequent meetings or calls with them, involve them in major decisions, and ensure there are no surprises. For example, if a client's representative is deeply invested in every detail, they might even have a desk in the project office to keep communication tight.

- **High Power, Low Interest:** Stakeholders who can greatly influence the project but aren't involved day-to-day. Often, this could be an executive sponsor, a city official, or the CEO of the contracting firm (someone who only wants to hear if things go awry). The strategy here is to keep them satisfied – without burdening them with minutiae, and make sure they hear about any big issues or milestones. For instance, provide monthly executive briefings or invite them to ribbon-cutting events, to keep them in the loop and supportive.

- **Low Power, High Interest:** People who care a lot but don't have much formal power to affect the project. Think of local community members or future end-users like a store owner in a retail development. They can't change the schedule, but their support can be valuable, and their discontent can cause noise. Keep them informed – through newsletters, community meetings, a project website, etc. – so they feel heard and know what's going on, even if they can't direct the project.

- **Low Power, Low Interest:** Folks who are barely affected and have little influence – maybe a distant utility that only has a minor coordination role, or a neighboring business that really doesn't care beyond "just don't block my driveway." These stakeholders typically require only minimal monitoring and engagement as needed.

By mapping stakeholders this way, the engagement strategy can be tailored. For example, if a critical subcontractor is absolutely key to success but seems not very engaged (perhaps they're stretched thin with other projects), the project team needs to elevate their interest – maybe by involving them in early planning sessions, making them feel like a true partner, or structuring performance incentives for on-time delivery. On the flip side, if a stakeholder is very interested but not powerful (like an enthusiastic community member), it's important to have channels to address their concerns without letting them derail decision-making.

## Clarifying Roles with RACI

A highly effective tool to clarify stakeholder roles and avoid confusion is the RACI chart. RACI stands for Responsible, Accountable, Consulted, Informed. It's essentially a matrix mapping key project activities or decisions against key roles, and assigning who is *Responsible* (performs the task), *Accountable* (the final decision-maker or owner of the task), *consulted* (those whose input is sought), and *Informed* (those kept apprised of the outcome) for each item. For example, for the activity "Develop Safety Plan": the Site Safety Manager might be Responsible for drafting it; the Project Manager is Accountable (approves it and owns its implementation); the Superintendent and a subcontractor representative could be Consulted (to ensure the plan is realistic on the ground); and the Owner and all field crews would be Informed once it's finalized.

Doing a RACI exercise for critical processes (design approvals, change order workflows, commissioning, etc.) flushes out ambiguity. If two people both think they're accountable for approving something, that conflict can be resolved by clearly defining roles in the RACI. If the exercise reveals that nobody was assigned as Responsible for a task, that gap can be fixed. RACI charts can even extend beyond the internal team. For instance, in one project's permitting process, the roles were charted as follows: the City Inspector was Responsible for inspections; the general contractor was Accountable for scheduling and facilitating those inspections; a third-party engineer was consulted for any special inspections; and the Owner was informed of all inspection results. Everyone knew the drill – no one could say "I thought you were calling the inspector?" because it was clearly defined in the RACI.

It's not enough to identify stakeholders; it's also important to consider each stakeholder's definition of success. Misalignment of goals can kill a project. A classic scenario: the architect may be dreaming of a stunning design, the contractor is worried about cost and buildability, and the owner might care most about opening by a certain date. If those priorities aren't reconciled, conflict is inevitable. On one project, the owner's operations team wanted a very robust (and expensive) mechanical system for ease of future maintenance, but the owner's finance group wanted to cut costs now. Two factions within the same owner organization had different priorities. The project leadership held a meeting to get everyone to agree on the project's guiding priorities – in that case, they reached a compromise of slightly higher up-front cost in favor of lower life-cycle maintenance. Exercises like this (sometimes formalized as a project charter or mission statement) should be done early on: have each key stakeholder voice their primary concerns (be it safety, quality, budget, schedule, legacy, etc.), and create a shared understanding of what the project is collectively trying to optimize.

Communication is a huge part of stakeholder alignment. As part of stakeholder planning, it's wise to develop a **communication plan:** who gets what information, how often, and via what medium. For example, the client's project manager might get weekly progress reports and a phone call for any major issue; the city council (if it's a public project) might get a monthly update on milestones; subcontractors get daily coordination meetings and weekly look-ahead schedules; the neighborhood might get a quarterly newsletter or have access to a project website with updates. Modern digital collaboration tools help as well – shared platforms where drawings, schedules, and RFIs are accessible to all stakeholders with the appropriate permissions. In one case, a shared online dashboard for stakeholders dramatically reduced "I didn't know about X" problems because everyone could see the latest plans, status updates, and even live camera feeds of the site. The key point is that no critical stakeholder should ever be surprised by information that affects them. If someone is left in the dark, that is a failure in planning.

## Dealing with Stakeholder Conflict

History is full of cautionary tales of stakeholder misalignment. One famous example is the Sydney Opera House saga – an iconic project plagued by a rift between the visionary architect and the government client. They had misaligned expectations: design changes piled up, budgets blew out from $7 million to over $100 million, and the schedule slipped by years. Ultimately, the architect quit and it took a decade longer to finish – a textbook case of stakeholder misalignment harming a project. Closer to home, consider a transit project where the transit authority was fixated on an immovable opening date (time-driven) while the contractors insisted they needed more time to ensure safety and quality after some scope changes. It became a tug-of-war that required high-level mediation to resolve – in the end, the owner relaxed the date slightly, and the contractors added resources, meeting in the middle. The lesson: when two powerful stakeholders have conflicting priorities (e.g. schedule vs. quality), they must come together, acknowledge the conflict, and negotiate a compromise or a realignment of expectations. Project leaders often have to act as honest brokers to facilitate this.

The best strategy, however, is preventative: involve stakeholders early and document decisions and responsibilities clearly. During project kickoff, it can be very effective to hold alignment workshops – bring everyone (owner, designers, key subs, consultants) into one room to hash out major concerns and how the team will work together. Create RACI charts for key processes (as described earlier) and perhaps even have everyone sign a partnering charter committing to shared project goals. Throughout the project, hold regular stakeholder meetings (e.g., weekly Owner-Architect-Contractor meetings, monthly executive committee meetings, etc.) to keep communication lines open. If something is changing – say the owner is considering a scope change – immediately discuss the impacts on each stakeholder (schedule, cost, risk) with all parties together rather than in silos. This transparency builds trust.

Another effective approach is to treat certain third parties as true stakeholders by integrating them into the team. For example, on a complex utility-heavy project, the power company and water authority can be treated as part of the core team – invite them to planning meetings, give them the schedules, and listen to their constraints – rather than only calling them when a conflict arises. This proactive inclusion can head off many delays (for instance, the team can avoid scheduling critical work when a utility crew knows they won't be available). Similarly, engaging the community early – holding town hall meetings to show local residents the construction plan and hear their concerns – can turn an adversary into at least a neutral party, if not a partner. The cost of a few meetings is minor compared to the potential of a lawsuit or a city council complaint if the community feels blindsided or disrespected.

Accountability is the other half of this equation, and it's crucial. It's not enough to map stakeholders and communicate; everyone must also own their part. Creating a culture and structure of accountability means each stakeholder (or team member) is answerable for their commitments. Some of this is enforced through contract mechanisms and project controls – for example, setting clear deliverable dates for the design team in the contract, or including required response times for submittal approvals by the owner's team. Other aspects are more cultural: fostering a sense of shared mission and professional responsibility. Project leaders should encourage an environment where anyone can speak up about an issue, even if it's outside their immediate scope. For instance, telling subcontractors in pre-construction meetings, "If you see a problem that isn't even yours, speak up – we're all accountable for this project's success." This must be backed up by acknowledging and rewarding proactive behavior (and by never punishing the messenger for raising a concern).

On some large projects, collaborative planning techniques (like the Last Planner System from Lean Construction) are used to create peer accountability among stakeholders. In Last Planner meetings, various trade foremen will literally stand up and promise to the group what tasks they will complete in the next week. If they don't meet their commitment, they have to explain to their peers why not. It's amazing how this dynamic – committing in front of fellow trades – motivates reliable promises. While not every project uses formal Lean methods, the principle remains: get commitments out in the open and track them. It's also helpful to keep an "accountability log" for key stakeholder actions – if the city is supposed to issue a permit by X date, or the owner is to deliver owner-furnished equipment by Y date, note it and follow up. No silent assumptions.

Stakeholder mapping and accountability planning ensure "the right people are involved at the right times" and that everyone knows their role in the project's success. By using tools like influence/power grids and RACI charts, and by actively aligning stakeholder goals through communication and early engagement, the costly miscommunications and conflicts that plague so many projects can be drastically reduced. Managing people's expectations and commitments must be treated as carefully as managing concrete and steel. When this is done right, the project runs more smoothly: decisions get made by the right players, potential opposition is addressed before it blows up, and the whole team moves in (roughly) the same direction. It's not easy – it's an ongoing effort throughout the job – but it's absolutely vital for a successful outcome.

## Balancing Time, Cost, and Scope (The Iron Triangle)

Every construction project operates under the shadow of the Iron Triangle – the classic trio of Time, Cost, and Scope. The idea is simple: you can't change one side of the triangle without affecting the others. In theory, this is "Project Management 101," but in the field it becomes a daily juggling act of trade-offs and tough calls. There's an old saying in the industry: "Fast, cheap, or high-quality – pick any two." That gets a laugh, but it's grounded in truth. If an owner suddenly wants to finish sooner (time), we'll probably need more money (cost) or have to cut some corners in scope/quality. If the budget gets slashed, scope might need to be reduced or the schedule extended. You cannot have everything at once – recognizing this early and managing it transparently is key to systematic planning.

## Let's break down the legs of this triangle:

- **Time:** This is the project schedule – deadlines, milestones, the completion date. In construction, time is often literally money because delays can incur liquidated damages or lost revenue for the owner. Also, there's usually a huge push to finish by certain dates (end of the year, tourist season, an announced grand opening, etc.).

- **Cost:** The budget or resources – including money for labor, materials, equipment, overhead, and so on. Contractors commit to deliver a defined scope for an agreed price, and owners have funding limits. Cost also ties into cash flow and financing considerations.

- **Scope:** The work to be done – defined by the plans, specifications, and the client's requirements. Scope also implies the quality or performance level. A change in scope might mean adding a new feature, increasing a performance spec, or conversely removing something from the project.

In a perfect world, we'd set a fixed scope, estimate a reasonable cost, plan a realistic schedule, and then execute exactly as planned. But in reality, changes happen. Perhaps the owner decides to add something (scope up), or unforeseen conditions force more work. Maybe market conditions drive material costs higher, affecting cost. Or delays occur that threaten the schedule. A systematic planner is always assessing the triangle: if one side changes, how do we adjust the other two to maintain a viable project?

## Understanding Trade-offs

To illustrate, here are some common scenarios and how the trade-offs are managed:

- **Scope–Time Trade-off:** Suppose a project is 80% complete and the client requests a significant addition or change (e.g., "Can we also add an extra elevator here?" or "We want to upgrade all the flooring material."). This is an increase in scope. If the completion date is immovable (maybe there's a grand opening event scheduled), then adding this scope will require compressing the remaining schedule – perhaps by adding crews, working overtime, or expediting materials. That will certainly increase cost (and needs to be negotiated as a change). Alternatively, if the budget is fixed and they still want the added scope, the team might plead for more time: "We can do it with regular hours if you give us an extra month." In practice, it might end up being a bit of both: a modest schedule extension plus a modest budget increase to accommodate a late scope change. The main point: added scope requires either more time or more cost (usually both).

- **Time–Cost Trade-off:** Now consider a case where the schedule is being squeezed. Perhaps the project is behind, or the owner suddenly needs an earlier completion. To save time, typical levers are: add more labor or shifts (overtime, second shift, etc.), bring in additional subcontractors, or overlap activities (fast-track the work). All of those usually add cost. More workers mean higher labor costs (with diminishing returns if they start tripping over each other). Overtime can boost productivity up to a point, but sustained overtime can also lower efficiency as people burn out. Sometimes, to hit a date, money is thrown at the problem – for instance, paying bonuses for early completion or premiums for suppliers to fabricate materials faster.

Conversely, if cost must be cut dramatically, one strategy is to slow down the project – a longer schedule with smaller crews to reduce the monthly burn rate (though that may incur other costs like extended supervision, so it's a tricky balance).

- **Scope–Cost Trade-off:** If the project budget is in trouble (say bids came in high or unforeseen issues ate the contingency), one response is to consider scope reduction or "value engineering." This could mean deleting some nice-to-have parts of the project, choosing cheaper materials or methods, or scaling back on quality. The intent is to maintain core functionality, but perhaps at a lower specification. This is common – for instance, substituting a cheaper flooring or lighting fixture to save money. Of course, reducing scope or quality might conflict with the client's expectations, so it requires their buy-in. Another approach, if the scope must be preserved, is to extend time (which owners often hate) because a slower, longer project might cost less per month (no overtime, maybe avoiding expensive winter conditions, etc.). In the end, something has to give: either reduce scope/quality, accept a longer timeline, or find more money.

What's important is transparency in these trade-offs. It helps to frame decisions explicitly in Iron Triangle terms: "*We can certainly add that feature, but it will either push the completion out by six weeks or cost an extra $200k – which is more acceptable?*" By laying it out like that, it forces a prioritization. The worst scenario is when a stakeholder expects to add scope, keep the deadline, and not spend more – that's magical thinking, and the team has to gently but firmly dispel it by referencing the Iron Triangle.

## Real-World Examples of the Iron Triangle in Action:

- **Accelerated Schedule (Time vs. Cost):** A dramatic example from the U.S. was the replacement of the I-35W Mississippi River bridge in Minneapolis after it collapsed in 2007. The public and government needed a new bridge fast. The normal timeline for such a project might be several years, but they set an aggressive target of roughly one year. How was that achieved? By pouring money and resources into it. The state highway department offered up to $27 million in bonuses for early completion – an enormous incentive. The contractor, Flatiron, assembled huge crews (around 400 workers by day and 200 at night) and essentially worked around the clock. In the end, the bridge was finished three months ahead of schedule, and the contractor earned the maximum bonus. That's a classic Time-Cost trade-off: they met the time goal by allowing cost to increase via bonuses and double shifts. (Notably, scope/quality were held constant – they didn't shrink the bridge or compromise specs; they just spent more to do it faster.)

- **Scope Creep and Overruns (Scope vs. Time/Cost):** The Sydney Opera House is also an iconic scope-change nightmare. The project's scope and design ambition kept growing during construction – new design ideas, higher quality demands – and neither time nor cost was adequately adjusted at first. The result was massive schedule overruns and budget explosions. Eventually, reality hit: to finish at all, a lot more money had to be injected and a much later completion accepted. It underscores that if scope changes dramatically, you must re-baseline the triangle; refusing to adjust time or cost will lead to failure.

- **Fast-Tracking (Time vs. Risk/Cost):** Many large projects employ fast-tracking – starting construction before design is fully complete – to shave time off the schedule. For example, foundation work might begin while the upper-level design is still being finalized. This can shorten the overall duration significantly (Time saved), but it raises risk and often costs. The risk is that design changes or coordination issues arise because you're building part of the project with incomplete information. The famous "Big Dig" project in Boston had elements of this – design was evolving even as construction proceeded, leading to costly rework and coordination problems (one of many issues on that project). In one segment of the Big Dig, a tunnel ceiling collapse occurred later, partly because of design/installation flaws with epoxy anchors – arguably, a fast-paced schedule and lack of redundancy contributed to that failure. Fast-tracking is an Iron Triangle decision: you gain Time, but you may pay in Cost (due to rework or inefficiencies) or even Scope/Quality (things might not be as optimal because of rushed coordination). The team can try to mitigate this with intensive coordination and contingency plans, but it's a trade-off to be aware of.

- **Small Project Triangle Example:** Even small projects encounter these issues. In one retail store interior build-out, late in the game, the owner decided they wanted a more elaborate lighting system (scope increase) just a few weeks before the grand opening date (time was fixed by a publicized event). The project team had a frank discussion: either extend the deadline (which wasn't really possible due to the event) or add resources and money to implement the changes faster. The owner ultimately authorized extra costs for expedited materials and overtime for the electricians, and the team managed to meet the original opening day with the new lighting installed – but it cost an extra chunk of money and required some very long nights for the crew. This everyday scenario shows the triangle clearly: the owner insisted on both scope and time, so cost had to flex.

To systematically manage the Iron Triangle, data and structured decision-making are crucial. Tools like Earned Value Management (EVM) provide early warning signs if the project is deviating from time or cost. For instance, the Schedule Performance Index (SPI) and Cost Performance Index (CPI) are useful indicators – if SPI is 0.80 it means only 80% of the planned work is being accomplished in a given time period (the project is falling behind schedule), so something's got to change: either pick up the pace (add resources = more cost) or consider trimming scope or accepting a delay. Similarly, if CPI is below 1.0 (over budget), the team might need to either inject more funds or find cost savings (scope reductions or process efficiencies). Running what-if scenarios is another useful technique. The team can take the schedule and simulate options – for example, "What if we add a second shift to the concrete crew – how much time would that save and what's the added cost?" Or, "What if we defer the landscaping scope to after opening – that saves time; is that acceptable?" By quantifying the impacts, the project leadership can make informed choices.

On large projects, it's wise to maintain formal contingency plans for each side of the triangle. For a time, this might mean keeping a bit of float in the schedule or a list of tasks that could be compressed if needed (sometimes called a schedule "reserve"). For cost, it means keeping a contingency fund (budgeted money) for unforeseen issues or changes. For scope, it can mean having pre-planned alternate deductive options agreed with the owner that could be exercised if needed (like a nice-to-have finish that could be omitted if absolutely necessary to save money). Think of these like shock absorbers: we hope not to use them, but they're there just in case. A classic scenario is near the end of a project when the schedule is tight – the team identifies a non-critical piece of scope that could be done after turnover (maybe a decorative element) and negotiates to push it out, effectively reducing the scope for the initial deadline to ensure opening on time, then completing it later. This actually happened on a hotel project: the rooftop terrace was not going to be 100% ready by opening day, so it was carved out and finished a month later, outside the critical path of opening the hotel. The owner accepted that trade so that revenue could start flowing from guests on schedule.

## Managing Stakeholder Expectations

This circles back to stakeholder alignment – the triangle reality must be communicated constantly. If an owner says, "Under no circumstances can the budget increase," the project leadership should clarify, "Okay, then if something unexpected comes up, we will have to either cut something, extend the schedule, or find other savings, because we have locked one side of the triangle." Conversely, if they say schedule is paramount ("we must be open by Black Friday for retail"), they should be warned that any issues will likely be solved by added cost or perhaps slight scope adjustments, and ensure they have some contingency funds available for that. In government projects, sometimes all three constraints seem fixed (scope by legislation, budget by funding, time by political promise) – in those cases, everyone needs to be braced for the possibility that quality or risk mitigation might suffer unless efficiencies are actively found; otherwise, it can end in a public debacle when something inevitably has to give.

Project teams should adopt a mindset of continuous monitoring of the triangle. Don't wait for a crisis. In regular meetings, it's useful to pose questions such as: "Right now, which of the three – time, cost, scope – is under the most stress? Are we trending behind schedule? Over the cost? Did the client's expectations on scope change?" Depending on the answer, adjustments can be made preemptively. For example, if the project is trending late (time issue), perhaps authorize some strategic overtime now rather than waiting until it's a full-blown crisis. Or if a design development comes back and it's more complex (scope creep), perhaps immediately start value engineering to offset the cost impact.

In essence, balancing the Iron Triangle is a continuous exercise of evaluation and negotiation. Successful construction managers are not those who stick stubbornly to one plan, but those who recognize early when one aspect is slipping and take decisive action to re-balance. The owner should be kept in the loop as well: for example, "We can still hit your date, but it will cost an extra $X – how do you want to proceed?" It's better to have that frank conversation in advance than to blow the date or budget without warning. This transparency also builds trust, as owners see that the team is actively managing trade-offs rather than hiding problems.

One more point: sometimes stakeholders will try to defy the triangle – for instance, insisting on adding scope without more time or money, hoping the contractor will just "find a way." In those moments, the best approach is to present concrete evidence or historical data to show why that's unrealistic. You might cite a reference or an industry rule of thumb, or bring up a past project: "On Project Z, we tried to compress 6 weeks of work into 4 without a budget increase and ended up with a mess and overtime anyway." Using data or past lessons can help convince them that a triangle trade-off isn't negotiable – it's essentially the physics of project management.

To sum up this section: Time, Cost, and Scope are the three primary levers we have, and systematic project planning means constantly steering the project by adjusting these levers in concert. We use tools (like EVM) to see when one lever is off track, communicate options in terms of trade-offs, and document any changes to baselines. The Iron Triangle is not just a theory from an exam – it's a daily reality on the jobsite, one that keeps everyone on their toes. The best outcomes come when the triangle is acknowledged openly and plans are made for its pressures, rather than pretending the constraints don't exist. By doing so, the team maintains control and can still deliver what's needed – maybe not everything at once, but the right balance of scope delivered at an acceptable cost and time.

## Planning for Temporary States

One of the more nuanced – but absolutely critical – aspects of systematic planning is planning for temporary states. What are temporary states? Construction is a process of transformation: we take a concept from drawings to a finished structure. Along that journey, the project passes through transient conditions that the final design never sees. Think about it: a building might be rock-solid when complete, but halfway through construction, it could be a half-framed skeleton vulnerable to wind. A bridge, once finished, can carry heavy traffic, but during erection, one span may still be unsupported and require temporary shoring. These intermediate conditions are often when the project is most at risk. Many experienced field leaders will admit they lose more sleep over what could happen during a critical lift, a scaffold erection, or a formwork pour than they do about the completed structure. The permanent structure has been rigorously engineered – but those temporary setups sometimes get less attention, and that's when accidents happen.

**Definition & Importance:** A temporary state in construction is any condition of the work that is not permanent but is needed to build the permanent structure. Examples abound:

- **Temporary Structures and Supports:** Scaffolding, shoring, formwork, bracing, temporary bridges or platforms, crane mats, cofferdams – all the stuff we build only to facilitate construction. They aren't part of the final building, but if they fail, the final building can't be built (and people can get hurt).

- **Partial Construction States:** When a structure is partially complete, it may not have all its strength yet. Freshly poured concrete is a classic example – until it cures to a required strength, it's an unstable state and relies on forms and shoring. A steel frame might be up, but until all the connections and bracing are done, it's in a precarious state if wind or load comes unexpectedly. An open excavation is another example – before the permanent retaining wall is built, you might have a hole held by temporary shoring; one heavy rain could spell trouble if not planned for.

- **System Testing/Commissioning:** Even turning on systems can create temporary risks. When first firing up an HVAC system for testing, you might bypass some safeties or not have all emergency systems active yet – you're in a temporary state where something could go wrong if it isn't controlled. The same goes for electrical systems being energized for the first time or a pipeline being pressure-tested.

- **Construction Sequencing Quirks:** Sometimes the construction sequence itself creates a temporary load path. For instance, consider removing a temporary support (stripping formwork, or removing a tower crane that's tied into a building) – at that moment, you have to ensure the structure stands on its own. These transitions are critical temporary stages.

Why is planning for these so important? Because many construction failures happen in temporary states, not in the final structure. The final building might be perfectly fine under design loads, but during construction, it might encounter a load it wasn't designed for, or lack a component, leading to collapse. There's a sobering statistic: the majority of major collapses on jobsites occur during erection, demolition, or some temporary condition, rather than after completion. That means builders must engineer not just the final product, but also the process of getting there.

Permanent designs assume everything is in place; for example, a concrete floor design assumes the supporting columns below are in place, and the concrete has cured to full strength. But when you're pouring that slab, none of those assumptions hold; the concrete is wet and green, relying on plywood and shores for support. The project team must plan for that scenario with the same rigor as an engineer designing the final building.

## Lessons from Tragedies

A clear modern example is the **Florida International University (FIU) pedestrian bridge collapse in 2018**. The bridge span had been placed, but the project was still in a temporary structural state. Significant cracking and distress were observed in the days and hours leading up to the collapse, yet traffic beneath the bridge remained open, and no temporary stabilization measures were implemented. When the structure failed, six people were killed. Investigations later showed that while the final structure may have met design intent once fully completed, the *temporary condition*—with incomplete load paths and unresolved distress—was dangerously unstable. The failure was not just an engineering issue, but a planning failure: the project lacked predefined criteria for when visible distress in a temporary state should trigger immediate load reduction, stabilization, or shutdown. The lesson is direct: temporary states require explicit decision thresholds and mandatory actions. If the plan does not define when work must stop and conditions must be stabilized, teams will default to optimism and momentum. That is not a safety system.

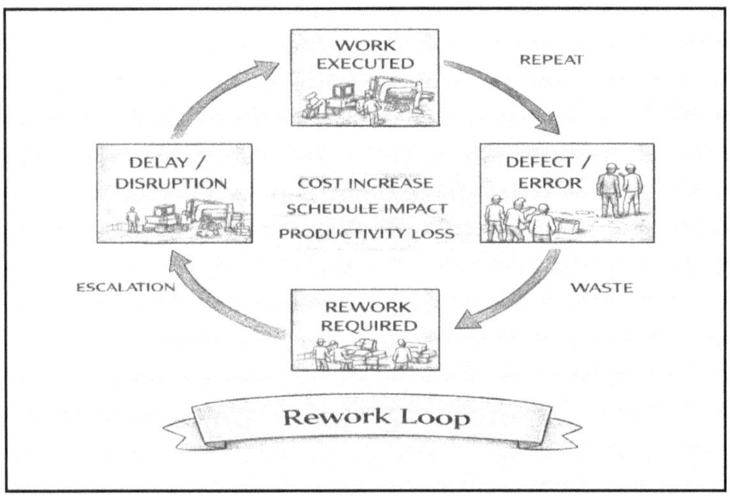


Another example comes from **formwork and temporary deck failures**, which continue to be a recurring source of fatal incidents. In **West Palm Beach, Florida, in 2016**, a formwork table collapsed during installation on a high-rise project, resulting in a fatal fall. The formwork system itself was intended to support loads safely once fully installed, but the failure occurred during a transitional phase—when the table was being positioned and not yet fully secured. As with many temporary-state incidents, the work was treated as routine rather than critical. The temporary configuration lacked redundancy, and the installation sequence did not adequately account for partial support conditions. The permanent concrete slab that the formwork was designed to support would have been stable once complete, but the temporary arrangement during installation was highly vulnerable. This incident reinforced an industry lesson that temporary works—formwork, scaffolding, shoring, and platforms—must be planned and engineered for every stage of use, not just their final configuration. Temporary does not mean informal, and installation phases often represent the highest risk period of all.

A third example illustrates how even common operations can become deadly when temporary load conditions are underestimated. In **New York in 2013**, a formwork collapse occurred during active concrete placement, killing two workers. The formwork system failed under the combined weight of wet concrete, placement sequencing, vibration, and worker movement. The permanent structure, once cured, would have been capable of carrying the design loads without issue. However, during the temporary state of concrete placement, the system experienced its most severe loading condition. Investigators found that the temporary works were not adequately braced for live placement loads and that changes made in the field altered load paths without engineering review. The lesson here is one the industry has learned repeatedly: wet concrete is not a static condition, and concrete placement is a critical temporary state that demands the same rigor as permanent structural design. Pour rates, lift heights, bracing, and inspection hold points must be planned explicitly. Once the pour begins, there is no margin for improvisation.

These events underline a consistent and uncomfortable truth: many construction failures do not occur because the final design was flawed, but because the **temporary condition was never fully planned, stabilized, or controlled**. The finished structure might have performed exactly as intended, but the project never safely reached that final state. We should not need tragedies to reinforce this lesson, yet history shows that temporary states—partial construction, installation phases, transitions, and live-load conditions—remain one of the most dangerous aspects of construction operations. Systematic planning must treat these states as engineered conditions with defined limits, responsibilities, and stop-work criteria. Temporary does not mean less important; in many cases, it is when the project is most at risk.

## Planning for Temporary States – What to Do
## Systematic planning for temporary states involves a few key actions:

- **Identify all critical temporary conditions in advance**. As the project plan and construction sequence are developed, deliberately flag points where the structure or workers will be at risk due to incomplete work or temporary support. This could be listed out: for example, erecting steel before final bracing is installed; an excavation open 15 feet deep until the permanent wall is built; using a temporary support beam to jack bridge girders; scaffolding for masonry work 5 stories high; a power switchover where backup systems are offline, etc. These should be listed and become focus points in planning meetings.

- **Engineer the temporary works.** For each identified temporary state, ensure there is an engineering design or calculation behind it. If a big concrete slab pour is planned, get a formwork engineer to specify the shoring layout, allowable loads, stripping sequence, etc. (Most jurisdictions now require engineered shoring/formwork plans for significant pours[28].) If doing a heavy lift with a crane, develop a rigging plan and use crane capacity charts that show safe loads with a margin. For an excavation, have a trench shoring design or a stamped sheet pile plan. The rule is: treat temporary structures with the same seriousness as permanent ones. That means proper design, review, and sometimes independent peer review. Bringing in specialty engineers can be very worthwhile – e.g., use a scaffold design specialist for an unusual scaffold configuration, or engage the crane supplier's engineer to help develop a critical lift plan. It's money well spent.

- **Procedures and Training:** Having a plan on paper isn't enough if the crew in the field doesn't follow it or understand it. Create method statements or step-by-step procedures for critical operations: how to erect and dismantle that scaffold, how to sequence the removal of shores, how to perform the crane lift (who communicates, who can abort the lift, weather limits, etc.). Then brief the team. Before a major temporary-state operation, the responsible supervisors should hold a crew talk: "Here's what we're about to do, here are the hold points, here's everyone's role, and here's what absolutely cannot be done (e.g., do not remove any brace without approval)." Also, ensure only competent, trained personnel handle these tasks. No one without experience should lead a high-risk temporary operation – if the crew lacks that expertise, bring in people who have it or get them properly trained under supervision.

- **Inspection and Verification:** Temporary works often need more frequent checks than permanent works because they can change or degrade quickly. Implement inspections – e.g., use a scaffold tag system where a competent person signs off daily that a scaffold is still sound, or have an engineer check trench shoring after heavy rain. Many codes require special inspections for things like bolting and welding during construction, which covers some temporary scenarios too (like checking that rebar and forms are correct before a concrete pour). Assign clear responsibility for these checks: who inspects the formwork before each pour? Who monitors the loads if a jack is applying force? Consider installing monitoring devices – e.g., load cells on a temporary jack to ensure it's not overloaded, or crack gauges on a partially demolished structure to see if it's moving. The goal is to catch any sign of distress in a temporary state before it leads to failure.

- **Build in Redundancy where feasible:** Often, temporary setups lack the redundancy of a finished building – if one support fails, the whole thing can go. Where possible, add a backup or safety net. For example, when using hydraulic jacks to lift a structure (like in a lift-slab or bridge lifting operation), put temporary cribbing stacks as a backup under the element being lifted so that if a jack fails, the member only drops an inch onto the cribbing, not the whole way. Another example: if steel is erected and left half-installed overnight, consider adding an extra cable brace as a backup in case a connection slips. In the Big Dig tunnel ceiling collapse, part of the issue was that there was no secondary support – heavy ceiling panels were held only by epoxy glue anchors, with no safety hanger wires, so when a few anchors failed, all the panels fell. Now, some designs add a secondary restraint to heavy elements if failure would be life-critical. Construction teams should apply the same logic: add secondary supports or fallbacks in temporary states wherever feasible.

- **Plan for Worst-Case and External Events:** Ask "What could make this temporary state go south?" and plan controls for those factors. Common ones are weather and human error. If a structure is half-built and a wind storm hits, could it topple? Have a wind action plan: for example, if forecast winds are> 40 mph, stop work and maybe even lower or secure cranes and add supports to any incomplete structures. If heavy rain is expected and there's an open excavation, ensure extra pumps are on standby and keep workers out of the pit until stability is assessed. Essentially, treat Mother Nature as a stakeholder too – one that can really mess with temporary conditions if we're not ready. Human error is addressed by extra supervision on critical operations and crystal-clear instructions (e.g., "Do not remove any formwork until concrete is at least X psi and approved by the engineer" – and actually test the concrete to confirm strength before stripping forms). Sometimes, just scheduling risky operations at the right time helps: avoid doing a major structural switch or heavy lift at the end of a long day when crews are tired. If it's critical, schedule it first thing in the morning when everyone is fresh and focused.

The industry has been improving on this front. Many companies now have dedicated Temporary Works engineers or require detailed Construction Engineering plans for high-risk activities. It used to be that contractors just "figured it out" in the field with timbers and nails; now, especially for large projects, it's expected to have engineered plans for all significant temporary structures. Checklists and internal permit systems are often used – for example, some projects require an internal permit before removing shoring or performing a critical lift, meaning a written permit must be signed by the responsible engineer and manager. This forces a deliberate review: "Have all criteria been met? Is everyone aware of the plan?"

**Temporary States and Safety Culture:** Beyond the technical planning, there's a cultural aspect. Every worker must understand that temporary does not mean lower standards. Just because a scaffold or trench is temporary doesn't mean anyone can be casual about it. Supervisors should tell crews: "That scaffold is as important as the building's frame – your life might depend on it." If any worker sees a makeshift support or something that doesn't look right, they are encouraged (in fact, required) to speak up, and work is halted and fixed. Many accidents happen when someone says, "Eh, it's just for now, it'll be fine." Part of systematic planning is battling that complacency. One effective practice is targeted training and toolbox talks specifically on past failures. Showing teams the photos and recounting the stories of Willow Island, L'Ambiance Plaza, etc., helps drive the point home that shortcuts in temporary stages kill. It's sobering, but it changes attitudes.

Technology can also help. Building Information Modeling (BIM) in 4D (time-sequenced) can be used to simulate construction stages. Animating the erection sequence allows the team to visualize what the structure looks like at each week of construction. This often helps spot "oops" moments – for example, realizing, "Wait, at week 20 that big precast beam will be in place, but the adjacent span won't be connected yet – is it stable by itself?" If not, the plan can include temporary support at that stage. On one high-rise project, 4D modeling was used to plan how a partially built concrete core would be affected by winds – the model showed that temporary braces were needed until the adjacent floor slabs were poured, so those braces were designed and installed. Catching that on the computer before it became an emergency in the field was invaluable.

Just to highlight, some of the critical temporary states that teams should always plan for include: heavy lifts (using cranes at high capacity – detailed rigging plans and weather protocols are a must), formwork/shoring for big pours (ensure the design can handle the weight of wet concrete, placement sequence, etc.), erecting tall prefabricated elements (ensure they are braced until permanent connections are in place), switchover operations (like transferring load from an old to a new structure during a renovation – plan the sequence so the load is always supported by something), and equipment assembly/disassembly (e.g. dismantling a tower crane – follow the manufacturer's procedures to the letter).

Planning for temporary states is about anticipating the "in-between" stages of construction and making sure those stages are as safe and controlled as the final state. It requires engineering know-how, yes, but also a mindset of vigilance: never assume "it's just temporary, it'll hold." Instead, engineer it, follow procedures, add safety factors, double-check it, and treat any deviation as a big deal. The painful lessons from the past – collapsing scaffolds, cranes, formwork – all say one thing: don't leave temporary conditions to chance. By being systematic in this regard, we protect our workers' lives and ultimately ensure the project can be completed successfully. After all, a project isn't truly successful if it collapses halfway, even if the final design was brilliant. The project is only a success if it reaches the finish line without catastrophe en route. Plan every stage and build it right at every stage, or don't build it at all.

## Checklist: Systematic Planning in the Field – Key Questions for Project Leaders

To wrap up, here's a checklist that can serve as a litmus test for whether the planning has been done thoroughly. As a project leader, ask yourself and your team these questions:

- **WBS & Scope Control:** Is our Work Breakdown Structure complete, clear, and understood by all? Does every team (design, construction, procurement) align their work plans to the WBS so that nothing falls through the cracks? Are we updating the WBS when changes in scope occur (so new work is accounted for and tracked)?

- **Stakeholder Alignment:** Have we identified all critical stakeholders, including those who aren't immediately obvious (regulators, community groups, future users)? Do we understand each stakeholder's interests or concerns? What's our plan to keep each stakeholder appropriately engaged or informed – e.g., regular meetings, reports, RACI assignments? Is it crystal clear who the decision-makers are for key issues, so we avoid confusion or paralysis?

- **Iron Triangle Monitoring:** What are the project's driving priorities – has the owner emphasized whether time, cost, or scope/quality is most critical if trade-offs are needed? Given that, do we have contingencies for the other two sides of the triangle (e.g., schedule float, budget contingency funds, scope alternates)? How are we tracking progress – will we recognize early if we're drifting off schedule or budget (using tools like EV metrics, variance reports)? And do we have a forum set with the owner to make quick decisions on trade-offs if, say, a delay or cost overrun looms?

- **Temporary Works & Critical Operations:** Have we listed out all high-risk construction activities and temporary conditions (e.g., major lifts, working at heights, critical concrete pours, structural swaps or tie-ins, demolition sequences)? For each, is there a documented plan or method statement? Have those plans been reviewed by a qualified engineer where needed? Are the responsible persons clearly identified and trained, and have we done a briefing with the crew for each critical operation? Do we have hold points in the schedule where we pause to inspect or sign off before proceeding (for example, do not remove shoring until concrete strength is verified by tests, or do not energize a system until lockout checks are done)? And do all workers know that they must never deviate from these plans or improvise without approval?

If you can confidently answer these questions, you likely have a strong systematic plan in place. If not, take the time to shore up those areas before you're in the thick of execution.

Systematic project planning might sound bureaucratic to some, but it is the very thing that enables a team to execute efficiently and safely. When the work is broken into well-defined pieces, everyone can see the roadmap. When stakeholders are aligned and clear accountability is assigned, the friction and fights that derail projects are minimized. By minding the Iron Triangle, the team can respond to change proactively instead of reacting with panic. And by planning for temporary states, the construction process can be navigated without falling victim to its inherent dangers.

A seasoned construction leader can often walk onto a site and tell within minutes whether a project was planned systematically or not. The signs are obvious: a well-planned project has an air of organized activity – crews know what they're doing that week (and why), materials show up when needed, problems are anticipated and met with Plan B or C rather than blank stares. In contrast, a poorly planned project is pure chaos – people scrambling, constant surprises, blame games among stakeholders, and often some safety close-calls to boot.

The principles in this chapter are not abstract; they're born from blood, sweat, and sometimes tears in the field. Many experienced managers have had their share of 3 A.M. emergency calls on projects that went sideways due to a lack of planning. The hope is that the next generation of field leaders can avoid those painful lessons by embracing a systematic approach from the start. Planning is not a one-time checkbox; it's a continuous discipline. Plans are refined as the project progresses, and teams stay proactive rather than reactive. When surprises come (and they will), a systematically planned project already has the structure and processes in place to absorb the shock – whether it's a sudden design change, an unexpected site condition, or a freak weather event during a critical pour.

Ultimately, systematic project planning is about making complex projects predictable and safe. It's about breaking a massive endeavor into controlled segments, aligning every person from the owner to the apprentices with the project's objectives, keeping the trade-offs transparent and managed, and never leaving critical transitions to luck. When this is done, what could be chaos is turned into something akin to an orchestra – different sections playing in harmony according to a well-composed score. That is how lasting projects are built: not by heroics and reacting to fires, but by diligent, disciplined planning and leadership that guides the team through every twist and turn toward success.

**Operational Takeaways for Today's Job:**

- Break work into clear, manageable chunks for your crew. At the start of the shift, outline the day's tasks in steps (a mini work breakdown) and assign specific people to each. Clarity in who does what by when reduces confusion and overlap.
- Identify your key constraints for the day (time, critical equipment, or limited crew) and communicate them. For example, if a concrete pour must be finished by noon due to the temperature, make sure everyone knows the hard deadline and plans work around it. Balancing time, cost, and scope on a micro scale keeps the project on track.
- Use a RACI mindset informally. Ensure everyone knows who is Responsible, who must approve, who to consult, and who to inform for today's activities (even if you don't call it RACI out loud). This prevents tasks from stalling due to "I thought someone else was handling that" syndrome.
- Plan for temporary conditions and transitions. If you're setting up something temporary (scaffolding, traffic control, trench support), have a plan for how it will eventually come down or change as work progresses. Likewise, address any handoffs (such as day-to-night shifts) with a briefing so nothing falls through the cracks.
- Manage stakeholder expectations in real time. If an inspector or owner's rep is visiting today, prep the site and crew on what they'll want to see. If a delay happens, inform the project manager or client early. Proactive communication of changes or issues prevents surprises and builds trust.

# Chapter 8

# Human Factors in Construction Operations

## Abnormal Event Critique

When something goes very wrong on a construction site – or even almost goes wrong – the response in the aftermath is crucial. Industry experience shows that a project's culture is revealed in how it critiques and learns from such incidents. One stark example is the April 2019 collapse of a tower crane in Seattle during disassembly. The 278-foot crane toppled while being dismantled, killing four people. Investigators later determined that crews had not followed proper procedures – they removed critical pins too early – and a sudden wind gust exploited this error, causing the crane to collapse. Importantly, the investigation did not pin the tragedy on one individual; instead, it identified multiple contributing factors: shortcuts in the disassembly process, inadequate supervision, and failure to account for high winds. The lesson is clear: major accidents are almost always the culmination of several compounding issues. Safety is the responsibility of everybody, and finding out "why" requires digging into all underlying causes, rather than just blaming the last person who touched something.

In a healthy project environment, an abnormal event is treated as an urgent learning opportunity. That means conducting a blameless but penetrating review. For example, after an incident, the team might gather to reconstruct the timeline of what happened and ask systematically: How did this slip through? What defenses failed? What can we change to prevent it? A common method is the "Five Whys" technique – asking why an event occurred, then asking why for each underlying cause, and so on, to peel back layers of causation. For instance, if a worker slips and nearly falls, ask why they slipped (was there a spill? improper footwear? rushing?), then ask why that condition existed (was housekeeping poor? training insufficient? a deadline pushing them to move too fast?), and continue digging. This process often branches into multiple factors rather than a single linear cause. In other words, there is rarely a single cause – it's usually a system of causes. Perhaps that worker slipped because a spill wasn't cleaned up, and they weren't wearing proper boots, and they felt hurried by a supervisor – all of those factors would need addressing.

Unfortunately, not every organization handles abnormal events in a productive way. There have been instances where a serious near-miss – say a suspended load swinging out of control and nearly hitting two workers – was essentially swept under the rug by a supervisor eager to "just move on." If no injury occurred, they figure there's nothing to report. That is exactly the wrong approach. As safety professionals often say, a near-miss is an accident that was narrowly avoided; it should be treated like a free lesson. The National Safety Council defines a near miss as "an unplanned event that did not result in injury, illness, or damage – but had the potential to do so." Because no one is hurt, it's easy for a crew to breathe a sigh of relief, say "well, that was close," and never speak of it again. But if the culture doesn't actively encourage reporting near misses, they tend to get swept under the rug as just an "oops." A strong safety culture pushes the opposite: report every near-miss, every anomaly, and analyze them just like actual accidents. By doing that, you can fix problems while they're still only warning signs, rather than waiting for a fatal outcome.

Indeed, accident investigations often find that workers were "surprised" an accident happened because "we've done it this way many times without trouble," or the exposure was very brief. This hindsight bias – it didn't seem dangerous until it caused a disaster – underscores why proactive critiques are needed. As one former OSHA investigator noted, it's not unusual to investigate a fatality that is very much like one that occurred not long before. In other words, the same patterns repeat when lessons learned aren't truly learned. If a formwork collapse in Texas looks just like one that happened a year earlier in California, it indicates the industry failed to share and act on knowledge from that first event. Thus, an abnormal event critique shouldn't just end with a report that sits in a file; its findings must be communicated broadly across the company and even the industry.

Let's consider another example of learning (or failing to learn) from an abnormal event. In 2015, a crew in Cincinnati was demolishing an old overpass bridge at night. Partway through, they noticed some steel girders had begun lifting off their supports unexpectedly, indicating potential instability. They wisely stopped work and brought in engineers, who recommended some remedial measures and then gave the go-ahead to continue demolition. Unfortunately, those fixes proved inadequate. Shortly after work resumed, the span suddenly collapsed onto Interstate 75 below, killing one worker and injuring another. The post-incident analysis found that while the team did recognize an abnormal condition (the shifting girders) and attempted a fix, the underlying instability wasn't fully understood or mitigated. This tragedy reinforced an important lesson: when you encounter an abnormal sign – even if you think you addressed it – double-check and err on the side of caution. In this case, a thorough critique afterward led the contractor and others to tighten procedures for demolitions, such as mandating a higher-level engineering review whenever something unexpected occurs mid-operation.

In summary, critiquing abnormal events means fostering an environment where problems are investigated relentlessly, and lessons are shared, rather than hiding mistakes or rushing back to work. It's human nature to want to put a bad incident behind us, but as construction leaders, we have to resist that impulse. Whether it's a dramatic crane collapse that makes national news or a near-miss nobody else hears about, the approach should be the same: gather the facts, identify all contributing factors (technical and human), and implement changes to prevent a recurrence. Done right, an incident review is not about assigning blame – it's about driving improvement. The only inexcusable mistake is the one we refuse to learn from. Treat every abnormal event as a critical learning opportunity, and the industry moves that much closer to an injury-free, failure-free jobsite.

## Shift Turnovers in Construction

Construction isn't always a 24-hour operation, but when it is, shift handovers become a potential weak link in the process. Even on single-shift projects, there are handoffs between crews or supervisors from one day to the next. A poorly executed shift turnover can lead directly to safety incidents, costly rework, or lost productivity. It's comparable to a relay race baton pass – a fumble in that handoff can derail the entire effort.

Other industries have famously suffered disasters due to poor shift communication. The 1988 Piper Alpha offshore platform explosion (which killed 167 workers) is one example: investigators found that a critical valve status was miscommunicated during a shift change, contributing to the catastrophe. In the chemical industry, the 2005 BP Texas City refinery explosion (which caused 15 deaths) provides a similar lesson – important steps were missed over the course of multiple shift turnovers during a unit startup, allowing dangerous conditions to build unnoticed. The takeaway for all of us is that whenever one crew steps in for another, there must be a clear and reliable transfer of information. Construction may not always run around the clock like a refinery, but the same risk exists: if the night shift doesn't tell the day shift that they removed bracing on a formwork, the day crew might overload it. Or if a trench was left unshored because work stopped at 3 AM, the morning crew might unknowingly send people into an unsafe trench. These handover slip-ups can be deadly.

So what does a good shift turnover look like in construction? In a word: deliberate. It should be a structured and unhurried process – not a rushed shout in the parking lot like "We're all set, good luck!" For projects that run multiple shifts, a best practice is to have a brief overlap meeting between incoming and outgoing supervisors (and key crew leaders if possible). Face-to-face communication is ideal whenever feasible. The outgoing team should prepare by jotting down vital information during their shift – equipment status, any incidents, unfinished tasks, measurements, and so on – so they're not relying on memory at the end of a long shift. The incoming supervisor should actively ask questions and, if possible, conduct a quick site walk with the outgoing supervisor to assess the status of the work in person.

This two-way exchange ensures responsibility truly transfers with understanding. Human factors experts note that effective handovers include preparation, a dialogue to exchange information, and a verification by the incoming person that they've understood everything. It's wise for the new shift to repeat critical points back (for example, "So those circuits are locked out and tagged at points A, B, and C, correct?") to catch any miscommunication before it causes harm.

It also helps standardize the information communicated at turnover. Some companies use checklists or digital log systems for shift handovers to ensure consistency. In construction, especially on fast-track projects, a checklist might include items like: safety issues (open permits, barricaded zones, any injuries or near misses from the previous shift), quality issues (any work that needs inspection or rework), progress updates (what was completed and what tasks remain in progress), equipment status (e.g. "Scissor lift #5 has a bad wheel – do not use"), and manpower or logistics notes ("the steel crew will start two hours late," or "delivery X didn't arrive and is rescheduled for today"). Using a written log or a dedicated app adds discipline and consistency to this process. In fact, industry analysis has found that poor shift handovers can cost operations tens of thousands of dollars per incident (due to safety hazards being missed or work needing to be redone), whereas organizations that implemented a digital handover system reduced miscommunication incidents by over 70% and significantly cut down equipment downtime. Those hard numbers demonstrate the value of doing shift turnovers right.

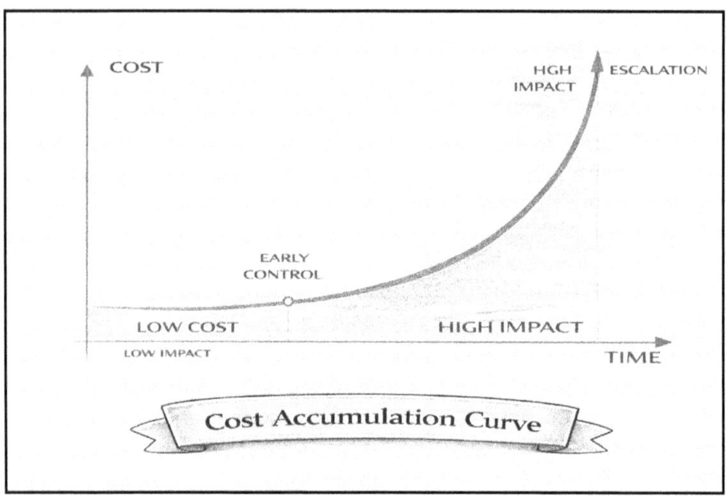

Cost Accumulation Curve

It's also important to identify particularly high-risk handovers in advance. Certain situations deserve extra caution at shift change – for example, transferring responsibility for a critical lift midway, handing over an active confined-space entry, or switching out crane operators in the middle of an operation. If a task is especially hazardous or sensitive, it's best to avoid doing it across a shift change at all; better to start and finish it on one shift if possible. But if a handover during such a task is unavoidable, plan a more extended joint briefing for it. The UK's Health and Safety Executive guidance on shift handovers emphasizes planning jobs so they don't bridge shifts whenever possible. In construction terms, that might mean not leaving the tensioning of a cable or the final bolting of a steel connection for the next crew – either complete the critical portion before the shift ends or get it to a clearly safe stopping point (with proper tags or markings...either complete the critical portion or at least get it to a clearly safe stopping point (with proper tags or markings) and ensure the next crew is fully aware of its status.

For example, consider how good and bad turnovers play out in practice. On a tunneling project that ran two 10-hour shifts, the team established a practice in which the night-shift foreman would fill out a simple written pass-down report and meet the day-shift foreman in person at the tunnel portal at the end of the shift. One night, the tunneling crew noticed unusual groundwater seepage at the heading – not a catastrophic amount, but more water than usual was coming through a fissure. They logged it and emphasized it during the morning handover. Because of that heads-up, the day shift immediately mobilized a grouting team and extra pumps before continuing excavation, and sure enough, the seepage increased later that day. Thanks to the thorough handover, a major inundation or work stoppage was avoided. Had the night crew not diligently passed on the warning, the day crew might have encountered a flood unprepared.

On the flip side, consider what happens when shift handover is handled carelessly. On a road project, an overnight crew removed some traffic barricades earlier than scheduled (to allow equipment through) but failed to inform the incoming crew. Come morning, the day crew assumed the lane closure was still in effect and nearly sent a team to work in what was now an open lane with live traffic. In reality, vehicles had already started coming through. The error was caught just in time, and the road was quickly re-closed, but it was a dangerously close call. The root cause was simply one crew failing to relay a critical change to the next. That near-miss was a wake-up call, and afterward the project instituted a policy that any change to site conditions (like barrier placements or power lockouts) must be written on a whiteboard in the site office and verbally communicated at every handover.

In short, smooth shift turnovers are a critical human-factor practice in construction. They require a bit of discipline and time – which might feel tedious at 6:00 AM or after a long night – but the cost of not doing them right can be immense. Many accidents and close calls ultimately trace back to someone not knowing what the previous shift did (or didn't do). A good rule to instill in crews is that if you're handing off a job, you still "own" that job until you've properly passed the baton to the next team. And if you're the incoming crew, never assume everything is okay – verify it for yourself. Those few extra minutes of communication at shift change are an investment that pays back in safety, efficiency, and peace of mind for everyone on the job.

## Night Shift Operations and Human Factors

Night work on a construction site is a different world. The jobsite at 2:00 AM is replaced by darkness and a hum of generators. The floodlights cast long shadows, and my crew moved carefully, aware that every step carried extra risk. As a construction leader, I learned early that night shift operations require heightened vigilance and thoughtful planning. In this chapter, we explore the unique challenges of running construction projects after dark and how to lead teams safely through them. Night shifts are often unavoidable – whether repaving highways while traffic is light or accelerating a project schedule – but they demand special attention to human factors. From limited visibility and worker fatigue to communication gaps and slower emergency response, night work tests a leader's foresight and commitment to safety.

Night shift construction is common in the U.S. – nearly 15 million Americans work evening or night shifts– and it offers advantages like reduced traffic and cooler temperatures. However, it comes with its own set of hazards that can turn routine tasks into high-risk operations. In my experience, success on the night shift begins with understanding these constraints and proactively addressing them in both planning and training. The following sections draw on real-world cases, industry research, and lessons learned in the field to illuminate best practices for safe and efficient night operations.

## Visibility Challenges and Increased Accident Risk

Perhaps the most obvious challenge of night work is limited visibility. Humans evolved to work in daylight; at night, even with artificial lighting, our depth perception and peripheral vision are diminished. Critical visual cues on a jobsite – a slight slope in the ground, a moving piece of equipment, a warning sign – can be missed or misjudged in the dark. This is one reason accident rates tend to be higher on night shifts. In fact, OSHA reports that injury and error rates are about 30% higher at night than on day shifts. Similarly, a CDC analysis found that accidents and errors increase by 28% on night shifts versus regular daytime work. The combination of darkness and artificial lighting can create glare and shadows that hide hazards. I've seen workers trip over materials that were plainly visible in daylight or nearly get struck by vehicles whose operators didn't spot them in time – all because of poor lighting conditions.

Proper job-site illumination is therefore a cornerstone of night-shift safety. U.S. regulations mandate minimum light levels; for example, OSHA requires at least 5 foot-candles of illumination in general construction areas at night (with higher levels for detail work or underground operations). In practice, this means deploying portable light towers, string lights, and task lighting to cover all active work zones. A detailed lighting plan should be developed before any night work begins. Such a plan identifies where to place lights to eliminate dark spots while minimizing glare for workers and passing traffic. It's a tricky balance: too little light increases accident risk, but poorly positioned, overly bright lights can blind workers or drivers. Some projects use balloon lights or glare-diffused LED floodlights to create more uniform lighting without harsh beams. In roadway construction, contractors often stagger light plants and use shielding to prevent oncoming drivers from being blinded, while ensuring crews are clearly visible.

Real-world incidents underscore the importance of visibility. One notable case occurred in a 2025 industrial project at an LNG plant in Texas: a crew was working on an elevated formwork system at 1:44 AM when the formwork collapsed, killing three workers. The contractor's investigation found that reduced visibility at night contributed to the crew not noticing a critical misalignment in the formwork. In response, the company immediately suspended all complex formwork "jump" operations at night, recognizing that verification of safety-critical connections is much harder under artificial light. This tragedy forced a policy change – certain high-risk tasks would only be done in daylight going forward. The lesson for us is clear: if a job requires extremely fine visual inspection or precise work, consider scheduling it during daylight whenever possible, or enhance lighting and double-checks at night. As one state DOT survey found, "poor visibility" was the number one problem cited with night work, and agencies stressed the need for detailed lighting plans and improved visibility of workers as key solutions.

In practice, every night shift should start with verifying that all lights are working and optimally placed. Crews should wear high-visibility PPE with retroreflective strips, so they "light up" in vehicle headlights or tower lights. Many contractors issue extra reflective vests or even LED-lighted armbands for night crews (indeed, nearly all contractors in one survey said they require additional reflective clothing at night). Paint and signage used at night jobs are often reflective or illuminated. These measures collectively help address the visibility gap, but the hazard can never be fully eliminated, which is why a night shift demands a more defensive mindset from workers and supervisors alike. "Expect to be unseen" is a rule I drill into night crews, meaning assume that others may not see you and act accordingly (such as making eye contact with equipment operators and never turning your back to oncoming traffic).

## Reduced Staffing, Skills, and Supervision

Running a project after dark often means fewer people on site – and not just fewer workers, but frequently fewer managers, support staff, and oversight personnel. This can lead to reduced availability of skilled workers and supervisors on the night shift, a challenge that has both practical and safety implications. Often, the "A-team" – the most experienced foremen, engineers, or craft workers – works during the day. At night, you might have a leaner crew, sometimes with less tenure, and a supervisor juggling multiple roles. In my career, I've seen night crews where one supervisor had to oversee several sub-teams spread over a large area, something we would never attempt during a busy day shift. Fewer hands and eyes on the job make it easier for mistakes to go unchecked.

There is also an industry skills gap at night due to worker preferences. Many seasoned workers are reluctant to take on graveyard shifts, so contractors may rely on younger or less experienced personnel for overnight work (or offer higher pay to entice veterans). Additionally, certain trades or subcontractors might not operate at night, limiting the pool of available expertise if an unexpected issue arises. For example, if a specialized welder or a surveyor is needed to solve a problem at 3 AM, you might simply be out of luck until morning. This can tempt crews to improvise fixes without the right expertise, which can introduce errors.

Another aspect is that supporting resources are scarcer. Site safety officers, quality inspectors, maintenance mechanics – often these roles are only lightly staffed at night, if at all. On one project, our night shift operated without an on-site safety manager (we had one on call by phone). Looking back, that was a mistake; we missed having that dedicated safety oversight when a series of near-misses occurred. The breakdowns in oversight and supervision during night operations have been cited as contributing factors in accidents. For instance, the internal review of the LNG project incident found that "moments to step in and take corrective action were missed" due to lapses in oversight, indicating supervision was not as effective during the night shift. A strong safety culture needs to extend through all shifts, but it can "go to sleep" at night if leaders aren't careful.

What can we do about this? First, it's crucial to staff the night shift with capable leadership and adequate crew levels, not just the minimum to get the work done. If the work is high-risk, consider having a project manager or senior superintendent also rotate onto nights periodically to support the night superintendent – a second set of experienced eyes. In some cases, owners or agencies require that a job's key personnel (like the site safety officer or a project engineer) be present for all night operations. Where that's not mandated, we as leaders should still evaluate whether we need to adjust our staffing plans. Remember that fatigue will reduce the effective manpower as well (a crew of 10 at night might only get the output of 8, for example, due to slower pace and breaks – see more on fatigue below). So plan crew sizes accordingly or break tasks into smaller steps.

Another best practice is to pair less experienced night crew members with veterans for mentoring, especially at the outset of a project. Some companies will rotate day-shift veterans onto nights for the first week or two of a night schedule to help train the permanent night crew on the unique hazards and workflow. Additionally, offer incentives or differentials for skilled workers to work nights, so you're not relying only on whoever is willing. And ensure your supervisors at night are well trained – often the night foreman is a step below the day foreman in seniority, so invest in their leadership training. They should be empowered to slow or stop work if conditions aren't safe, since senior management might not be on site immediately.

Finally, acknowledge human limits. One supervisor cannot manage the same span of control at night as they might in daylight. Communication is tougher (more on that next), and the supervisor might be tired too. Therefore, adjust supervisor-to-worker ratios on night shifts – you may need more supervisors or shorter inspection rounds. The goal is to prevent scenarios where an overstretched night supervisor overlooks a brewing problem.

Consider the safety risk of "lone work" as well; avoid having individuals working in isolation at night with no nearby help or oversight. Implement a buddy system or regular check-ins if lone work can't be avoided. In summary, ensure your night team has the right people and enough people to do the job safely, even if that means spending a bit more – it's a worthy investment to prevent accidents.

## Fatigue, Circadian Rhythm, and Productivity Loss

Fatigue is the constant companion of night shift workers. No matter how tough we think we are, humans are biologically wired to sleep at night. Working through the small hours disrupts our circadian rhythm – the body's internal clock – leading to reduced alertness and slower reaction times. Research confirms what any bleary-eyed night worker knows: night shifts dramatically increase the risk of errors and accidents due to fatigue. The Occupational Safety and Health Administration bluntly states that non-traditional night schedules "intensify fatigue and reduce alertness" and that workers "generally will not fully acclimate to night work". In fact, even after many years of nights, most people never achieve the same level of cognitive sharpness at 3 AM as they have at 3 PM. Studies suggest it can take up to 10 days of a consistent night schedule to partially adjust, but few construction projects keep workers on nights that long without rotation.

The effects of fatigue are wide-ranging and insidious. Physically, tired workers may experience weariness, headaches, slower motor skills, and even dizziness. Mentally, fatigue causes impaired concentration, memory lapses, irritability, and poor decision-making. In construction, this can translate to someone forgetting a step in a procedure, misreading a measurement, or not noticing a safety hazard. From a safety standpoint, the risk of injury shoots up when workers are sleep-deprived. For example, a study in the American Journal of Industrial Medicine found that working 12-hour days is associated with a 37% higher risk of injury. Extended hours combined with night work compound the problem. It's no coincidence that some of the worst industrial disasters had fatigue as a contributing factor – the 2005 BP Texas City refinery explosion, for instance, occurred on a unit where operators had been working prolonged shifts and was partly attributed to fatigue-related errors. Likewise, transportation accidents (like the 2009 Colgan Air crash) have been linked to pilot fatigue from night flying. In construction, while catastrophes may be less publicized, we have our share of fatigue-induced mishaps, such as vehicle crashes, equipment collisions, or workers falling asleep at the wheel on the drive home.

Another consideration is productivity and quality on night shifts. Expecting the same output at night as during the day is often unrealistic (though some owners try to demand it). Studies and field surveys indicate that productivity usually drops at night – one reason being that workers operate more slowly and cautiously (which isn't bad for safety) and need more breaks. Certain fine-detail tasks suffer in quality under artificial light and tired eyes.

In one survey of highway contractors and engineers, respondents noted that while some heavy operations (earthwork, paving) can be done effectively at night, they avoided doing "detailed finish work" or precision tasks on the night shift because quality would likely be lower. I have followed that guidance in my projects: for example, we scheduled critical weld inspections and complex concrete finishing during early morning daylight, even if pours were done at night, to ensure quality wasn't compromised by bleary-eyed inspectors.

Given these issues, managing fatigue is a top priority for night shift planning. Shift length and scheduling are key tools. Whenever feasible, keep night shifts shorter than day shifts – the traditional 8 hours is ideal. Pushing up to 10 or 12 hours overnight exponentially increases fatigue risk toward the end of the shift. OSHA's guidance warns that working longer than 8 hours, especially at night, generally reduces alertness and productivity. If long shifts are unavoidable, build in additional break periods and consider a napping policy (even a 20-minute power nap during a midnight lunch break can refresh a worker, though policies on this vary by company). Also, limit consecutive night shifts – many companies use a rotation (e.g., 4 nights on, 3 off) to prevent cumulative sleep debt. Ensure workers have adequate time to rest between shifts; at least 8 hours off, but more is better if one includes commute and winding down. A common practice is day-evening-night rotations, but quick rotations can be even harder on the body. Some experts suggest keeping crews on a consistent night schedule for a set period, so their sleep patterns aren't constantly changing.

Beyond scheduling, train and equip workers to manage fatigue. This includes education on sleep hygiene (dark, quiet places to sleep during the day, avoiding caffeine or alcohol before rest, etc.) and on recognizing signs of fatigue in oneself and crewmates. Supervisors should be taught to spot signs of severe fatigue – e.g., yawning, nodding off, slowed reactions – and remove or reassign a worker who is too tired (the OSHA general duty clause would back this in the name of safety). Encourage a culture where workers can speak up and say "I'm not safe to continue right now" without fear. For example, On a tunnel jobs implementing a simple fatigue-monitoring protocol: every two hours, the crew took a brief pause for a quick self-assessment quiz (asking whether they felt alert or were struggling) and a peer check (crew members gauged each other's alertness). If anyone scored poorly, shuffle them to a less dangerous task or get them some rest. Technology is emerging here, too – wearables and smart sensors can track signs of fatigue (such as head droop or microsleeps). Some oil and gas companies have used wearable monitoring for night workers, which has helped reduce fatigue-related incidents and even cut sick time by 40% by ensuring workers get needed rest.

Finally, be mindful of health and long-term impacts. Chronic night work has been linked to higher rates of heart disease, digestive issues, and even certain cancers (as NIOSH and others have reported). As managers, we have a duty to rotate people out of long-term night assignments when possible or implement health monitoring for those who must remain on nights. Bottom line: Fatigue is an enemy lurking in the shadows of every night shift. A proactive approach – reasonable shift scheduling, breaks, worker education, and a no-fault policy for reporting fatigue – will go a long way to keeping our crews safe and productive when the midnight oil is burning.

## Communication Barriers and Shift Handoffs

One often underestimated human factor in night operations is communication, especially communication *across shifts*. A construction project that operates 24 hours effectively has multiple teams, each picking up where the other left off. If information isn't conveyed accurately during shift turnovers or in daily coordination, mistakes and omissions can occur. Shift handover time is a vulnerable moment: the outgoing day crew is tired and eager to head home; the incoming night crew is gearing up but may not be fully up to speed on what transpired earlier. It's all too easy for a critical piece of information – say, that a certain circuit was left de-energized, or that an inspection result is pending – to "fall through the cracks" during this transition. Miscommunications or a lack of information flow can lead to rework, conflicts, or safety incidents. For example, if the day shift doesn't tell the night shift that a scaffold was left incomplete, a night worker might unwittingly climb it and be put in danger.

At night, communication on site also faces hurdles. Radios can have more static at longer range, hand signals are harder to see in the dark, and not everyone from support departments (engineering, management) is immediately reachable if an issue arises. Here is a scenario where a night crew encountered an unexpected underground obstruction. The day team's excavator operator had gone home, and the night crew wasn't aware of exactly what had been located earlier that day. Without clear instructions, they debated whether to dig around it or stop work. It took an hour to reach the project engineer (woken up at home) to clarify what the obstruction was and how to proceed. This kind of delay or confusion can be costly. It illustrates why detailed handover notes and communication protocols are crucial for night work. In fact, effective handovers are considered an essential risk management strategy in any 24-hour operation.

To mitigate these issues, implement a structured shift handover process. This might include a short overlap period (e.g., 30 minutes) where day and night supervisors walk the site together and go over a checklist of status items. Many projects use a shift log or handover report – a written log (or digital app) that captures all pertinent information: work completed, any incidents, equipment or systems status, materials needed, safety concerns, etc. Standardized documentation ensures nothing important is left out. For example, a handover report should clearly state if a lock-out/tag-out is in place on a piece of equipment, so the next shift doesn't remove it blindly. Today, some teams use digital tools accessible on tablets or phones so that handover information is recorded in real-time and easily accessible. In our company, we introduced an evening "tailgate" call: before the night shift starts, the night superintendent joins a quick call with the day superintendent and project manager to hear what happened during the day and what priorities or hazards to be aware of that night. This supplements the written log and allows questions to be asked live.

It's also wise to establish clear communication channels and emergency contacts for the night shift. Create and distribute a call list of key personnel (project manager, safety manager, engineers, utility contacts, etc.) who can be reached after hours if needed. The night crew should not be left guessing whom to call in a given situation. I recommend conducting a brief training or simulation for night shift leaders on emergency communication: for instance, if there's a utility strike at night, they should know exactly which manager and authorities to notify, even if those occurred at 2 AM. Similarly, ensure the night shift has access to up-to-date plans and documents – if the site office is closed at night, provide them with needed drawings or a way to retrieve information (some projects issue iPads with cloud access to plans).

Language barriers can be exacerbated at night as well, due to fatigue. Workers for whom English is a second language might have extra difficulty communicating when they're tired. Encourage night supervisors to double-check understanding when giving instructions and perhaps use more visual aids (like pointing with a flashlight, or sketches in toolbox talks). Speaking of toolbox talks: always do a nightly safety briefing at the start of the shift, covering what the night's tasks are and any handover notes from the day. Even if a toolbox talk was done in the morning with another crew, the night crew needs their own. It's an important forum to clarify any questions from the handover.

Another good practice is using modern communication to keep night workers in the loop with company safety news or changes. Some firms send out regular safety updates via email or text to night shift employees – things like incident learnings or safety reminders. This can help night crews feel connected to the wider project and not like the "forgotten shift." However, be careful not to overload them with information during work hours when focus is needed.

Treat information as something that must be actively managed in a 24/7 operation. Solidify your handover routines, document everything, and foster a culture where outgoing crews conscientiously set up incoming crews for success. When I finish my day shift, I often pause and think: if I were about to start working now with no sunlight, what would I need to know? Then I make sure the night foreman gets that information. This extra effort can prevent costly errors and keep everyone safer by ensuring continuity of knowledge around the clock.

## Emergency Response and After-Hours Support

An often overlooked constraint of night construction is the slower response time for emergencies and the limited availability of outside support services after hours. On a daytime site, if a serious injury occurs or if you need urgent assistance, you can typically count on a prompt 911 response, nearby hospitals fully staffed, and perhaps even an on-site medic during peak shift. At 2 AM, those safety nets are stretched thin. Emergency services can be delayed at night – fewer ambulances on the road, perhaps a longer wait for paramedics, and hospital ERs running skeleton crews. If your site is in a remote area or spread out (like a highway project), it might take even longer for first responders to navigate to your location in the dark. As one safety firm noted, responding to accidents at night is more challenging: poor visibility combined with reduced staffing can delay emergency response, increasing the risks of an incident escalating or a medical issue becoming more serious.

This means that night shift teams must be even more self-reliant in the event of an incident. It's critical to have a robust emergency action plan (EAP) tailored to night operations. Key elements include: ensuring that first aid trained personnel are present on each night crew, having fully stocked first aid kits and trauma kits on site, and readily accessible emergency equipment (like fire extinguishers, spill kits, etc.) positioned for the night work areas. In some high-risk night operations (e.g., highway lane closures), contractors arrange for an on-site ambulance or EMT unit dedicated to the work zone, recognizing the risk of vehicle strikes or other trauma. While not always feasible, consider this added precaution for particularly dangerous night jobs.

Communication with emergency services should be planned in advance. At night, identifying your location for responders can be tricky ("We're 3 miles down County Road X, 500 yards east of the old barn…"). To streamline this, I advise informing local fire/EMS before a night operation begins – provide them a map of the site, the best access routes at night, and even have signage or lighting that can guide them in (e.g., a flashing amber light at the gate). On one project, we arranged a meeting point at a nearby intersection where an employee would stand by to flag down the ambulance and escort them to the injured person, since the site entrance was hard to spot at night. Little steps like that can shave minutes off response time, which could be life-saving.

Additionally, consider alternative emergency transport. If it's safe to do so, having a plan for a site vehicle to transport an injured worker to meet an ambulance or go directly to a hospital might be faster than waiting. But weigh the risks – never delay proper care or transport someone unsafely. The key is to plan those decisions ahead of time in your EAP rather than improvising under duress.

Another aspect of after-hours operation is equipment or infrastructure emergencies. Suppose a piece of heavy machinery breaks down in the middle of the night. During the day, you could call the equipment dealer or your maintenance shop and likely get a tech on site within an hour. At night, you might not reach anyone until morning. Similarly, if you suddenly need an extra load of materials (concrete, steel plates, etc.), suppliers are likely closed. This can lead to unexpected downtime and pressure to "make do" without the right tools or parts, which can be dangerous. One classic example: a crew might be tempted to operate machinery with a jury-rigged fix through the night because the proper part isn't available until the day – this can obviously be a safety hazard.

The way to mitigate these issues is through thorough pre-planning and contingency planning for night work. Before starting night shifts, double-check all equipment is in good working order (conduct those maintenance inspections in daylight if possible). Have backup equipment or spare parts on site for critical items. For instance, if you have one lighting tower that's absolutely essential, have an extra generator or light tower on standby in case the first fails – you don't want to be plunged into darkness. If a mobile crane or excavator is mission-critical and would halt work if it breaks down, maybe arrange for a second unit to be available or plan other work that could continue while repairs are underway. This ties back to the earlier advice in Arrowhead's safety tips: preventive maintenance and equipment checks are crucial before and during night shifts, because a mechanical failure at night can "quickly escalate into a dangerous situation," and fixing it may not be as simple as it is during the day.

For materials, schedule deliveries strategically. Many highway jobs requiring asphalt or concrete paving at night will coordinate with suppliers to run overnight batches. But if you're relying on standard suppliers, make sure you have enough material on hand by the end of the day shift to last the night. I once managed a pour that ran past midnight only to realize we were short on rebar ties; our supplier was closed, and we lost precious hours until we found a workaround. Lesson learned: stock critical consumables in advance for night work.

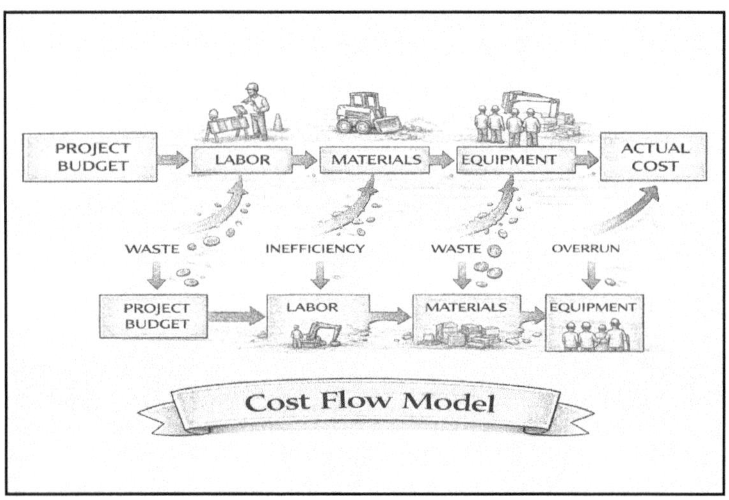


Coordination with external stakeholders is also important. If your night work could impact utilities (e.g., tie-ins that might require an electric company on standby) or trigger alarms (noise, security systems), arrange those ahead of time. Some urban projects hire off-duty police officers for traffic control and first response on night shifts – ensure they are properly briefed on the site EAP too.

In summary, night shift leaders must operate with a mindset of "*What if something goes wrong right now, and who can help?*" The answer, often, is "*mostly just us*". Thus, we bolster our internal readiness: trained first-aiders on the crew, clear emergency procedures, pre-planned communication with 911 and hospitals, backup equipment and supplies at the ready, and perhaps reduced risk-taking (maybe it's better to postpone a particularly perilous task if emergency support isn't adequate at that time). One positive development is that many OSHA and NIOSH guidelines now encourage having a Fatigue Risk Management Plan and ensuring "workers know the limitations of protective measures during extended hours," which includes planning for emergencies when people are tired, or help is far. By anticipating the worst-case scenarios and having plans to address them, we improve the resilience of our night operations and protect our crews when the unexpected strikes in the dead of night.

## Case Studies: Night Shift Pitfalls and Successes

Learning from real-world examples helps underscore why these night shift challenges matter. Let's look at a few U.S.-based cases across different sectors that highlight failures and best practices during night operations:

Highway Work Zone Accidents: Night road construction is common to avoid daytime traffic jams, but it has led to tragic incidents. A study of California freeway projects found that the presence of nighttime construction was associated with an 87% increase in the accident rate on those road segments. In one grim incident from California, a DOT maintenance worker was struck and killed around 1 AM by a speeding driver who did not see the work zone in time. Such incidents are far too frequent – impaired or drowsy drivers at night combined with work zones can be deadly. Research by Arditi et al. (2007) even suggested that nighttime highway construction can be five times more hazardous in terms of fatal crashes than daytime work. On the flip side, some states have implemented successful countermeasures. For example, Virginia DOT in the late 1990s revamped their night work zone protocols, mandating better lighting, advance warning signs, and police presence. Many jurisdictions now double traffic fines in work zones and run public awareness ("Slow for the Cone Zone") campaigns to get drivers to pay extra attention at night. The takeaway: robust traffic control and defensive setup (barriers, rumble strips, flaggers or automated flagger devices) are key best practices. One Federal Highway Administration statistic is worth noting: although fewer work zones operate at night, about half of work zone fatalities nationwide occur during night hours, underscoring the risk and need for continued improvements.

Industrial Project Fatigue Disaster: We've mentioned the BP Texas City refinery explosion (2005) – while not a construction project, it's a case study in night shift fatigue. Operators had been pulling extended 12+ hour shifts, some at night, leading up to a startup procedure that went awry. Investigations by the U.S. Chemical Safety Board found fatigue and communication lapses (e.g., handover issues between shifts) contributed to the catastrophe. In construction, an analog might be a long shutdown turnaround where crews work continuous nights. Recognizing this risk, some industrial contractors now impose work hour limits – for instance, no more than 14 days straight of night shift and mandatory rest days – and employ fatigue monitoring during intensive periods. A positive example comes from an Alaskan oil facilities project: management noticed quality and safety declining on night shifts during a winter fast-track schedule. They responded by reducing night shift length from 12 to 10 hours and providing a warm, quiet sleeping camp for workers to rest during the day. The result was a measurable drop in incidents and rework. The lesson: listen to the warning signs (near-misses, worker complaints of tiredness) and adjust schedules before an accident forces you to.

Vertical Construction and Urban Night Work: Building construction at night is less common (due to noise ordinances), but it does happen for pouring concrete during cool temps or to hit schedule milestones. One case that illustrates multiple night shift issues was the construction of a high-rise in a dense urban area (let's call it Project X in a big U.S. city). The contractor chose to do a major concrete mat foundation pour at night to minimize traffic disruption from the fleet of concrete trucks. They obtained a city noise variance permit and had everything lined up – or so they thought. On the night of the pour, a key concrete batch plant suffered an equipment failure at 1 AM. Because it was after hours, it took extra time to reroute trucks from a farther plant, and the pour was nearly compromised as some concrete started to set cold. Engineers on site had to add extra working additives to the mix and scramble to finish. Fortunately, quality was maintained, but the incident highlighted the risk of relying on just-in-time supplies at night. On the same project, neighbors called in multiple complaints about noise and light glare. The contractor learned and for future night work (tower crane jumps) they installed noise dampening blankets on fencing and directed floodlights away from residential windows, plus communicated schedules to the community in advance. This calmed community relations and avoided city fines. The success story here is that with proper planning – coordinating backup material sources, controlling noise/light, working closely with city regulators – the team was able to execute critical night operations in a busy city without incident. But it required treating night work almost as its own project, with its own risk assessment and mitigation plan, rather than just "another shift."

Port Arthur LNG Tank Collapse (2025): This recent incident has become a case study on multiple night shift hazards converging. As described earlier, a night crew was performing a "climbing formwork" operation (jumping the formwork to the next level on an LNG storage tank) when the form's support gave way. It was the middle of the night, and three workers fell to their deaths. The post-incident analysis revealed several contributing factors: inadequate inspection of the bracket connection (likely worsened by low visibility at night), possibly fatigued workers towards the end of their shift, and a breakdown in supervision/

oversight at that critical moment. Bechtel, the contractor, responded robustly – beyond just pausing night work, they conducted company-wide training on the specific operation, added triple-check verification steps, and as noted, banned that task at night indefinitely. The broader industry takeaway is that some high-risk tasks may simply not be appropriate for night execution without special precautions. If you must do them, then you need extraordinary measures (e.g. floodlighting like daytime, redundant inspections, etc.). This case also shows a positive aspect: transparency and learning. The contractor publicly shared the findings to help others learn, emphasizing that safety culture must be constantly reinforced, especially on night shifts where it can vary from crew to crew.

Each of these examples – highways, refineries, high-rises, industrial tanks – teaches us something about managing night work. They show how things can go wrong (fatal crashes, explosions, near-miss quality issues, structural failures) and also highlight best practices (enhanced traffic control, hour limits, community planning, specialized training). A common thread is that night shift problems are usually foreseeable: fatigue was predictable, visibility issues were known, communication gaps were avoidable. With diligent planning and strong safety leadership, we can anticipate and mitigate these issues. As a seasoned superintendent, I often tell my peers: *the challenges of night work are not surprises – darkness, tiredness, logistics – we know they're coming, so we have no excuse not to address them head-on.* When we do, as many projects have shown, we can achieve safe and productive night operations.

## Strategies for Safe and Efficient Night Operations

Having examined the challenges and cautionary tales, let's compile a toolbox of practical strategies to manage human factors and safety on night shifts. A successful night operation doesn't happen by accident – it is the result of deliberate planning and proactive measures. Here are key strategies and best practices:

- Comprehensive Night Shift Planning: Before night work begins, conduct a thorough risk assessment specifically for nighttime conditions. This should cover lighting needs, traffic control, noise control, emergency access, and worker scheduling. Develop a Night Operations Plan – some DOTs require contractors to submit one, detailing how you will handle the known challenges (visibility, fatigue, etc.). Include the plan as part of your overall project safety plan so it's not an afterthought. Identify high-risk activities that will occur at night and outline extra precautions for each. It's helpful to run a nighttime site walk with key team members before work starts: go to the site at night, simulate the setup, and see what hazards jump out (you might discover, for example, that existing streetlights blind your truck drivers as they enter your site, leading you to install temporary shielding).

- Optimal Lighting and Visibility: As discussed, invest in a detailed lighting plan. Use adequate portable lights to meet or exceed OSHA's minimum illumination standards. Position lights to eliminate dark corners but also minimize glare and deep shadows. Consider new technologies like balloon lights or LED light towers that provide broad, even lighting. Ensure work areas at heights (scaffolds, forms) are well-lit too – personal headlamps on hardhats can help workers see in nooks and crannies. Improve worker visibility by mandating high-reflectivity PPE. For road crews, reflective vests (Class 3 at night) and even active LED lighting on clothing can dramatically increase how far away a driver spots a worker. Mark work zone perimeters with steady burn lights or glow sticks so everyone knows site boundaries in the dark. Regularly inspect your lighting throughout the night – lights can drift or generators can sputter out fuel, and you don't want to find out only when someone is working in the dark.

- Noise and Community Measures: To comply with laws and keep community goodwill, implement noise reduction methods. Use equipment with sound mufflers, avoid idling heavy machinery near residences, erect temporary sound barriers if needed, and schedule the loudest tasks earlier in the evening (or during day, if possible) when it's less disruptive. Always secure required permits for after-hours work. For example, New York City prohibits construction noise at night without an approved variance – working without one can lead to shutdowns and fines. So get your paperwork in order and communicate with neighbors. A pro-tip: distribute flyers or notices to nearby residents/businesses with a schedule of night work, a contact number for complaints, and what you're doing to mitigate impacts. Being transparent can reduce frustration and calls to authorities.

- Fatigue Management and Shift Design: Design your night shifts with human limits in mind. Limit shift lengths – 8 or 10 hours max for continuous night duty is a good rule (remember, accident risk jumps after about 8 hours on task). Enforce rest breaks; a common practice is at least a 30-minute meal break plus additional 10-15-minute breaks every 2-3 hours on nights. Encourage micro-breaks to stretch and move, which fights off drowsiness. If using rotating shifts, try forward-rotating (day → evening → night) as it's slightly easier on the circadian rhythm than backward. Provide resources on proper sleep – some companies even give night workers blackout curtains or hotel rooms to get quality daytime sleep. Consider split shifts or naps for extremely long operations: e.g., for a concrete pour that goes 16 hours overnight, rotate crews such that each individual might work 6-8 hours then rest. Monitoring fatigue should be part of supervisors' duties. One practical tool is a simple questionnaire or a handheld device that workers use to self-report alertness level at start and mid-shift; if someone flags high fatigue, proactively rotate them to a less hazardous role or have them take a break. As a leader, be mindful not to schedule physically or mentally heavy tasks at the very end of a night shift – plan critical lifts, tricky installations, etc., for earlier in the night when everyone is fresher (per OSHA guidance to perform heavy labor at the beginning of extended shifts).

- Training and Awareness for Night Work: Don't assume workers automatically know how to handle night conditions. Provide specialized training or orientation for night shift crews. This training should cover the specific hazards of night work (reduced visibility, fatigue, etc.), emergency procedures at night, and any differences in protocol (like different PPE requirements or communication methods). Emphasize situational awareness: for example, teach them how machinery behavior can seem different at night (e.g., depth perception when signaling a crane is harder – use extra caution and clear signals). If possible, conduct a safety drill on a night schedule – e.g., a fall rescue drill in darkness to practice with headlamps and see the challenges. Also, train day shift folks on how to coordinate with night shift – for instance, engineers should know to leave written instructions for night crews if they won't be there in person. The goal is to embed night-shift awareness into the project's culture. One revealing survey found that while contractors stress safety at night, only 1 in 20 surveyed contractors provided advanced training specifically for night shift work. That's a gap we can close. Even a short pre-shift briefing each night (15 minutes to go over hazards and plans) can keep safety top-of-mind despite the late hour.

- Robust Communication Protocols: Establish clear communication methods for night operations. Equip the crew with reliable radios (and have spare batteries). Use a lot more verbal confirmation at night – what might be a simple hand wave in daytime should be a verbal "All clear!" at night. Standardize shift handovers with checklists and written logs as discussed earlier to ensure continuity. If language barriers exist, have bilingual supervisors on nights or use translation tools as needed; miscommunication in the dark is a recipe for disaster. Consider implementing a "buddy system" where workers operate in pairs or small teams that communicate frequently, especially in hazardous areas – no one should be out of earshot of others.

- Emergency Preparedness: We covered this in depth in the earlier section, but in summary, beef up your night emergency plans. Ensure quick response capability – e.g., have an on-site vehicle always fueled and ready as a potential ambulance. Provide personal flashlights or beacons to each worker in case of power loss or if they need to signal for help across a dark site. Nighttime EAP should include alternate means of communication (if radios fail, do we use air horns or sirens for alarms?). Do a nighttime emergency drill if possible – it's very eye-opening for a crew to try, say, an injury evacuation or fire drill in the dark; it often exposes issues like "the path to the muster point isn't lit," which you can then fix proactively.

- Staffing and Supervision Adjustments: Treat the night shift as equally important as days in terms of management attention. Don't skimp on supervision – assign strong leaders to nights. If the project is large, have multiple foremen or a safety officer present, not just one person in charge of everything. Management should also conduct unannounced night visits occasionally. When I was a project manager, I made it a point to come by the site in the middle of the night once a week. Not only did it show the crew that leadership cares about their shift, but it also allowed me to see conditions firsthand and spot things the day team might not realize (like poor housekeeping or lighting issues). These visits help enforce that safety standards should be uniform across shifts – night shift isn't the "wild west." Additionally, provide adequate manpower: ensure the crew size at night accounts for the slower pace – one guideline I use is to schedule about 10-20% more personnel at night for the same task you'd do in the day, to counteract the productivity dip and provide redundancy for safety. Also, consider specialized roles like a Night Shift Safety Lead who only focuses on safety monitoring and not production, if your project complexity warrants it.

- Technology Aids: Leverage technology to assist with night work challenges. Some examples: install CCTV or drone surveillance with night-vision to monitor widespread sites (helpful for security and safety oversight). Use wearable tech that can detect microsleeps or unusual inactivity – some systems can alert a supervisor if a worker hasn't moved for a long time (potentially having collapsed or fallen asleep). Anti-collision and proximity warning systems on equipment (back-up alarms, flashing strobes, radar proximity sensors) are extra valuable at night to compensate for low visibility. Even simple laser pointers can help a signalperson indicate where a load should land when guiding a crane in darkness. Newer "connected worker" apps allow crews to quickly report hazards or near-misses with a photo from their phone, even on night shift, so issues are logged and can be addressed by management promptly.

- Policy and Compliance: Adhere to all relevant safety regulations and go beyond them for night work. OSHA has guidelines but not specific hours-of-work rules, so it's on management to create internal policies. For instance, many companies have policies like "No worker shall work more than 16 hours in any 24-hour period" or "At least 8 hours rest between shifts" – enforce these strictly on jobs with night shifts to prevent burnout. Keep records of hours to ensure compliance. If working on a federal or state road project, be aware of agency specifications for night work (some DOTs require additional lighting units on standby, higher reflective gear classes, or special signage for night work). Comply with local laws on noise and lighting nuisance – this is not only legal but also part of operating responsibly in the community. It also circles back to safety: a community upset by noise might try to shut down your job, causing rushed work later, etc. Good community relations can indirectly improve safety by reducing external pressures.

In implementing these strategies, leadership must set the tone. Night shift crews often feel like second-class citizens compared to the day crew – a dangerous mindset that can breed corner-cutting ("nobody's watching anyway"). To counter this, leaders should actively involve night crews in safety planning, solicit their feedback, and recognize their work. Simple things like providing decent lighting, hot coffee, and a food truck on the night shift show that the company cares about their well-being. And when a night crew achieves a milestone safely, celebrate it just as you would a daytime achievement.

By weaving together all these measures – from hard controls like lights and barriers to soft controls like training and culture – you create a robust net of defenses against the inherent risks of night work. In my experience, when night operations are well-planned, crews actually appreciate the cooler, quieter environment and can be very productive. They take pride in knowing they are tackling the "tough shift" and management has their back. Ultimately, the goal is that everyone who heads to the jobsite under the stars goes home safely to see the sunrise – and that takes diligent, compassionate leadership guided by the strategies we've outlined.

## Policy and Legal Considerations for Night Work

Operating at night introduces not only safety challenges but also policy, legal, and regulatory considerations that construction managers must navigate. Ignorance of these can result in project delays, fines, or liability issues. Here we summarize the key considerations in the U.S. context:

- OSHA and Labor Regulations: There is no specific OSHA standard limiting night work or shift length in construction (OSHA does not mandate, for example, a maximum number of night hours). However, under the General Duty Clause, employers must provide a workplace free of recognized hazards – which arguably includes managing fatigue and lighting hazards. OSHA has published guidance acknowledging that extended or irregular shifts pose increased risks and recommends employers implement fatigue management programs. So while you might not get cited for having workers on a 16-hour night shift per se, if an accident occurs and it's found you ignored obvious risk factors (like extreme fatigue or darkness without proper lighting), OSHA could cite the employer for a violation of the General Duty Clause. Additionally, OSHA's construction standards do have relevant provisions: Illumination standard 29 CFR 1926.56 (which we cited earlier) legally requires maintaining minimum lighting levels on construction sites. Failure to provide adequate lighting (e.g., a worker injury attributed to poor lighting) could result in an OSHA citation under this standard. There are also standards for things like signs, barricades, and flagging in work zones (incorporating MUTCD guidelines) which effectively require high-visibility apparel at night, etc. Compliance with these is not optional.

- Noise Ordinances and Permits: Most cities and counties have noise ordinances restricting construction noise during night hours. For example, in New York City, general construction is only allowed 7 AM to 6 PM on weekdays without special permission. To work at night, one must apply for an after-hours variance permit and adhere to noise mitigation rules (such as limits on decibel levels at the property line). Similar rules exist in many jurisdictions (Los Angeles, Chicago, etc.), often with specific allowed dB levels or requiring community notice. Failing to get a permit can lead to stop-work orders or fines. So, part of night work planning is engaging with local authorities well in advance: submitting applications detailing why night work is necessary, what dates/hours, and how you'll minimize noise. Ensure your project's scheduling accounts for the time to get these approvals. Also, check if there are any union or labor contract stipulations about night shifts (for instance, some union agreements mandate higher pay for nights – which is a cost factor – or limit consecutive night shifts for workers).

- Transportation and Roadway Regulations: If your night work is on or near roadways, additional legal requirements come into play. The Manual on Uniform Traffic Control Devices (MUTCD) has specific guidelines for nighttime work zone setup – e.g., use of retroreflective signage, warning lights on barricades, temporary roadway lighting. Many state DOTs incorporate these into contract specifications. As a contractor, you're typically obligated by contract (and by law for public safety) to follow those standards. Some states like Florida or Texas may require submitting a Traffic Control Plan that specifically addresses night conditions (like requiring law enforcement presence for certain high-speed road work at night). Non-compliance can not only be a safety issue but breach of contract.

- Worker Health and Fair Labor Standards: Outside of OSHA, consider general labor laws. The Fair Labor Standards Act (FLSA) doesn't prohibit night work but remember to pay proper overtime if night shift pushes hours beyond 40 per week, etc. Some states mandate certain break periods – for example, a state might require a 30-min meal break after 8 hours, which could impact an overnight schedule that might otherwise try to compress time. Additionally, there are recommendations from agencies like NIOSH or CDC on shift work and worker health – while not law, these can influence your company policies. NIOSH in particular has studied shiftwork in healthcare and other industries; one notable stat from CDC's recommendations is that night shift workers should have health assessments (some jurisdictions or companies require periodic health checks for night workers to ensure the unusual hours aren't causing severe health issues). It's worth being aware of such guidance because it can become part of industry standard of care. In the event of a lawsuit (say a worker claims the employer negligently overworked them and caused an injury), these guidelines can be used to argue what a reasonable employer should have done.

- Contract and Liability Considerations: If you're a contractor, check your contract clauses about night work. Some owner contracts explicitly require owner's approval for night shifts (since it can raise costs or complaints). Others might have clauses transferring certain risks to you during off-hours. Also, ensure your insurance coverage (workers' comp, general liability) doesn't have any exclusions or special requirements for night operations – typically it doesn't, but sometimes insurers want notification if a project will involve extensive night work (due to elevated risk). An interesting angle: if working at night reduces one kind of risk (like public traffic interference) but increases another (worker injury risk), you may need to justify the trade-off in any incident investigation or claims.

- Community Relations and Environmental: We touched noise, but also consider any environmental permit constraints. For instance, some projects have environmental permits that restrict work at night to protect wildlife (sea turtle nesting season lights off on beaches, or limiting nighttime construction near endangered bird habitats, etc.). Make sure to review any environmental impact requirements. Culturally, remember that night work can draw public attention – a failure at night (like an accident with bright lights and sirens) might look particularly bad in media. Hence, the emphasis on transparency and proactively communicating with stakeholders. Some municipalities require community boards to be informed or even to approve nighttime construction schedules for major projects.

- Fatigue Management Policies: While not enshrined in law, there's a growing trend for clients or regulators to expect formal fatigue management policies on projects that involve long or night hours. For example, the Federal Railroad Administration and Federal Motor Carrier Safety Administration have strict hours-of-service rules for train operators and truck drivers. While those may not directly apply to most construction workers, it shows a norm. The National Safety Council and other safety organizations publish voluntary standards that could be considered best practice. Keeping logs of worker hours and having management sign off if anyone exceeds certain thresholds is a good practice that might later become mandated if fatigue-related accidents continue.

- Learning from Industry Standards: Bodies like the American National Standards Institute (ANSI) have standards for high-visibility apparel (ANSI/ISEA 107) which effectively are law when working at night on roads, since OSHA and DOTs incorporate them. Also, NIOSH recommendations (like "Plain Language about Shiftwork" or the newer Center for Work and Fatigue Research updates) provide practical steps that, while not law, should inform your internal policies. For example, NIOSH suggests employers limit consecutive night shifts and provide training on fatigue – if you ignore these and something bad happens, regulators or courts might view it as negligence. On the other hand, following such guidance can be a shield: if inspected, you can demonstrate you've implemented a Fatigue Risk Management Plan as OSHA's additional resources suggest.

In essence, the legal landscape for night work in construction is a patchwork of OSHA requirements, local ordinances, and contract rules rather than one unified law. As a construction leader, you need to be aware of all these layers. Before starting night operations, consult with your safety managers, legal team, or consult OSHA resources to ensure compliance. It's wise to document your efforts – e.g., keep records of lighting measurements to show you met OSHA foot-candle requirements, document worker training on night hazards, and archive any permits or community outreach communications. This not only keeps you compliant but also provides defense if there's ever a claim that you shortcut safety at night.

Remember, safety and legality go hand in hand. Many of the legal requirements (good lighting, proper rest, noise control) are in place to protect workers and the public. By diligently following them, you create a safer jobsite. And beyond just following the letter of the law, strive for the spirit of protecting your team: if that means doing more than the minimum – like voluntarily capping night shift lengths or investing in better lighting than required – it's worth it. In my experience, those investments pay off by preventing accidents and the far greater costs they entail.

## Guiding the Night Shift

Leading construction operations at night is truly a test of management skill and commitment to safety. There's a certain pride that comes with a night crew – a camaraderie in knowing you're doing tough work under tougher conditions. As a leader, tapping into that pride while safeguarding your team is the ultimate goal. Human factors – fatigue, visibility, communication, stress – are the critical variables. Throughout this chapter, we've highlighted how those factors can either be mitigated through thoughtful strategies or, if neglected, can compound into serious incidents.

The narrative examples and statistics drive home a simple reality: night shifts are inherently riskier. Accident rates climb, mistakes lurk in the shadows, and workers face strains beyond the norm. But we've also seen that with proactive leadership and planning, these challenges are manageable. Whether it's a highway lane closure at midnight or a pour on a high-rise before dawn, success comes from anticipating the hazards and doubling down on safety protocols.

In practical terms, always ask yourself: "*What if I were a worker out here at 3 AM – what would I need to stay safe, alert, and effective?*" That perspective will guide you to provide the right lighting, reasonable schedules, clear information, and a supportive culture. Make it standard to include night shift considerations in every project plan and safety meeting. When training your teams, share lessons from night shift incidents (the case studies in this chapter are a good start) so everyone learns vicariously and remains vigilant.

One point to emphasize in closing is the value of teamwork and communication across shifts. A day crew and night crew might hardly see each other, but they must function as one unit handing off a baton. Fostering respect and smooth interaction between them is part of the human factors equation. I've seen projects flounder when day and night teams blamed each other for problems ("the night guys left a mess," "the day guys didn't leave us the equipment") – and I've seen projects excel when they treated each other as partners. As a leader, set that tone. Encourage day crews to set up night crews for success, and night crews to deliver quality progress for the day shift to continue. Celebrate achievements of the night team in company newsletters or meetings just as you would daytime achievements, so they feel equally valued.

Lastly, be mindful of your own limitations. Managing night operations can mean long hours for supervisors and managers too. Don't burn yourself out; ensure you have a rotation or sufficient rest, or delegate trusted deputies, so that leadership remains sharp. A tired boss can make poor decisions that affect everyone.

Running a construction project at night is like conducting an orchestra in the dark – you need every player to be in tune, and you must trust the systems you've put in place because you can't see everything. With the knowledge and strategies outlined in this chapter, you are better equipped to shine a light – both literal and figurative – on the path to safe, efficient night work. Lead by example, stay vigilant, and your crew will take their cue that safety doesn't punch out at sunset. In the morning, when the sun rises and the day shift arrives to see a job well done through the night, you'll know that thorough planning, respect for human factors, and strong leadership made it possible. That is the hallmark of excellence in construction management – around the clock.

## Communication Failures and Their Costs

Effective communication on a construction site is like the oil in an engine – it's seldom the main focus, but everything can grind to a halt (or worse, suffer a catastrophic failure) if communication breaks down. We've already discussed shift handovers specifically, but communication failures can occur at any moment: a signaler's unclear hand gesture, a radio call that wasn't heard, a language barrier between coworkers, or a critical warning that never reaches the people who need to hear it. The consequences of these communication breakdowns range from costly rework and delays to life-changing injuries or worse.

Consider a heartbreaking example of how a simple communication failure can lead to disaster. In 2024, a precast concrete crew was installing large wall panels on a multi-story building. The panels were being held in place with temporary bracing (bolts) until the final welds and connections were completed. At the end of a shift, one worker mistakenly believed that a colleague had finished welding a panel permanently and that it was safe to remove the bracing. He was wrong – the panel was not yet secure. Because of this misunderstanding between crew members, the brace bolts were removed too early, and the entire five-story concrete panel suddenly toppled over. A young laborer working from an aerial lift in the fall zone was struck; the falling panel knocked over his lift, and he was killed. Investigators of the incident explicitly concluded that a communication breakdown had led to the premature removal of the supports. It's a stark reminder that even a small misheard or assumed instruction – the difference between "Yeah, go ahead and take those bolts out" and “Hold on, don’t remove them yet” – can be the line between an ordinary day and a fatal catastrophe. In this case, that line was the collapse of a wall.

Communication failures also carry high financial and schedule costs. Mistakes stemming from bad information often result in rework or delays. For instance, if an excavation crew doesn’t get the memo about a buried electrical line in their dig area and they strike it, you’re suddenly dealing with repair costs, potential OSHA fines, and investigations – not to mention the immediate danger that the crew was in. Even if nobody is injured, the resulting downtime and rework can be very expensive. There’s an old rule (originating from NASA studies on engineering errors) that the later a mistake is caught, the more exponentially expensive it becomes to fix. On a construction site, miscommunications follow a similar pattern – catching a misunderstanding before any action is taken is critical. If a crane operator and a rigging foreman take a moment to clarify a confusing hand signal before a lift, that’s ideal; if they don’t and a load gets dropped due to miscommunication, you might be looking at an insurance claim and a wrecked piece of equipment.

We should also acknowledge language barriers, which are a very real communication challenge on many U.S. construction sites. The workforce is diverse, and crews often include both native English speakers and others who might be more fluent in Spanish, Polish, Vietnamese, or other languages. In some regions and trades, Spanish in particular is prevalent. Unfortunately, language barriers, coupled with a lax work culture regarding safety standards, lead to alarming accident statistics. If a worker can’t fully understand safety warnings or instructions because they’re only given in English, that worker is at greater risk of harm. Or, if a worker is not equipped in a culture of safety, then they are at greater acccedent risk.

For example, consider a skilled concrete finisher on a crew who speaks very limited English. He might be highly proficient in his trade when you demonstrate tasks visually, but would he understand an urgent shouted warning like “Watch out – run!” if it’s only given in English? To mitigate this, companies should provide bilingual safety briefings and signage whenever needed, but not all do. A 2024 investigative report noted that while some construction firms hire translators for safety meetings and toolbox talks, others do not, meaning that vital warnings and instructions may be missed by workers who aren’t fluent in English. The same report observed that many immigrant workers also hesitate to speak up about hazards – either due to the language barrier or out of fear of losing their job if they “make waves.” So the communication breakdown goes both ways: safety messages from management might not reach some workers, and workers’ concerns might not reach management. Both failures can have deadly consequences. The human cost is enormous – not only for the injured workers themselves, but also for their families and communities. And for companies, beyond the clear moral imperative to fix this, failing to address language gaps ultimately means higher incident rates and all the costs associated with accidents.

Even among crews who all speak the same language, everyday miscommunications are common. Think about how we direct equipment, for example. We use hand signals, radios, and sometimes just shouting over the noise. It's very easy for signals to get crossed. In one incident, a spotter was guiding an excavator operator near some underground utilities. The spotter intended to signal "slow down/stop," but the operator misinterpreted the hand signal and continued digging, nearly striking a gas line. They had no radio communication, and the environment was noisy – a classic recipe for misinterpretation. Luckily, in that case, disaster was averted. Afterward, the team established a clearer set of standard hand signals and started using two-way radios with headsets so the spotter and operator could talk directly. It was a small investment that could prevent extremely costly mistakes.

Industry data backs this up: when communication is garbled or unclear, the risk of accidents increases – and companies often pay the price in damages, injuries, and even legal penalties. One safety publication put it bluntly that communication failures on construction sites can cost hundreds of thousands of dollars in project losses and fines, not to mention that they can end in injury or death in the worst cases.

Noise is another major obstacle to communication on-site. Construction environments are loud – heavy machinery, generators, saws, road traffic – and on top of that, we require hearing protection, which can further impede hearing each other. Often workers end up pulling out their earplugs to talk, which then exposes them to harmful noise levels. It's a tough trade-off to manage. However, some crews are now using wireless communication headsets that have built-in hearing protection. These systems allow real-time, hands-free communication without the need to shout and without sacrificing anyone's hearing. A team equipped with such headsets can carry on a normal conversation ("Heads up, I'm swinging the load toward you from the left") even if a concrete pump or other loud equipment is running nearby. This technology was once rare and expensive, but it's becoming more affordable and prevalent, especially on large projects. The payoff is improved safety and efficiency – coordination is faster and clearer, and there's less frustration. When workers can easily talk and hear each other despite high background noise, they reduce errors caused by misheard directions and also cut down on the stress and fatigue that constant loud noise can cause.

So how do we prevent communication failures? A few best practices can be highlighted:

- Standardize critical commands: Make sure the crew shares a common "language" for important directions. For example, establish a set of standard hand signals for crane operations (OSHA provides a standard set) and ensure all signalers and operators are trained in them. Likewise, for radio communications, agree on clear wording – if someone says "Stop," it means cease operations immediately, and the operator should repeat back the command to confirm. (Pilots and air traffic controllers use this kind of read-back protocol to avoid misunderstandings; construction crews can do the same, e.g., on the radio: "Confirm, you want me to shut down the pump?" – "Yes, shut it down now.")

- Use interpreters and bilingual materials: If a significant portion of your workforce isn't fluent in English, invest in bilingual safety training, signage, and translation. Provide safety manuals, toolbox talk handouts, and warning signs in the languages your workers understand. Whenever possible, have a bilingual supervisor or safety officer who can translate instructions in real time. Even simple steps like posting hazard signage in multiple languages or using universally understood pictograms go a long way. Some companies also identify bilingual crew members and designate them to help bridge communication gaps on mixed-language teams.

- Encourage questions and double-checks: Create a culture where workers feel free to say, "Wait, can you clarify that?" or "I'm not sure I understood, could you repeat that?" Often, miscommunication goes uncorrected because people are hesitant to admit they missed something. Supervisors should explicitly tell crews that if they are not 100% sure what's being asked, they must ask for clarification – and they will never be punished or ridiculed for doing so. It's far better to repeat an instruction than to have someone guess wrong and get hurt or cause an error.

- Improve communication tools and methods: Provide the right tools and protocols for the situation. Don't rely on shouting over noise or vague gestures when better options exist. For example, if teams are working out of each other's line of sight (say, a crane operator and a rigging crew on a rooftop), equip them with two-way radios or headsets so they have direct communication. If mandatory hearing protection is making it hard for crew members to hear each other, look into technological solutions like noise-canceling communication headsets or adjust the workflow (for instance, pause noisy equipment during critical briefings so everyone can hear). The goal is to match the communication method to the environment.

- Plan communications for high-risk tasks: For especially complex or hazardous operations (such as a multi-crane lift or an energized electrical switchover), take the time to explicitly plan how the team will communicate throughout that task. This might mean writing a short "communication plan" as part of the work plan or lift plan. Designate one person to be the sole signaler or the single point of contact on the radio. Establish what channels or hand signals will be used and have a backup method in case the primary communication fails. When things are high-risk and high-complexity, you cannot afford communication chaos – so choreograph it in advance.

At the end of the day, communication is a human factor that underpins everything else on the project. When it flows well, you hardly notice it – the job runs like a well-rehearsed orchestra. When it breaks down, the discord becomes apparent immediately: work crews accidentally doubling up on the same task, critical steps left undone because someone assumed someone else did them, or worst of all, emergency sirens wailing after an accident that could have been avoided with one timely message. Field leaders often find they must devote a significant portion of their time to facilitating communication between different trades, between the field and the office, and between day and night shifts. It may not be the most glamorous part of the job, but it absolutely saves lives and prevents costly mistakes. Remember, you can have the best safety equipment and the most detailed plans in the world, but if the right people don't get the right information at the right time, it can all unravel. Keeping those information channels clear and effective is a core responsibility for everyone, especially leaders.

## Embedding Human-Factor Awareness into Culture

In construction, you can implement all the safety rules and procedures imaginable, but real, lasting improvement happens only when attention to human factors becomes part of the culture. By "culture," we mean the collective values, beliefs, and habits of the organization – the unwritten rules about how things are done. We want a culture where workers naturally look out for one another, anticipate potential human errors, and feel empowered to speak up about concerns. Building this kind of culture is arguably one of the most important jobs for leadership, because a strong culture continues to guide behavior even when no one is watching or enforcing rules.

What does a human-factor-aware safety culture look like? For starters, it goes beyond mere compliance. Some companies focus only on ticking the OSHA boxes – that's a compliance mindset (doing the bare minimum, like wearing PPE only when the supervisor is watching, or filling out paperwork just to satisfy requirements). In contrast, the best contractors foster a culture where safety is a core value, not just a regulatory obligation. Such companies often exceed the legal requirements simply because it's the right thing to do and because they truly believe every injury is preventable. For example, instead of saying "We have to do this training because OSHA requires it," a culture-driven approach says "We do this training because we want everyone to go home safe; it's part of who we are as a team." In practical terms, companies with a strong safety culture treat safety initiatives as investments and sources of pride. They may even highlight their safety culture in marketing – indeed, some clients now ask for evidence of a strong safety culture before awarding projects, knowing that it often correlates with high quality and reliable performance.

Leadership commitment is the foundation of any cultural change. Workers pay far more attention to what leaders do than what they say. If management loudly promotes safety but then quietly cuts the safety budget, or if a supervisor is the first to ignore the hard hat rule, the crew will take note of those actions as the "real" message. Leaders at all levels need to walk the talk. That means, for example, showing up to the daily safety huddles, wearing the required PPE diligently, and even personally donning a harness and climbing up to inspect a tricky location if that's what the crew is being asked to do. It also means never pressuring the team to cut corners or rush in ways that compromise safety. Top executives should demonstrate a similar commitment. At one large general contractor, the CEO himself conducts unannounced monthly safety walks on various projects. Imagine the impact on the workforce when the top boss is out there in work boots and a hard hat, personally checking safety conditions. Likewise, forward-thinking companies hold their management accountable for safety performance – treating it not just as a field issue but as an organizational one. One effective tactic is tying a portion of managers' bonuses or performance evaluations to safety metrics. This approach certainly grabs the attention of office personnel and motivates them to take a hands-on interest in jobsite conditions. Most importantly, true leadership commitment means that when push comes to shove, safety takes precedence over production. If a serious hazard emerges, the best companies will stop the job – even if it means a schedule delay or additional cost – until the issue is resolved. In high-level project meetings at such companies, you might hear an executive say something like, "I don't care if we're a week behind schedule, fix that scaffolding issue before anyone goes back up there." That kind of backing from the top empowers everyone on down the chain to do the right thing without fear.

Another essential element is empowering workers and creating a sense of psychological safety. Every person on site – from the most veteran superintendent to the newest laborer – should feel both responsible for safety and authorized to take action if they see something unsafe. Many organizations formalize this through a "Stop Work Authority" policy: any worker, at any level, has the right (in fact, the obligation) to stop work if they believe there's a safety hazard, no matter what, and they can do so without any fear of reprimand. This is a game-changer for safety culture. No manager can have eyes everywhere at all times; you rely on those closest to the work to catch issues. But workers will only speak up if the culture supports it. So management must actively encourage and positively reinforce this behavior. Sites that truly embrace this often see a dramatic shift: crew members will respectfully interject during an operation with comments like, "Hold on, let's double-check our fall protection," or "I'm not comfortable with that rigging – let's pause and fix it." When that happens, supervisors thank them for speaking up, recognizing that they may have just prevented an accident. In fact, some companies go as far as to celebrate near-miss reports or hazard identifications. For example, one firm hands out a small award each month for the best "Good Catch" – recognizing a worker who spotted and corrected a potential hazard. This approach sends a clear message that finding and reporting problems is a positive contribution, not a mistake to hide. In a strong safety culture, near-miss reporting is welcomed and never punished. Workers are treated as partners in safety, not as children who need to be controlled.

Embedding human-factor awareness also involves focusing on behaviors – often referred to as Behavior-Based Safety (BBS). BBS has a mixed reputation because if it's implemented poorly, it can feel like a punitive "Big Brother" system that blames workers. But when done correctly, it can truly help instill safe habits on the job. The idea is to observe work in progress, give constructive feedback, and reinforce the positive behaviors you want to see. For example, on one project, a peer-observation program was set up: each week, a different crew member was assigned to spend a couple of hours observing the work practices of their teammates. That person would then share any safe practices they noticed or any minor risky behaviors observed (all anonymously, without naming anyone) at the next toolbox talk. Initially, workers were skeptical about this process. But over time, it became routine and even a bit of a friendly competition – crew members would half-jokingly try to get "caught" doing something safely so they could be highlighted as a positive example. This kind of proactive focus keeps everyone more conscious of their actions.

One technique borrowed from psychology is the "ABC model" – Activator, Behavior, Consequence. The idea is not to wait for bad outcomes (like injuries) to occur and then punish them, but to actively reinforce good behaviors as they happen. This can be as simple as a supervisor giving verbal praise: "I noticed you took the time to re-secure that ladder – good job, that's exactly the kind of attentiveness we need." It could also involve more formal incentives or recognition programs for safe work practices. Research has shown that positively reinforcing safe behavior helps strengthen the safety culture overall. Of course, there still need to be consequences for willfully unsafe behavior – for example, if someone repeatedly disregards safety rules even after coaching and warnings, some form of disciplinary action is appropriate. However, in a mature safety culture, such punitive measures are rarely needed because peer pressure and personal pride tend to maintain standards. When safety is ingrained, no one wants to be the only person on the crew not wearing their safety glasses when everyone else treats it as second nature.

Training and education are also crucial for ingraining human-factor awareness. This goes well beyond the basic OSHA 10-hour class or other mandatory courses. Those cover the regulations and common hazards, which are important, but we also need to train workers and supervisors on things like communication skills, situational awareness, fatigue management, and teamwork – the so-called "soft" factors that have very hard consequences if neglected. Many leading contractors provide robust orientation programs and ongoing training that emphasize why certain practices are important, not just what to do. For instance, an orientation might include a module on human factors where the group discusses real incidents that were caused by factors like rushing, complacency, miscommunication, or wrong assumptions. They often use case studies from within the industry to make these lessons concrete. Additionally, top companies conduct periodic refresher trainings or safety stand-downs to refocus everyone on lessons learned. After any significant incident or even a near-miss on a project, they will call an all-hands toolbox talk to review what happened and how to prevent it in the future. Some even share lessons from incidents across the whole company and sometimes with industry peers, because nobody has to suffer an accident that someone else has already learned how to prevent.

Embedding awareness into daily practice also means institutionalizing certain habits. A few examples of effective habits are:

- Daily pre-task safety plans: Also known as Job Hazard Analyses (JHAs), these are brief discussions (5–10 minutes) that the crew holds before starting a task to identify what could go wrong and how to prevent it. This habit forces everyone to slow down and think about human factors and potential hazards each day. For example, a pre-task plan for a lifting operation might include questions like, "Who is watching the backup alarm while we move this truck? Do we all know the hand signals for the lift?" By addressing such details up front, crews are more alert to the risks and the precautions needed.

- Interactive safety meetings: Regular safety meetings or "safety huddles" should be more than a perfunctory checklist. They need to be engaging and relevant to the day's work. On some projects, each morning begins with a brief safety sync-up where every crew or crew leader shares any hazards that others should know about for that day. For instance, one crew might announce, "We'll be working on the scaffolding above the east entrance today, so no one should enter that doorway below us." These quick exchanges ensure that safety is integrated into the daily work plan, not treated as an afterthought. It sets a tone that every day starts with safety.

- Active field supervision: Continuous monitoring and coaching by supervisors helps embed safe practices into the routine. A superintendent or foreman who stays out in the field, doing frequent walkabouts, can catch small issues before they become big problems and also recognize good practices on the spot. It's important that this oversight is not seen only as nitpicking. Good supervisors will just as readily say, "Thank you for wearing those harnesses properly, guys – I appreciate it," as they will correct a lapse. Recognizing positive behavior reinforces it. When crews see that management truly cares about how the work is done (not just that it gets done), they are more likely to uphold those standards themselves.

A culture that truly values safety also demonstrates care for the well-being of its people. This means if someone is clearly exhausted, the culture encourages them to speak up and maybe take a break or switch to a less strenuous task for a while. If someone is new to the crew, they get paired with an experienced mentor who looks out for them. If someone is dealing with a personal crisis or stress, supervisors and teammates discreetly keep an eye out that day. These are the human-factor considerations that a tight-knit crew adopts instinctively. A simple motto often heard is, "Look after your buddy." In practice, that means if you see a coworker struggling or on the verge of doing something risky, you step in to assist – not to scold them, but to help them get back on track safely. For example, an older carpenter on one crew noticed a younger laborer was zoning out in the afternoons. He found out the younger man had a newborn at home and was severely sleep-deprived. So the crew quietly adjusted their plan: they shuffled some tasks around so that the tired worker was not doing anything especially dangerous (like operating machinery) while he was groggy. That kind of peer support only happens in a culture where people feel genuinely responsible for each other's safety.

Embedding human-factor awareness into the culture means making safety and mindfulness an integral part of how the team operates every day. It is achieved through visible and consistent leadership, empowering workers to act and speak up, continuous education and training, and daily routines that reinforce the message. Over time, you can tell that it's working when you observe new hires naturally absorbing the norms from veterans ("Around here we always tie off—no exceptions") and even being comfortable reminding seasoned workers if they slip ("Hey, don't forget your goggles – you know we never cut without eye protection"). In a strong culture, safety isn't viewed as an external requirement or a drag on productivity – it's simply the way work is done. In fact, such sites often become more productive and deliver higher quality, because the same discipline and care that the team applies to safety carries over into every aspect of the job. The ultimate goal is that every person on site genuinely believes that no deadline, no cost savings, and no convenience is worth putting a life or limb at risk, and they act accordingly as a unified crew. When that point is reached, human factors are truly embedded in your construction culture – and the results will show in improved safety records, higher morale, and even better overall project performance.

**Operational Takeaways for Today's Job:**

- Do a thorough shift turnover. When handing off work to the next crew or shift, take 10 minutes to walk them through what was done, what's left, and any issues to watch out for. A face-to-face handover or a written log with a conversation can prevent miscommunication that leads to errors.
- Be mindful of fatigue and adjust accordingly. If workers pulled a long shift or are on nights, consider rotating tasks to keep people alert, provide a bit of extra break time, or use a buddy system for monitoring each other's well-being. Alertness is a safety tool – treat it like one.
- Improve night shift safety with simple tweaks. Ensure the site is well-lit (add extra lighting towers if needed), keep noise down for communication clarity (or use radios/headsets), and have supervisors more visible during off-hours. Small changes counteract the higher risks that come with darkness and reduced staffing.
- Acknowledge the human limits openly. Remind the crew it's okay to speak up if they're feeling too tired or stressed to work safely – better to rearrange work than to push through and have an accident. Create an atmosphere where asking for help or a breather is encouraged, not seen as a weakness.
- Adapt communication for the conditions. During night or loud operations, use clear hand signals or repeat-backs to ensure messages are understood. Don't assume a nod means yes – verify instructions. Effective communication bridges human factor gaps, especially when conditions are less than ideal.

Chapter 9

# Event Investigations and Learning Systems

When an incident strikes on a construction site – be it an injury, equipment failure, or a near miss – how you respond as a leader can mean the difference between repeating the mistake or preventing it for good. A project executive or site superintendent should turn these unplanned events into opportunities to improve. On a construction site, every **incident is a learning opportunity** if approached correctly. This chapter provides a practical guide to structuring investigations, assigning investigative roles without casting blame, and converting hard lessons into lasting organizational knowledge. It rests on a simple belief: very few incidents are truly "just bad luck." Almost every mishap can be prevented with proper foresight and by strengthening our systems. By treating incidents as symptoms of deeper issues rather than one-off flukes, construction teams can improve their processes and safety culture.

The sections that follow outline a structured investigation process, clarify the roles and responsibilities of investigators, examine how to critique performance **before, during, and after** events, and discuss methods for recording and sharing lessons learned. The goal is that when things go wrong, the company doesn't just fix the immediate problem – it learns how to avoid all such problems going forward.

## Structured Investigation Process

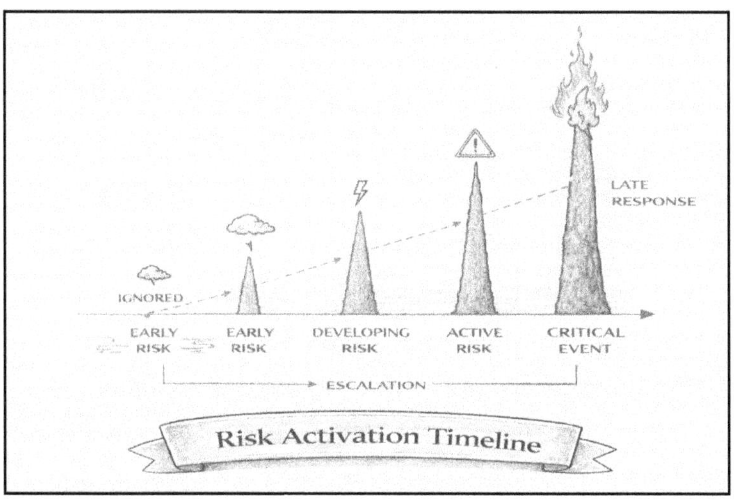


A structured investigation process is the backbone of learning from any jobsite event. It brings order and rigor at a time when emotions may run high and pressure is on to assign blame or "just get back to work." Without a defined approach, an investigation can easily descend into chaos or finger-pointing. It's all too common for companies to treat an inquiry as a hunt for a culprit and stop once someone is blamed. An effective leader avoids that trap. Instead, they ensure the investigation digs into underlying causes and system factors, not just the immediate trigger. In other words, the focus is on what happened and why, rather than who messed up. Making this shift in mindset – from blame to curiosity – is critical. A thorough, fact-focused investigation will not only reveal how to prevent a repeat of the same incident but also often uncovers hidden issues that could cause other accidents if left unaddressed.

**Investigate Everything, Big or Small:** In principle, every incident and significant "near miss" should be investigated, no matter how minor it seems. For example, a small scaffolding slip with no injury might reveal that workers have gotten into a bad habit with their harnesses – a lesson that could avert a serious fall later. Investigating even the small stuff sends a powerful message to the crew (and to regulators) that the company takes safety seriously and is always looking to improve. It also reinforces an important cultural point: reporting problems is valued, not punished. Workers will only speak up about close calls if they see management routinely examining issues in a fair, problem-solving manner rather than looking for someone to blame. By investigating near misses, teams basically capture free lessons – no one paid the price in blood or broken equipment yet, so you learn from it and fix the issue before it causes real harm.

**Prepared, Not Improvised:** Smart construction organizations set up their investigation procedures long before an incident ever occurs. They have clear guidelines and even forms or checklists ready in advance. The worst time to figure out how to investigate is in the middle of a crisis. If every investigator has to "make it up as they go," the consistency and quality of investigations will be all over the place. Instead, a standard framework should kick in automatically when an event happens. The severity of an incident will, of course affect how deep or expansive the probe is, but the basic process should remain the same for any event. This consistency ensures that important steps aren't skipped and that results from different investigations can be compared and trusted. It also means everyone – from the crew to management – knows what to expect when an incident occurs. They've essentially practiced for it (much like a fire drill), so they can step into their roles calmly and methodically, even under pressure.

**Immediate Response and Scene Control:** The moment something happens, the first priority is always emergency response: make sure people are safe. That means providing first aid to the injured, calling 911 or site medics if needed, and eliminating any immediate dangers (e.g. shutting down equipment, isolating energy sources) to prevent further harm. Right after that, whoever is in charge (usually the superintendent or site safety officer at first) needs to secure the scene for investigation. This might involve stopping work in the area, roping off or barricading the incident site, and preserving any physical evidence (equipment, materials, etc.) so it isn't moved or tampered with. For instance, if a trench wall collapsed, you would cordon off that section and leave it as undisturbed as possible until the investigation team can examine it. Preserving the scene is crucial because memories fade fast and physical clues can disappear if you wait too long.

During this initial phase, management also determines the scope of the investigation and what resources are needed. A serious incident might warrant assembling a larger investigation team or bringing in an outside expert, whereas a minor mishap might be handled by one person. But in all cases, the predefined process kicks in. Notifications are part of the immediate response as well: the project manager and safety manager should be informed, and if it's a severe incident that triggers regulatory reporting (for example, a hospitalization or fatality), you may need to contact OSHA or other authorities within a specified timeframe. In sum, the first stage is about making people safe, then securing information, stabilize the situation, then lock down the scene so you can figure out what happened.

**Fact Gathering (Evidence Collection):** Once the scene is stable and secured, the investigation turns into a fact-finding mission. This phase is like detective work: you want to gather every piece of information that could shed light on what happened. It should begin as soon as possible – ideally within minutes or hours of the incident – while details are still fresh. Investigators (who might be the supervisor, a safety officer, and/or others assigned) will conduct interviews with witnesses and those involved. A good practice is to have each person recount what they saw or did, and even write down and sign their statement if possible. Even if someone didn't see anything relevant, it's useful to note that too, because it helps establish who was present and can help cross-check others' accounts.

Importantly, these interviews are not about grilling people or assigning blame. They should be done in a calm, respectful manner. Ask fact-based questions like, "Can you describe what you were doing right before the incident and what you observed?" rather than accusatory ones like, "Why on earth did you do that?" The tone matters: people will speak more openly if they know the goal is to understand the problem, not to punish them.

Besides interviews, investigators collect physical and documentary evidence. Take photographs or video of the scene, any damage, and the surrounding conditions. Secure any relevant equipment or tools (for example, keep the broken scaffold piece or failed sling for analysis). Pull records such as equipment maintenance logs, inspection checklists, permits, or training records for the people involved. If technology is available, save data from machinery or site sensors (perhaps a crane's black box or security camera footage). For example, if a crane tipped over, you'd gather the crane's load charts, the operator's certification and training records, the crane's maintenance history, and maybe the weather reports or wind readings at the time. In short, **grab every piece of the puzzle** you can find.

The investigator should be thinking about what information might be missing and chase down those leads. If a key witness left the site, give them a call. If a tool that failed was sent off-site, track it down or preserve it in storage. In practice, especially on a busy project, you can't hold the site frozen forever. Once you're confident that you've documented everything important – photos taken, measurements done, witness statements collected – you will eventually "release" the scene so work can resume in that area. The goal is that before normal operations start again, nothing crucial to the investigation has been lost or overlooked. It's a balancing act: you want thorough evidence collection, but you also want to minimize unnecessary downtime. Good planning (as mentioned, having checklists and roles defined in advance) helps a lot here, so the team can efficiently gather what they need.

**Analysis and Root Cause Determination:** With a trove of information in hand, the next step is to figure out **how and why** the incident happened. This stage is often the most intellectually demanding. A useful way to start is by constructing a timeline of events in as much detail as possible. List out exactly what happened and in what order – for example, "7:45 AM: crew began pouring concrete on the second floor; 8:10 AM: formwork beam X failed; 8:11 AM: section of formwork collapsed." Laying out the sequence helps ensure everyone understands the progression of events clearly.

After establishing the timeline, the team will analyze the data and evidence to identify causes. Crucially, the analysis must **dig deeper than the obvious**. We don't stop at "what broke?" or "who made a mistake right before it happened?" We ask "**Why** did this happen?" – and then keep asking "why" multiple times to uncover deeper causes. Often, an incident has multiple contributing factors.

For example, imagine a scenario where a pre-cast concrete wall panel toppled over during installation. The immediate cause might be that a brace was removed too early by a worker. But if you ask why that happened, you might discover the crew was rushing because they were behind schedule (possibly due to management pressure to finish the job quickly). You might also find that the bracing plan wasn't clearly detailed in the method statement, so the crew improvised (a procedural gap). Ask "why" again: maybe the supervisor overseeing the lift wasn't trained on the proper bracing techniques (a training deficiency). And perhaps the engineering review of the lift plan missed this potential failure mode (an organizational oversight). In nearly every incident, if you keep peeling back the layers, you'll uncover a chain of events and conditions that all contributed to the outcome.

There are formal techniques to aid this analysis – for instance, the "5 Whys" method (literally asking why repeatedly to go deeper into cause and effect), fishbone (Ishikawa) diagrams that map out cause categories, or more specialized methods like TapRooT. The specific tool is less important than the mindset: **look beyond the proximate cause** (e.g., a worker's action) to the systemic issues that set the stage (e.g., why was the worker in a position to make that mistake? what organizational factors were at play?). It's usually misleading and counterproductive to pin everything on "operator error" or "carelessness" without further context. In almost every case, factors such as equipment problems, unclear procedures, insufficient training, or flaws in the safety management system are part of the story. If we fail to identify those, we fail to fix them – meaning the underlying problem could strike again. In other words, failing to find the root causes is like repairing a broken part without addressing why it broke; you're likely going to have the same failure in the future.

As the team pieces together their analysis, they should also reflect on why existing safeguards didn't prevent the incident. Was there a rule, checklist, or safety system in place that should have caught this issue? If so, why didn't it work this time? This might reveal, for example, that a particular safety rule exists only on paper but isn't practical in the field (so people routinely bypass it), or perhaps that supervision was stretched thin at the moment and missed a critical check. Essentially, you're also critiquing your overall safety program's performance under stress (more on this kind of systemic critique later on in the chapter).

The output of the analysis phase is a set of **findings** – these are the key facts and root causes the team identified – and from those findings, the team will formulate **recommendations** for corrective and preventive actions.

**Documentation – The Investigation Report:** An investigation isn't truly complete until its findings are documented clearly and comprehensively. A written report is vital for several reasons. First, it provides one official, factual record of the event that everyone can reference, rather than relying on rumors or fading recollections. Second, it becomes a key piece of documentation for accountability and potential legal or regulatory needs – if months or years later someone has questions (insurance, OSHA, lawyers, etc.), a well-prepared report shows what the investigation found and what was done about it. Third, the report is a tool for organizational learning, feeding into the "lessons learned" system we'll discuss at the end of this chapter.

A good investigation report typically includes: an executive summary (a brief overview of what happened), a detailed narrative of events (the timeline and analysis of causes), a list of the findings/root causes, and the recommended actions or corrective measures. All the supporting evidence – photos, diagrams, witness statements, equipment inspection results, calculations – can be attached or referenced in appendices so that if someone needs to dig into the details, they can.

When writing the report, the tone should remain **objective and precise**. Stick to the facts and well-supported conclusions, and avoid emotional language or finger-pointing. For example, instead of writing, "Joe Smith was negligent and didn't check the harness," the report would state, "The investigation found that the worker did not inspect his fall harness prior to use, contrary to company procedure; the worker reported that he felt rushed by the tight schedule that morning." This way, the report documents *what* happened and *why* in a factual manner, noting both the action (no harness inspection) and the context (felt rushed due to schedule). It avoids casting moral judgment like calling someone "negligent" – which adds heat but not light – and it provides a basis for understanding and improvement.

It's often wise to have a fresh set of eyes review the draft report – maybe another manager or a safety professional who wasn't involved in the incident – to ensure clarity and that nothing important has been missed. A peer review can catch any bias or gaps and strengthen the report's credibility.

**Implementing Corrective Actions:** The ultimate test of an investigation's value is what the company does with it. A brilliant analysis means nothing if it just sits in a binder on a shelf. Therefore, a structured process demands a step of action and follow-up. Usually, the investigation team (in consultation with senior management) will develop an **action plan** based on the recommendations from the report. This might involve engineering fixes (e.g., redesigning a scaffold connection detail that failed), changes to procedures (for example, introducing a new checklist or altering a work sequence), additional training for workers or supervisors, disciplinary action if it turns out someone willfully violated a critical safety rule, or even broader policy changes at the corporate level.

Each action should be assigned to a responsible person or department, with a target date for completion. For instance, if one finding is that "Crane operators were not aware of the updated wind-speed restrictions in the new load chart," then an action might be: "**Equipment Department:** Conduct refresher training with all crane operators on the new load chart and wind limit guidelines by March 1." That way it's clear who must do what, and by when, to address the issue.

After deciding on actions, it's important that leadership follows up to **ensure they are carried out and that they are effective**. It's a good practice to schedule a review (say 30 or 60 days later) to verify that all the assigned corrective actions have been completed and to check if the changes are indeed solving the problem. Part of this follow-up might involve observing work in the field or auditing the area of concern to see if the new measures are being implemented and working. For example, if a new pre-work checklist was added because of the incident, you might observe a couple of crews using it to confirm that it's practical and that it's actually preventing the type of oversight that led to the incident. This ongoing verification **closes the loop**, making sure the time and energy spent investigating truly lead to improvements. Many firms will incorporate the new lessons into their regular safety audits or inspections going forward – essentially institutionalizing the change into everyday practice.

**Communication and Sharing:** Another critical element of the process is sharing the findings and lessons with everyone who needs to know. On the immediate level, that means communicating to the project team – and often the whole company – what was learned from the incident and what will change going forward. Being transparent about this builds trust.

When workers see that management isn't sweeping a problem under the rug but instead is openly saying, "Here's what went wrong and here's how we're fixing it," it reinforces that the goal of the investigation was to improve the work, not to hide issues or simply punish someone. It also ensures that everyone on the team is aware of any new safety measures or procedures that resulted from the investigation.

Sharing lessons shouldn't be limited to the site where the incident occurred. If your company has multiple projects or crews, you want those other teams to learn proactively from this event so it doesn't happen to them. Companies often issue brief "**Lessons Learned**" **bulletins or safety alerts** describing the incident (in general terms) and key takeaways, which are distributed company-wide. These might be discussed in safety meetings or posted on notice boards. The idea is to spread the knowledge gained so that others don't repeat the mistake.

A word of caution: if there are legal sensitivities (for example, ongoing litigation or an OSHA investigation), you might need to have any written communications vetted by your legal team before sharing broadly. Even so, you can almost always communicate the core safety lesson in a way that doesn't jeopardize any legal position – for instance, you might not name individuals or admit fault, but you can still say "We had an incident with a trench collapse, and as a result we are implementing these new precautions...". The bottom line is that the knowledge gained must not remain siloed with the investigation team or management; it needs to reach the people on the ground.

**Figure:** A simple flowchart illustrating a standard incident investigation procedure, from the initial response through forming an investigation team, establishing the facts, identifying immediate causes and underlying root causes, determining remedial actions, writing the investigation report, and following up. Having a predefined process like this helps ensure no critical steps are missed in the heat of an incident.

A structured investigation process flows from the initial emergency response (making the scene safe and preserving evidence) through disciplined fact gathering, analysis of root causes, documentation of findings, implementation of fixes, and follow-up – all culminating in **organizational learning**. This disciplined approach ensures that even when an accident happens, something constructive comes out of it. Investigations become less about assigning guilt and more about strengthening the project's defenses. As one safety manager put it, "*We find causes, not culprits*." By rigorously investigating every incident and near miss, construction leaders demonstrate a commitment to a safer workplace and often boost morale and productivity in the process. Workers see that the company cares enough to do things right rather than brush problems under the rug. Over time, this approach creates a positive cycle: a robust investigation process catches issues and drives improvements, those improvements prevent future incidents, and fewer incidents mean a safer, more productive jobsite for everyone.

## Investigator Roles & Responsibilities

Even the best-designed process won't work without the right people executing it. In construction incident investigations, having clear roles and responsibilities is key to getting a fair and thorough outcome. This section examines who should be on the investigation team, what each person's role should be, and how to assign responsibility in a way that avoids blame games. For a field leader like a superintendent or foreman, the focus is on how to staff and empower an investigation effort that has credibility with the crew and that gets to the real truth of the matter.

**Who Should Investigate?** On many sites, the default approach is for the crew's supervisor or the site manager to lead an incident investigation. Indeed, supervisors often have first-hand knowledge of the work and what was going on at the time, so they are a logical choice to spearhead the inquiry. However, relying solely on a single supervisor can be limiting. In fact, many companies have found that the most effective investigations involve a team of people – typically a mix of management, safety staff, and even some representation from the crew – working together. Each member brings a different perspective and expertise: a project manager or superintendent might understand scheduling pressures or design issues; a safety officer knows regulatory requirements and investigation techniques; and a veteran crew member can provide practical insight into site conditions or work methods that management might not be aware of.

Including an employee from the field (for example, a well-respected foreman or senior worker who wasn't directly involved in the incident) can also increase trust in the process. Workers will see that "one of our own" is part of finding solutions, not just management behind closed doors. In some cases, companies involve a safety committee member or a union safety representative for this reason. Ideally, the investigation team should be **cross-functional and impartial**.

**Avoid Conflicts of Interest:** A fundamental principle in selecting the team is to avoid any conflicts of interest. Team members should not be in a direct reporting relationship with each other when possible, and especially not with anyone whose actions are under scrutiny. In other words, the person leading the investigation shouldn't be the direct boss of someone who is being interviewed about the incident – that situation could intimidate the interviewee or subconsciously bias the leader to defend their crew or area. Likewise, you wouldn't want someone on the team who might feel pressured to cover up information to protect themselves or their department.

For example, if the incident involved a piece of equipment failure that falls under the maintenance department, it might be wise to have someone *outside* of the maintenance department take the lead on examining that aspect (or even bring in an external expert), rather than the maintenance manager investigating his own area. Objectivity is paramount. Everyone on the team should be able to follow the facts wherever they lead without fear or favor.

Also, ensure that team members can dedicate the time needed for a proper investigation. If a manager is put on the team but is too busy with other duties and can only spend an hour here or there, crucial details might be missed. Serious incidents may warrant temporarily backfilling someone's regular job so they can focus on the investigation. In some cases – particularly high-severity incidents – companies bring in external investigators or consultants (say, a forensic engineer for a structural collapse, or an outside safety auditor) to support the team with specialized expertise or just to provide an independent eye. These outsiders can be very helpful, but they should *supplement* the internal team, not completely replace it. It's important that line management remains involved and accountable for the investigation's outcome; if you just hand everything to a third party, workers might think management is trying to distance themselves or, conversely, that management isn't interested in learning from it. A good balance might be having a site manager or project manager co-lead the investigation with assistance from a safety consultant or specialist as needed.

**Investigation Team Composition:** For anything beyond a very minor incident, it helps to formally designate an investigation team. For a moderate-to-major incident, a typical team might include several roles:

- **Team Leader:** Often the project manager, construction manager, or site superintendent fills this role – basically someone with the authority to direct the investigation and allocate resources. The Team Leader coordinates the overall process: they set the scope of the investigation, assign tasks to team members, ensure everyone knows their responsibilities, and keep the effort on schedule. They'll also act as the point of contact with outside parties, which could mean communicating updates to upper management, interfacing with OSHA or other regulatory inspectors on site, or briefing the client if the client has their own parallel investigation. The Team Leader needs enough clout to gather information and cooperation from all involved, and should be trusted by both workers and management. Often, the team leader will oversee writing the final report and will brief the company's leadership on the findings. Strong communication skills (both upward to executives and downward to workers) are a big plus for this role.

- **Safety Officer (HSE Advisor):** This is the safety professional who brings methodical investigation experience to the team. Their job is to ensure the investigation is thorough and follows best practices. The safety officer is usually skilled in techniques like root cause analysis, evidence preservation, and interviewing, which help keep the investigation on the right track. They will also make sure that any regulatory reporting (for instance, OSHA notifications or required documentation) is handled properly and that the investigation considerations (like not disturbing evidence until certain analyses are done) are respected. In many cases, the safety officer may take detailed notes, help with analysis techniques, and serve as a technical advisor on the team.

- **Technical Expert(s):** Depending on the incident, it's valuable to have one or more people who are experts in the type of work or equipment involved. This could be a project engineer, a structural engineer, an equipment superintendent, a crane operator, a master electrician – whoever has deep knowledge relevant to the incident. They can help the team understand technical factors and evaluate whether something failed due to design, usage, environmental conditions, etc. For example, if a scaffold collapsed, having a scaffold competent person or an engineer on the team would be important; if an electrical accident happened, an experienced electrician or electrical engineer should be involved. Technical experts might analyze physical evidence, run calculations, or identify what further tests might be needed (like a lab analysis of a broken part).

- **Employee Representative:** This might be a senior crew member, like a foreman or lead worker, who was not directly involved in the incident but can represent the perspective of the workforce. Their presence helps ensure the investigation team doesn't become too insulated from how things actually operate on the ground. An hourly craft worker on the team can sometimes spot practical issues or causal factors that managers might overlook ("You realize nobody on night shift could hear that alarm in the compressor room – it's too far away," etc.). It also, as mentioned, builds trust with the rest of the crew: they know someone who does the same kind of work they do is in the room when causes and fixes are being discussed.

- **Management/Legal Oversight (optional):** For incidents with potentially serious liability or broader implications, a representative from senior management or even an in-house attorney might be included in an oversight capacity. They might not dig into the nuts and bolts on site, but they'll be aware of what the team is finding and help ensure that the investigation is adequately resourced and that any legal considerations are minded (e.g., preserving attorney-client privilege for certain analyses, if applicable). Senior management involvement also reinforces that the company takes the incident seriously. However, this person typically wouldn't interfere with the technical investigation; they're there to support and observe, and perhaps to help communicate findings into company-wide decisions.

Once the team is formed, roles should be clarified so everyone knows who is doing what. The **Team Leader** will coordinate the activities: for instance, they might say, "Alice (safety officer) and Bob (engineering) will handle the scene measurements and photos, Carlos (maintenance manager) will collect the equipment records, and I'll interview the foremen and witnesses along with Danielle (safety rep). We'll reconvene at 3 PM to review what we have." Dividing tasks plays to people's strengths and also speeds up the process.

The **team members' responsibilities** include all the hands-on work we described earlier: gathering evidence, conducting interviews, analyzing data, and so on. Often, they will split up duties – one person can sketch and photograph the scene, another can focus on interviews, another can dig through documents and logs. They then come together to share and synthesize the information. Team members should be chosen with an eye to their skills: someone detail-oriented and familiar with equipment might be great at examining a failed part and reviewing maintenance records, whereas someone with good people skills might be ideal for interviewing witnesses who may be shaken or defensive. All members should contribute to formulating the findings and recommendations, not just passively collect data. It's a collaborative analysis. One benefit of a diverse team is that it guards against "tunnel vision." If one person says, "It must have been worker error," another might counter, "But why would a competent worker make that error? Was something confusing or different than usual?" The team can then explore multiple angles together and ensure they aren't jumping to the first conclusion prematurely.

**Training and Competence:** Being an effective investigator is a skill of its own – just as operating a tower crane or welding pipe requires training, so does leading an investigation. It's a good idea to train key personnel in investigation techniques before they ever need to use them. Ideally, at least one member of any investigation team has formal training in incident analysis (for example, they took a root cause analysis course, or learned a methodology like ICAM – Incident Cause Analysis Method). This person can help bring structure to the process, ensuring that the team uses tools like "5 Whys" or cause-and-effect diagrams correctly and doesn't skip steps.

Experience is also a great teacher. If possible, the Team Leader should be someone who has participated in a few investigations previously. There's a big difference between theory and practice when it comes to investigations – experience teaches you how to manage interviews (like knowing when to give someone a break or how to build rapport), how to handle sensitive information (e.g., what to document in writing versus what to communicate verbally if there are liability issues), and how to avoid common pitfalls (like rushing to judgment too soon). Many large construction firms run internal workshops or even mock drills to practice investigations. They might set up a fake "incident" (such as a staged equipment failure or safety scenario) and have participants go through the process of investigating it, then critique how they did. These simulations mean that when a real event occurs, the individuals involved aren't doing it for the very first time under high stakes. They've had at least some rehearsal and feedback, which builds confidence and competence.

**Integrity and Fairness:** Perhaps the most important qualities for anyone leading or participating in an investigation are integrity and objectivity. The team must approach the task with a mindset of uncovering facts and understanding problems, not with an agenda to vindicate or blame anyone. In essence, investigators should be like neutral fact-finders – almost scientific in their approach. They follow the evidence, wherever it leads.

It's vital that investigators resist any pressure to steer the findings or sugarcoat anything. If an investigation starts getting influenced by politics (say, someone in management subtly implies the report should downplay a particular issue) or by personal loyalties ("I don't want my buddy to get in trouble, so I'll omit that detail"), then the whole purpose is defeated. The truth is going to come out eventually, if not on this project then on a future one, possibly in a worse way. So investigators must have the courage to report what **actually** happened and why, even if it's uncomfortable.

For example, suppose during an investigation, a foreman on the team realizes that an unrealistic schedule acceleration imposed by upper management created a scenario where the crew was rushing and took unsafe shortcuts – and that directly contributed to the incident. It might be awkward for that foreman to say, "Management pressure was a cause here," but integrity demands that he bring it up. A good company will want to know that, so they can fix it (maybe by adjusting schedules or communicating expectations better). Leadership should encourage this honesty by explicitly stating that **no one will be punished for truthfully reporting findings**. In fact, one of the best ways to set the right tone is at the very start of an investigation to say to the team (and anyone being interviewed): "*This is a fact-finding exercise, not a blame-assigning exercise. We care about understanding what happened and how to prevent it going forward, nothing else.*" When workers and investigators hear that and see it backed up by actions, they are far more likely to be candid.

Building trust in the investigation process is essential. Other employees will be watching how investigations are handled. If they see that someone who speaks up about a management issue in an investigation gets ostracized or punished, you can bet people will shut down in the next investigation. Conversely, if they see that the company openly accepts tough findings (like "we didn't train our people well enough for this task") and uses them to improve, they'll trust the system more.

**Assigning Responsibility Without Blame:** In a mature safety culture, there's a clear understanding of the difference between **accountability** and **blame**. An investigation will often find that certain actions or decisions – made by specific individuals or groups – played a role in the incident. Acknowledging that is part of finding the truth. But it should be done in a constructive, non-accusatory way. A concept known as "just culture" is often cited here: it means the organization strives to be fair and just, balancing learning with accountability.

Instead of asking "Who screwed up?" in an angry way, the team asks "What went wrong and why did it make sense for people to do what they did?"

This doesn't mean individuals are never held accountable. If someone knowingly violates a safety rule or is reckless, there will be consequences (re-training, discipline, etc.). However, the emphasis of the investigation is on understanding the **context** of people's behavior and the weaknesses in the system. Many mistakes that workers make are not purely random or willful – they often happen because the system allowed them or even inadvertently encouraged them. Maybe a procedure was too complicated, so people skipped steps. Maybe they were under pressure and took a shortcut. Maybe they weren't properly trained and didn't realize the risk. The investigation team should document these contributing factors in a neutral way.

For instance, rather than writing "Operator was careless and didn't lock out the machine," a report might say: "The operator did not lock out the machine before performing maintenance (At-Risk Behavior). He stated that he was trying to save time because the crew was behind schedule. The lockout procedure was available but not actively enforced or routinely practiced on site." This kind of finding points out the action (didn't lock out – which is a serious lapse), but it also gives the why (felt time pressure, culture of lax enforcement). It avoids loaded terms like "careless" which don't help fix anything. Instead, it provides a basis for solutions: perhaps adjust the schedule or staffing so lockout isn't seen as a time-consuming burden, and definitely start enforcing the lockout procedure through training and supervision.

A **just culture** approach means the recommendations will usually focus on strengthening systems – like improving training, simplifying procedures, adjusting incentives – while still addressing individual behaviors fairly. If a rule was knowingly violated, appropriate action will be taken, but it's done in the context of also fixing whatever allowed or prompted that violation. The leadership must back the investigation team in this. If the investigation finds, say, that a management decision or lack of a safety program element was a major contributing factor, the top leaders have to be willing to accept that finding and not scapegoat a worker to avoid owning the problem. Conversely, if the team finds truly egregious misconduct by someone (for example, a worker was caught using drugs on the job, or a supervisor was actively covering up safety defects), a just culture doesn't mean those actions get a free pass – it means you handle them consistently and fairly (everyone knows that being impaired at work is a no-go, and that person is removed according to policy), and you also ask, "How did our system allow that situation to happen and go unnoticed?"

The bottom line is that the investigation team's role is to be truth-tellers for the improvement of the organization. They shine light on whatever problems they find, whether it's human error, equipment failure, management system gaps, or all of the above. It's up to leadership to ensure that truth is heard and acted upon, not buried because it's inconvenient. When investigations are handled in this forthright, fair manner, workers actually become more supportive of them. They see that the company is using incidents to learn and improve, not just to point fingers. Over time, this leads to more reporting of near misses (because people trust they won't be unfairly blamed) and better overall safety outcomes.

**Communication and Empathy:** Another often under-appreciated responsibility of the investigators – particularly the team leader – is communicating with the workforce and any affected individuals in a compassionate and clear way. If someone was injured in the incident, the investigation team should show empathy towards that person and their family. Let them know (through appropriate channels) that the reason for the investigation is to prevent anyone else from suffering the same fate, not to conduct a witch-hunt. Details of the investigation may need to remain confidential until it's done, but the tone can still be open and caring.

Additionally, investigators will likely be interviewing co-workers who might be shaken or traumatized by what happened. Conducting those interviews with care is crucial. Choose a private setting and a time when the person is composed enough to talk. A good interviewer might start with something like, "I know this is difficult to talk about, and I appreciate your time. We're doing this to understand what happened so we can make sure it doesn't happen again. How are you feeling today?" Acknowledging the human element goes a long way. The person being interviewed then understands this isn't just a cold procedure – it's about protecting their team in the future. If they become emotional during the interview, the investigator should be patient, maybe take a short break, or offer to continue later. The information they have is valuable, but their well-being is important too.

Throughout the investigation, the team should **communicate periodically with key stakeholders**. For example, the project director or a senior manager back at headquarters will want to know the general status ("We've interviewed 8 people so far, collected the physical evidence, and are starting the analysis phase…"). If regulatory inspectors (like OSHA) are also investigating, it's wise to have one team member act as the liaison to them – sharing factual information as appropriate and ensuring the two investigations don't interfere with each other. Typically, companies will cooperate with OSHA investigations while still protecting certain information (like attorney-privileged analysis). It's often beneficial to keep an open line with any client representatives, too, if the client is concerned or conducting their own review.

It should also be clear who speaks for the company to outside parties. If news media get wind of a serious incident, usually a designated corporate communications person handles any press releases or statements – not the investigation team. The investigators' findings, however, will inform what the company ultimately says publicly or to the client. For example, if the cause was determined to be a certain equipment failure, the company will frame its message around that once it's confirmed.

Finally, once the investigation is completed, it's the team's responsibility (through the team leader or a senior manager involved) to **present the findings** to different audiences in the company. This could involve a formal debrief with executives, a written summary shared with all project managers, and a safety meeting with the crew on site. In those presentations, the team must walk a delicate line: be honest about what went wrong – even if it's embarrassing or inconvenient – but keep the focus on solutions and *improvements*. For example, instead of saying "John messed up and didn't follow the procedure," they might say "The procedure wasn't followed in this case – we've identified why that happened and here's how we are going to ensure it's followed going forward." It's not about avoiding the truth (if John didn't follow procedure, that will be made clear), but it's about framing it in a way that emphasizes fixing the problem rather than shaming the individual. This approach reinforces to everyone listening that the company cares about **preventing future issues**, not simply punishing past mistakes.

When an investigation team fulfills these roles effectively – with objectivity, fairness, technical rigor, and good communication – the site and the company emerge from an incident stronger. Workers see that incidents are handled professionally and that the company learns and adapts, which in turn encourages them to keep participating in the safety program. Management recognizes that even bad events can yield valuable lessons that improve the organization. In short, the investigation becomes a driver of continuous improvement, not a dreaded ordeal.

## Before, During, and After Event Critiques

Up to now, we've focused on reacting to an incident after it happens (investigating, etc.). But effective construction leaders know that learning opportunities **surround** an incident – not just afterward, but also *before* it ever occurs and *during* the incident's unfolding. In this section, "**event critiques**" refers to structured reflections or reviews that take place at three stages: **before an incident** (proactively critiquing plans and conditions to prevent accidents), **during and immediately after an incident** (real-time management and quick "hot debriefs"), and **after the incident** (formal post-mortems and continuous improvement processes). By embracing all three stages, a team makes sure they capture the full spectrum of lessons. It's about creating a continuum of learning: not only reacting to failure after the fact, but also anticipating potential failure before it happens, managing it effectively in the moment, and then drawing improvements from it afterward.

## Before an Event: Proactive Critiques and Foresight

Long before an incident occurs (and ideally to prevent it from ever occurring), construction teams should be engaging in *pre-event critiques* – essentially, critical reviews of plans, processes, and risk factors **while work is ongoing and before anything goes wrong**. In practice, this can take several forms:

- **Pre-Task Planning and "What If" Analysis:** Whenever you're about to perform a high-risk activity – say a complex crane lift, work at a significant height, or entry into a confined space – it's crucial to pause and think through how it could go wrong before the work starts. Many job sites hold Pre-Task Plan (PTP) or Job Hazard Analysis meetings each day or week, where the crew reviews the job steps and associated hazards. To get real value from these, you need a critical mindset, not just a checklist mentality. Encourage the team to actively question the plan. A great technique is sometimes called a "**pre-mortem**" (the opposite of a post-mortem): ask the crew, "Imagine it's the end of the day and something went wrong – what do you think it would be? What could cause this task to fail or someone to get hurt?" For instance, before lifting a large concrete panel into place, a superintendent might pose, "Alright, let's play devil's advocate – how could this lift go sideways on us?" The crew might bring up issues: "*Well, the crane is set up pretty close to that soft shoulder; the ground could settle." Or "If the wind picks up, this panel might start to swing out of control.*" These are exactly the kinds of concerns you want to hear beforehand. By soliciting these thoughts, the team can then adjust the plan or add precautions: improve the crane pad or add matting to stabilize the ground, set a wind-speed limit, and maybe delay the lift if it's too gusty, etc. Using the collective experience and imagination of the crew in this way shores up defenses before any incident occurs. It's so much better to find a weakness in a meeting or tailgate talk than to find it by experiencing an accident.

- **Safety Observations and Audits as Ongoing Critique:** Another form of before-event critique is the continuous process of safety observations and audits. Whenever a safety officer or supervisor walks the site and points out a concern – say they spot a scaffold that looks a little unstable, or they notice a worker handling a sharp piece of metal without gloves – they are essentially critiquing current operations in real time. A robust safety culture will empower everyone on site, not just supervisors, to do this kind of observation. This might come through formal programs (like behavior-based safety observations, where workers are trained to observe and give feedback) or simply through encouraging an open "see something, say something" attitude. Each near miss or small hazard that someone points out is an opportunity to fix something before it causes an injury. For example, suppose a crane operator radios in and says, "Hey, that last load was awkward – I nearly lost control for a second. We should revisit the rigging method we're using." That near-miss report is effectively a proactive critique of the plan: something isn't ideal, and if we don't address it, the next time we might not be so lucky. High-performing teams treat these near misses and safety suggestions as **free lessons** and respond with the same seriousness as they would to an actual incident. Supervisors investigate them (albeit on a smaller scale) and correct the underlying issues. Over time, cultivating this proactive critique culture – where identifying and addressing issues early is the norm – drastically reduces the number of actual accidents. You're catching and fixing problems while they're still small and no one's gotten hurt.

- **Training and Drills:** Rehearsals and drills can be viewed as a pre-event critique of your emergency preparedness. When you conduct a drill (for example, a fall-arrest rescue drill or a simulated fire evacuation), you're essentially *testing* your response plan and then critiquing how it went so you can improve it before a real emergency happens. If a fire drill reveals that workers on the far end of the site couldn't hear the alarm, that's a critical finding – it means you might need additional alarm speakers or a different notification method (maybe text alerts to phones) to reach everyone. If a fall rescue exercise shows it took 15 minutes to rig the rescue gear to lower a hanging worker, you might realize you need to pre-rig anchor points or provide additional training to do it faster. These drills are invaluable because they often expose weak points in emergency plans that look fine on paper. The important thing is to take them seriously (no "checkbox" drills just to satisfy a requirement) and to promptly fix any shortcomings they reveal. Good leaders foster an environment where practice runs are treated as real opportunities to improve, not as annoying exercises. That way, if and when a real emergency occurs, the crew is much better prepared and the response will be smoother.

*In summary, "before-event" critiques are about being a step ahead – thinking critically before something bad happens.* It requires a mindset of hyper-vigilance and a bit of humility, essentially assuming "we might have missed something" and actively looking for it. This is where seasoned construction leaders often shine: they've seen a lot, so they know what can go wrong, and they aren't shy about asking the tough "What if?" questions up front. To some, this approach might feel like pessimism or over-cautiousness, but it's actually just prudent planning. If teams make a habit of this kind of proactive questioning and improvement, many potential incidents can be caught in the hypothetical stage and defused. As the saying goes, "*Plan for the worst, and you'll likely prevent it.*"

## During an Event: Real-Time Management and "Hot Debriefs"

When an incident does occur, the immediate priority in that moment is, of course, managing the situation – controlling the incident, helping any injured people, and preventing things from getting worse. However, even **during** the event and its immediate aftermath, there are opportunities for quick learning and adjustment. In this stage, two aspects stand out:

how you manage the incident in real time (while it's happening) and how you conduct an immediate "**hot debrief**" right after the situation has been stabilized.

**Real-Time Incident Management:** Even as an emergency is unfolding, the person in charge (and others around) should keep their eyes open for what's working and what's not. Think of it like this: you're the incident commander dealing with a crisis, but part of your brain is observing and noting any snags or unexpected issues. There's obviously no time for a detailed analysis at the moment, but there is a need to stay situationally aware and gather mental notes for later.

Take a scenario: say a section of formwork collapses on site, and there's a concern someone might be trapped. The site superintendent immediately initiates the emergency response plan – she halts all work in the area, has someone call 911, and organizes a headcount and search of the affected zone. While doing all this, she might notice certain things: maybe the site's two-way radio channel is flooded with chatter, and it's hard to communicate, or perhaps some workers didn't know where the muster point was, and there was confusion about who should do what. A savvy leader will recognize these as **issues to fix later** – essentially, she's performing a mental critique of the emergency response even as she's directing it. It requires composure to both manage the crisis and simultaneously observe it, but it can be done.

Another part of real-time management is being **adaptive**. No matter how well you plan, real incidents often don't unfold exactly as per the book. Equipment might not work as expected, or people might be too shaken to carry out their assigned roles. Good leaders make quick adjustments – and those adjustments themselves should be noted and later questioned. For example, perhaps your emergency plan said the lifting crew should assist in rescue operations, but in this collapse scenario, the lifting crew members were in shock or minorly injured and weren't able to help. So another crew stepped in to assist. Later on, when things are calm, you should ask: "Did our emergency plan assign the wrong people to the task? Should we have a backup for that role? Why did we need to deviate from the plan, and was our original plan missing something?" Each deviation from the plan is a learning opportunity to make the plan better for next time.

During an incident, the leader should also encourage **clear, two-way communication**. In a chaotic situation, information is gold. Encourage everyone involved in the response: "If you see any other hazard or problem, speak up immediately!" In effect, everyone on site becomes an extra set of eyes and ears. For instance, maybe a worker notices that a certain machine wasn't shut off amid the emergency, and it's creating a secondary hazard – if they shout that out and someone kills the power, a second accident is prevented. This kind of collective vigilance acts as a live critique of the unfolding event, catching things that the primary responders might not see. It helps to keep the incident from escalating further.

**Hot Debrief Immediately After:** Once the immediate danger is over and any injured persons have been cared for, it's extremely valuable to conduct a short, informal "**hot debrief**". This concept is borrowed from emergency services and military practice, but it works on construction sites too. A hot debrief is basically a quick gathering (often just 5–15 minutes) of the people directly involved in or affected by the incident, held right afterward to capture initial impressions and feedback while it's fresh.

Emotions may still be high – people might be relieved, shaken, upset, or adrenaline-charged – so the person facilitating the debrief (perhaps the site manager, foreman, or safety officer) should handle it with sensitivity. Emphasize that this is not about blame. The purpose is simply to hear from everyone: *What happened from your perspective? What did you think went well in how we responded? What didn't go well?* It gives each person a voice and often reveals immediate clues or issues that the formal investigation will examine in greater depth.

Imagine there was a small fire on site – perhaps a piece of equipment caught fire, but it was extinguished quickly. In the hot debrief afterward, one crew member might say, "We had trouble finding a fire extinguisher at first because it was hidden behind some stored materials." Another might add, "Once we got it, it worked well, and John's quick thinking to shut off the fuel supply really kept it from getting worse." These are golden nuggets of information. The first comment tells you there's a housekeeping or placement issue with fire extinguishers that needs to be *fixed immediately*. The second comment highlights a positive: someone did something right that could be reinforced as a good practice.

The facilitator of the hot debrief can jot down these points (or just mentally note them if writing seems too formal for the moment). This session also serves as an emotional release for the team. It's a chance for people to express concerns or even just say "I'm glad everyone's okay" and hear each other out. Management or the supervisor should thank the crew for their handling of the situation – especially if their actions prevented a worse outcome. That kind of immediate positive reinforcement ("Hey, everyone did a good job evacuating calmly – that made a big difference, thank you") builds trust. The workers see that management acknowledges their effort even when things go wrong, which makes them more likely to respond well (and report accurately) in the future.

I recall a construction superintendent describing his approach to hot debriefs: after any incident, once the area is secured and everyone is safe, he gathers the crew in the break area. He'll say something like, "Alright, we just went through a serious situation. Let's take a few minutes – grab some water, catch your breath – and then I want to hear from you about what happened." He makes sure to frame it as a conversation, not an interrogation. In those few minutes of open talk, crew members might mention things they would hesitate to bring up later after the lawyers and big bosses are involved. For instance, someone might admit, "Honestly, I wasn't sure what that alarm was – I've never heard it before, so I froze for a second." That indicates maybe training on alarms or emergency signals is lacking. Another might say, "We've had a lot of near-misses with that machine sparking, and we kind of ignored them," which is incredibly useful information ("why were they ignored?" becomes a question). Or someone could commend a coworker: "If Diego hadn't yelled for everyone to get back, I might not have realized what was happening." That tells you Diego noticed something others didn't – maybe he smelled something or saw a flash – which could lead to understanding how the incident developed.

These details can get lost as time passes and memories get edited, so capturing them "while the fire is still hot" is important. However, **hot debriefs should remain brief and focused**. They're not the full investigation. If someone starts venturing deep into speculation ("I think the pump failed because the mix was wrong…"), the facilitator can note it but also gently steer away by saying, "That's something we'll definitely look into. Right now, I want to focus on what you experienced and observed in the moment." The immediate questions to answer are: *What can we learn or fix right now? and Are we okay to continue work, or do we need a pause?*

If the hot debrief reveals that a fire extinguisher station was blocked, you don't wait for a report – you go out and clear that area **right away** before work resumes. Or if someone says, "Honestly, I'm pretty shaken up and can't focus right now," the supervisor might decide to pause work for the rest of the day for that crew (or reassign them to non-hazardous tasks) because continuing in an emotionally rattled state could lead to another incident. Sometimes one key takeaway from a hot debrief is, "We're not in a safe headspace to keep working at the moment – let's regroup later or tomorrow."

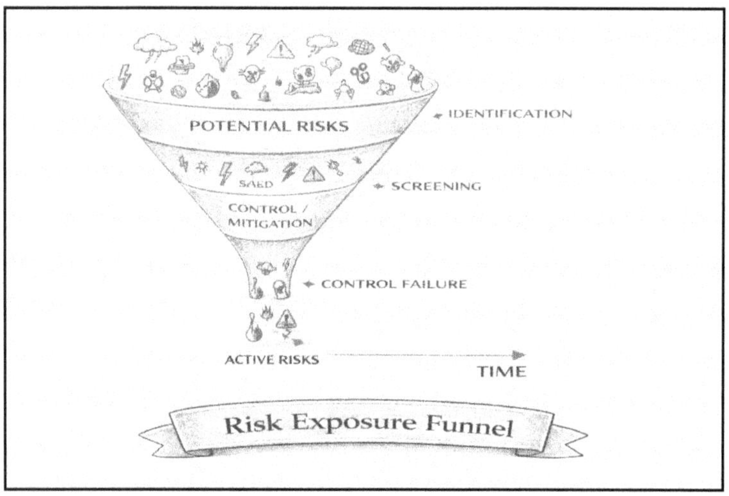


In sum, **during an event** and immediately after, leaders should simultaneously manage the crisis and critique what's happening. By capturing lessons on the fly and conducting a quick hot debrief, you begin the learning process even before the dust settles. This not only can improve the eventual investigation (because you've gathered those first-hand impressions and issues), but it also signals to the team that even when something goes wrong, the response is to learn. Additionally, any obvious fixes that come out of the hot debrief can be implemented right away, potentially preventing further problems that day. It's an approach of "stabilize, observe, learn, and improve" all in short order, making the best of a bad situation.

## After an Event: Formal After-Action Reviews and Continuous Improvement

The most extensive and in-depth critiques happen after the incident, once the situation is fully stabilized, the dust has settled, and the initial investigation results are coming in. This is where the organization truly digests what happened through structured reflection sessions and by integrating the lessons into future practice. We've already covered the investigation process, so in this section, we'll focus on the **critique and learning activities that follow or accompany the investigation** – essentially, how to cement the lessons learned and make sure they lead to real improvements.

**After-Action Review (AAR) Meeting:** One powerful tool for post-incident learning is the **After-Action Review (AAR)**, a concept borrowed from the military that has also been adopted in many industries and project management practices. In construction, you can conduct an AAR shortly after the investigation is complete (or even after the conclusion of an entire project or major phase, especially if that phase experienced some incidents or major challenges).

An AAR is an open and blame-free discussion centered around a few key questions: 1. **What was supposed to happen?** (What was the plan or expected result?) 2. **What actually happened?** (What occurred during...Leaders listen actively to such admissions rather than scolding – this is valuable feedback that something in training or communication failed and needs improvement.

Crucially, an AAR should drive home the lessons learned and lay out how the team will change going forward. It's an opportunity for leadership to reinforce their commitment: for example, "*We found our inspection process was lacking on this job. Going forward, here's what we will do differently on every project...*" and assign specific people to carry out those changes. Documenting the discussion is useful because many good ideas can surface. In serious cases, the AAR might include executives or even involve an outside facilitator to ensure tough questions are asked. But whether it's a formal, large-scale meeting or a quick on-site wrap-up, the real test of an AAR's effectiveness is whether participants leave with a clear understanding of the **lessons learned** and see those lessons turn into action. If folks can say, "We had an incident, we talked it through openly, and here are the concrete things we're now doing differently because of it," then your AAR did its job.

**"Hot Wash" and Phase-End Reviews:** On long projects, it's often beneficial not to wait until the very end to reflect. After any major phase or milestone – especially one that had some problems – the team can hold a phase-end critique (what some call a "hot wash," similar to an AAR but done periodically). For example, after finishing the deep excavation phase of a project (which may have had a few trench collapses or utility strikes), the project team might pause and evaluate: *What safety issues did we encounter in this phase? How can we manage this type of work better next time?* By doing so before moving on to the next phase (say, foundation work), they can apply fixes immediately to the same project, rather than waiting to do better on the next project. In essence, you create a **continuous improvement loop** within the project itself. This practice of regular reflection is a hallmark of high-performing teams – you're constantly learning and adjusting as you go, rather than only learning in hindsight.

In fact, institutionalizing these post-project or post-phase reviews is a best practice. Safety organizations often recommend treating incident investigations as part of an ongoing **cycle of improvement** – investigate, implement changes, then later review if those changes are effective. Many companies, at project completion, will include a section in the close-out report on safety lessons learned. They might hold a meeting with key stakeholders to discuss insights like, "We had a lot of minor electrical shocks on this job. Why was that? Maybe our approach to temporary power and GFCIs needs an overhaul before the next project." Doing these post-mortems ensures that the mistakes of one project inform the plans for the next, so the same issues don't crop up again (or are better mitigated if they do).

**Critiquing the Response and the System:** After an event, it's also important to critique how well the **organization's response** and overall safety management system functioned. This goes beyond the direct causes of the incident and looks at how our systems performed under stress. For instance, suppose a worker fell from a scaffold and was injured. The investigation finds the cause (maybe guardrails were missing and a lapse in training). But separately, ask: *How did our emergency response go?* Did the team respond quickly and correctly? Was first aid administered right away, and was the injured worker safely and promptly transported to medical care? Did we notify the worker's family and OSHA in a timely and compassionate manner? By answering these questions, you might find areas to improve in your emergency plans or communication protocols. Maybe during the confusion of the incident, several minutes passed before 911 was called because everyone assumed someone else had called – that indicates a need for a clearer assignment of who makes the call. Or maybe the process for notifying the family had hiccups, suggesting you should refine that procedure.

Likewise, at a higher level, treat the incident as a "**stress test**" **of your safety program**. Leaders should ask, "What weakness in our safety management system allowed this incident to happen?" Perhaps you discover that supervision was spread too thin due to cost-cutting – that's a management decision to rethink. Or maybe a near-miss a week earlier foreshadowed this event, but workers didn't report it – that might point to a trust or communication gap (so you know to work on your reporting culture). By critiquing the system in this way, you view the incident not just as an isolated failure but as evidence of potential broader issues. This can be uncomfortable ("maybe we in management contributed to this"), but it's necessary for true improvement. High-reliability organizations in other industries do this routinely: after a failure, they examine both the technical causes and the organizational causes. Construction should do the same. If a subcontractor was involved in an incident, for example, the general contractor might reflect on how they pre-qualified that sub or communicated expectations to them – and whether doing that better could prevent similar incidents. This kind of introspection – looking in the mirror – is the hallmark of a learning organization.

**Maintaining a Learning Mindset:** It's worth noting that the "after" phase of learning isn't just a one-time meeting or report. It should evolve into an **ongoing learning system** for the company. A lesson identified might lead to a new standard or training program, which then should be revisited months later to ensure it's working. In essence, the aftermath of an incident has a long tail: initial fixes, monitoring, adjustments, and knowledge sharing. Good leaders keep the topic alive as long as needed. They might say, "Three months after that fall incident, we'll do a special audit of all scaffolds to make sure our new scaffolding checklist is being followed." If that follow-up audit finds issues, that feedback loops into further action (maybe we need additional training, or maybe the checklist needs tweaking).

One safety director shared that after any serious incident investigation, he personally visits the site (or the successor project) a few weeks later and leads a follow-up discussion: "*What have we changed here, and how is it working? Is there anything we might have missed?*" This not only reinforces that management has not forgotten about the event, but it also shows that they genuinely care to learn everything they can from it. Employees see that lessons aren't just declared and then shelved; they're actively being integrated and tested in real operations.

*In summary, after-event critiques involve formalizing the learning through reviews and making sure the organization as a whole absorbs it. Whether it's through an AAR meeting, a post-project "hot wash," or a systemic analysis of your safety program, the goal is to turn one project's hard-won lesson into common knowledge for all projects. The tone in these after-event critiques stays blame-free and forward-looking: we're examining processes and decisions not to punish past mistakes, but to improve future performance. It's akin to a sports team watching the game replay after a loss – not to shame the player who missed a catch, but to understand how to catch it next time. With that mentality, teams emerge from accidents wiser and more resilient.*

## Recording & Disseminating Lessons Learned

A lesson that isn't captured and shared is a lesson lost. Construction is an industry of transient projects and rotating teams – knowledge can easily slip through the cracks when one project finishes or when key people move on. That's why developing robust systems to **record and disseminate** lessons learned is the capstone of event investigations. In this section, we explore how organizations can document lessons in a usable way and how they can ensure those lessons reach the people who need them, leading to lasting improvements.

**Documenting Lessons in a Structured Way:** The immediate output of an investigation is often a detailed report, but a thick report filed away is not the most user-friendly way to inform future projects. Many companies, therefore, maintain a "**Lessons Learned**" **register or** database, where each incident (or significant near miss) is distilled into key points. A typical entry might include: a brief description of the event (*what happened*), the root causes identified (*why it happened*), and the corrective actions taken (*what was done to prevent it next time*). It often also has tags or categories (e.g., "Scaffolding," "Electrical," "Crane Ops") so that others can search by topic. The goal is to create an accessible knowledge base that anyone planning similar work can reference. For example, before starting a new high-rise construction project, a project team could search the database for "tower crane incidents" and find write-ups from past projects, learning about pitfalls and best practices discovered from those events.

It's important to be *disciplined* in how you record a lesson. A vague note like "Worker was injured due to tool misuse – remind everyone to use tools properly" is too superficial and will likely be ignored. Instead, a well-documented lesson reads like a mini case study with a clear takeaway. For instance: "*Incident: Laceration injury when a grinder's cutting disc shattered. Causes: The grinder disc had expired and had developed cracks, and the worker was not wearing a face shield. Lesson: Always inspect grinder discs for cracks/expiration before use, and always wear a face shield when grinding. Actions: After the incident, we updated the PPE policy to require face shields for grinding and held a toolbox talk on inspecting abrasive wheels. Result: No similar injuries in the 12 months since.*"

An entry like that tells a story and gives a concrete preventative measure. It's specific enough that someone else reading it can take action ("inspect your grinder discs and enforce face shields, because here's proof of what could happen if you don't"). The process of writing it out also forces the team to articulate exactly what they learned.

**From Reports to Organizational Knowledge:** Simply writing a lesson in a database doesn't mean the organization has truly learned it. A lot of companies dutifully file lessons learned, but then fail to actually change anything – the "lesson" becomes just a line in a report instead of a change in behavior. To avoid this, each lesson must be tied to tangible follow-ups within the organization: updating training materials, revising standard procedures, changing design guidelines, etc., as appropriate. For example, if a lesson is "Type X form ties can fail in cold weather – use Type Y in winter conditions," that insight should prompt your engineering department to update the design standards or approved tie list accordingly. The lesson might live in the database, but more importantly, it needs to live in day-to-day practice going forward.

Often, it helps to assign an "**owner**" to each significant lesson – someone responsible for carrying that lesson into the broader organization. If Project A learns something, you might assign the Safety Manager or Operations Manager to ensure that the lesson is communicated and applied on Projects B, C, and D. Some firms have a safety committee or senior management group review all new lessons on a quarterly basis to make sure the follow-up actions have been implemented company-wide.

**Dissemination: Getting the Word Out:** A lesson does no good if the people on other sites or future crews don't hear about it. So spreading lessons learned is just as critical as the investigation itself. There are several effective channels to do this in construction:

- **Safety Bulletins and Alerts:** Many organizations issue quick safety bulletins or "alerts" after a noteworthy incident or near miss. These are usually one-page notices (often with a photo or graphic) summarizing what happened and highlighting the key lessons or precautions. They can be posted on jobsite bulletin boards or emailed to project teams. For example, if there was a near miss where a heavy pipe slipped from a crane sling, the company might send out an alert: *"Near Miss: Pipe Drop During Hoisting – Lesson: Always use tag lines and check sling angle when lifting long objects."* It would briefly describe the incident and the recommended safe practices. These alerts get information out quickly across the company and keep everyone vigilant. Even industry groups sometimes share anonymized alerts among different companies, because everyone benefits from broader awareness.

- **Toolbox Talks and Safety Meetings:** Incorporating recent lessons into routine safety meetings is a great way to reach the crew level. A foreman or superintendent can start a morning huddle or weekly toolbox talk by recounting a lesson from another project. "*I want to share something that happened on one of our sister jobs last month – a worker got a bad shock because an extension cord's insulation was damaged. So today, let's all double-check our cords and remember to get rid of any that are worn out. Here's what to look for...*" By doing this, crews learn from incidents they weren't directly involved in. Even lessons from other companies or from news reports can be discussed: "*Did you hear about that scaffolding collapse in the news? Let's talk about how to make sure we never have that here.*" When workers see that these aren't abstract stories but real events that could happen to them if they're not careful, it resonates. The idea is to foster a mindset of **learning from others** – we don't have to touch the stove ourselves if we can learn from someone else who did.

- **Lessons Learned Workshops/Roundtables:** Some larger firms organize periodic meetings (quarterly, biannually, etc.) where project managers, superintendents, and safety professionals come together (in person or via video) to share recent lessons. Each project might present one challenge or incident, and what they learned from it. These roundtables create a culture of openness – you're essentially saying, "Here's a mistake we made and how we're fixing it." It can feel a bit humbling, but it is incredibly useful. Others in the room can ask questions or offer their experience if they have dealt with something similar. This prevents siloed knowledge and builds a network of problem-solvers across the company. The key is that management must encourage this open sharing and not penalize projects for admitting problems. In fact, it's good to praise teams for bringing up issues in these forums, because it shows they're proactive about improving.

- **Digital Platforms and Apps:** In today's connected world, technology can significantly aid lesson sharing. Companies might use an intranet portal, a shared drive, or a custom safety app where lessons learned are uploaded and easily searchable. Some go further and push notifications to employees. For example, a mobile app could send a brief alert: *"Safety Flash: Two cases of scaffolds not being fully planked – ensure yours are 100% planked!"* That way a field supervisor or crew leader sees it on their phone immediately. Some organizations create short e-learning modules or videos reenacting an incident and showing what went wrong, which can be viewed during safety meetings or individually. The idea is to integrate learning into daily workflow – maybe at the next company-wide call or meeting, the topic of a recent lesson is on the agenda. With data analytics, some companies even monitor trends: if multiple projects report, say, near misses with a certain power tool, the system flags it, and the safety team can issue an alert or inspect that issue across all sites.

**Ensuring Lessons Stick:** Dissemination isn't just a one-way broadcast – we need to make sure the lessons are understood and adopted. One effective approach is to engage workers in conversations about the lessons. Encourage foremen and supervisors to ask their crews after sharing a lesson, *"Has anyone here seen something like this? What would you do if you encountered this situation?"* By discussing it, people process the information actively rather than just hearing it passively. You know a lesson has truly "stuck" when you start hearing it come up later on the jobsite without prompting. For instance, a crew member might say, *"Let's use extra cribbing under this beam – remember that lesson from last year about the beam that slipped?"* When past incidents become part of the crew's decision-making vocabulary, that's proof of a learning culture. It means the organization's memory is working: people are internalizing the lessons and applying them.

However, as time passes or personnel turn over, even well-learned lessons can fade. That's why it's important to periodically **refresh** key lessons. Annual safety refresher trainings are a good opportunity to highlight, "*Here are the top 5 lessons we learned last year,*" reminding everyone of those stories and takeaways. Orientation for new hires is another opportunity: rather than just reading rules to them, also share a few significant historical incidents from the company and the lessons they taught. "*For example, we had a tragic accident with a trench collapse a decade ago. Since then, we have had a strict policy that no one enters an unshored trench over 4 feet deep. Let me tell you how that accident happened so you* *understand why this rule matters." Real stories like these stick with people much more than rules on paper, and they convey the seriousness behind the safety policies.*

It's also valuable to **verify** that lessons learned are actually being put into practice. This can be done by targeting certain audits or inspections on areas where you know you've had issues before. If multiple past incidents and lessons relate to, say, lockout-tagout procedures, then make it a point in the next quarter to do focused inspections of lockout practices on all projects. If those audits find that people are still making mistakes or skipping steps, it tells you the lesson didn't truly take hold – maybe more training is needed, or a different approach to enforcement. Essentially, you "trust but verify." When you find a gap, you circle back and reinforce the lesson until it sticks.

**Sharing Beyond the Company:** Some forward-thinking contractors even share key lessons with the broader industry – through trade associations, safety conferences, or published case studies – especially if they learned something that could benefit others. For example, the U.S. Army Corps of Engineers maintains a public database of construction lessons learned; a company could submit a write-up of an incident and its lessons to contribute to industry knowledge. While construction is competitive, safety is one area where everyone wins by collaborating. By sharing lessons (even anonymously) with peer companies, you help raise the bar industry-wide. It also enhances your reputation as a safety leader. Clients and partners appreciate when a contractor isn't just fixing their own problems but also helping others avoid them. Of course, any external sharing should be done carefully (usually anonymizing project names or individuals), but the overarching idea is that **knowledge should flow freely when it comes to keeping people safe**.

**Making Lessons Easy to Find and Use:** Despite good systems, sometimes lessons don't spread simply because people are busy or don't know where to look. It's important to make your lessons learned system user-friendly. Ensure entries are concise and well-tagged so that an engineer or superintendent can quickly search and find relevant info without wading through hundreds of pages. Some companies highlight a particular "Lesson of the Month" – e.g., a short email or poster focusing on one notable incident and its takeaway – to keep lessons in the spotlight one at a time. Embedding lessons into daily tools and procedures is also a clever strategy. For instance, if you update a crane lift plan template, you might add a small note in the margin: "*Remember: Past incidents have occurred due to high winds. Always check wind speed and don't lift if >20 mph.*" That way, whenever someone is planning a lift, they see a built-in reminder of a lesson learned, right at the point of use.

Finally, leadership must foster an attitude that **learning never stops**. The moment a company grows complacent – thinking "we've seen it all" – something new will catch them off guard. A true learning system treats each project, each incident, and even each near miss as data to get better. It also treats success as data: if a project went off with zero incidents, what did we do right and how can we repeat it? Lessons learned shouldn't only be about failures; capturing what worked well helps propagate best practices, too. In other words, the learning culture is not about blame; it's about constant curiosity and improvement.

To conclude, recording and sharing lessons learned is about creating a "memory" for the organization – one that outlives individual projects and people, and prevents the same mistakes from happening again. It closes the learning loop that begins with an incident and an investigation. When an incident's lessons are documented clearly, shared widely, and woven back into training and procedures, the pain of that event is transformed into wisdom for the future. Over time, a construction company with a strong lessons-learned culture will find that it repeats mistakes less often, responds to new challenges more effectively, and earns a reputation (inside and outside the company) as a high-reliability organization. The leadership challenge is to champion this culture: dedicate time and resources for capturing lessons, insist that those lessons be integrated into everyday work, and celebrate team members who contribute to the company's collective knowledge. In doing so, leaders ensure that every event – even the bad ones – ultimately makes the team **stronger, smarter, and safer**.

## Operational Takeaways for Today's Job:

- If an incident or near-miss happens, secure the area first. Pause work, rope off the spot if needed, and preserve any evidence (broken parts, positions of equipment) until it can be reviewed. This not only protects the scene for investigation but also prevents a secondary incident.
- Begin fact-finding immediately while memories are fresh. Have the involved crew and witnesses briefly write down or tell what happened as soon as it's safe to do so. Capturing honest accounts on the day of the event provides valuable lessons that might otherwise fade or get skewed over time.
- Assign someone to lead the follow-up. Whether it's the safety officer or a competent supervisor, put one person in charge of investigating the cause, gathering reports, and coordinating any analysis needed. Clear responsibility ensures the investigation doesn't fall by the wayside once work resumes.
- Hold a post-incident toolbox talk. After an event (minor or major), gather the team to discuss what went wrong and how to prevent it in the future. Focus on the facts and improvements, not blame. This reinforces a learning culture and helps everyone take away something actionable.
- Close the loop on every investigation with action. If an investigation finds that, say, a ladder slipped because of oil on the rungs, implement a new cleaning check or buy grip tape – and communicate that change. Ensure lessons learned translate into a concrete change on-site or in procedure.

Chapter 10

# Independent Verification and Quality Checks

Quality in construction is everyone's responsibility, but the most effective teams don't just rely on individuals "doing it right" – they build in verification and checks to catch mistakes before they become costly problems. This chapter explores how to systematically verify work quality through a mix of informal and formal methods, independent reviews versus in-process checks, focused verification of critical path activities, and ways to guard against confirmation bias in quality assurance (QA) and quality control (QC)

## Informal vs. Formal Verification

Every construction team uses some level of verification to ensure work meets requirements. Informal verification often happens continuously on the jobsite: a foreman double-checking a measurement, a crew member eyeballing a weld, or a superintendent doing a quick walkthrough to spot obvious issues. These checks are usually undocumented, quick, and based on experience. They catch many day-to-day errors in real time. In contrast, formal verification involves structured inspections, tests, and sign-offs – often with documentation like inspection reports, test results, photographs, or checklists. Formal verification tends to occur at defined milestones or completion points and follows established standards or procedures (for example, building code inspections, specified hold points, laboratory material tests).

The key differences between informal and formal verification come down to rigor and record-keeping:

- **Rigor**: Informal checks are generally visual or based on the craftsperson's judgment; formal checks use specific criteria or instruments. For example, a worker might simply glance at a concrete surface, whereas a formal check would involve using a calibrated device or measuring against a tolerance (reviewing concrete strength test reports, for instance). Informal methods might catch obvious defects, while formal methods aim to prove compliance with requirements.

- **Record-Keeping:** Informal verification is usually not recorded – if a foreman spots a misaligned anchor bolt and has it fixed on the spot, it may never be written down. Formal verification produces evidence and documentation. Test reports, inspection photographs, and non-conformance records serve as proof that work meets the standards (or not), creating a trail of accountability. If a city inspector signed an approval sticker (the classic "green tag") on an electrical panel, it is a formal record that the work passed inspection. Such records are crucial if disputes arise later, showing that the project team took the required steps to check quality.

Both types of verification are valuable. Informal verification is immediate and frequent, forming the first line of defense against errors. Crew leaders and skilled workers should be encouraged to self-check and peer-check work continuously – this fosters a "right-first-time" culture where workers take ownership of quality. On the other hand, formal verification provides an objective assessment and a pause to ensure everything is correct before moving forward. Formal inspections often bring a fresh set of eyes (like a building inspector or QA manager) who isn't as close to the work and may spot issues others overlooked. Formal checks also typically use established checklists and criteria. For example, a formal pre-pour inspection checklist for a slab might include verifying rebar size, spacing, and clearance against the structural drawings, whereas an informal check might simply be the foreman scanning that the rebar "looks about right."

**The Value of Formal Checks:** A commercial concrete contractor once planned a major slab pour at dawn. The superintendent did an informal evening walkthrough and thought the rebar was all in place. However, just 30 minutes before the pour, an independent inspector performing a formal pre-pour check discovered that an entire section of top-mat reinforcement was missing – a detail easy to overlook without a thorough, systematic check. The pour was halted until the steel fixers corrected the issue. In this case, the formal verification prevented a costly repair, since pouring concrete over missing rebar would have meant jackhammering out the slab later. The lesson is clear: informal checks alone might not catch subtle or hidden defects, especially when the team is under schedule pressure or simply assumes everything is fine. Formal verification acts as a safety net to ensure nothing critical slips through.

**Superintendent's Perspective:** "Early in my career I trusted my guys so much that I skipped a formal rebar inspection on a footing. We poured and later found out we had missed some reinforcement. It became a nightmare of drilling and epoxy doweling to fix. Never again. Now I always do that formal check with a second set of eyes – it's not about trust, it's about making sure we didn't overlook something obvious."

**Balancing Informal and Formal:** Effective projects use a layered approach: encourage crews to catch issues informally as they work, but also schedule formal hold-point inspections and tests for critical work. Pipefitters might informally pressure-test a line themselves first (to see if gaskets hold) before calling in the third-party inspector for the official pressure test. By doing so, the crew can fix any obvious leaks (informal correction) and then confidently undergo the formal test, which becomes part of the project records. This saves time and ensures that formal inspections are more likely to pass on the first go.

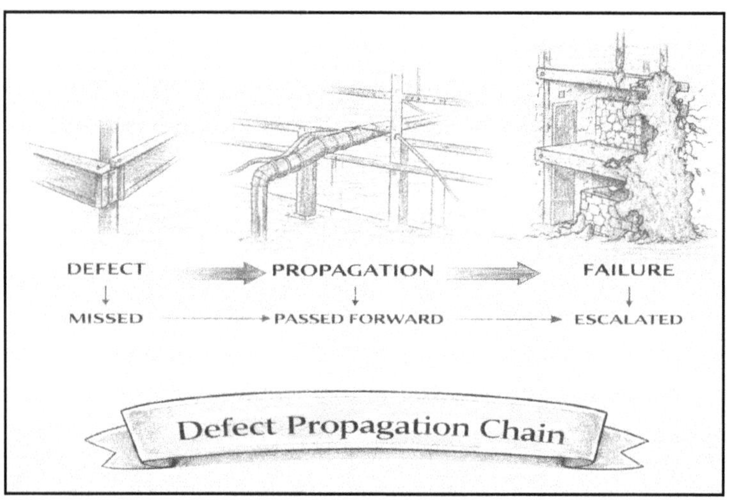


Informal verification is the day-to-day vigilance of the field team – it's flexible, fast, and done continuously. Formal verification is the structured quality gate – it's thorough, objective, and documented. A successful quality management system uses both in tandem: "trust, but verify." Trust the craftsmanship of the team (and have them double-check each other), but verify through formal checks that the work truly meets project requirements before it's considered complete.

## Independent vs. In-Process Checks

Quality checks can be categorized by who performs them and when. In-process checks (also called first-party checks) are those performed by the team doing the work as part of their workflow. Independent checks (sometimes called second-party or third-party checks) are done by someone not directly involved in executing that particular work – this could be a separate internal QA/QC staff member, a consultant, or an owner's representative or inspector. Both approaches have important roles on a project, and using them together provides a more robust quality assurance program.

**In-Process Checks (Self-Control):** These are the checks that crews and subcontractors perform on their own work as they go. It includes things like a mason frequently using a level to ensure walls are plumb, or an electrician using a checklist to verify all terminations are tight in a panel before closing it up. In-process checks are essentially quality built into the process – often guided by the contractor's Quality Control plan or standard operating procedures. A good crew has a culture of not waiting for an outside inspector to find a mistake; they prefer to catch mistakes themselves. Many contractors implement a "three-step" quality process (similar to the Army Corps' Preparatory, Initial, and Follow-up phases): before a task begins, the team reviews requirements (prep); once work starts, they inspect the first piece of work (initial); and they continuously monitor ongoing work (follow-up) to make sure it stays on track. This is all in-process QC done by the people performing or supervising the work. The advantage of in-process checks is that issues are identified immediately – ideally before they become bigger problems or before a lot of defective work is put in place. It's also efficient because the feedback loop is short: the person who made a mistake finds it and fixes it on the spot.

However, in-process checks have limitations. The crew or subcontractor might develop "tunnel vision" or have a bias of thinking their own work is correct (after all, they just installed it, so they believe it's fine). There's a risk of confirmation bias here – the installer may unintentionally overlook their own errors. Additionally, some contractual or code requirements stipulate that an independent party must verify certain items regardless of the contractor's own checks (common for safety-critical items such as structural connections, welding, and electrical systems).

**Independent Checks:** These are verifications by personnel not directly responsible for doing the work. This could be an on-site QC inspector who works for the general contractor but floats between trades, verifying work; a third-party testing agency (for example, for soil compaction tests, concrete cylinder breaks, and weld x-rays); a consultant hired by the owner to conduct quality audits; or city/code inspectors. The idea is to bring an unbiased perspective to evaluate the work. Because they are independent, they are less likely to have the "pride of ownership" that might cause someone to gloss over an issue. Independent verifiers also often have specialized expertise or certifications (for example, a certified welding inspector checking weld quality, or a certified firestop inspector ensuring fire caulking is installed correctly).

**A multi-tier quality system often uses both approaches:** the contractor (and subs) perform their own QC checks, and an independent party performs overarching QA oversight. This dual layer provides a built-in check-and-balance. Top-performing teams might review inspections internally and then have an independent verifier cross-check those inspections. On one utility project, inspections were reviewed by both the contractor's QA engineers and independent verifiers, ensuring bias-free oversight and cross-validation before closeout. In other words, the contractor's QA/QC team might conduct the first review, and then an independent auditor or owner's rep conducts another pass to confirm that nothing was missed and that the inspection process itself was effective. This significantly reduces the chance of a systematic oversight, because an independent reviewer might catch something that the project team accepted in error.

**When should you use independent checks (in addition to in-process checks)?** In general, every task should have some level of in-process verification – that's just good practice. Independent checks are not needed for absolutely everything, but there are certain situations where they add tremendous value. Key scenarios include:

- **Critical or High-Risk Work:** If a failure in a particular work item could be catastrophic for safety or extremely costly to rework, an independent check is prudent. For example, have a structural engineer or third-party inspector verify reinforcing steel placement before a major concrete pour (since you cannot see the rebar after pouring). In fact, construction inspection hold points are particularly important when work cannot be inspected later because a problem could be covered up. A hold point before a concrete pour allows verification of the steel reinforcing before it is embedded in concrete. The crew will do their own checks while installing rebar (in-process), but a hold-point inspection by an independent party double-checks it formally before the pour. Another example: before a large electrical switchgear is energized, have an independent testing firm perform equipment testing and infrared scanning – catching issues at that stage averts failures later when the system is live.

- **Compliance and Code Requirements:** Many codes and contracts mandate independent testing or inspection. Building codes, for instance, often require special inspections by licensed professionals for structural steel connections, concrete strength, soil compaction, and so on. Even if your team feels confident through in-process QC that the work is done right, you must still plan for these independent verifications. Embrace them as an extra quality gate rather than a hurdle. Also, ensure the schedule accounts for these third-party checks so they don't become a bottleneck.

- **New or Complex Work, or Team Familiarity Bias:** If the crew is doing something new or especially complex, or if there's concern that familiarity may breed complacency, bring in independent eyes. A fresh perspective can catch issues a habituated team might miss. If a subcontractor has been installing hundreds of identical prefabricated wall panels, they might start assuming each one is perfect and stop paying close attention. An independent QA visit might spot a subtle manufacturing defect in a panel that the install crew overlooked in their routine. (Independent audits can also be unannounced spot-checks to see if in-process checks are truly effective.)

- **When Objectivity is Needed:** On some projects – especially design-build jobs or private projects – the general contractor's own QA staff might serve as an "independent" checker relative to the subcontractors. They roam the site doing spot checks. The point is to have someone evaluate the work who isn't the person who installed it. This prevents the scenario of "marking your own homework." A painting subcontractor might inspect their own surface preparation (an in-process check), but the GC's QC officer then does a tape adhesion test or measures dry film thickness as an independent verification of the paint job quality before final acceptance.

- **Building It into the Process:** Incorporate independent checks into the project plan by identifying hold points and witness points in the Quality Control Plan or Inspection and Test Plan (ITP). A hold point means work pauses until an inspection is passed – for example, don't pour concrete until the rebar inspection (by an engineer or inspector) is signed off. A witness point doesn't stop the work, but alerts an independent party that they have the opportunity to witness the activity or inspection if they choose. Communicate these clearly to all parties. List hold points on the short-term schedule and ensure the responsible inspector or engineer is notified in advance. Hold points should be clearly listed in the inspection/test plan so everyone knows when a hold point is in effect, and use a visible hold status indicator (like the green tag for city inspections) to show when the hold point is released. The project team should treat an independent inspector as an ally in delivering quality – involve them early and share relevant information (drawings, specs, approved submittals) so their check can be efficient and useful.

**Project Manager's Insight:** "On a recent job, our paving crew was confident the soil subgrade was compacted well. They had rolled it and it 'felt solid.' But we still brought in a third party to do density tests – and thank goodness. The first tests showed we were under spec in a few spots. The crew re-compacted those areas and we avoided what could have been future pavement failure. Since then, I've never skipped independent testing on critical work, no matter how sure we are in the field."

**In-Process + Independent = Quality Assurance Synergy:** In-process checks by crews ensure that quality is built in and issues are caught early, while independent checks ensure objectivity and catch what the team may have overlooked. When both are in play, quality issues are dramatically reduced. Independent checks also provide a form of "verification of the verifiers." If an independent QA audit consistently finds no issues, it's a good indication that the in-process checks are working well. If the independent party finds a lot of defects, it may signal that the subcontractor's self-check process needs improvement. Project teams should use those insights to adjust training or increase oversight as needed.

Note that independent verification can also extend to verifying processes and data, not just the physical work. Some owners will even perform an independent audit of the contractor's entire quality management system – checking that inspections are being done, paperwork is in order, test equipment is calibrated, etc. This meta-level verification keeps the QA/QC program honest and continuously improving. It might sound burdensome, but on large projects, it ensures that the quality system itself is functioning properly and not being taken for granted.

## Verifying Critical Path Activities

In project management, the critical path represents the sequence of activities that directly determine the project's finish date. Delay any critical path activity, and the project is delayed. From a quality perspective, any rework or quality failure on a critical path task is doubly damaging: you not only incur the cost to fix it, but you also likely blow the schedule. Therefore, critical path activities demand special attention for verification and quality control. The goal is "do it right the first time" on critical path tasks so that quality issues do not derail the project timeline.

As a field leader, you should always know which activities on the site are critical (or near-critical) at a given time. These are the ones with little or no float. Common examples might be completing the foundations in time for structural steel erection, or pulling primary electrical feeders that must be done before power-on and commissioning can start. Once identified, treat these tasks with heightened QA/QC focus. Some practical steps to manage quality on critical path activities include:

- **Front-Load Planning and Checks:** Before a critical activity begins, conduct a thorough preparatory meeting with all stakeholders (foremen, superintendent, QA representative, subcontractor PM, and perhaps the designer if needed). Review the plans, specs, and quality requirements in detail and ensure everyone understands the acceptance criteria. This is essentially a Preparatory Phase meeting (as discussed in earlier chapters) to prevent errors before they happen. Prior to a critical concrete shear-wall pour, review the formwork design, rebar details, embed plate placements, and concrete mix requirements. From this meeting, develop a specific checklist or Inspection & Test Plan (ITP) tailored for that pour.

- **Assign Your "A-Team":** Critical path work is not the time for on-the-job training or skeleton crews. Ensure your most competent supervisors and crew members are assigned to these tasks. If needed, bring in additional supervision or specialist inspectors for that activity. If a certain weld on a steel moment frame is critical (perhaps due to seismic design importance), you might have both the subcontractor's Certified Welding Inspector (CWI) and an independent CWI from the owner verify those welds. Yes, it's "belt and suspenders" – but the cost of discovering a bad weld later (or worse, experiencing a failure) far outweighs the cost of extra inspection up front.

- **Institute Hold Points on Critical Tasks:** Do not allow work to progress past key milestones without verification. This is where formal hold points, as introduced earlier, come into play. A classic example: don't start structural steel erection until the foundation inspection is complete, because fixing anchor-bolt layout errors after the steel is up is extremely painful and costly. In scheduling terms, consider inserting a "quality hold" activity after a critical installation and before its successor activity. That indicates no further work can continue until the quality check is done and passed. After a waterproofing membrane is applied on a foundation wall (critical path for starting backfilling), schedule an inspection and flood test as a hold point before backfill. If the membrane leaks and it's not caught until after backfilling, you'd have to excavate again – a nightmare scenario for both schedule and budget. By verifying at the hold point, you ensure the critical path isn't unknowingly jeopardized.

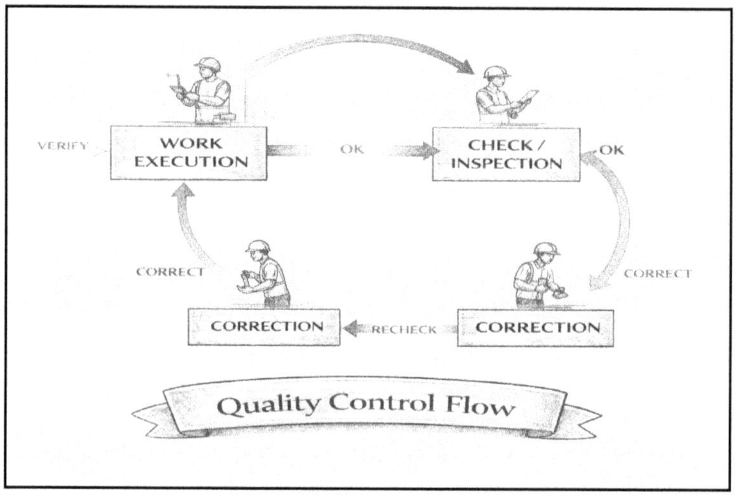


- **Increase Frequency of Checks During Execution:** For critical tasks that span time or volume, don't wait until the end to inspect everything. Verify as the work progresses. If a critical pipeline installation runs 1000 feet long, you might mandate that every 200 feet, a hydrostatic pressure test is performed (or every joint is visually inspected by QA) instead of only testing at the very end. In this way, if a problem is found, you localize the rework to a small section, and you catch it early enough to possibly resequence other work or mitigate delays. Continuous monitoring can also be done via daily walkthroughs by the superintendent or QA manager, focusing specifically on that activity's quality.

- **Leverage Technology for Critical Work:** Take advantage of digital tools to help verify work in real time. Some projects use laser scanning or photogrammetry to check critical dimensions or alignments immediately after installation (such as confirming precast concrete panels are within tolerance). Drones can be used to inspect the critical connections of tall structures that are hard to reach manually. Even project management software can be set to flag any quality issues on critical path tasks with high priority. If your project has access to such technology, use it to augment human inspections on critical path items.

- **Document and Communicate Verifications:** When a critical path activity is verified successfully, make it visible. Issue a quick verification report or at least send an email update to the project team (e.g., "All 50 caissons drilled, inspected, and passed – no defects noted, ready for concrete"). This builds confidence and also records that quality checks didn't delay the task (or if they did, that fact is documented for transparency). Conversely, if an issue is found, escalate it immediately through the proper channels and log it as a non-conformance or issue in the project's tracking system. The earlier project management knows about a critical path quality problem, the more options there are to mitigate the schedule impact (like resequencing other work, adding resources, etc.).

**Case Example – Preventing a Schedule Disaster:** On a fast-track industrial project, installing large-diameter underground pipes was on the critical path leading up to system commissioning. The team implemented a rigorous verification plan: for each pipe segment, the crew performed an in-process check (correct gasket placement, bolt torques), then an independent QA inspector observed a pressure test for each segment before burial (a hold point). Partway through, one segment failed the pressure test – a flange hadn't been tightened properly, and a gasket was pinched. Because this was caught at the segment level, only that portion had to be reworked and re-tested (a half-day fix). If they had waited and tested the entire system at once at the end, they would have had to dig up multiple segments to find the leak, possibly delaying the project by weeks. This example underscores how verifying critical path activities in smaller chunks and early can isolate issues and keep the overall project on track.

Not all tasks can receive the same level of scrutiny – it wouldn't be efficient – but critical path tasks warrant a higher dosage of QA/QC. Think of it like a surgeon taking extra precautions for a high-risk operation: additional pre-operative checks, more monitoring during the procedure, and careful post-op follow-up. As a general contractor or field leader, you should allocate your "quality energy" in proportion to the importance of the activity. Use risk-based thinking: focus on what would hurt the project most if it went wrong, and verify those things to a degree that makes failure highly unlikely.

Also, remember that sometimes near-critical activities deserve similar treatment, because a quality issue could consume all the float and suddenly push them onto the critical path. The project schedule is dynamic; review it regularly (say, at each schedule update meeting) and adjust your quality focus if new critical activities emerge.

Finally, one benefit of rigorous verification on critical path work is improved client confidence. Owners and stakeholders are very sensitive about anything that could delay the project. When they see that you have robust quality checks on the key activities, it boosts their trust that the project will finish on time and meet the required quality. It's common for clients or third-party project monitors to specifically audit critical path activities – so being proactive with independent verification can save you headaches during those audits. For example, an owner's rep might ask, "How do we know the tensioning of those post-tension cables (a critical path item) was done correctly?" If you can pull out records of calibrated jack certifications, elongation measurements, and an independent engineer's sign-off, you've answered the question and proven your team's diligence.

Treat critical path activities with extra care in QA/QC: plan them thoroughly, verify prerequisites, inspect during and after execution, impose hold points for independent checks, and respond to any issues with urgency. This proactive stance will significantly reduce the chance that a quality problem causes a schedule slip. It's far better to spend an extra hour today preventing a rework than to lose ten days later scrambling to fix something that could have been caught earlier.

## Preventing Confirmation Bias in QA/QC

Humans are wired with cognitive biases – systematic ways our brains filter and interpret information. In construction QA/QC, one of the most dangerous biases is confirmation bias. This is the tendency to seek out or interpret information in a way that confirms pre-existing beliefs or expectations, while ignoring or downplaying information that contradicts them. In simple terms, we see what we expect to see. For a field team, confirmation bias can cause you to miss defects because you assume everything is okay, or to explain away warning signs because they don't fit your mental model of "we did this right."

Why is confirmation bias hazardous? Construction teams often develop a strong belief in the quality of their work and their processes – especially if they've had success in the past. Confidence is good, but overconfidence can blind us. A QA manager might firmly believe that a trusted subcontractor always delivers high-quality work. That belief might lead to quicker, less thorough inspections on that subcontractor's work – and potentially cause missed problems. If an issue later surfaces (say, a waterproofing defect), someone might say "I never saw that coming," when in fact subtle signs were there but overlooked. Another example: if the schedule is tight, project leaders might convince themselves that "there's no time for mistakes," and thus subconsciously ignore signs of trouble to maintain the narrative that the project is fine. This has happened in some famous project failures – early warning signs were dismissed because the team was biased to believe in their plan.

Be on the lookout for signs of confirmation bias in QA/QC. Red-flag attitudes or remarks include: "We've never had a problem with this type of work, so I doubt anything's wrong now," or "This inspection is just a formality; everything should be good." Another warning sign is filling out checklists with all "Yes/OK" without real verification (sometimes called pencil-whipping the checklist). Also, if quality reports never find any issues at all – a 100% clean record – it could mean your inspection process isn't rigorous enough or is unintentionally only finding what it expects to find.

The good news is that once you acknowledge the risk of confirmation bias, you can put measures in place to counteract it. Here are strategies to prevent confirmation bias from undermining quality checks:

- **Embrace Skepticism and a Questioning Culture:** Cultivate a mindset in yourself and your team that actively looks for what might be wrong, not just proof that everything is right. Encourage inspectors and foremen to approach each check with healthy skepticism: ask "What if something isn't as per the plan – how would I know?" Rather than walking the site trying to confirm that "yes, it all looks good," deliberately search for anomalies or deviations. Create an environment where anyone can question the work without fear. Team members should feel free to challenge assumptions and ask for evidence. If a junior engineer says, "I'm not sure that bolt was tightened," respond positively: "Thank you for flagging that – let's double-check it," rather than "Of course it was tight, the crew knows what they're doing."

- **Diversify Perspectives:** People from different backgrounds or with different experiences may spot issues others overlook. If the same person checks the same work every day, they might become blind to certain faults. Rotate inspection duties if possible, or conduct cross-team audits. Bringing in someone from outside the project for a fresh look can break internal "groupthink." Encouraging diverse perspectives can significantly reduce the likelihood of confirmation bias. Have a safety officer join a quality walk – they might notice an issue (like improper concrete curing that could also pose a safety hazard) that the regular QA inspector took for granted.

- **Independent Verification to Counter Bias:** This ties back to section 10.2 – use independent checks specifically to counter internal blind spots. An outside inspector has no emotional investment in the work and thus can be more objective. Even within your team, you can simulate this by swapping duties. Have one superintendent review another's area of work. Some organizations even take it a step further and use "blinded" re-checks, where a second inspector tests or examines something without knowing the first inspector's conclusions, to see if they independently come to the same result. If a QA inspector marked an entire checklist as all OK, a peer might later do a spot re-check of a few items without being told what the first inspector found – if the second finds issues, it indicates possible bias or oversight by the first. This kind of redundant check ensures that the verification process itself is under scrutiny.

- **Clear Criteria and Checklists:** A well-designed checklist or inspection test plan can serve as a guardrail against bias by forcing the inspector to verify objective points rather than rely on intuition. Make sure your checklists have specific, measurable items (e.g., "Gap around pipe penetration sealed with approved fire caulk – Yes/No?") rather than vague prompts like "Firestopping acceptable?". Objective criteria make it easier to identify something that doesn't conform. Inspectors should gather evidence for each item (notes, photos, measurements). Also, periodically audit the checklists themselves – ensure they cover the key potential failure points and don't skip known trouble spots.

- **Pay Attention to Contradictory Data:** Often, there are hints when something is wrong – maybe one test out of five had a borderline result, or a crew member voiced a concern that was dismissed, or a subcontractor's report doesn't quite align with your own logs. Instead of ignoring these outliers, treat them as valuable clues. If one concrete cylinder break result came back low while others passed, don't just average it in and forget it – investigate why. Was there something different about that batch or that test? Sometimes teams explain away anomalies to maintain the belief that "our process is fine," but those anomalies might be early warnings. One quality auditor suggests documenting any contradictions as valuable leads rather than nuisances. In practice, if an independent test result or observation conflicts with your expectation, dig deeper until you're satisfied that the discrepancy is resolved or understood.

- **Regular Reality Checks (Quality Reviews):** Just as project managers hold regular schedule or budget reviews, conduct periodic quality review meetings. In these meetings, look at recent QA/QC data: How many inspections passed vs. failed? Are there patterns in the issues found? Are there areas where you aren't finding any issues at all (maybe "too good to be true")? By reviewing quality results with the whole team, you might spot blind spots. Perhaps no one has been checking a certain specification because each person assumed someone else was doing it. Or maybe the team has been focusing on structural quality and overlooking finish work. Regular reviews combat the complacency that feeds confirmation bias. Experts suggest implementing regular and rigorous review processes where plans and quality data are reassessed with an open mind. If something seems off, do not sugar-coat it – face it and adapt. Use these meetings to recalibrate the QA/QC focus and to remind everyone that finding problems is a good thing (as long as they can still be fixed). It's far worse to have a "perfect" record during construction, only to discover later that issues were simply not detected.

- **Training and Awareness:** Make confirmation bias a topic in your team training or safety/quality stand-downs. Many people exhibit bias without realizing it. Simply educating the team about what confirmation bias is and giving examples of how it can manifest on-site, will make individuals more self-aware. For example, train inspectors to consciously ask themselves during inspections: "Am I objectively looking for defects, or am I assuming it's fine because I trust the installer or I'm in a hurry?" Even a quick mental reset—pause for 30 seconds and look again—can help an inspector check their own assumptions. Encourage this practice of slowing down briefly to question one's assumptions. You can pair this training with lessons learned: share stories (from your projects or industry case studies) where biases led to issues. (*For instance, a case from Chapter 3 described how a* *team's assumption that "the existing structure was fine" led them to overlook critical cracks – an expensive lesson in the importance of verifying assumptions.*)

- **Encourage Reporting of Bad News:** Teams often hide or downplay problems because of pressure to only deliver good news. Create an environment where reporting a potential quality issue is seen as a positive, proactive act – not a failure. If people know that management welcomes hearing about issues early (rather than "shooting the messenger"), they will be less likely to self-censor important information. Establish clear channels for feedback and concerns, and encourage workers at all levels to speak up if they notice potential issues. Crucially, respond constructively when they do. If a field worker says, "Boss, I think the concrete mix looks off, maybe the slump is high," don't reply, "It'll be fine, keep going." Instead, thank them and check it out – maybe perform an extra slump test to be sure. One superintendent even implemented a system where any crew member could halt a concrete pour if they suspected a quality issue, no questions asked; it empowered the team to be vigilant rather than assume "it's probably okay."

**Quality Manager's Tip:** "I have a personal rule on site: if I hear myself or someone else say, 'It should be okay,' that's a red flag. 'Should be okay' is not good enough – I want to know it's okay. That forces us to get the facts. Sometimes that means taking a closer look, doing another test, or asking for an outside opinion. It's saved us from plenty of near-misses. Good quality management is a bit like detective work – assume nothing, verify everything."

By actively countering confirmation bias, you add another layer of quality control – a mindset layer. Your QA/QC processes won't just be going through the motions, but will truly be effective at finding and fixing issues. The result is a project where decisions and assessments are based on evidence, not assumptions. In the long run, this leads to safer, higher-quality, and more successful projects. It also fosters a culture of openness and continuous improvement, where the team is always asking "Is there anything we're missing?" – and thereby catching issues while they're still manageable.

## Practical Field Checklists and Templates

The following checklists are practical tools that field leaders can use to implement the concepts from this chapter. They can serve as quick on-site guides or be incorporated into the project's quality plan. (These examples can be adapted and expanded into project-specific forms, and included as needed in an appendix or project templates.)

## Checklist: Independent Verification of Work Quality

Use this checklist to plan and ensure independent QA/QC checks are effectively integrated into the project workflow:

- **Identify Critical Inspections:** List key hold points for each definable feature of work (DFOW) – for example, pre-pour rebar inspections, post-installation pressure tests, weld X-rays, etc. Ensure these hold points are on the schedule and that everyone knows work must stop for these checks.

- **Assign Responsible Verifier:** For each hold point or independent check, assign who will perform it (a specific third-party inspector, QA manager, or engineer). Schedule them in advance and confirm their availability to avoid delays.

- **Prepare Reference Materials:** Before the inspection, verify that the latest drawings, specifications, and criteria are on hand for the verifier. (For example, have the rebar shop drawing on site for the inspector, the torque specifications ready for a bolt inspection, etc.)
- **Ensure Pre-Check by Crew:** Have the crew or subcontractor perform their own in-process check before calling in the independent verifier. This way obvious errors can be caught and corrected internally first, preventing wasted time for the inspector and keeping the process efficient. (Example: the crew fills out an internal checklist and fixes any issues, then invites the independent inspector for the formal check.)
- **Document the Outcome:** For each independent verification, require a brief report or sign-off. This could be the inspector's report, a signed checklist, a test certificate, or even a simple email confirmation. File these records in the project's QA log – they are crucial for closeout and for protecting all parties in case of future disputes or claims.
- **Track Non-Conformances:** If the independent check finds a problem, log a Non-Conformance Report (NCR) or issue immediately. Assign responsibility and a deadline for corrective action. Do not consider the hold point cleared until re-inspection is completed and the item passes.
- **Monitor Timely Closure:** Ensure that any required rework resulting from independent inspections is done promptly so it doesn't stall the project. (It helps to keep a log – e.g. "Welding inspection on 5/12 failed on 2 joints – rework completed 5/13, re-inspected and passed on 5/14.")
- **Review and Improve:** Periodically review the effectiveness of your independent checks. Are they catching issues? Are some independent checks consistently finding nothing (which might indicate the process could be streamlined, or conversely that the checker might be too lenient)? Adjust the plan if needed – add or remove checks based on where the risk actually lies.

## Checklist: Preventing QA/QC Confirmation Bias

Use this short checklist as a reminder for inspectors and managers to keep bias out of the quality process:

- **Approach each inspection with fresh eyes:** Don't assume "it's fine because it was fine last time." Look at the work as if you're seeing it for the first time, every time.
- **Seek disconfirming evidence:** Actively look for anything that does not meet the standard, rather than trying to confirm that everything is okay. (For example, deliberately inspect the hard-to-reach areas or less visible details where errors could hide, rather than only the obvious areas.)
- **Cross-check with a colleague:** For critical inspections, have a second person perform a quick independent check, or swap checklists with a colleague and see if you both get the same results.
- **Use objective measurements:** Whenever possible, use tools (levels, gauges, meters) and numeric criteria instead of just eyeballing. Numbers don't lie – a conduit either has the required clearance or it doesn't; a concrete test cylinder either met the strength requirement or it didn't. Let data inform you, especially if your gut is telling you "it's probably fine."
- **Encourage every team member's voice:** If anyone on the team (no matter their role or level) expresses a concern or observes a potential issue, take it seriously and investigate. Document what they reported and what you found. This creates a culture where raising a hand is welcomed, and biases are less likely to silence dissenting observations.

- **Double-check "clean sweeps":** If an entire checklist or inspection comes back with perfect results across the board, do a random spot audit of a few items to verify. (For example, if all 50 drywall sections were signed off by their installers as "no defects," randomly pick one or two sections to re-walk and ensure there truly are no defects.) This isn't about undermining trust – it's about verifying that the process is honest and thorough.

- *Reflect on assumptions:* Before closing out an inspection, ask yourself: "Am I assuming anything? Would I come to the same conclusion if I didn't know who built this, or if this weren't my own work?" If you feel even a hint of uncertainty, take a second look or get a second opinion.

By using checklists like these and instilling the practices discussed in this chapter, general contractors and field leaders can significantly strengthen their quality management processes. The result: work that passes inspections the first time, fewer costly fixes, and projects that finish on time with a quality product – exactly what a systematic construction operations handbook is all about.

### Operational Takeaways for Today's Job:

- Treat critical tasks with extra scrutiny. Before a major pour, lift, or critical installation, get a fresh pair of eyes on the setup. For example, have a colleague double-check formwork bracing or re-calc a lift plan – independent verification often catches things the primary team missed.

- Incorporate "holding points" for quality. Identify points in the work (like after rebar is tied, before concrete pour) where you stop and verify the work with a checklist or third-party inspector before proceeding. These deliberate pauses for quality checks can save enormous rework later.

- Rotate inspection duties if possible. One day, assign one foreman to inspect another's work and vice versa. This cross-check reduces confirmation bias – we tend to overlook mistakes in our own work but spot them in others', so structured peer review improves overall quality.

- Use instruments and technology to verify, not just eyeballs. Calibrate and use tools (laser levels, torque wrenches, multimeters) to confirm critical measurements or torques meet specifications. Don't rely on "looks good to me" when an exact value matters.

- Document your checks. When you verify something important – like that bolts are torqued or a trench shield is in place – make a quick note or take a photo. This not only proves it was done but forces a moment of mindfulness to truly confirm the quality/safety item is correct.

## Part III - *Regulatory and Professional Frameworks*

**Roles:** *Compliance-focused leaders* – project managers, safety managers, design leads, and anyone responsible for regulatory compliance.

**Focus:** Maps out the laws, regulations, and professional standards that govern construction work. Part III serves as a reference for "the rules of the game," covering topics like OSHA requirements, building codes, permitting, and licensing.

**Operational Decisions Informed:** Compliance and ethical judgment calls. Equipped with Part III's insights, a manager or superintendent will better understand when a task requires a certified professional or permit, how to schedule inspections and audits, and how to anticipate regulatory scrutiny. This ensures field leaders can navigate red tape and keep their projects on the right side of safety laws and industry standards, rather than learning the hard way after an incident.

Chapter 11

# U.S. Regulatory Standards in Construction



Construction projects in the United States operate under a network of safety, health, and environmental regulations. Field leaders must navigate federal OSHA requirements, consensus standards, environmental laws, and local building codes on a daily basis. This chapter provides a plainspoken guide to these U.S. regulatory standards. The goal is to translate the legal framework into practical actions for construction managers and site supervisors. We'll cover OSHA's construction safety rules, the role of ANSI/NFPA/ASCE/AISC standards, environmental compliance, building code enforcement, and how global best practices can elevate safety beyond mere compliance. The tone is grounded and pragmatic - focused on what field professionals and project leadership need to know to keep jobsites safe, lawful, and productive. Real cases and patterns will illustrate key points (without repeating case studies used elsewhere). Where helpful, suggestions for visual aids or diagrams are noted to reinforce complex concepts. The chapter is organized as follows:

## OSHA Core Standards (1926, 1910)

Federal OSHA (Occupational Safety and Health Administration) is the cornerstone of construction safety regulation in the U.S. Created by the OSH Act of 1970, OSHA's mission is to ensure safe and healthful working conditions by setting and enforcing standards, and by providing training and assistance. In construction, OSHA's influence is felt through the 29 CFR 1926 standards (Construction Standards) and related parts of 29 CFR 1910 (General Industry Standards). This section explains OSHA's role on projects, when construction vs. general industry rules apply, which hazards drive the most citations and injuries, and how site leaders can effectively comply day-to-day. It also covers the concept of "competent persons," the importance of documentation, multi-employer worksite policy, what to expect during OSHA inspections, how violation severity and penalties are determined, and tips for integrating compliance into the daily rhythm of field supervision.

## OSHA's Role on a Construction Project

OSHA plays both a proactive and an enforcement role in construction. On one hand, OSHA sets nationwide safety standards that construction employers must follow – covering everything from fall protection to scaffolding to machine guarding. These rules in 29 CFR 1926 are essentially the "safety code" for construction sites. On the other hand, OSHA is the enforcement authority: OSHA compliance officers can inspect job sites, investigate accidents and complaints, and issue citations and fines for violations. In practice, this means every construction project in the U.S. operates under OSHA's oversight, whether or not an inspector ever actually visits the site.

OSHA's Mission and Approach: OSHA's core mission is to ensure that every worker goes home safe at the end of the day. The agency does this by *"setting and enforcing standards, and providing training, education and assistance".* Practically, OSHA publishes regulations (the standards), conducts outreach (like free on-site consultation services and training programs), and performs enforcement inspections. A construction field leader should view OSHA as the baseline setter for safety – the minimum requirements that must be met. Many companies choose to go above and beyond OSHA rules (which we'll discuss later in the chapter), but no responsible contractor can ignore the OSHA standards.

OSHA Inspections: OSHA typically does unannounced inspections on construction sites deemed high-risk or where complaints/incidents occur. An inspector (Compliance Safety and Health Officer, or CSHO) will arrive on site, show official credentials, and hold an opening conference to explain the scope and reason for the visit. They then conduct a walkaround inspection of the work areas, looking for hazards that violate OSHA standards. The inspector may interview workers and review records like injury logs or safety training documents. After the walk, a closing conference is held to discuss any findings or likely citations. Citations and penalties (if any) are later mailed to the employer, typically within a few weeks, and the employer has the right to contest or discuss them in an informal conference.

Worker Rights and OSHA: It's worth noting that OSHA also empowers workers: employees have a right to a safe workplace and can file complaints with OSHA if they believe there are serious hazards or violations. Retaliation against workers for raising safety concerns is illegal, and OSHA can protect whistleblowers. On a well-run job, the culture encourages workers to speak up internally about safety issues so that fixes can be made before OSHA involvement is ever needed. Field leaders should foster this openness and not simply rely on the fear of OSHA inspections.

OSHA's role on a project is both standard-setter and watchdog. They provide the playbook of safety rules and can show up to audit compliance. A field leader should not view OSHA as an adversary, but rather as an ever-present authority whose requirements must be woven into the project's planning and daily execution. OSHA compliance is the starting point for jobsite safety.

## 1926 vs. 1910 and When Both Apply

One source of confusion in construction safety is whether to follow the OSHA construction standards (29 CFR 1926) or the general industry standards (29 CFR 1910) for a given activity. In principle, OSHA's rules are segmented by industry: Part 1926 applies to all "construction work," while Part 1910 covers non-construction workplaces. OSHA defines "construction work" as *"work for construction, alteration, and/or repair, including painting and decorating".* This means that if the task is part of building or altering a structure, the construction standards (1926) generally apply. Section 1910.12(a) of OSHA's regulations explicitly says that the construction standards apply "to every employment and place of employment of every employee engaged in construction work".

**General Industry vs. Construction:** OSHA uses the term "general industry" to refer to all industries not in construction, agriculture, or maritime. Most construction projects clearly fall under the 1926 rules. However, there are situations on a jobsite where 1910 standards might still kick in. For example, equipment maintenance or fabrication work done on-site could be considered general industry work if it is routine and not part of building the structure. OSHA has long distinguished between construction and maintenance: generally, if you're doing anticipated, routine upkeep of equipment (maintenance), 1910 standards apply, whereas building or significantly repairing something is construction (1926). The line can be gray – often it depends on the scope and scale of work. A one-for-one replacement of a small component might be deemed maintenance, but replacing a major structural element is construction by OSHA's view.

**When Both Sets Apply:** On some projects, an employer may need to comply with both 1910 and 1926. A common example is hazard communication – the OSHA HazCom standard (1910.1200) technically falls under general industry regulations, but it applies to all industries including construction for handling chemicals. In fact, many of OSHA's general requirements (like PPE standards, first aid, fire extinguishers) apply across industries. OSHA clarifies that many employers (especially large contractors) "may need to comply with both general industry standards and construction standards depending on the work being performed". If a construction crew is operating a forklift, OSHA's Powered Industrial Trucks standard (1910.178) applies (there's an equivalent in 1926 by reference). If they set up a workshop on site for custom fabrication, general industry machine guarding and lockout/tagout rules can apply inside that workshop.

From a practical standpoint, construction OSHA standards (1926) should be the field leader's primary reference for typical site operations – fall protection, scaffolding, trenching, etc., are all covered there. But be aware of gaps: if 1926 doesn't have a specific standard for a hazard, the General Duty Clause of the OSH Act might be used, or OSHA might look to a 1910 standard for guidance. For instance, 1926 has no comprehensive standard for confined space work for years – OSHA would apply the general industry confined space rule or the General Duty Clause. (Note: In recent years OSHA added 1926 Subpart AA for confined spaces, drawn from the 1910 rule.)

**Key Takeaway:** Determine whether an activity is construction or maintenance, and know which OSHA rules govern it. When in doubt, err on the side of the more stringent requirement. It's often safest to comply with the relevant

1910 standard even on a construction site if 1926 doesn't address the hazard, because OSHA can and will hold employers accountable for recognized hazards using general industry standards as evidence. For example, lockout/tagout (LOTO) procedures for controlling hazardous energy are only explicitly spelled out in 1910.147, but if you're servicing equipment during construction, implementing LOTO per 1910.147 is expected as a best practice (and OSHA has cited construction companies under the General Duty Clause for not doing so).

1926 rules are your primary playbook on a construction project, but 1910 rules fill in many specifics and still apply to work that is maintenance-like or not explicitly covered in 1926. Smart field leaders keep a working knowledge of both. Think of 1926 as the specialization for construction hazards, and 1910 as the general safety rulebook that underpins everything.

## The Citation Drivers and Fatal Four Patterns

Not all OSHA rules are violated equally – some hazards are far more likely to draw citations (and cause harm) than others. Year after year, the same core issues top OSHA's list of most frequently cited standards in construction. By understanding these "citation drivers," a field leader can focus attention on preventing the most common (and deadly) problems.

**The "Fatal Four":** OSHA has identified four hazard categories that are responsible for an outsized share of construction worker deaths. These "Fatal Four" are: Falls, Struck-By, Caught-In/Between, and Electrocutions. Together, they consistently account for roughly 60% of U.S. construction fatalities. Falls from height are the single leading cause, often about one-third to nearly half of all construction deaths in a typical year. The other three categories (being struck by equipment or objects, being caught in collapsing materials or equipment, and electrocution by contact with live electricity) collectively add up to another 20-30% or more of fatalities. The pattern is clear: if you eliminate the Fatal Four hazards, you eliminate the majority of construction deaths.

Because these hazards are so severe, OSHA inspections focus heavily on them. For example, an inspector on a random jobsite sweep will almost certainly look for fall protection issues first – knowing that falls are the #1 killer. If workers are on roofs or edges without proper fall arrest systems, expect a citation. If trenches are unshored or missing a safe means of egress, expect a citation (trenching hazards fall under caught-in/between). If heavy equipment is operating, they'll check for backup alarms and swing radius barricades (struck-by controls). If electrical panels are open or temporary power isn't GFCI-protected, they'll key in on electrocution risks. The Fatal Four drive OSHA's enforcement priorities because they drive injuries and deaths.

**Top OSHA Violations:** Looking at OSHA's annual Top 10 violations list for construction further illustrates the point. In recent data for fiscal year 2024, the most cited standard in all of OSHA (across industries) was Fall Protection – Construction (29 CFR 1926.501), with over 6,000 citations. In construction-specific rankings, other top citations included: Ladder safety (1926.1053), Scaffolding (1926.451), Fall Protection Training (1926.503), Eye and Face Protection (1926.102), along with some general industry standards that apply widely (Hazard Communication 1910.1200, Respiratory Protection 1910.134, Lockout/Tagout 1910.147, Machine Guarding 1910.212). We see a pattern – the citation "hot spots" are directly related to the Fatal Four and other common construction hazards:

- Falls: Cited via fall protection requirements, ladder usage, scaffolding safety, and training deficiencies. Fall hazards also tie into roof work, leading edges, floor openings, etc.
- Struck-by: Can manifest in citations under PPE (eye/face protection), vehicle safety standards, crane operations, or material storage (to prevent object falls).
- Caught-in/between: Often cited through trenching (1926 Subpart P), equipment guarding (machine guarding or lockout), or scaffold collapse potential.
- Electrocutions: Cited through electrical standards – e.g. lack of GFCIs, improper extension cord use, working too close to power lines, missing lockout on electrical systems.

It's no coincidence that those areas are heavily policed by OSHA – they are exactly where workers get killed or hurt if corners are cut. A plainspoken way to put it: if you address fall hazards properly, maintain trench safety, prevent equipment strikes, and control electrical risks, you've tackled the big ones. You'll not only save lives but also avoid the citations that cost companies thousands in fines and lost reputation.

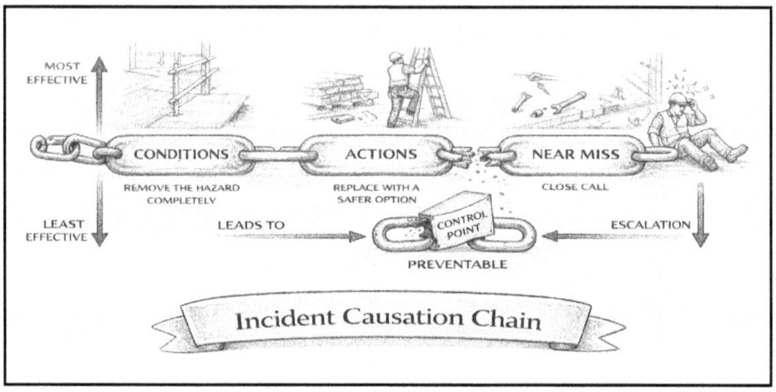


**Focus on Falls:** Falls deserve special mention because they are by far the leading cause of death and the most cited violation. OSHA requires fall protection (guardrails, safety nets, or personal fall arrest systems) at elevations of 6 feet or more in construction. Yet, violations are rampant – residential construction sites with workers scrambling on roofs or edges without harnesses, commercial jobs without guardrails on open floors, etc. Given that more than 300 workers die from falls each year and thousands are injured, OSHA has made fall protection a national emphasis area. The takeaway for field leaders: ensure strict compliance with fall protection rules (have a plan for every work at height scenario). This includes proper scaffold assembly, using the right ladders and using them correctly, covering holes, and enforcing 100% tie-off with harnesses and lanyards where required. Leading contractors often implement fall protection that exceeds OSHA (for instance, requiring tie-off at 100% above 4 feet, instead of OSHA's 6 feet) – an example of moving from compliance to best practice.

**Other Common Citation Traps:** Aside from the Fatal Four, some other issues frequently lead to OSHA citations in construction: - Housekeeping and Material Storage: Debris, scrap wood with nails, or materials haphazardly piled can violate housekeeping standards and create struck-by or tripping hazards. - Hazard Communication: Not having Safety Data Sheets (SDS) on site or not training workers on chemical hazards is a common finding (and easy citation to avoid by maintaining a simple HazCom binder and training). - Personal Protective Equipment (PPE): Missing hard hats in areas with overhead work, not wearing safety glasses, or inappropriate footwear – these are basics that, if neglected, signal poor safety culture and invite inspector scrutiny. - Tools and Equipment: Damaged electric cords without ground pins, missing guarding on power saws, or not securing compressed gas cylinders – each of these is a specific OSHA violation that shows up often in citations.

**The patterns in OSHA citations tell a story:** the same fundamental hazards cause the majority of problems. Thus, a field leader can prioritize efforts on these fronts. It's impossible to memorize every OSHA rule, but you don't need to – if you focus on controlling the well-known high-risk activities (working at heights, excavation, heavy equipment, electrical, etc.) and keep a tidy, well-organized site, you will inherently cover the major compliance bases. It's a bit like the Pareto principle (80/20 rule) in effect: a minority of hazard categories account for the majority of risk and citations. Put your daily energy there.

## Competent Persons and Daily Field Control

OSHA regulations frequently refer to a "Competent Person" – a term with a very specific meaning. A competent person is not just an experienced worker; it is defined by OSHA as *"one who is capable of identifying existing and predictable hazards in the surroundings or working conditions which are unsanitary, hazardous, or dangerous to employees, and who has authorization to take prompt corrective measures to eliminate them"*. In simpler terms, a competent person is someone on site who both knows what to look for and has the authority to fix it. Many critical construction activities explicitly require a competent person to be involved or present, according to OSHA standards.

**Examples of OSHA-Required Competent Persons:** - Excavations: OSHA's trenching standard mandates that a competent person inspect trenches daily and after rainstorms. This person must be able to identify signs of potential cave-in (like tension cracks) and is empowered to stop work and call for additional shoring or sloping as needed. - Scaffolding: A competent person must supervise scaffold erection, inspect scaffolds and their connections before each shift, and after any event that could affect integrity. They'll know how to calculate loads, identify if cross-braces are properly installed, etc., and if something's unsafe, they can get it corrected immediately. - Fall Protection Systems: When workers use safety harnesses, a competent person might be required to inspect anchor points or supervise the installation of lifelines, ensuring the system will actually arrest a fall safely. - Demolition: A competent person must survey the structure for hazards (like posted load limits, potential collapse zones, utility locations) prior to demolition and continually monitor during demolition. - Cranes and Steel Erection: While "qualified" persons (with specific credentials) often handle rigging plans or crane assembly, a competent person is typically on deck to inspect crane conditions (e.g., soil support, setup) and steel connection points during operations, with the authority to pause work if safety is in question. - Confined Spaces (in construction): The entry supervisor could be considered a competent person who understands the hazards (like low oxygen) and can take action if conditions become unsafe.

In essence, competent persons are OSHA's way of pushing real-time hazard control to the field level. OSHA can't have inspectors everywhere all the time; instead, the expectation is that the contractor designates sufficiently trained and experienced individuals on site who will police safety and make immediate corrections. As a field leader (superintendent, foreperson, site safety officer), you are very likely the competent person for many operations. It's not just a title – it's a daily duty. You must be able to recognize hazards (which means you need knowledge of applicable standards and the particular job conditions), and you must have the clout to implement fixes (which means management must back you when you say work stops until a safety issue is resolved).

**Daily Field Control Routines:** A competent person's impact is most effective when integrated into the daily routine. Here's what that often looks like in practice: - Morning Prep & Inspection: At the start of each day, the competent person should walk the site (often as part of a daily safety inspection). This might include checking that overnight conditions didn't create new hazards – e.g., rain didn't accumulate in an excavation, wind didn't loosen scaffold ties, etc. They check that fall protection systems (guardrails, covers) are in place where needed for the tasks planned that day, that machinery inspection checklists (like crane or forklift daily checks) are done, and that no new debris or tripping hazards have appeared. - Pre-Task Safety Meetings (Toolbox Talks): Many sites hold quick safety huddles each morning. The competent person often leads or participates in these. They discuss the tasks of the day and key hazards to watch for. For example, "Today we're stripping formwork on level 3 – watch for overhead loads and wear your hard hats, and remember the fall protection requirements at the open edge." This is also a chance to get feedback: workers might mention a concern ("that stairway light is out, it's pretty dark"). The competent person notes it and ensures it's addressed. - Monitoring Work in Progress: Throughout the day, the competent person should circulate and keep an eye on high-risk operations. It's a balancing act with other duties, but the idea is to catch unsafe conditions or behaviors before an incident or an OSHA inspector does. Did someone remove a guardrail to load material and forget to put it back? Time to intervene. Is a crew working too close to another group using power tools overhead? Re-route them or enforce barricading. This on-the-fly hazard spotting and correction is core to the role. - Empowering Others: A competent person can't be everywhere, so a strong safety culture encourages all workers to act competently in their own realm. The formally competent person might train crew leaders to also recognize hazards. For example, a foreman of the trenching crew learns from the safety competent person how to identify when soils look unstable, so they too become watchful. Nonetheless, OSHA will hold the designated competent person (and the company) accountable if obvious hazards were not identified or corrected in areas under their watch. - Stopping Work When Necessary: Importantly, the competent person must be ready (and unafraid) to stop work if an imminent danger is present. It could be as simple as halting a crew about to enter a 12-foot-deep trench that has no shoring or telling an equipment operator to shut down a machine that's leaking fuel near an ignition source. This authority to pause or modify work is what separates a "competent" person from just a knowledgeable person – you have the backing of management to make safety calls. The project leadership should make it clear that production goals never override the competent person's safety authority.

OSHA's requirement for competent persons in many standards is essentially a mandate for daily field-level safety control. It pushes safety responsibility down to the site where hazards unfold in real time. For a field leader, embracing the competent person role means constantly learning (staying up to date on standards and safe practices) and communicating (with your crew, management, and subcontractors) about hazard recognition and correction. It's an active role – not a title you put on a resume and forget. If an accident happens and OSHA comes in, one of the first questions often will be: "Who was the competent person supervising this operation?" You want to be able to demonstrate that you (or your designee) were not only named as competent, but behaved competently – through inspection logs, through training records, and through actions taken to fix issues.

Competent persons are the lynchpin of OSHA compliance on the ground. They turn the thousands of pages of OSHA regulations into actual safe conditions by daily vigilance and quick action. Every morning when a field leader steps onto the site, a useful mental exercise is: "What hazards am I competent to spot today, and am I authorized to correct them?" If there's any gap in being able to answer that, it needs shoring up – either through training or through management granting the needed authority. The project's safety depends on it.

## Documentation That Protects the Job

In construction, if it isn't documented, it's almost as if it didn't happen. This saying holds especially true for safety and regulatory compliance. Good documentation is a field leader's best ally in both preventing issues and defending the project and company if things go wrong. Here we'll discuss the key types of safety documentation and how they "protect the job" – from demonstrating due diligence to helping with legal defenses in case of an OSHA citation or claim.

**Why Documentation Matters:** Imagine an OSHA inspector arrives after a reported incident. They ask for proof that you've been training workers, inspecting the site, maintaining equipment, etc. If you can pull out well-organized records – training sign-in sheets, daily inspection checklists, equipment maintenance logs, toolbox talk minutes – it paints a picture of a safety-conscious operation. If you have nothing to show, OSHA is more inclined to believe safety was neglected. Documentation provides evidence of compliance. It can show that the company had rules and was actively enforcing them, which can mitigate citations or even form the basis of a defense. For example, one recognized defense to an OSHA citation is the "unpreventable employee misconduct" defense, which essentially argues that the employer did everything reasonably required (set rules, trained, supervised, disciplined) and a rogue employee still violated a rule. But that defense "hinges on what you do well before, not after, OSHA shows up". The employer must have established safety rules, trained employees on those rules, conducted regular jobsite inspections, and documented any incidents and corrective actions. Only with such documentation can you credibly claim an employee's violation was an unpreventable anomaly and not a symptom of a lax safety program.

## Key Safety Documents to Maintain:

- **Safety Program/Plan and Orientation Records:** Most projects start with a site-specific safety plan or at least the company's safety manual. Ensure you have a copy on site. When workers go through safety orientation, have them sign off. Keep those sign-in sheets or acknowledgement forms. They show OSHA that workers were introduced to rules (e.g., PPE requirements, fall protection policy) from day one.

- **Toolbox Talks / Safety Meetings:** Weekly (or daily) toolbox talk sign-in sheets are simple but powerful documents. They prove that you are regularly engaging workers on safety topics. Make sure the topic and date are noted, and collect signatures of attendees. If a specific hazard was discussed (say ladder safety on May 3rd) and later an issue arises, you can show you did proactive education.

- Job Hazard Analyses (JHAs) or Pre-Task Plans: Many sites use daily JHAs where crews plan out their tasks and identify hazards/controls. If you have these documents filled out and signed by crews, they demonstrate foresight. They are also great tools for the field leader to review each day, to ensure crews actually implement the identified controls.

- **Inspection Checklists:** Document your daily or weekly inspections. Whether it's a simple site walk checklist or a digital inspection report, note the date, who did it (likely the competent person or safety rep), and observations. If you find issues and fix them, write that down – e.g., "Found extension cord with damaged insulation – removed from service and replaced." This shows a pattern of hazard identification and correction. Should an

OSHA inspector say, "How do you ensure electrical safety?" you can provide several past inspection reports illustrating that process.

- **Equipment and Tool Inspections:** Keep records for cranes, forklifts, excavators, etc. Typically, operators do a daily check (fluid levels, brakes, etc.) – those should be logged. Similarly, fire extinguishers need monthly checks (tag them), harnesses and lanyards need periodic inspections (record it). If a scaffolding tag system is in place (green/yellow/red tags), maintain the scaffold inspection forms. These all demonstrate proactive maintenance of safe conditions.

- **Training Certifications:** Many operations (forklift driving, aerial lift use, OSHA 10/30 cards, first aid, etc.) require trained personnel. Keep copies of certifications on site. If a question arises, "Has this person been trained to operate that boom lift?" you can pull the proof. It's also wise to maintain a log of all employees on site with their training and certs – this helps ensure only qualified folks do certain tasks.

- **Incident Reports and Investigations:** If, despite best efforts, an accident or near-miss occurs, document it thoroughly. Write an incident report, investigate root causes, and document what corrective actions were taken to prevent recurrence. This not only helps internally to improve, but if OSHA investigates, they will want to see that the company took the incident seriously and learned from it. It can reduce the likelihood of a willful or repeat citation if you can show the incident was an anomaly and steps were implemented afterward.

- **Disciplinary Records:** It may sound harsh, but keeping records of safety warnings or disciplinary actions for safety violations can protect the company. If an employee repeatedly violates rules and is reprimanded (and you document those write-ups), it demonstrates enforcement of your safety policies. In OSHA's eyes, this supports that any violations were against company policy (unpreventable employee misconduct, as mentioned). Conversely, if everyone breaks the hardhat rule and no one is ever corrected on paper, it undercuts any claim that you truly require hard hats.

- **Permits and Plans:** Certain activities require permits or plans – for instance, a trench deeper than 20 feet needs a professional engineer's protective system design. A hot work (welding) permit system might be in place. If so, retain those permits and any associated checklists (like fire watch logs). For confined space entries, keep the permits that show atmospheric testing results and supervisor sign-offs. These show compliance with procedural safety requirements.

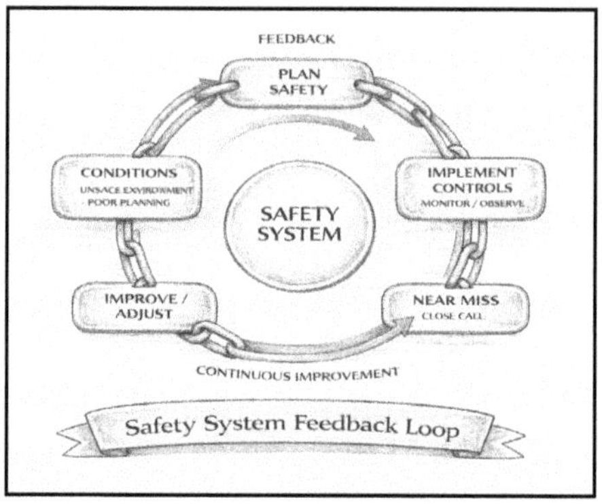


- **Environmental Compliance Records:** Keep logs of stormwater inspections, waste disposal manifests, etc., as these are also regulatory documents that can protect you if environmental regulators come knocking.

All this documentation might sound like bureaucratic busywork, but it serves a real purpose on the job: 1. For Prevention: The very act of keeping records forces a routine. If you know you have to fill out a daily trench inspection form, you will actually go inspect the trench – thereby catching hazards early. Documentation creates accountability and habit. 2. For Communication: Written reports, like a safety inspection that is copied to the project manager, communicate issues that need resources or higher-level attention. They leave a paper trail that management was informed of a hazard (which motivates management to address it, because if an accident happens and it was documented but ignored, that looks bad). 3. For Defense: Should OSHA issue a citation, documentation can be a lifesaver in negotiations or legal proceedings. For example, if cited for lack of training after an incident, producing a

roster of all training sessions, including the involved employee's signature, can lead OSHA to withdraw or reclassify the citation. Likewise, comprehensive inspection logs can sometimes get penalty reductions by showing good faith efforts. Attorneys often say, "If it's not documented, it didn't happen" – which is admittedly cynical, but underlines that an assertion ("we always inspect scaffolds") carries little weight compared to a file of scaffold inspection forms.
4. For Insurance and Claims: In case of worker compensation claims or third-party lawsuits, safety documentation can demonstrate the contractor's diligence. It might not prevent a claim, but it could reduce liability or show that an injury was due to a worker deviating from clear instructions (which can shift how a case is viewed).

**Efficient Documentation:** Modern projects often use digital tools (mobile inspection apps, online safety management systems) that can simplify the paperwork burden. A superintendent might do a walkthrough with a tablet, check off items, take photos of issues, and instantly generate a report sent to all stakeholders. Whether digital or paper, the key is consistency and organization. It's no use having documents if they're scattered or inaccessible. Set up a safety binder or digital folder system. Tab it out: training, inspections, incidents, permits, etc. Update it weekly at a minimum. This way, if someone higher-up or an inspector asks for something, you can retrieve it within minutes.

**Documentation vs. Actual Safety:** Of course, paperwork alone doesn't make a site safe – action does. There is a danger of getting so caught up in forms that one loses sight of the field. The balance to strike is using documentation as a tool, not an end in itself. A checklist is only as good as the thoroughness and honesty of the inspection behind it. A training sign-in sheet is only meaningful if real training occurred. As a field leader, don't let the paperwork devolve into pencil-whipping or retroactively filling things in just to have it. It should be a reflection of genuine safety efforts. When done right, documentation and safety performance go hand in hand. When done poorly (like falsified records or cookie-cutter JHAs that no one reads), it's just window dressing that won't hold up under scrutiny.

Document what you do and do what you document. It protects the job by ensuring critical safety tasks aren't missed and by providing evidence of those efforts. Should regulatory agencies come by or if an accident triggers an investigation, you'll be glad to have a trail showing that you took safety seriously. It's part of the professional approach to construction management in the modern era – no longer is it just about getting the work built, it's also about keeping records that show you built it safely and responsibly.

## Multi-Employer Worksite Accountability

Walk onto any sizeable construction site, and you're likely to find multiple companies working shoulder to shoulder – a general contractor, a dozen subcontractors, maybe staffing firms, specialist installers, etc. With so many employers sharing one worksite, a common question is "Who is responsible for safety?" The answer, in OSHA's eyes: potentially everyone. OSHA's Multi-Employer Worksite Policy allows them to cite more than one employer for the same hazardous condition if certain criteria are met. This is a crucial concept for field leaders, especially general contractors, to understand. You can't assume "that's the sub's problem, not mine" when it comes to safety. Let's break down how accountability is distributed on multi-employer projects.

**OSHA's Roles on Multi-Employer Sites:** OSHA identifies four categories of employers that could be held responsible for a hazard on site: - Exposing Employer: The company whose workers are exposed to the hazard. (E.g., a subcontractor whose employee is working below an unguarded skylight – that sub is an exposing employer to the fall hazard). - Creating Employer: The company that caused the hazard to exist. (E.g., a sub who removed a guardrail, thereby creating a fall hazard for everyone). - Correcting Employer: A company responsible for fixing hazards (often one hired specifically for site safety or a sub tasked with site cleanup, etc.). - Controlling Employer: The company with general supervisory authority over the site, usually the General Contractor or Construction Manager, who has the power to prevent or abate hazards through its coordination role.

A single company can have multiple roles. But importantly, OSHA can cite any or all of these employers if a hazard is not addressed, depending on the circumstances. The policy essentially ensures that responsibility isn't dodged via contracting. For instance, if a GC oversees a site where a sub's workers are routinely exposed to obvious dangers, OSHA expects the GC to have exercised its authority to get that corrected – if not, the GC can be cited as a controlling employer even if none of its own employees were directly at risk.

General Contractor (Controlling Employer) Responsibilities: As the controlling employer, the GC holds a lot of accountability. OSHA requires employers to exercise "reasonable care" to prevent and detect violations on the site. This doesn't mean the GC has to babysit every moment or inspect the site with a microscope, but it must implement a safety management system that includes: - Orienting subs to site safety rules. Many GCs hold mandatory safety orientations for all subcontractor workers and issue site-specific safety handbooks. - Requiring sub's safety plans or Activity Hazard Analyses before work begins and reviewing them. - Regular safety audits of subcontractor work areas. A GC superintendent or safety manager should be walking the site, and if they see a sub's employees doing something hazardous (like not tying off on a scaffold), they must address it – not ignore it because "it's not my crew." Reasonable care might include weekly documented safety meetings with subs or using a safety checklist while touring the site each day. - Enforcement and coordination: The GC should enforce safety requirements through the contract (stop-work authority, removal of noncompliant workers, etc.). Also, the GC often coordinates a "safety committee" or similar group with representation from each major subcontractor to discuss and resolve safety issues collectively.

If a GC can show they have a robust system of oversight and they acted in good faith to enforce safety, it may avoid citation or get a lesser citation if a sub still has a lapse. However, simply saying "that's the subcontractor's job, not ours" is a sure way to earn an OSHA citation as a controlling employer.

**Subcontractor (Exposing/Creating) Responsibilities:** Subcontractors, for their part, are directly responsible for protecting their own employees (and anyone affected by their work). If they create a hazard – e.g., leave out a live electrical wire – they are on the hook as the creating employer to fix it, even if, say, the hazard could hurt another trade's workers. If their employees are exposed to a hazard another created, the sub is expected to take steps to protect its workers – that could mean complaining to the GC or even stopping work until the GC or creating employer fixes it. If a subcontractor knows about a hazard and continues work without protection, OSHA can cite that sub as the exposing employer.

There is an exception: if the exposing employer (sub) can show they had no authority or ability to correct the hazard, informed the controlling employer, and took steps to keep their workers away from the danger, they may avoid a citation. But that's a narrow path. Generally, subs cannot be passive. For example, if a painter sees that the framers removed a floor cover leaving a hole, the painter's supervisor should not let their crew near it and should immediately inform the GC to correct it. If they do nothing and someone falls, the painter's company can be cited along with the framer and the GC.

**Communication and Coordination:** Multi-employer safety is all about communication. A practical measure is holding coordination meetings where upcoming work is discussed – this helps identify if one trade's work could endanger another (e.g., overhead work that requires a drop zone). The GC's role is to facilitate these conversations and timing (perhaps schedule certain high-risk tasks off-hours or sequence so that no one is below another crew).

**Contractual Indemnification vs. OSHA Responsibility:** It's worth noting that while contracts often spell out that each subcontractor is "independently responsible" for safety compliance and may even indemnify the GC for safety issues, this does not limit OSHA's ability to cite the GC. OSHA doesn't care about the internal contract – they care about who had control and knowledge. So from a legal liability standpoint, a GC might later pass costs to a sub via contract clauses, but in terms of citations both can get fined. In practice, this should motivate GCs not to be hands-off, and subs not to rely solely on the GC for safety. Everyone must keep their own house in order and also be their brother's keeper to some extent on a construction site.

**A Real-World Example:** A telling case was reported through a safety consortium (CROSS) about a steel erection accident. A trade contractor had installed a column with only two anchor bolts as allowed by the engineer's detail, and it fell, injuring a worker. The investigation noted that both the steel contractor and the engineer failed to recognize OSHA's requirement that all columns have at least four anchor bolts. Here, OSHA could cite the steel contractor (creating and exposing employer) for not following the standard, and also potentially the GC (controlling employer) for not ensuring the steel sub adhered to OSHA rules. The learning point: the GC should have procedures to review critical erection plans for compliance, and the sub, even though the design showed two bolts, should have known the OSHA standard and sought to correct that detail. Shared responsibility – each party had a duty to catch the hazard, and neither did, so accountability was shared.

**Multi-Employer Citations are Common:** Especially on large commercial projects, OSHA often issues citations naming both the GC and one or more subs for the same condition. For instance, if an open shaft is left unguarded, the GC might get cited as the controlling employer, the concrete sub who left it open as creating employer, and any sub whose workers were exposed as exposing employers. Penalties might be apportioned or duplicated. It can become a finger-pointing mess post-incident, which is why it's far better to handle hazards proactively together than to deal with them in OSHA conferences later.

**Preventive Strategies:** As a field leader, what can you do to manage multi-employer accountability? - Pre-qualify subcontractors for safety. Don't bring on subs with poor safety histories or no program. - Kick off safety meetings with each subcontractor before they start work on site. Make safety expectations crystal clear and let them know the GC will be monitoring. - Include safety in every coordination meeting agenda. It keeps it front-of-mind that safety isn't separate from schedule or logistics – it's part of them. - Joint safety inspections: Sometimes the GC's safety manager will do a walkthrough with each sub's foreman weekly. This not only finds issues but also educates the subs on what to look for. - Enforce PPE and site rules universally: A GC superintendent should correct any worker on site (regardless of employer) if they see, say, no hard hat or a misuse of a tool. Consistency matters; if one sub sees another getting away with lapses, they may slack too. - Documentation of communication: If a hazard is noted involving a sub, write it up (say, in a daily report or a safety memo to the sub's supervisor) so there is a record that it was identified and the sub was instructed to fix it. OSHA will ask, "When you noticed this violation by your subcontractor, what did you do?" Having an answer (and proof) is important.

To sum up, on a multi-employer job, safety is a team sport with shared accountability. OSHA's policy ensures that no employer can completely wash their hands of a hazard just because it didn't directly hurt their own employees. For general contractors, the duty is to actively manage and police the overall site conditions. For subcontractors, the duty is to protect their workers and not create dangers for others – and if they do, to fix them fast. The best outcomes occur when all parties foster a culture of "we're all in this together" for safety. The GC is often the quarterback of this effort, but every employer has a position to play. If everyone does their part, the site is safer, and the risk of anyone being on the wrong end of an OSHA citation (or worse, an injured worker) goes way down.

## How Inspections Unfold

Even the most safety-conscious construction site can receive an OSHA inspection. It's important for field leaders to know what to expect when an inspector arrives and how the inspection process typically unfolds. Being prepared can ensure the inspection goes as smoothly as possible and that the company's interests are protected while also addressing any genuine safety concerns.

**Unannounced Arrival:** In most cases, OSHA inspections are unannounced. The compliance officer (CSHO) might show up at the site gate or trailer during working hours without prior notice (advance notice is actually illegal except in rare special cases). As a site supervisor, you or your gate/security personnel should have a protocol: welcome the inspector, check credentials (OSHA inspectors carry federal IDs), and notify the project manager or safety manager immediately. You have the right to verify their credentials; genuine CSHOs will willingly show their government ID and business card.

**Opening Conference:** After introductions, the OSHA inspector will hold an opening conference with the employer representatives (this could be in the jobsite office). They will explain why the site was selected for inspection and the scope of what they plan to cover. Selection reasons could include: a programmed emphasis program (e.g. random inspection focusing on fall hazards in construction), a recent serious accident or fatality, a worker complaint, a referral from another agency, or just being in a high-hazard industry in a target area. During the opening, the inspector will outline the process: they'll mention that they will do a walkaround, possibly interview employees, review certain records, etc. They will also ask for an employer representative to accompany them (this should be someone knowledgeable about the site – often the safety manager or superintendent). Additionally, if there's a union on site, a worker representative has the right to accompany the inspector as well.

**Records Review:** Early in the inspection (often during the opening or just after), the inspector may request to see certain records. Common requests: - The OSHA 300 Log of injuries and illnesses (to see if there are any recordable incidents and patterns). - Training records (e.g. "show me proof of OSHA 10-hour training for employees" if relevant, or specific certification records). - Equipment inspection logs, safety meeting minutes, etc., especially if related to the focus area (they might say, "do you have your crane inspection records?"). - They will check the posting of the OSHA workplace poster and that the OSHA 300A annual summary is posted in season (Feb-April). - If it's a focused inspection (like falls), they might review the site's fall protection plan or JHAs for elevated work.

Make sure these records are organized and accessible. If an inspector asks for something reasonable and you say "I don't know where that is" or it takes an hour to find, it doesn't give a good impression.

**The Walkaround Inspection:** This is the heart of the OSHA visit. Following the opening conference, the compliance officer, accompanied by the company rep (and possibly an employee rep), will tour the worksite. They will go to the areas relevant to the inspection scope (and sometimes beyond – an inspection can expand if they notice other serious hazards). During the walk: - The inspector will observe work in progress. They are looking for violations of OSHA standards or the OSH Act's General Duty Clause (for hazards not covered by specific rules). If they see an apparent violation, they might point it out immediately or take notes and pictures. - Photos and Measurements: Expect the CSHO to take photographs or video of what they consider hazards or compliance issues. They may also take instrument readings (for example, a light meter for checking the welding flash screens or a sound level meter for high noise, or air monitoring pumps if they suspect overexposure to dust/chemicals). - Interviews: The inspector will likely talk to some employees privately. They have the right to do so. They may ask workers if they've been trained, if they know about certain procedures, or get the workers' perspective on a hazardous situation. As management, you cannot interrupt private interviews of non-supervisory employees. However, if the inspector talks to a supervisor (whose statements legally bind the company), that is effectively an interview of the company – it's wise to have your designated rep present for those conversations to clarify any issues. - Correcting Hazards On the Spot: If the inspector identifies a minor hazard that can be immediately abated, they might mention it during the walk. For instance, "That extension cord is damaged – that needs to be removed." It is in your best interest to correct issues on the spot when possible. OSHA notes if you abate a hazard promptly, it can be a point in your favor (though they may still cite it, showing "abated during inspection" it demonstrates good faith). For serious imminent dangers, the CSHO can even ask that work be halted until fixed (and if not, they can initiate an immediate action to remove workers or get a quick court order, though that's rare if you cooperate). - The inspector will try to minimize work interruption, but obviously, there will be some, especially if they need to see something in detail. Work can generally continue unless it's creating imminent danger.

During the walkaround, the company representative's job is to observe and take notes on everything the inspector notes. Keep a parallel record: if the CSHO takes a photo of something, you take a photo of it too. If they use an instrument reading, note what it was. This information is invaluable later if citations come in – you'll know what they're based it on. It's also fair to politely ask the inspector clarifying questions: "Can you explain what the concern is here?" Sometimes they might misinterpret something, and you can provide clarifying info. For example, they might see what looks like asbestos insulation when it's actually fiberglass – letting them know could shift their approach. Don't be argumentative or deceitful; focus on facts and showing that you're responsive.

**Scope of Inspection:** OSHA has priorities for inspections when resources are limited. If this is a "focused inspection" (a policy for construction sites with effective safety programs), the CSHO may limit the scope to the project's most hazardous activities (often related to the Fatal Four: falls, electrical, struck-by, caught-in). There was guidance allowing focused inspections if a site had a competent safety program in place. In contrast, if it's a comprehensive inspection (such as after a fatality), they will go through the entire site and all operations with a fine-tooth comb.

**Closing Conference:** Once the inspection portion is complete, the inspector will hold a closing conference with the employer (and the employee rep, if present). In the closing, they will discuss the findings – essentially telling you what violations they believe were observed and which standards might be cited. They often also inform you of your rights and options post-inspection. They might not list every little item (and sometimes they discover more when writing the report later), but you should get a good sense of the issues. Take careful notes during this meeting, too.

**During the closing, the inspector will typically cover:** - Possible Citations: They'll say something like "We observed an apparent violation of 29 CFR 1926.501(b)(1) – employees exposed to falls of 15 feet without fall protection on the west scaffold." They might not give you the classification (serious, other-than-serious, etc.) on the spot, but sometimes they indicate the severity. - Courses of Action: They will explain that after the OSHA area office reviews the report, citations (if any) and penalties will be proposed and mailed. They'll remind you that OSHA has up to 6 months from the inspection to issue citations. - Employer Options: The inspector will mention that once citations are received, the employer has the right to an informal conference with OSHA's area director to discuss them, or to contest the citations formally within 15 working days. They often encourage informal resolution. They might hand you brochures on your rights. - Abatement and Next Steps: If they found things you can fix immediately, they might encourage you to do so even before citations arrive. For things not immediately fixed, they may ask, "What will you do about X?" – sometimes to gauge your willingness to comply. If citations are later issued, each will come with an abatement deadline (when you must certify the hazard is corrected). The closing conference is a good time to clarify what corrective actions are expected and to ask any questions. For example, if a machine lacked a guard, you can confirm the type of guard needed. - Consultation Programs: Often, CSHOs will mention OSHA's consultation services or other programs that can help, emphasizing that OSHA's goal is hazard abatement and compliance (not just citations). Take that for what it's worth – at this point, the inspection is basically done, and it's about wrapping up professionally.

**After the Inspector Leaves:** You have a bit of a waiting game until the formal report and any citations arrive (typically a few weeks). Use this time wisely: immediately address any noted hazards. Don't wait for the paperwork. If guardrails were missing, install them; if training was lacking, do it. Not only is this morally right to protect workers, but if you fix things, it can potentially reduce penalties or even change a citation's classification when OSHA sees you acted in good faith. - Debrief internally. Gather your management and safety staff and go over what happened, what was found, and what needs to be improved. It could expose gaps in your program you weren't aware of. It's far better you learn and improve, regardless of the citation outcome. - Prepare for citation response. If you anticipate citations, plan who will be involved in reviewing them (safety director, perhaps legal counsel for serious cases). Prepare documentation or evidence that might refute or mitigate issues.

**Types of Violations and Penalties:** When citations arrive, they'll categorize each violation and assign a proposed penalty. Violations can be: - Other-than-Serious: a safety issue that is not likely to cause death or serious physical harm. These often carry smaller fines or even $0 fines, but they are still citations. - Serious: where there's a substantial probability of death or serious harm if an accident were to occur (even if no accident did occur). Most construction hazards (falls, electrocution risk, etc.) end up classified as Serious. These carry higher fines. - Willful: an employer knowingly failed to comply or acted with plain indifference to safety – this is very bad and carries very high fines (and potential criminal referral if someone was killed under a willful violation). - Repeat: a violation of the same (or very similar) standard that the company was cited for in the past 5 years. Repeat penalties can be up to the same level as willful. - Failure to Abate: if you don't fix a cited issue by the deadline, OSHA can fine you per day beyond the abatement date.

The closing conference usually won't label things as willful or repeat on the spot (unless conditions were egregious). Those decisions come from higher-ups. But they will likely indicate if something is serious. If you strongly disagree with something the inspector noted, you can voice it diplomatically at closing ("We believe those employees were tied off, and perhaps the lanyards weren't easily visible from the ground – we have a policy and can provide records"). They won't argue on the spot usually, but it signals that you might contest that item.

**Attitude and Cooperation:** Throughout an inspection, maintaining a respectful, cooperative attitude can influence the outcome. That doesn't mean roll over – you should ensure your rights are observed (e.g., if an inspector wants to wander into unrelated areas, you can ask if that's within the scope). But outright obstruction or hostility is a mistake. It sets a negative tone and could even escalate the situation (e.g., the inspector might seek a warrant if denied entry, which delays things and sours relations). OSHA notes employer good faith in their assessment; demonstrating care for safety and a proactive approach during the inspection can sometimes result in penalty reductions for "good faith" (except they don't give that reduction for willful violations). Conversely, being dismissive of safety or hiding information can lead to no breaks given.

What If You Disagree? If you truly think an inspector is wrong (perhaps citing a standard that doesn't apply, or misidentifying a condition), it's okay to respectfully discuss it at the closing conference. Provide evidence if you have it ("Actually, that material was not asbestos; here are lab test results we had done"). They may or may not take it on board immediately, but it could influence their final report or how the area office views it.

**Trade Secrets and Confidential Info:** If certain processes or materials on your site are trade secrets, you can inform the inspector, and they are supposed to label their notes accordingly so that if someone FOIAs the report, those sections can be redacted. This is more common in manufacturing than in typical construction, but just in case you're building something with proprietary methods, be aware.

**End of Inspection:** After the closing, the inspector leaves. There is typically a bit of relief, but don't waste the learning. If they pointed out even small issues, those were real hazards that could cause harm. Use the experience to bolster your safety efforts.

An OSHA inspection is a structured process – Opening, Walkaround, Closing – within which the inspector will identify hazards and potential violations. By knowing this process, a field leader can navigate it calmly: ensure the inspector sees a cooperative and safety-conscious site, fix what can be fixed promptly, and gather all information. Inspections are a fact of life in construction; they shouldn't be feared, but they should be taken seriously. And the best "defense" against a bad inspection is consistently good safety practices long before OSHA ever shows up.

## Severity Levels and What Raises Penalties

When OSHA issues citations after an inspection, not all violations are viewed equally. The agency assigns a severity level to each violation and calculates penalties based on several factors. As a construction manager, it's helpful to understand what makes one violation a slap on the wrist and another a heavy fine (or even a potential criminal case). Here, we break down the types of OSHA violations, how penalties are determined, and what factors can raise or lower the penalties. Knowing this can guide how you prioritize corrective actions and handle citation processes.

## Types of Violations:

- **De Minimis Violation:** These are extremely minor technical violations that have no direct impact on safety or health. OSHA doesn't issue fines or even citations for de minimis issues; they might just mention them. (Example: a ladder rung spacing off by half an inch beyond spec might be de minimis if it doesn't create a hazard).
- **Other-than-Serious Violation:** This violation is related to job safety/health but is not likely to cause death or serious physical harm. OSHA can cite it, but the penalties are usually lower. Sometimes, OSHA may levy zero monetary penalty for an other-than-serious, especially if it's quickly fixed or the employer has otherwise a good record. Example: failing to post the OSHA poster is an other-than-serious violation.
- **Serious Violation:** This is the most common type in construction. OSHA defines a serious violation as one where there is a substantial probability that death or serious physical harm could result from the hazard, unless the employer did not or could not have known of the violation. Essentially, if someone could realistically be badly hurt by the condition, it's serious. Falls, trench hazards, electrical risks, etc., are almost always classified as Serious. Serious violations carry higher penalties – OSHA is required by the OSH Act to consider a mandatory penalty for serious violations (though they have discretion on the amount).
- **Willful Violation:** This is the gravest category. A willful violation is one committed with intentional disregard of, or plain indifference to, OSHA requirements. It means the employer knew something was required or hazardous and consciously didn't comply. Willful citations have very high fines (into six figures per violation in current times). If a willful violation results in a worker fatality, OSHA may refer the case to the Department of Justice for criminal prosecution under the OSH Act (which could lead to fines and jail for individuals responsible).
- **Repeat Violation:** If OSHA had cited the company (at any of its sites nationwide) for a similar condition in the past five years, and they find it again, they may classify it as Repeat. Repeat violations can carry penalties up to the same maximum as willful (because Congress treats a repeat as showing the employer didn't learn or prioritize correction). Example: If you were cited in 2023 for not having guardrails on a roof project, and in 2025 OSHA finds lack of guardrails on a different project of yours, that could be a Repeat violation of 1926.501.
- **Failure-to-Abate:** This comes into play after a citation is issued. If OSHA comes back for a follow-up (or you have to certify abatement and fail to), and the hazardous condition was not corrected by the deadline, they can fine you per day beyond the abatement date. These fines add up quickly and can exceed the original fine.

## Penalty Amounts and Calculation:

OSHA penalties are proposed based on a formula considering the severity of the violation and the employer's size, history, and good faith. Each year (and now each January), OSHA adjusts penalty maxima for inflation. As of early 2025, for example, the maximum penalty was about $16,550 per Serious or other-than-serious violation, and $165,500 per Willful or Repeat violation. (Those numbers creep up annually with inflation adjustments.)

However, OSHA rarely hits the max for every serious violation. They use factors: - Gravity of the violation: This is the primary driver. Gravity is a combination of severity (how bad an injury could be – e.g., death, serious injury, minor injury) and probability (likelihood of an incident – high, medium, low). OSHA basically rates each violation on these and gets a gravity-based penalty. For instance, an unprotected 30-foot fall hazard has high severity (death likely if fall) and may have medium likelihood (if workers are frequently near it), yielding a higher gravity than, say, an unguarded 6-foot trench that's 2 feet deep (which might be lower severity). - Small Employer Penalty Reductions: OSHA gives penalty discounts for smaller businesses as an acknowledgement that huge fines could cripple them. For employers with 1-10 employees, penalties can be reduced by as much as 80%. For those with, say, 11-25 employees, maybe 60%, and so on, scaling down to no discount for very large firms (over 250 or 500 employees, etc.). Construction often involves many small subs, so this can be significant. - History: If the company has no citations in the past 5 years, OSHA might reduce the penalty (e.g., 10% off) as a "history" credit. If they have past issues, no reduction, or if they have a severe past history, sometimes they treat it more harshly (especially if related). - Good Faith: If OSHA sees that the employer had a written safety and health program and was making earnest efforts to comply, they often give a penalty reduction (commonly up to 25% off for good faith). However, no good faith reduction is given for willful violations, because if it's willful, by definition they weren't acting in good faith. Also, if a fatality occurred or if the inspection was initiated by a severe incident, they may not give a good faith discount. - Quick Fix: OSHA sometimes offered a quick-fix reduction if the hazard was corrected immediately during the inspection (an additional incentive to fix it on the spot).

All these factors are applied to come up with a proposed penalty for each item. For example, OSHA might start with a $14,000 base for a serious violation (if gravity is high), then if the company has 30 employees, give 50% size reduction (-$7,000), had no history (-$0 because history credit given only if no previous citations, let's assume none so maybe -10% of remainder), and had a good safety program (-$? maybe 15% of remainder), ending up with, say, a $5,000 proposed fine. These calculations happen behind the scenes, but you see the end number on the citation.

Willful and Repeat Penalties: These are where numbers skyrocket. As of 2024, a single willful or repeated violation could be over $160,000. And that's per violation. If OSHA finds, say 5 willful fall hazards, theoretically, that could be 5 × $160k = $800k. They often group related instances into one combined violation, but they have discretion. Under current trends, OSHA has been more aggressive in citing on a per-instance basis for egregious cases – for example, if 10 workers were exposed, they might cite 10 separate violations. Parker Poe (a law firm) noted that OSHA increased penalties and sometimes splits one hazard into multiple items to boost fines. This is part of an effort to make fines bite more, as OSHA leadership felt old fines were too low to deter violations.

Criminal and Severe Cases: It's rare, but if an employer willfully violates an OSHA rule and that leads to a worker's death, individuals can be criminally charged (though under federal law it's a misdemeanor). There have been cases where construction company owners or superintendents faced criminal penalties (and even jail time) for egregious negligence (e.g., instructing workers into obviously dangerous situations that led to fatalities). Environmental and other laws can also trigger criminal charges (as we'll see in 11.3). The key is: a willful violation is not only expensive but puts one in a very bad light legally.

## Factors that Raise Penalties:

- **Injury or Fatality Outcomes:** While technically OSHA cites the condition, not the outcome, the reality is that if an accident happened (especially a death), OSHA is more likely to classify violations related to it as Serious or Willful. They might also do a broader inspection under the "fatality/catastrophe" protocol, yielding more citations. Public and internal pressure is higher to assign accountability.
- **Willful/Repeat Classification:** As described, if OSHA sees you had a chance to fix this from past experience and didn't, or you plainly ignored a known requirement, they crank up the classification. Example: You've been cited before for not providing cave-in protection in trenches, and they find it again – expect a Repeat citation with a much bigger fine.
- **Lack of Cooperation or Attitude:** If, during the inspection, the employer was uncooperative or showed indifference, OSHA might be less inclined to reduce penalties for size or good faith. They have discretion. On the flip side, being very cooperative and showing a good safety program could lower perceived severity or get the good faith reductions.
- **Number of Exposed Employees and Duration:** A violation involving many workers or lasting a long time might be considered more severe. For instance, two companies might each have an excavation with no shoring. If one was a 6-foot-deep trench that one person was in for 10 minutes, and another was a 12-foot trench with a whole crew working all day, the gravity and willfulness (if they clearly should've known) are higher for the latter. OSHA might even cite per instance.
- **OSHA Emphasis Programs:** If your violation falls under a national or local emphasis (like falls, silica, trenching), OSHA may be less lenient because these are priority hazards they are targeting.

**Abatement and Follow-Up:** Once citations are issued and final (after contest or settlement), the employer must abate by the dates given. Sometimes OSHA will do a follow-up inspection specifically to check abatement on serious or willful citations. If they find you didn't fix it, those failure-to-abate fines can add up daily. So, ignoring a citation is definitely not an option.

**Contesting and Informal Conferences:** Employers have the right to formally contest citations, which sends the case to an independent Occupational Safety & Health Review Commission. Often, before doing that, there's an informal conference with the OSHA Area Director. In that meeting, you can negotiate – maybe you show evidence that some facts were wrong, or you've fixed everything and have a better program now. OSHA has latitude to adjust citations: they can reclassify (e.g., drop a willful to serious if convincing argument is made), combine or delete items, and reduce penalties (sometimes significantly) in a settlement. It's a bit of a negotiation process balancing your appetite for fighting versus just resolving.

From a field perspective, avoiding the violation in the first place is ideal. But if you get cited, swiftly fix the hazard and learn from it. If you disagree with it, use the channels available to contest. The key is to not repeat the same mistakes – a repeat citation can be 10 times as costly, and beyond cost, it indicates to everyone (workers, clients, regulators) that the safety culture is lacking.

Focus on eliminating serious hazards – especially those that can kill or maim – because they carry the highest human and regulatory cost. If an issue is pointed out, fix it before someone gets hurt or OSHA finds it, and you not only save lives but save the company from financial and legal pain. If you do get a citation, don't take it personally – take it as a prompt to improve. And absolutely avoid ever letting a known hazard go unaddressed (that's flirting with willful territory). Build a track record of good faith efforts; it will serve you well in the long run, both in safety outcomes and in regulatory relations.

(Figure 11.4: A table showing types of violations vs. penalty range, e.g., Other-than-serious: up to ~$16k; Serious: up to ~$16k; Willful/Repeat: up to ~$165k, with notes on reductions.)

## ANSI, NFPA, ASCE, AISC Codes and Their Applications

OSHA may be the law of the land for construction safety, but it is not the only set of rules influencing a project. The construction industry relies heavily on standards developed by professional and industry organizations – ANSI, NFPA, ASCE, AISC, and others – to fill in details, define best practices, or even serve as de facto regulations. Understanding when these consensus standards become requirements and how to apply them, is a mark of a mature field professional. In this section, we explore the role of key standards and codes beyond OSHA: when voluntary standards effectively become mandatory, and practical examples of using ANSI, NFPA, ASCE, and AISC standards on site. We'll also highlight a common blind spot: temporary conditions that codes don't always adequately address.

## When Standards Become Requirements

Not everything that governs construction safety and quality is printed in a government regulation. Many standards originate from professional organizations (like ANSI – American National Standards Institute, NFPA – National Fire Protection Association, ASCE – American Society of Civil Engineers, AISC – American Institute of Steel Construction, and others). These are developed by experts and often updated faster than laws can be. They start as recommendations or technical guidelines. However, they often become requirements by adoption or reference in contracts, codes, or regulations.

Here are several pathways by which a "voluntary" standard attains mandatory power on a project:

- **Incorporation by Reference in OSHA or Other Regulations:** OSHA sometimes weaves consensus standards directly into its regulations. For instance, OSHA's rules for eye and face protection (29 CFR 1910.133 and by reference in construction) incorporate ANSI Z87.1 standards for safety glasses. Likewise, OSHA's head protection rule mandates hard hats that meet ANSI Z89.1. This is called "incorporation by reference" – OSHA doesn't rewrite the whole ANSI standard in the CFR, but it cites it, making compliance with that ANSI spec mandatory under OSHA law. So what was a private standard (like ANSI Z89.1 for hard hat performance) becomes effectively a legal requirement on the job, effectively because OSHA says any head protection must meet that ANSI criteria. Another example: OSHA's scaffold standard references certain design provisions that align with ANSI/SSFI (Scaffold & Access Industry Association) guidelines for capacity and safety factors.

- **Adoption by Building Codes or Fire Codes:** Building codes and fire codes (often based on ICC or NFPA model codes) frequently reference standards for specific details. For instance, the National Electrical Code (NEC), which is NFPA 70, is adopted as law in all 50 states for electrical installations. So while NEC is technically an NFPA consensus standard, it becomes law when the state or city adopts it as part of their building code. That means on your job, the NEC's detailed requirements for wiring, grounding, etc., are not optional – they must be followed. Similarly, model building codes reference ASCE 7 (Minimum Design Loads for Buildings) for structural loading requirements – so ASCE 7's specs become enforceable. NFPA 241, a standard for safeguarding construction and demolition sites from fire, is increasingly required by local jurisdictions and fire departments (for example, Massachusetts now requires an NFPA 241 fire safety plan at permit application for construction projects). So if the city building code or fire code says "submit a fire safety plan per NFPA 241," complying with that standard's guidance (on things like temporary fire standpipes, hot work protocols, etc.) becomes a project requirement.

- **Contract Documents (Specifications):** Often, the project specs or owner requirements will mandate compliance with certain standards. A project manual may state "All scaffolds shall be erected and dismantled in accordance with ANSI/ASSP A10.8 standard" or "All welding shall conform to AWS (American Welding Society) D1.1 code," etc. These contract requirements are binding – failing to meet them can be a breach of contract. For safety, owners sometimes require contractors to meet additional consensus standards beyond OSHA. For example, an owner might specify that all fall protection equipment must comply with the latest ANSI Z359 standards (which could be stricter or more up-to-date than OSHA's rules). On federally funded projects, there may be references to Army Corps of Engineers safety manual EM 385-1-1 (a very stringent safety standard). If it's in the contract, it's effectively a requirement for the job.

- **General Duty Clause and "Recognized Hazards":** OSHA's General Duty Clause (Section 5(a)(1) of the OSH Act) requires employers to keep the workplace free of serious recognized hazards, even if there's no specific OSHA standard. How do you establish something is a "recognized hazard" and what should be done about it? Often OSHA will point to national consensus standards as evidence. For example, before OSHA had a specific silica dust standard in construction, they could cite employers under the General Duty Clause by referencing ACGIH or NIOSH guidelines on silica exposure or ANSI ventilation standards. Or if there's no specific OSHA rule on how to handle building structural modifications during construction, but an ASCE standard or manufacturer guideline exists, ignoring that might be considered failing to address a recognized hazard. In short, even if not codified, consensus standards can create a benchmark for what a reasonably prudent employer would do. Not following any known standard could be seen as negligence.

- **Industry Best Practice and Liability:** Beyond regulatory enforcement, if an accident happens and lawyers get involved, compliance with industry standards (or lack thereof) will come up. For instance, suppose a formwork collapsed. There might not be a precise OSHA detail for every aspect of formwork design, but there are ANSI/ASCE standards and guidelines. If the contractor deviated from those and an injury occurred, in court that consensus standard can be used to demonstrate they weren't following accepted practice. Conversely, if you followed the consensus standards, that can be a strong defense that you were being responsible. So in a more indirect way, these standards become "required" to protect oneself legally and professionally.

**Practical Impact:** For a field leader, the key is to know which standards apply to your work beyond OSHA. This could involve: - Checking the project specifications for any referenced standards (there might be a whole list in the spec intro section). - Knowing your trade's go-to standards: e.g., if you are dealing with steel erection, know AISC's Code of Standard Practice; if dealing with concrete, know ACI (American Concrete Institute) standards referenced by code; if doing demolition, ANSI A10.6 on demolition might have guidance, etc. - Ensuring equipment and PPE meet the latest ANSI standards. While OSHA might lag (for instance, OSHA's regulation might still reference an older ANSI version), many owners or states require the newer ones. For example, some state OSHA plans or contractors require high-visibility apparel that meets ANSI/ISEA 107 class 2 or 3, even if federal OSHA doesn't explicitly mandate ANSI 107 except in highway work. It's on you to proactively use the highest standard applicable.

**When Standards Conflict or Update:** Sometimes you'll encounter conflicting guidance. Say OSHA references ANSI Z89.1-2009 for hard hats, but the current ANSI standard is Z89.1-2014, which might have some differences. Best practice is to meet or exceed both – typically the newer standard is as stringent or more. In rare cases of conflict, generally the regulation or code in force takes precedence, but you'd want to resolve that via an RFI or asking the authority having jurisdiction. Most of the time, consensus standards complement regulations, filling gaps rather than opposing them.

**Examples of Standards Becoming Requirements:** - Scaffold Design: OSHA says scaffolds must support 4 times intended load. ANSI A10.8 might provide more detailed guidance on design and testing. Many times, scaffold suppliers design to ANSI and OSHA recognizes that via reference or practice. - Cranes: OSHA's crane rule (1926 Subpart CC) references standards from groups like the ASME (American Society of Mechanical Engineers) for signals, etc. Many cities require cranes to be erected per the manufacturer's specs and ASME B30 series standards. If an accident occurs and you weren't following those, it's trouble. - Fire Prevention: OSHA has basic fire protection rules, but NFPA standards (like NFPA 51B for hot work) are often specifically required by owners or local rules. A contractor ignoring NFPA's guidance on having a fire watch after welding might not violate a specific OSHA rule if no fire occurs, but if a fire does occur, that standard will be the lens through which their actions are judged.

Consensus standards often carry the force of requirements on modern construction projects, either directly through law or indirectly through contracts and professional expectations. A diligent field leader treats them with the same respect as OSHA rules. Keep a library (even if digital) of relevant standards for quick reference – and ensure you're using them when planning site safety and quality. It's part of the due diligence that elevates a project's safety from just legal compliance to industry best practice (which we'll explore more in section 11.5 on global best practices).

## ANSI Practical Applications

The American National Standards Institute (ANSI) doesn't itself write standards, but it accredits various organizations (like ASSP, ASTM, etc.) to develop "American National Standards" across industries. In construction, many ANSI standards (often prefixed by the developing organization's acronym, like ANSI/ASSP, ANSI/AWS, ANSI/ISEA, etc.) provide detailed guidance on equipment and safe practices. Let's look at some practical applications on site where ANSI standards come into play:

**Personal Protective Equipment (PPE) Standards:** Perhaps the most visible ANSI influence is in PPE: - Hard Hats: ANSI Z89.1 sets performance classes (Type I or II for impact, and Classes G, E, C for electrical protection). When you buy hard hats, you ensure they're labeled as meeting ANSI Z89.1 (most will be). OSHA requires this, and it's good practice to use the appropriate class (e.g., Class E for electricians to withstand higher voltage). Field application: If a hard hat is damaged or over 5 years old (per manufacturer recommendation), replace it – it may no longer meet ANSI's criteria due to wear. - Safety Glasses and Face Shields: ANSI Z87.1 defines standards for impact resistance and optical clarity. On site, you ensure all safety eyewear worn has "Z87" stamped on it, indicating compliance. If doing work with high impact risk or lasers, refer to Z87.1 for appropriate ratings (like Z87+ for high impact). When an OSHA inspector checks a construction site, they often look for that Z87 marking on glasses – it's become shorthand for "okay, these are legit safety glasses." - High-Visibility Vests: ANSI/ISEA 107 classifies hi-vis apparel (Class 1, 2, 3) based on amount of reflective material and background color. Many work sites near traffic or with crane operations require Class 2 or 3 vests. A field leader might say: "All workers must wear Class 2 vests at all times." That's not explicitly in OSHA (except by reference for flaggers), but it's an ANSI-driven best practice. It's practical to specify because it ensures a minimum visibility level for workers. - Gloves and Hand Protection: ANSI/ISEA 105 is a standard for glove cut resistance, puncture, etc. Some contractors now specify cut-resistant gloves meeting a certain ANSI cut level for tasks like material handling or using knives. This reduces hand injuries. Field example: requiring gloves with ANSI Cut Level A3 or higher when handling sharp metal or rebar. - Hearing Protection: ANSI S3.19 (older) or ANSI/ASA S12.6 are methods for measuring Noise Reduction Rating (NRR) of earplugs/muffs. While OSHA sets exposure limits, the effectiveness of hearing protectors is rated by these ANSI standards. Ensure that hearing protectors have an NRR labeled (which is per ANSI method). If you implement a hearing conservation program, you might use ANSI standards to evaluate if a protector brings noise down to safe levels.

**ANSI/ASSP A10 Series – Construction Safety Standards:** The American Society of Safety Professionals (ASSP) publishes the A10 series (accredited by ANSI) covering various construction safety topics: - ANSI/ASSP A10.1 – Pre-Project & Pre-Task Planning: Provides guidelines on how to plan safety into projects and tasks. Practical use: adopt a pre-task planning form (essentially an A10.1 concept) where crews identify hazards before work. Many companies use some form of this standard's principles to structure their job hazard analyses (JHAs) or toolbox talks. - ANSI/ASSP A10.2 – Safety, Health, and Environmental Training: Could guide how you develop training programs for your crew. For example, A10.2 suggests topics and frequency for construction safety training beyond OSHA 10/30. - ANSI/ASSP A10.8 – Scaffolding Safety Requirements: While OSHA has scaffold rules, A10.8 gives additional best practices for scaffold design, erection, use, and dismantling. Field application: maybe using A10.8's guidance to implement a scaffold tagging system, or to ensure proper training for scaffold erectors. - ANSI/ASSP A10.32 – Fall Protection Systems: This covers fall protection in construction in detail. Practically, you might refer to A10.32 when designing a fall protection plan for unique situations (like leading edge work, or using safety nets) that OSHA mentions but doesn't detail. A10.32 can provide guidance on selecting anchorages or devising rescue plans. - ANSI/ASSP A10.33 – Safety and Health Program Requirements for Multi-Employer Projects: This is directly relevant to section 11.1.6's topic – it outlines how to coordinate safety among GC and subs. A field leader on a large project could use A10.33 as a template to set up a site safety committee, subcontractor safety prequalification, etc. In fact, the ASSP offers such standards to help manage exactly what OSHA expects in multi-employer scenarios[48]. - ANSI/ASSP A10.38 – Basic Elements of Employer's Program to Provide Safe Work Environment: Essentially guidelines for a construction safety program. A savvy project manager might check their project safety plan against A10.38's elements to ensure completeness. - ANSI/ASSP A10.44 – Control of Energy Sources (Lockout/Tagout) for Construction: OSHA's LOTO standard is in general industry, not detailed in construction. A10.44 fills that gap with specifics for construction scenarios (e.g., when setting up temporary power or working on equipment). If you need to lockout a piece of machinery on site, A10.44 provides the protocol to do it safely in construction context.

**Equipment and Tool Standards (ANSI/UL/ASME):** Many tools and pieces of equipment adhere to ANSI or related standards: - Ladders: ANSI ASC A14 standards cover portable ladders design. So when you buy ladders, they likely meet ANSI A14. Field tip: using ladders within their rated capacity (Type IA, etc.) which is based on ANSI testing. Ensure the ladder's duty rating matches the use (don't have 300-lb person plus tools on a 250-lb rated ladder). - Aerial Lifts and MEWPs: ANSI A92 standards (now SAIA/ANSI A92) govern design and safe use of Mobile Elevating Work Platforms (boom lifts, scissor lifts). Recently, new ANSI A92.22/.24 standards require more rigorous training and annual inspections. Many rental companies now enforce these – e.g., they will not rent unless you show proof of operator training as per ANSI. So as a field leader, you arrange that training because effectively the ANSI standard has become the practice. The new standards also changed some equipment features (like load sensing). Understanding those changes, which came from ANSI, helps in using the lifts properly. - Cranes: ASME (American Society of Mechanical Engineers) B30 series standards cover cranes, rigging hardware, etc. While not "ANSI" branded (ASME is accredited by ANSI though), these consensus standards are heavily used. For instance, many contractors follow ASME guidelines for critical lift planning (like requiring load charts, rigging inspections, etc. per ASME B30.5 for mobile cranes). If your site has a tower crane, you might refer to ASME B30.3 for operations and erection practices. Practical application: using proper hand signals as shown in ASME standards (often posted on site as charts), or inspecting hooks and slings to criteria that come straight from ASME/ANSI standards (like remove a hook from service if throat opening > X% of original – those criteria are from consensus standards). - Power Tools and Machines: Underwriters Laboratories (UL) standards or ANSI machine tool standards ensure equipment is built safe (with guards, interlocks). As a user, ensure you don't defeat those safety features. For example, circular saws built to ANSI/UL have guards – train workers not to pin them back. If you're using a powder-actuated tool, it likely meets ANSI A10.3 standard – you comply by ensuring only trained operators use it and PPE is worn (A10.3 provides those guidelines). - Temporary Traffic Control: ANSI/ITE and MUTCD (federal guideline incorporating ANSI standards) control how to set up traffic signs, cones for road work near construction. If you have road interface, you follow those standards (like taper lengths for lane closures, sign spacing, etc.). Field example: using retroreflective cones, advance warning signs, flagger practices per the Manual on Uniform Traffic Control Devices (which is based on national standards).

**Communication of ANSI Standards on Site:** Often, you won't pull out a thick ANSI book in the field. Instead, the practical bits of ANSI standards are distilled into checklists, toolbox talks, or company safety protocols. For instance: - A safety manual might state: "All safety harnesses and lanyards must comply with ANSI Z359.1 and be inspected prior to each use. Anchor points shall support 5,000 lbs per person or be designed by a Qualified Person (per ANSI)." This language is straight from ANSI fall protection standards. - A scaffold tag might say "Scaffold built to ANSI A10.8 and OSHA standards" indicating compliance. - A toolbox talk on "Rigging Inspection" might list removal criteria for slings (like broken wires for wire rope) that come from ASME B30.9 (which is ANSI-approved).

As a field leader, you might not memorize all these, but it's good to have cheat sheets or quick references. Many contractors produce laminated cards or phone apps with key ANSI values (for example, "Max arresting force allowed by ANSI Z359 – 1,800 lbs, so our shock lanyards must limit force to that").

**Staying Updated:** ANSI standards get revised periodically. OSHA regs, however, often reference older versions until they update their rule (which can take decades). Best practice is usually to follow the latest standard. Manufacturers will anyway – you can't hardly buy a non-ANSI-compliant hard hat or lanyard today because manufacturers design to the latest. But training materials and site rules should also update. For example, if ANSI changes the rating system (like they did for hi-vis apparel classes some years ago), make sure your requirements reflect the current terminology so there's no confusion.

In summary, ANSI standards find their way into daily construction practice through PPE requirements, tool specifications, and safety program elements. While they might seem like behind-the-scenes technical guidelines, they have very real practical applications: from the label on your hard hat to the way you plan a complex lift. Embracing those standards helps ensure you're not just compliant, but using quality equipment and methods that the broader industry agrees on as safe. It's part of moving from minimal compliance to best practice, one step at a time, using the knowledge embedded in these standards.

## NFPA Practical Applications

The National Fire Protection Association (NFPA) is known for codes and standards that go well beyond "fire" in the narrow sense. Many NFPA standards directly impact construction activities, especially with regard to temporary facilities, building system installations, and fire prevention during construction. Here's how some key NFPA codes come into play practically on a construction site:

**NFPA 70 – National Electrical Code (NEC):** Perhaps the most universally applied NFPA code in construction. The NEC dictates how electrical systems are installed safely. Practical construction applications: - Temporary Power: When setting up temporary electrical panels, distribution boxes, lighting, etc., the NEC's rules on wiring methods, grounding, GFCI protection, etc. all apply. For example, using proper outdoor-rated cords, installing GFCI outlets or breakers on all circuits (OSHA also requires that), properly sizing cables for the load, ensuring overcurrent protection. A field electrician will run temp power per NEC Article 590 (Temporary Installations) which allows some leeway but still has requirements (like clearance around panels, protection of cords from damage). A site super should coordinate with a qualified electrician to set up and routinely check the temporary power system is code-compliant and safe. - Permanent Installations Testing: As systems get installed, say the fire alarm or sprinkler system, the NFPA codes require certain acceptance tests. NFPA 72 (fire alarm code) requires testing of all devices, NFPA 13 (sprinkler) requires hydrostatic testing of pipes. The construction team must plan and execute these tests to pass final inspections. That's further down the line, but definitely part of construction. - Lockout/Tagout and Electrical Safety: NFPA 70E (Electrical Safety in the Workplace) is an NFPA standard that's widely used by contractors to protect workers from arc flash and shock when working on electrical systems (especially hot work on live panels). While OSHA has some rules, NFPA 70E gives detailed practices: boundary distances, PPE categories, etc. A construction company may adopt 70E rules whenever their people are doing electrical tasks. For instance, requiring that any work on live circuits above 50V requires an energized electrical work permit (a concept from 70E), and appropriate arc-rated PPE per the incident energy analysis or category method in 70E. So on a site, if an electrician needs to do a hot tap, the safety plan might say "comply with NFPA 70E – de-energize if possible; if not, wear Category 2 arc flash gear, face shield, insulated gloves, and have a safety watch." - Equipment and Wiring Methods: Field engineers and inspectors ensure that things like the wiring in HVAC units, elevator controllers, generators, etc., meet code (e.g., proper conduit types, supports, clearances). For a field leader, you might not know all those details, but coordinating inspections and not covering up work before inspection sign-off is vital.

**NFPA 241 – Construction & Demolition Fire Safety:** This is a key standard often overlooked until an incident happens. NFPA 241 provides guidance to prevent fires in buildings under construction or renovation. Its practical applications include: - Fire Safety Plan: NFPA 241 recommends a written plan covering things like hot work protocols, temporary fire protection, alarm systems during construction, storage of flammables, etc. Increasingly, jurisdictions require a formal NFPA 241 plan for large projects. For example, Boston and NYC require site-specific fire safety plans and a fire prevention program manager on certain projects. So practically, a field leader might have to develop this plan (often with safety officer or a consultant), get it approved by the fire department, and implement it. Implementation means training crews on the plan, posting emergency contact info, etc. - Temporary Standpipes: For buildings over a certain height, NFPA 241 says you should install a standpipe (often the permanent one, even if the building isn't finished) as construction progresses, and it should be operational (charged or able to be charged) typically once the building reaches a certain number of floors. Many locales enforce something like "standpipe must be up to within 1-2 floors of highest work deck and operational." Field implication: coordinating the install of standpipe risers floor by floor, and not delaying it because it might be easier to do later. Also ensuring that the siamese (fire department connection) is accessible as building construction might obscure it, and keeping it clear of debris. - Hot Work Protocol: NFPA 241 and NFPA 51B (the Hot Work standard) require a hot work permit system. In practice, the GC designates a permit authorizing individual. Whenever welding, cutting, grinding or other flame/spark producing work is to be done outside designated shops, a permit must be filled out, site inspected, fire watch assigned, and a 30-minute fire watch maintained after. Many contractors have their own hot work permit forms based on NFPA templates. The superintendent or safety manager often signs off on them daily. Additionally, NFPA requires fire extinguishers present and, in some cases, possibly a charged hose line if significant fire risk. - Housekeeping and Combustible Waste: NFPA 241 emphasizes removing combustibles (wood scraps, paper, trash) regularly to reduce fuel for any fire. So a field leader should enforce daily cleanup and not allow large accumulations of debris. Also storing flammable liquids properly (in UL-approved safety cans, in flammable storage cabinets, etc.) – that's an OSHA requirement too, but NFPA codes give specifics on quantities and storage conditions. For example, having more than 5 gallons of gasoline on site outside a cabinet violates code; or not storing propane properly (should be in a secure cage away from ignition sources). - Temporary Heaters: In winter construction, use of salamanders or other heaters is common. NFPA 241 would require using listed safe heaters, keeping them clear of combustibles, and having ventilation. As a sup't, you ensure any open flame heaters are far from tarps or wood, there's a person monitoring them, and they're shut off properly. If using LPG (propane) heaters, you have to follow NFPA 58 (LP-Gas code) on how to store and use cylinders. - Fire Watch During System Impairment: If you turn off a permanent fire protection system (like during a sprinkler tie-in), NFPA 241 says you need a fire watch patrol or other compensatory measures. So coordinate such shut-downs with the building owner (if existing part of building) and maybe local fire dept if required, and assign someone to walk the area looking for fire until systems restored. - Security Against Arson: NFPA 241 also talks about site security – maintaining fencing, locking doors, perhaps even surveillance – because construction sites can be targets for vandalism or arson. For instance, there have been cases of fires started by trespassers at night. Keeping the site illuminated and secure is a preventive measure.

**NFPA 101 – Life Safety Code:** Though mostly for finished buildings, if you are renovating or doing phased occupancy, NFPA 101 might dictate requirements for exits, emergency lighting, alarms even during construction. In a hospital renovation, NFPA 101 might require that exits remain available or that temporary partitions are fire-rated to separate construction from occupants. A practical step is building temporary enclosures with fire-retardant plywood or poly and maintaining clear egress paths (with exit signs) for any occupied part of the building. Also, if you disable an alarm in one zone, you might need to set up a fire watch per NFPA guidelines.

**NFPA 30 & 30A – Flammable Liquids and Fueling:** Storing fuel on site? NFPA 30 is the Flammable and Combustible Liquids Code. For instance, it will say that a certain amount of gasoline above X gallons should be in a flammable liquid storage cabinet, and no more than Y gallons in one cabinet, etc. It also defines how to set up fuel storage tanks (with secondary containment, distance from buildings). NFPA 30A covers refueling areas. On site, if you have a diesel tank for equipment, you follow those guidelines (like having a double-walled tank or a containment berm, fire extinguisher nearby, no smoking signs). Many of these are echoed in environmental rules too.

NFPA 10 – Portable Fire Extinguishers: This standard guides how many extinguishers and of what type should be available. OSHA gives a bit on this, but NFPA 10 is the source for things like "one extinguisher within 75 feet travel distance for ordinary hazards." In practice, a field leader ensures extinguishers are distributed on each floor or work area, especially near hot work or fuel, and that they are inspected monthly (tagged) and maintained. It also says personnel should be trained in their use – a good toolbox topic.

**NFPA 90A – HVAC Ducts and Air Handling:** During construction in an existing building (renovation), if you're doing hot work or generating dust, you might need to cover or shut off HVAC ducts to prevent fire spread or smoke migration. NFPA 90A deals with fire dampers and such. So perhaps during a fit-out, you ensure any duct penetrations are sealed with fire-resistive material temporarily if fire partitions are breached.

**NFPA 3 and 4 – Commissioning of Fire Protection and Integrated Testing:** As projects near completion, NFPA has standards for commissioning fire protection systems (NFPA 3) and integrated fire protection testing (NFPA 4). Practically, a commissioning plan might be executed where all life safety systems are tested together – fire alarm triggers fire doors to close, elevators to return to ground, etc. A project lead coordinates these final tests often with authorities present.

The overarching practical theme with NFPA in construction is fire and life safety risk management during the building process. Fire is a lesser day-to-day concern than falls or other immediate hazards, which is why sometimes it gets neglected – but a single fire can be catastrophic (to workers and the project timeline). By applying NFPA standards, you impose controls like you do for other hazards: plan for it, equip for it (extinguishers, alarms), control ignition sources (hot work permits), limit fuel (good housekeeping), and maintain emergency response readiness (standpipes, alarms, drills maybe).

A good field leader liaises with the local Fire Marshal – they often inspect mid-project for compliance with these things, especially on large builds. Building a good relationship by voluntarily following NFPA 241, etc., can make those interactions smoother. Some GCs even assign a "fire prevention manager" on big projects who focuses on these tasks.

NFPA standards provide the roadmap to prevent fires and ensure life safety in construction. Practically applying them means integrating fire safety into site logistics (like how you'll handle hot work, where you put that temporary generator, how you phase installing fire systems, etc.). It's about looking at your site from a fire risk perspective and not just hoping a fire won't happen, but actively preventing it and being prepared if it does.

## ASCE and Temporary Conditions

The American Society of Civil Engineers (ASCE) issues standards and guidelines that are crucial when dealing with structural and geotechnical aspects of construction, especially for temporary conditions. Temporary conditions include any non-permanent state of the structure or site: excavations, partially erected structures, formwork and falsework, scaffolds, temporary supports, etc. These conditions often govern construction safety and stability but may not be fully addressed by the final design or building codes. ASCE has stepped in with guidance to make sure these in-between stages are safe.

**ASCE 37 – Design Loads on Structures During Construction:** This is a standard specifically for considering loads on partially built structures. Key ideas: - Reduced Load Factors for Short Duration: ASCE 37 recognizes that some loads (like wind) have a lower probability of occurring during a short construction period than over a 50-year building life. It allows using reduced wind load factors for structures under construction for limited time frames. For example, a steel frame that's only half complete might not need to be designed for a 100-year wind if it will be finished in 6 months, but it still needs to handle a decent wind that could reasonably occur in that period. - Considering Construction Stages: It addresses loads like the weight of construction equipment (cranes on a partially built floor), stacking of materials, asymmetrical loads when only part of the structure is built, etc. It basically guides engineers to check "what's the worst that could happen to this structure while it's incomplete?" and design for that. - Field Use: How do we apply ASCE 37 on site? Typically, the construction engineer or contractor's engineer uses it to design things like: - Temporary bracing: e.g., if a steel frame is up but floors aren't in, it might need extra bracing to withstand wind. ASCE 37 gives the wind pressure to design that bracing for. Field leader ensures those braces are installed as per that plan. - Concrete early strength: The loads from upper levels of concrete or equipment might exceed capacity if concrete below hasn't fully cured. ASCE 37 might be used to plan formwork stripping times or reshoring spacing. - Non-uniform loading: Say you plan to stockpile bricks on one bay of a slab during construction – that bay sees higher load than final code design assumed (because in final use, it's uniform offices, not a concentrated pallet of bricks). You either avoid that or shore under it. ASCE 37 gives some guidance on such scenarios. - In practice, a field team may not directly calculate ASCE...(Continued)

... designers and contractors had to consider temporary wind loads and bracing for structures under construction, but often these intermediate stages didn't get the same attention as the final design. ASCE 37 helps formalize that attention.

**ASCE/SEI 7 – Temporary Structures and Loading:** While ASCE 7 is mainly for permanent structures, it has begun incorporating provisions for temporary structures. For instance, the 2022 IBC and proposed ASCE 7-28 are adding more on temporary loading requirements. Historically, building codes lacked explicit mention of loads on temporary works, which was indeed a blind spot. Going forward, field leaders can expect plan reviewers or inspectors to ask for engineered designs of temporary works (shoring, stage platforms, etc.) with appropriate load considerations.

**Temporary Retaining Structures and Excavations:** ASCE's Geo-Institute often publishes guides on excavation bracing, cofferdams, etc. While OSHA gives some excavation slope and shoring tables, deeper excavations often need a geotechnical engineer's design. In practice, a site may have an engineered plan for sheet pile or soldier pile with tie-backs. The field role is to install according to that design and monitor (via surveying or inclinometers) any movement, as often specified by the engineer. If struts or tie-backs are part of the system, ensure they're installed on schedule to maintain support.

**Temporary Bridges and Structures:** ASCE manuals or state DOT standards cover temporary bridges or pedestrian walkways on construction sites. For instance, a temporary bridge for construction equipment crossing a creek will likely need to handle heavy loads – an engineer designs it, maybe using ASCE guidelines. The field crew must build it to spec (appropriate beams, decking, guardrails) and not overload it beyond design (e.g., don't send a 150-ton crane over a 50-ton rated bridge).

Formwork and Falsework (ACI and ASCE): The American Concrete Institute (ACI) has standards for formwork (ACI 347) and ASCE has worked on formwork/falsework guidance too. Many tragic failures (scaffold or formwork collapses) happen due to ignoring loads during concrete placement. Practical steps: - Ensure formwork drawings are reviewed and followed. These often come from specialty engineers and include required shore spacing, bracing, pour rates, etc. - Before a big concrete pour, inspect that formwork/falsework is built per plan (all required ties, wedges tight, etc.). Have the engineer of record sign-off if needed. - During the pour, control how fast and where concrete is placed to avoid overloading one area. (For example, ASCE/ACI might say "don't exceed 4 feet of height per hour" for wall pours to limit pressure). - Don't remove shores too early. Often the engineer will specify a concrete strength or time before stripping. Use field-cured cylinders or test results to verify if needed.

**Partially Constructed Structures Stability:** Many structures aren't self-supporting until completion. Eg: Precast concrete parking garages – stability might rely on all pieces and connections being in place. If you erect partially, you may need temporary bracing or sequencing to maintain stability. ASCE's Temporary Works Committee emphasizes clarity of responsibility among designers and contractors for that stability. Field application: - Hold pre-erection meetings to discuss stability at each step. Identify when a frame might be vulnerable (like a long-span truss lifted in place but not yet connected at both ends). - Use temporary guys, struts, or support towers as recommended by the design engineer. For steel, OSHA Subpart R mandated things like minimum 4 anchor bolts per column and evaluating need for guying – that came from hard lessons and AISC/ASCE input. - Don't deviate from approved erection sequence. If the structural drawings or erection plan say "Do not remove bracing until all floor diaphragms are poured," heed that.

**ASCE's Temporary Works Design Philosophy:** "Temporary works" refers to any temporary construction used to build the permanent work (formwork, scaffolds, cofferdams, temporary ramps, etc.). ASCE stresses that these need proper design and oversight, just like permanent works. They often aren't explicitly checked by building officials (the blind spot) – so it falls to contractors and their engineers to self-regulate. Many large contractors have in-house "temporary works engineers" or hire consultants. The practical upshot: field leaders should never brush off temporary structures with "it's just temporary." They should ensure someone competent has engineered or evaluated it.

**Monitoring Temporary Conditions:** Another aspect – monitoring is crucial since temporary states can be less robust. If using a temporary crane base or scaffold in high winds, have a wind action plan: at X wind speed, stop work, at Y speed, evacuate scaffold. Keep an eye on weather forecasts. - Monitor excavations after heavy rain or if adjacent structures show cracks – bring engineer if needed to reassess shoring. - If a building is only partially connected (like precast panels temporarily bolted), restrict access on windy days if stability is borderline.

The temporary condition blind spot Is often due to everyone focusing on final design compliance while assuming contractors will "figure out" the construction means and methods safely. ASCE documents basically say "you need to actively engineer and manage those means and methods too." Field leaders implementing ASCE recommendations will, for example: - Insist on design calculations for a large trench box or excavation support, not just seat-of-pants installation. - Provide adequate lateral support for walls or frames under construction – e.g., don't remove temporary bracing from a masonry wall until the concrete tie beams are in place and cured. - Consider construction loads like stacked materials or construction equipment weight on partially built slabs – and if not sure, ask a structural engineer to okay it.

ASCE's contributions guide us to treat temporary construction phases with the same rigor as permanent conditions. They push us to ask, "Is the structure safe right now, not just when finished?" Practical application means getting engineering input for critical temporary works, following engineered plans for those works, and continuously monitoring safety of partially completed structures. When this is done, collapses and failures during construction can be avoided, protecting both lives and the project itself.

## AISC and Steel Erection Reality

The American Institute of Steel Construction (AISC) sets the standards for steel design and construction. While AISC's Specification and Code of Standard Practice deal with the final product, they also profoundly influence steel erection in the field. There's often a gap between what the design drawings show and what is needed for safe, efficient erection. This is where AISC guidelines and the "steel erection reality" on site meet.

**AISC Code of Standard Practice (COSP):** This document (AISC 303) defines roles and expectations between engineers, fabricators, and erectors. Key points that affect field work: - Anchor Bolt Responsibility: The Code states that anchor rods (bolts) are part of the overall design; but once placed, if they're out of tolerance, how to handle it? It used to be gray. Now, OSHA's steel erection rules (derived in part from AISC input) require columns to have at least 4 anchor bolts and prohibit arbitrary field fixes to anchor bolts without engineer approval. So field practice: if anchor bolts don't line up, do not torch them or bend them without engineer sign-off. Instead, call the engineer; often they'll approve a fix like drilling new anchors or a bracket, but that's formal. This change came after collapses caused by anchor bolt issues. - Erection Aids and Stability: The Code acknowledges that some pieces (like temporary bracing, safety cables) are not shown on design drawings but are necessary for erection. It's understood the erector will add these as needed. Field implication: use common erection aids – for example, when setting long-span steel joists, OSHA and AISC expect you to install "row of bolts" or bridging before releasing the hoist cable. In practice, have crews bolt a minimum pattern (often two bolts per connection) to stabilize before letting go. - Sequence and Means/Methods: AISC leaves the erection means and methods to the erector but expects a reasonable plan. Complex sequences (like heavy lifts, tandem picks) should be planned and sometimes submitted for approval. If a design is particularly sensitive to sequence (e.g., require a specific order of removing temporary supports), the engineer should note it. If not noted, the erector plans but should still follow good practice (like not removing all temporary bracing until final connections are made and tightened).

**OSHA Subpart R – Steel Erection (inspired by AISC):** The 2001 update to OSHA's steel erection standards was heavily informed by AISC, ironworkers, etc. That standard is law, but understanding the rationale (often AISC consensus) helps in the field: - Example: "No load on open web steel joists until bridged" comes from AISC guidance to prevent joist collapse. Field: ensure all bridging is installed and joist connections made before landing deck bundles on them. - Example: "Shake-out" of steel on deck: There's an AISC practice to not overload a floor with too many steel pieces at once. OSHA says if you landed deck bundles, you can't set another on until distributed. Use AISC tables or engineer input for how much weight a partially constructed floor can take. - Column Stability: OSHA requires 4 anchor bolts and other measures because in "real life" (the steel erection reality) columns sometimes got set with 2 bolts and collapsed. Now field crews ensure all columns have 4 anchors (fabricators and foundation crews too). If a column has base plate issues, they must be fixed with engineer oversight, as seen in CROSS report where a two-bolt temporary connection failed and caused a collapse. - Double Connections: AISC and OSHA addressed the hazard when two beams connect at opposite sides of a column flange using common bolts – swapping one out could let the first beam fall. Field solutions per standards: use a seat or stub angle so first beam is supported when bolts are removed, or use a moment clip that keeps one bolt. As a field leader, ensure crews follow the method – many modern steel connections include a safety seat detail because of this. - Torquing and Tensioning: AISC specifications say how bolts should be pretensioned for final structural integrity. On site, ensure ironworkers properly tighten bolts (snug tight for shear connections, fully tensioned for others as required). While it's a quality issue, it's also safety: a fully plumbed and tightened frame is stable. Partially bolted frames left overnight should be evaluated – often the erector will secure enough bolts to hold until final torquing. - Field Fixes: AISC CoSP Option 3 allows connection design to be delegated to fabricator's engineer (with approval). But any deviation during erection (drilling extra holes, cutting a member) technically requires approval. Real-world, ironworkers sometimes want to "fix" misfits with flame or force. The AISC approach: don't – get the engineer to approve modifications. We saw that in CROSS case: they field-welded smaller bolts to an embed plate in lieu of proper anchors, which failed. If they had paused and consulted an engineer, a safer fix (like add braces or redesign base plate) would be made. Field takeaway: stop and ask on significant fixes. - Communication and Roles: The CROSS report highlights blurred lines of responsibility between erector, fabricator, and engineer. AISC code tries to clarify: the erector is responsible for means and methods (safety of erection), the engineer is responsible for the final design adequacy. But when a temporary state (like that two-bolt base connection) is inherently part of design, someone must take charge. Practically, if something doesn't look or feel safe, even if "approved", the field must speak up and perhaps get a specialty engineer to assess erection stability. Better a small delay than a collapse.

**Steel Erection Reality – Things Field Crews Do:** - Use Crane Swing and Coverage: Place the crane to minimize moving heavy steel by hand. Sequence steel delivery so members needed first aren't buried in the laydown.

This isn't AISC per se, but good planning that prevents workers from excessive manual handling or unsafe lifts. - Connecting Practices: Ironworkers often work without fall protection until they get decking in (they are allowed by OSHA up to 30' or 15' in some cases to free climb, though many contractors require 100% tie-off at 6' regardless). AISC doesn't set that, OSHA does. Field management can go beyond OSHA to require fall protection at all heights, providing they supply the means (like moveable anchors on steel). This is a safety culture choice aligning with best practice beyond just compliance. - Bolt Tossing and Falling Objects: On a real steel job, bolts get thrown up to connectors, tools can fall. The reality is controlled by measures like bolt bags, tie-offs for spud wrenches, and exclusion zones below. No specific AISC standard, but industry practice. As a field leader, ensure barricades below steel work, require tool tethering when practical, and keep site clear of unnecessary personnel under steel work. - Welding Sequence and Distortion: When field welding (like moment connections), sequence can cause warp or locked-in stresses if done thoughtlessly. AISC provides guidelines to stagger welds. The field should follow welding procedures not just for quality but to avoid things like a flange suddenly warping out and causing a dangerous release of stored energy.

**Case Example – Column Collapse (Revisited):** In the CROSS case, a tube column on two temporary bolts toppled with a worker on it. Lessons aligning with AISC/OHSA: - Always have columns supported by 4 full-strength bolts (so they can't pivot). - If only 2 bolts are in place temporarily, guy the column or brace it until remaining bolts are installed. The erector in the case failed to do so. - Ensure welds or other "fixes" for connections have known capacity – in the case, the weld of the temp bolts likely wasn't sufficient. AISC would have mandated a proper engineered fix. - Communication: Fabricator/erector should have gone back to engineer when anchor bolt placement wasn't per design to figure a safe solution.

AISC standards and the reality of steel erection safety are intertwined. The standards provide a framework and engineering backbone (properly designed connections, tolerances, etc.), while field "reality" requires careful sequencing, on-the-fly problem solving within the limits of standards, and sometimes more conservative measures than the bare minimum. When field crews respect AISC guidelines, and management ensures compliance with OSHA's steel erection rules (largely derived from industry best practices), steel erection can be done without incident. The mantra could be: Don't take shortcuts with steel – every piece and bolt plays a role in stability. Use the proper pieces (no missing braces), follow the plan, and if something goes off-plan, get the experts involved rather than improvising dangerously. That's how you reconcile the sometimes messy reality of steel erection with the safety standards that AISC and OSHA expect.

## EPA and Environmental Compliance for Construction

Construction projects not only must protect workers and create structures, but also must safeguard the environment. The Environmental Protection Agency (EPA) and state environmental agencies enforce regulations that can profoundly impact construction operations. Environmental compliance is sometimes seen as separate from construction's core tasks, but failing to integrate it can lead to costly project delays, fines, and harm to community relations. This section covers how environmental risks translate to schedule/cost risk, and practical approaches to managing stormwater, erosion, dust, air quality, noise, hazardous materials, waste, and spills. It also outlines roles and accountability for environmental compliance on site.

## Environmental Risk as Schedule and Cost Risk

Ignoring environmental requirements can stop a project in its tracks as surely as a safety shutdown or structural failure. Environmental risk is schedule and cost risk. Construction managers must view compliance with environmental laws not as red tape, but as integral to project success.

**Permits and Approvals**: Before work starts, most projects require environmental permits – stormwater discharge permits, wetlands permits, tree removal approvals, etc. These come with conditions (like implementing a Stormwater Pollution Prevention Plan, or SWPPP). If you break those conditions, agencies can issue stop-work orders or hefty fines. For example, a contractor building a subdivision might need a state NPDES (National Pollutant Discharge Elimination System) permit for stormwater. If muddy water is leaving the site and polluting a creek, the state can shut down earthwork until fixes are in place. That's a direct schedule hit. In one case, Swinerton, a large contractor, had to pay $2.3 million to settle EPA allegations that it failed to use proper stormwater controls and inspections, leading to sediment discharges. Beyond the fine, imagine the work stops and remediation efforts that cost time.

**Project Delays from Violations:** An example scenario: a project clears land and doesn't install silt fence; a heavy rain sends silt into a nearby stream. A local environmental official issues a cease and desist until cleanup and proper controls are established. That could mean earthwork halts for weeks. Rainy season might extend that, cascading into missing a foundation deadline, etc. Or consider discovering contaminated soil unexpectedly – if not handled right, regulators can halt excavation until a remediation plan is approved.

Treat it as Part of Construction, Not an Afterthought: Successful projects integrate environmental planning into the master schedule. For example: - Plan grading and earthwork phases around rainy seasons to minimize risk. - Build in time for installing and maintaining erosion controls (it's as much a task as formwork or rebar – schedule crews to do it and check it). - Ensure long-lead environmental items (like getting an off-site disposal facility lined up for hazardous debris) are on the procurement schedule. - If you need to do environmental testing (say, test soil for contaminants), schedule it early to avoid surprises when you have a hole open and find out you can't reuse the soil.

Think of Boston's "Big Dig" or other mega-projects – environmental oversight was huge. They had full-time environmental compliance staff to coordinate with agencies. While smaller projects might not have that luxury, the principle holds: assign someone knowledgeable to keep eyes on enviro issues (maybe the safety officer doubles as environmental officer). Regular internal audits help catch issues before the regulator does.

**Opportunities in Compliance:** It's not all doom – managing environmental aspects well can sometimes speed things up or save money: - Good planning might let you reuse materials on site (if tested and clean) instead of hauling to landfill – saving tipping fees and trucking time. - Avoiding violations means you don't have unplanned work stoppages. - Some sustainable practices (like water reuse for dust control) can reduce resource costs too. - A project with a good compliance record might get faster sign-offs from agencies – inspectors build trust with you and perhaps focus less scrutiny, indirectly easing and speeding approvals.

So, view every environmental requirement (be it silt fence, a noise barrier, or a hazardous waste manifest) as a critical path item. The cultural shift is to treat "legal and ethical" obligations as non-negotiable parts of the plan.

When scheduling excavation, budget crew days for environmental setup and monitoring. When reviewing costs, consider the fine you're avoiding as an ROI for the compliance efforts.

Environmental risk management is project risk management. Recognize that a dirty site can muddy your timeline and bottom line. By being proactive environmentally, you keep regulators and neighbors satisfied, which keeps your project on track and on budget.

## Stormwater and Erosion Control

Few things can shut down a construction site faster than a mudslide into a storm drain or a neighbor's property. Stormwater runoff and erosion control are therefore top priorities, especially for sites with significant earth disturbance. The goal is to prevent soil, sediment, and pollutants from leaving the site when it rains or from wind – a requirement under the Clean Water Act enforced via NPDES permits.

Stormwater Pollution Prevention Plan (SWPPP): Almost all projects over a certain size (disturbing 1 acre or more, or even smaller in sensitive areas) must have a SWPPP as part of their construction general permit. This plan identifies potential sources of sediment and pollution and prescribes Best Management Practices (BMPs) to prevent them from reaching waterways. As a field leader: - Know your SWPPP. Keep a copy on site. It will have a site map showing drainage patterns and where BMPs (like silt fences, basins) are to be installed. It also lists who is responsible for inspections and maintenance. - Implement BMPs per the plan and permit timeline – usually, before clearing or grading begins, erosion controls must be in place. For example, silt fence around downslope perimeters, stabilized construction entrance (rock pad to scrape mud off truck tires), sediment basins or traps for large sites, inlet protection (fabric or barriers around storm drain inlets). - Preserve existing vegetation as long as possible and only clear what you need when you need, to minimize exposed soil at one time. - Schedule earthwork strategically: e.g., avoid leaving large areas unseeded during rainy season; phase grading so that one area is stabilized (with straw, seed, mulch or temporary cover) before opening another.

**Erosion Control BMPs:** Some common ones and field tips: - Silt Fence: A basic barrier for sheet flow. Ensure it's trenched in (6 inches buried) and the stakes on downhill side. Many failures come from improper installation or not enough fence. Space fence runs with overlaps at low points (j-hook pattern) to actually catch sediment rather than funnel it. Remove sediment that builds up when it reaches about half the fence height. - Straw Wattles or Fiber Rolls: Good for small slopes or around inlets. They need to be staked and sometimes trenched slightly. Replace when they degrade or if a big event moves them. - Check Dams: Small barriers of rock or wattles in drainage channels to slow flow and let sediment settle. Space them so the toe of the upstream one is at the elevation of the top of the next, to create a step-down pool effect. - Sediment Basins: For bigger sites, you dig a basin to hold runoff, letting sediment drop out. Maintain as required; often you might pump from it (after letting sediment settle or through a filter) to discharge water. Some states require specific sizing (e.g., volume for a 2-year storm). - Construction Entrance/Exit: Essentially a pad of large rock (2-3" stone, maybe with a geotextile underlayment) at site exits to knock mud off tires. Ensure trucks actually use it – possibly requiring a wheel wash on really muddy sites. Replenish rock when it becomes packed with mud. - Dust Control (also stormwater related): Dry soil can blow and then settle in storm drains. Water down haul roads or use a water truck regularly in dry conditions (but not to the point of creating runoff). Some projects use soil binders or cover stockpiles with tarps to prevent erosion by wind or rain. - Slope Stabilization: For exposed slopes, especially steeper ones, quickly apply straw mulch (often with a tackifier) or erosion control blankets (rolled matting, sometimes jute or coir fiber) to hold soil until vegetation grows. The SWPPP may dictate seeding and mulching if an area will be inactive for more than 14 days, for instance. - Ditches and Conveyances: Any channels meant to carry water should be stabilized – e.g., line with riprap or erosion control blanket – to prevent the channel itself from eroding.

**Inspections and Maintenance:** Under EPA's Construction General Permit and similar state permits, you must inspect the site at regular intervals (often weekly and after significant rain events ≥0.5 inch). The inspector (often a certified person) looks at all BMPs: is silt fence intact? Are inlets protected and not bypassed? Is there mud on the street (if so, sweep it immediately – many cities fine if you track mud offsite)? They fill out a report and note any fixes needed. Crucially, you must fix issues promptly – usually the rule is within 7 days or before next rain, whichever sooner, any noted problem should be addressed.

**Sediment Tracking and Turbidity:** Many jurisdictions not only check for visible mud but also do turbidity tests in nearby waters. If your runoff increases water turbidity beyond certain threshold, you're in violation. For example, in some states, if a sample from your site's discharge is more than, say, 50 NTU above background, you have to report and take corrective actions. The easiest way to avoid that is to keep muddy water on site (through ponds or filtration).

**Concrete Washout:** Wet concrete and cement-laden water is highly alkaline and a pollutant. The SWPPP should specify a concrete washout area – a lined pit or container where concrete trucks and tools can be washed, preventing wash water from soaking into ground or storm drains. Field practice: clearly mark the washout, inform drivers (they often try to wash out wherever if not directed), and when it's 75% full, clean it out (hardened concrete can be hauled to recycling or landfill).

**Spill Prevention (as part of stormwater):** Fuel, oil, paints, chemicals – if they spill and get in water, big trouble. The SWPPP often overlaps with a spill control plan. Keep chemicals in a designated, contained area (like a bermed and covered storage). Refuel away from waterways, have spill kits. Even something like hydraulic oil leak from a machine should be contained (with drip pans or absorbents) before rain spreads it.

**Mindset for Crews:** Erosion control isn't just the site super's job – equipment operators, laborers, etc., should all be mindful: - If you see a silt fence knocked down, report it or fix it if you can. - Don't dump spoil or materials on top of BMPs. - Keep dumpster lids closed (rain in dumpsters can create polluted leachate). - Don't wash out concrete or paint brushes except in designated area. - If a pump is dewatering an area, ensure it's not discharging dirty water directly offsite – often you pump through a filter bag or to a silt basin.

**Documentation:** Keep logs of all inspections and maintenance performed. If an agency questions you, showing a thick folder of inspection reports, actions taken, and dated photos of BMP upkeep can demonstrate your good faith (and possibly stave off penalties by showing you are trying, and any issue was maybe from an extreme event).

In essence, stormwater and erosion control are about keeping your dirt on your site and preventing any concrete, paint, or chemical from hitching a ride with runoff to waters. It's one of the most visible environmental compliance areas on a construction project. Doing it right not only avoids violations but also keeps the project literally out of the mud (less rework, better access when site isn't a soupy mess). It's a win-win: the environment is protected and your site is more efficient and safer (less slip hazards, etc., when mud is managed).

## Dust, Air, Noise

Construction sites can be unruly neighbors when it comes to dust, air emissions, and noise. Beyond being considerate, there are regulations and local ordinances that require control of these nuisances and hazards. Effective management protects worker health, keeps the community happy, and maintains compliance.

**Dust Control (Fugitive Dust):** - Sources: Excavation, grading, traffic on unpaved roads, demolition of structures, cutting of concrete/masonry (which also raises silica concerns for workers), and stockpiles of soil or sand can all generate dust. - Regulations: Many states (especially arid ones like California, Arizona, Nevada) have strict dust control rules. For example, Maricopa County (Phoenix area) requires dust permits for earthmoving and mandates specific measures (water trucks, gravel pads, vehicle speed limits on site) with fines for non-compliance. The federal Clean Air Act also has standards for particulate matter (PM10 and PM2.5). If your dust significantly contributes to local PM levels, you could be violating air quality standards. - Practical Controls: - Watering: The go-to method. Use water trucks or sprinklers to keep soil damp during active earthwork or when trucks are driving on dirt. The SWPPP often ties into this for erosion too – but not over-water to cause runoff. Fine misting can also control dust during concrete cutting (plus it controls silica). - Chemical Stabilizers: On longer-term haul roads or stockpiles, spraying a dilute solution of soil binder (like magnesium chloride or organic polymers) can crust the surface and reduce dust. - Speed limits: Enforce, say, 10-15 mph speed on unpaved areas. High speeds kick up exponential dust. - Gravel roads/entrances: Laying gravel on frequently traveled paths reduces dust from those surfaces. Also prevents kicking up dust onto public roads. - Covering loads: Require dump trucks leaving site with soil or debris to cover their loads (usually law) so debris/dust doesn't fly out in transit. - Street cleaning: If dirt does get on adjacent pavement, use street sweepers (preferably with water spray or vacuum) to clean it promptly. Dry sweeping just stirs more dust. - Stockpile management: Keep soil piles low, cover them with tarps or plastic sheeting if they'll sit unused. If very large, consider hydroseeding them temporarily. - Demolition dust: Use misting cannons or fire hoses to spray water during building demolition or implosions to suppress dust clouds. Many cities require water spray during demo of concrete/brick structures. - Monitoring: Some projects use dust monitors (like particulate monitors on site perimeters) especially near sensitive areas like schools or hospitals. These can trigger alarms if dust exceeds thresholds, prompting immediate action (like increase watering).

**Air Quality (Equipment Emissions, Odors):** - Equipment Exhaust: Diesel equipment emits particulate (soot) and NOx. In some metro areas (e.g., California), ordinances or contract requirements mandate using newer Tier 4 engines or retrofit filters on older equipment to reduce emissions. Field enforcement: ensure subcontractors use compliant equipment or provide proof of emissions controls. Limit unnecessary idling (many places have 5-minute idling laws). Not only does this reduce air pollution, it saves fuel. - Odors and VOCs: Activities like asphalt paving, roofing with hot tar, painting, applying sealants or spraying foam can release smelly or hazardous volatile organic compounds (VOCs). While usually short-term, neighbors might complain or workers might get headaches. Control by scheduling such work for times of least impact (e.g., early morning paving when few neighbors out), using low-VOC products when possible (e.g., low-VOC paint), and ensuring good ventilation for indoor work. If doing something like fiberglass layup or solvent use in an enclosed area, use localized exhaust fans or carbon filter air scrubbers if needed. - Asbestos/Lead: If demolishing or renovating older structures, asbestos and lead paint become airborne hazards if not properly abated. There are strict EPA and OSHA rules for removal (e.g., wet methods, negative pressure enclosures, not letting visible emissions escape during asbestos removal). Non-compliance can release fibers or lead dust, harming workers/neighbors and resulting in severe fines or even criminal charges. Always survey for these materials pre-demo and hire licensed abatement contractors. Don't start demo until you have clearance that hazardous materials are removed or contained. - Concrete/Cement Dust (Silica): When cutting or jackhammering concrete, the fine dust contains silica, harmful when inhaled (can cause silicosis). OSHA requires dust control (like wet cutting or using saws with vacuum dust collection). This also prevents clouds of dust leaving the immediate area. In urban sites, uncontrolled concrete dust can settle on cars, houses, etc., leading to complaints and potential citations under air quality or worker safety rules. Use water or HEPA vacuums consistently. - Site Fires/Smoke: Burning waste on site is typically illegal (and dangerous). Not only does it produce smoke (air pollution), it's a fire hazard. Have a strict no-burning policy; use proper disposal. If a piece of equipment catches fire or a welding spark causes a small blaze, that smoke is not a violation per se but avoid such incidents by good practices (fire watch, maintenance on equipment to prevent fuel leaks). If one happens, notify authorities as required.

**Noise Control: -** Ordinances: Almost every locale has noise ordinances. Typically, they set permissible working hours (e.g., no construction noise before 7am or after 6pm, and perhaps none on Sundays or holidays) and sometimes maximum decibel levels at property lines. On big projects, especially if working 24/7 is needed, you may need a noise variance or to implement noise mitigation to get permission. - Community Relations: Engage neighbors early if noise is going to be an issue. Provide a hotline or contact they can call with concerns. Sometimes just giving notice ("We'll have to do early morning concrete pours for the next 3 weeks, here's why...") helps manage expectations. - Noise Mitigation Techniques: - Use quieter equipment or methods when possible (e.g., hydraulic pile pressing instead of pile driving hammer if feasible). - Maintain equipment – lubricate and repair mufflers. A bad muffler on a generator can double noise. - Stationary loud equipment (generators, compressors) should be located as far from noise-sensitive boundaries as practical, and/or behind barriers. - Temporary noise barriers: These can be plywood walls or specialized noise curtains erected between the work and neighbors. For instance, a 8-foot plywood fence with sound blankets attached can significantly reduce ground-level noise for adjacent homes. For higher work like jackhammering a tall structure, hanging sound curtain tarps on scaffolding around the work area can dampen noise transmission. - Limit certain activities to daytime: If you must jackhammer or use impact tools, do those during permitted hours; at night, stick to quieter tasks if any. - White noise backup alarms: Instead of the high-pitched beepers on equipment, many places allow (or encourage) broad-spectrum "white noise" alarms that are less annoying and localize to the vehicle area. - Sequencing: avoid having multiple loud operations concurrently if possible (like two jackhammers at opposite ends of site vs. doing sequentially). - Worker Hearing: Remember, high noise isn't just an offsite concern – onsite, provide hearing protection and implement a hearing conservation program for your crew as needed (OSHA requires if 8-hr average > 85 dB). Tools like air hammers, circular saws, and heavy machinery often exceed 90 dB. Rotate workers or enclose noisy machines to reduce exposure. This also coincidentally reduces how much noise leaves the site.

**Monitoring Noise:** For critical projects, you might do periodic sound measurements at site boundary during operations to ensure compliance with decibel limits. Some community concerns can be objectively addressed ("We measured 68 dB at your front door, which is within the city's 70 dB limit").

**Enforcement and Penalties:** Repeated dust or noise complaints often bring inspectors. For dust, visible dust crossing property line can result in notices of violation. For example, some California air districts fine per day of violation; it can be thousands of dollars and even a shut-down order until a dust mitigation plan is approved. For noise, the city may issue stop orders for offending activities or police can cite you. Better to be proactive: if neighbors or inspectors see you earnestly controlling dust and scheduling noisy work appropriately, they are more lenient or even complimentary.

Dust, air, and noise control is about being a respectful and law-abiding neighbor while also protecting health. Water down the dust, tune down the noise, and keep the air as clean as practicable. It's part of the logistics of running a clean site. And it pays off: less cleanup, fewer complaints, and no surprises from regulators. As one superintendent quipped, "Keep the site's noise and dust down, and we stay out of the news and in the good graces of City Hall." That sums it up well.

## Hazardous Materials and Waste

Construction projects often encounter or generate hazardous materials and waste that require special handling. Whether it's an old building with asbestos, contaminated soil on site, or simply the paints and fuels used during work, mismanaging these can pose serious health risks and lead to legal violations. Properly dealing with hazardous materials protects workers, the public, and avoids expensive cleanup liability down the road.

**Asbestos:** Commonly found in older buildings (insulation, floor tiles, mastic, roofing, popcorn ceilings, cement siding, etc.). Federally, any facility demolition or renovation must follow EPA's NESHAP (National Emission Standards for Hazardous Air Pollutants) for asbestos: - Asbestos Survey: Before disturbing any suspect materials, get a licensed inspector to survey. If asbestos is present above a minimal threshold, you must have it abated by licensed asbestos contractors prior to general demo. - Abatement Process: Involves sealing off area, workers in protective suits, wet removal techniques to prevent fibers, and proper disposal in leak-tight containers with asbestos warning labels. It's expensive, but non-negotiable. Cutting corners can expose many to cancer-causing fibers and result in heavy penalties or jail. An Illinois contractor was fined $1.7 million by OSHA for willfully exposing workers to asbestos – a career-ending event. - During Construction: If you unexpectedly find asbestos (e.g., buried pipe wrapping or unforeseen material in walls), stop work in that area and get professionals to test and remove it. Educate workers on what it looks like and to report suspicious materials (like old insulation or 9"x9" vinyl tiles). - Disposal: Asbestos waste goes to approved landfills. Maintain waste shipment records. Ensure your subcontractor does – you as GC could be held liable if asbestos waste "disappears" illegally.

**Lead-Based Paint:** In structures built pre-1978, lead paint is likely. Sanding or demolishing that can create toxic lead dust. - Renovation (especially homes, schools, daycares): Must follow EPA's RRP (Renovation, Repair and Painting) Rule. That means certified renovators, containment of work area (plastic sheets, sealing vents), specific cleaning and verification after to ensure no lead dust remains. - Demolition: If doing whole building demo, the concern is less occupant exposure (since it's coming down entirely) but more about worker exposure and environmental release. Keep it wet to avoid dust and handle debris appropriately. Large-scale demo is usually exempt from RRP, but OSHA lead standards still apply to protect workers (they may need respirators or blood lead monitoring if levels high). - Soil contamination: Exterior lead paint can flake into soil; if you have to excavate around an old painted structure, test the soil for lead. If above hazardous levels (RCRA threshold is 5 mg/L by TCLP test), that soil has to be managed as hazardous waste, not just dumped. - Waste: Lead painted debris often isn't "hazardous waste" unless very lead-rich, but some states have special rules. It may go to construction debris landfill, but check local regulations. Do not abandon debris; provide proper disposal receipts.

**PCBs (Polychlorinated Biphenyls):** These were in old electrical transformers, capacitors, light ballasts, some caulks and paints (1970s era). If your project involves old industrial sites or even old fluorescent lighting, you could encounter PCBs. - Transformers/Capacitors: Utilities usually handle these. But if you find an old transformer in a building, assume it may have PCB oil unless labeled otherwise. Specialized removal and disposal is needed (per TSCA – Toxic Substances Control Act). - Fluorescent Light Ballasts: Many pre-1980 ballasts contain PCBs in their potting material unless marked "No PCBs." Collect and dispose of as PCB waste (through hazardous waste channels) rather than tossing in dumpster. - Soil or Caulk: If PCBs are in caulk (like around windows in old schools) or in soil from a spill, EPA has strict remediation requirements. You'd need environmental engineers to test and devise a cleanup plan, often with specified disposal methods. This can be a project in itself.

Contaminated Soil and Groundwater (Site Remediation): - If the site was previously industrial (gas station, dry cleaner, factory), a Phase I & II Environmental Site Assessment should have been done. If contamination is known, there might be a Remedial Action Plan to follow during construction. For example, digging up old fuel tanks (USTs) or hotspots of polluted soil. - Handling contaminated soil: Often, you'll segregate it, test it, and haul to appropriate disposal (hazardous waste landfill, or special treatment facility). Workers must be trained (potentially HAZWOPER 40-hr if significant exposure potential). There may need to be air monitoring at perimeter if volatile chemicals are present (to protect neighbors). - Vapor issues: If building on a site with solvent contamination, you might need to install vapor barriers or venting systems under the new building to prevent vapor intrusion. - Water dewatering: If you have to dewater an excavation and groundwater is contaminated (e.g., with gasoline), you likely need to treat that water (like running through carbon filters) before discharging, under a permit. - Unexpected finds: If you smell strange odors or see stained soils during digging – stop and get it tested. It's cheaper to address properly than to have regulators find you "spreading contamination" later. In many places, you're legally required to report discoveries of certain contamination (like free product petroleum or hazardous waste) to state agencies.

**Construction Hazardous Wastes (from materials we bring):** - Common hazardous wastes generated: spent solvents (paint thinners), waste paint, excess adhesives, waste welding rods (sometimes cadmium), used oil (though that's recyclable in many cases), old batteries, aerosol cans, fluorescent lamps, mercury thermostats, etc. - RCRA Requirements: If you generate hazardous waste above certain quantities, you need an EPA generator ID and must follow storage, labeling, and disposal protocols. Most construction sites try to be Conditionally Exempt (generate very little), but large jobs can tip into Small Quantity Generator territory if disposing of lots of paint or solvent. - Storage on site: Hazardous wastes must be kept in labeled containers (e.g., "Hazardous Waste – Paint Solids, start date...") and stored safely (usually with secondary containment for liquids, protected from rain). There are time limits (typically 180 or 90 days depending on generator status) to ship off site. It's not uncommon for contractors to neglect this until end of project – that's risky if an inspector shows up mid-job. - Disposal/Recycle: Use licensed transporters and facilities. Keep manifests (legal shipping documents) for all hazardous waste shipments. For example, waste paint and solvent rags might go as F-listed hazardous waste to a fuel blending facility. If quantities are small, sometimes local household hazardous waste programs might accept them, but a professional approach is better. Some materials like aerosol cans can be depressurized and recycled as scrap metal with special equipment (if you invest in that). - Special wastes: Pressure-treated wood, for example, contains arsenic or other chemicals. Some states treat sawdust or scraps as hazardous. Most treat it as solid waste but with restrictions. Similarly, gypsum drywall isn't hazardous, but if left to get wet and rot it produces hydrogen sulfide gas (smells like rotten eggs) – which can be a safety hazard in enclosed areas and a nuisance.

**Worker Protection: When handling hazardous materials:** - Provide proper PPE (respirators, Tyvek suits, gloves) and training (like OSHA's HAZCOM, or full HAZWOPER if appropriate). - Have Safety Data Sheets (SDS) for chemicals on site. Workers should know how to read them and precautions. OSHA's HazCom standard requires labeling and accessible SDS. - For tasks like torch cutting coated steel (could release lead, PCBs if old paints), create an exposure control plan – air monitoring, ventilation, PPE. - Use engineering controls e.g., negative pressure enclosures for lead paint removal (with HEPA filtration of exhaust).

**Reporting and Plans:** - Many projects require a Hazardous Materials Management Plan or Waste Management Plan as part of specs (especially sustainable projects under LEED or Owner requirements). These outline how you will store fuels, handle unexpected contamination, segregate and document waste streams (even how much construction debris is recycled). - If a spill or release happens, some must be reported immediately to agencies (see next Spill section). Know your reportable quantities (e.g., oil spills over a certain gallon amount to water must be reported to the National Response Center).

**Long-term Liability:** If your project doesn't properly dispose of hazardous waste and, say, leaves containers buried or sends to an unlicensed dump, the generator (which could be your company or the owner) can be held liable under Superfund law to clean it up later – often at enormous cost. Always do it by the book now to avoid those nightmares.

Treating hazardous materials and wastes with respect is part of running a responsible site. It means planning for proper abatement of legacy hazards (asbestos/lead), safe usage and cleanup of current materials (fuel, chemicals, etc.), and lawful disposal of wastes generated. If unsure, hire an environmental consultant for guidance – a small expense compared to the consequences of mishandling. Emphasize to the crew: "If you see something weird (odor, barrel of unknown goo, etc.), don't ignore it – we handle it safely." Through vigilant management, you ensure hazardous substances don't harm people or the environment, and your project doesn't get hit with compliance violations or cleanup costs.

## Spill Response Expectations

Despite best efforts, spills happen – a hydraulic hose bursts, a fuel truck overflows a tank, a solvent drum tips over. How a construction site prepares for and reacts to spills can mean the difference between a minor incident and a major environmental problem. Regulatory expectations for spill response are clear: contain it, report it if required, and clean it up promptly and thoroughly.

**Spill Prevention, Control, and Countermeasure (SPCC) Plan:** - Under EPA rules, if you store more than a certain amount of oil on site (above 1,320 gallons in above-ground containers, or a single container over 660 gal), you're required to have a formal SPCC plan. This plan is often prepared by a professional engineer. It outlines the site layout, oil storage locations, potential spill pathways, and specific measures to prevent and respond to spills. - Even if you don't hit the SPCC threshold, it's wise to have a simplified spill plan. Construction projects typically have fuel for equipment (diesel tank, jerry cans of gas), oils (form release oil, vehicle maintenance fluids), and perhaps other chemicals (formwork chemicals, concrete admixtures, etc.). - A typical plan includes: - Regular inspection of tanks and equipment (look for leaks, rust). - Secondary containment (like double-walled fuel tanks or berms around single-walled tanks that can hold 110% of the volume). - Procedures (e.g., never leave filling operations unattended, keep valves locked). - Training for workers on fueling procedures and emergency actions.

**Spill Kits and Equipment:** Every site should have readily accessible spill control materials: - Spill Kits: Typically containing absorbent socks (to confine spill spread), absorbent pads or granular absorbent ("oil dry"), protective gloves, disposal bags, and perhaps putty for plugging leaks. Place these kits near high-risk areas (fuel storage, generators, near water bodies). Also, each piece of heavy machinery often carries a small spill kit. - Drain Covers: If you have storm drains on or near site, have rubber drain mats or covers in your kit. In event of a spill, slap those over drains to prevent oil entering. - Tools: Shovels and containers to scoop up contaminated soil or absorbents once you soak up the spill.

**Immediate Response Steps (The 3 C's):** 1. Control/Cut off the source – stop the flow. That might mean shutting a valve, uprighting a container, plugging a hole (some kits have epoxy putty that hardens quickly to patch tank holes). 2. Contain – use absorbent snakes to encircle the spill, creating a dam. For larger spills on soil, build a dirt berm around it with a shovel. If on a slope, block downhill progress. Protect any waterways or drains aggressively – that's priority one. 3. Clean up – once contained, start recovering the material. If it's oil pooled, use absorbent pads or a vacuum truck for larger amounts. If it soaked into soil, dig up the saturated soil. Place waste (used absorbents, oily soil) into drums or heavy-duty bags for disposal as hazardous or special waste.

**Notification:** Determine who needs to be notified: - Site management: Always inform the project manager or superintendent immediately, even for small spills (so they can help and also be aware in case of any further action). - Environmental/safety officer: They may coordinate the response and any reporting. - External Reporting: This depends on quantity and type. For petroleum, SPCC rules require reporting to the National Response Center (NRC) if a spill could reach water and causes a sheen, or if over a certain amount (like any amount to water, or >1,000 gal to land in one go, or smaller repeated spills). Many states also have hotlines (for example, Florida requires reporting any spill >25 gallons, some states any amount that could harm environment). It's better to report if in doubt – penalties often escalate if you fail to report. - When reporting, be ready with info: location, substance, estimated amount, what you've done so far, and if any waterway is impacted. They'll advise if more action is needed. - Client/ Owner: Many owner contracts require immediate notice of environmental incidents. - Local authorities: If a spill is threatening public property (like oil flowing to a city street or sewer), the fire department or local environmental agency should be contacted. Fire Dept hazmat teams might respond if it's something large or particularly hazardous (like a tanker fuel spill or unknown chemical). - OSHA: Only if it's also a safety incident (like a worker overcome by fumes or a fire), typically environmental spills themselves aren't OSHA-reportable unless there's an injury or exposure.

Overall, regulators expect construction sites to be spill-ready – that means having a plan, materials, and trained personnel to handle spills immediately. The first minutes count: if you can keep a spill out of the water or drain, you've massively reduced the impact. Emphasize a culture of "see a spill, own the spill response" – any worker who spots one initiates action and alerts others, rather than walking by. By meeting these expectations, sites not only avoid environmental harm and fines, but also demonstrate professionalism and responsibility.

## Global Best Practices in Safety Standards

Achieving compliance with U.S. regulations (OSHA, EPA, building codes) is foundational, but leading construction organizations strive for safety excellence beyond mere compliance. Around the world, many best practices and frameworks have been developed to foster a strong safety culture and nearly eliminate serious incidents. In this section, we explore how one can climb the ladder from compliance to excellence, translating global standards like ISO management systems into the construction context, adopting High-Reliability Organization habits, maintaining safety culture even under schedule pressure, and what the "best practice ladder" looks like for continuous improvement.

## Compliance vs. Excellence

There's a significant mindset shift between doing only what is legally required (compliance) and pursuing the highest level of safety performance (excellence). Compliance means meeting the letter of laws and codes – you avoid penalties, you tick the boxes. Excellence means proactively creating an environment where incidents are not just avoided by rule, but by ingrained culture and systems that seek out and eliminate risks continuously.

**Characteristics of Compliance-Driven Approach:** - Tends to be reactive. Safety measures might come mainly after an incident or citation ("We got fined for poor fall protection, so now we use guardrails"). - Focus on minimum standards. For example, doing training only because OSHA requires 10-hour cards, conducting inspections only at the frequency OSHA mandates, etc. - Often manager-led only. Rules are enforced top-down ("because OSHA says so" or "because company safety officer says so"), with less buy-in from workers. - Documentation-centric – heavy on paperwork to prove compliance, sometimes at expense of actual engagement (filling out forms vs. having meaningful discussions). - Goals = Zero Citations. If a project got no OSHA citations, they consider it a success, even if there were some near-misses that got lucky.

**Characteristics of Safety Excellence:** - Proactive and preventive. The project team anticipates potential hazards and addresses them before they cause harm. There's an emphasis on near-miss reporting and lessons learned – finding weaknesses in the system when no one's hurt yet and fixing them. - Beyond the rules. Voluntarily implementing controls more stringent than required if it makes the job safer. For instance, even if OSHA doesn't require a full-time safety manager on a smaller project, an excellence-minded firm might dedicate one because they want that oversight. Or enforcing 100% hardhats even in areas where it "might not be strictly required" because it's part of their culture. - Worker empowerment and involvement. Excellence sees safety as everyone's responsibility. Workers are encouraged to speak up, even stop work, if they see a hazard – and are praised for it, not reprimanded. They are involved in planning (like crew participation in job safety analyses) which leads to better adherence and ideas. - Continuous Improvement. Such organizations set internal safety performance indicators (leading indicators like number of safety suggestions, or trailing like total recordable incident rate far below industry average) and strive to improve them year over year. They audit themselves against best practices, not just regulations. - Ownership from the Top. Senior leadership not only supports safety, they visibly champion it. Budget and schedule are structured to never sacrifice safety – e.g., building in time for safety training, rewarding project teams for safety performance (not just productivity). - Goals = Zero Injuries / World-Class Safety Culture. They measure success in terms of lives protected, not just fines avoided.

A compliance-oriented site might do the legally required weekly toolbox talk because it's mandated. An excellence-oriented site will have daily safety huddles because they find value in constant vigilance and crew feedback. - Compliance might put guardrails on obvious edges to avoid fines; excellence might also implement an internal "Fall Hazard Awareness Week" where they re-evaluate all elevated work and even invite fall protection equipment vendors to demonstrate new devices – aiming to reduce even minor fall-related incidents, not just prevent citations. - When an incident occurs, a compliance-driven firm might blame a worker for violating a rule. An excellence-driven firm uses a no-blame investigation to find systemic causes (maybe training was inadequate or the procedure was flawed) and fix those systemic issues across the company, not just discipline the individual. - Compliance stops at following OSHA or building code. Excellence looks at best practices across the industry or globe: e.g., adopting a Behaviour-Based Safety program (which might not be required by law but has proven to reduce incidents by influencing worker behavior). - Compliance tends to be siloed (safety as separate function), whereas excellence integrates safety into every decision – it's a value, not just a priority that can shift. On an excellent project, a planning meeting for a big lift will inherently involve safety staff and considerations from the outset, not as an afterthought because "we have to file a lift plan per OSHA."

It's worth noting that compliance is the foundation – you cannot be excellent if you don't even meet basic compliance. So first ensure compliance, then build upon it.

**Business Case for Excellence:** Companies sometimes question the cost of going above and beyond regulations. But numerous studies and case studies show that excellent safety correlates with improved productivity and quality (fewer disruptions from accidents, better morale, less turnover, etc.). Also, the worst accidents can be catastrophic to budget and reputation – preventing them is priceless. Many clients now evaluate contractors' safety records and programs when awarding jobs, so excellence can be a market differentiator.

**Transitional Strategy:** For a project or company trying to move from compliance to excellence: - Start by plugging any compliance gaps (to stabilize). - Then introduce higher standards gradually: perhaps set internal rules like "100% eye protection on site" even if some tasks wouldn't legally require it, with clear communication on why. - Encourage reporting of hazards and near-misses without fear – this shifts focus to prevention rather than blame. - Provide positive reinforcement: celebrate milestones like X days injury-free not as complacency but as recognition of team effort. - Train supervisors in leadership safety skills, not just rule enforcement. They should be coaches and mentors in safety. - Involve workers in safety committees or hazard hunts (where crews go find and fix hazards each week). This fosters a sense of ownership beyond compliance.

Compliance is doing things right; excellence is doing the right things (even when not explicitly told to). It's a journey that requires cultural change and continuous effort. But projects that aim for excellence often see the benefits in safer work environment, better morale, and often better overall project performance.

## High-Reliability Habits that Work

The construction industry can learn a lot from High-Reliability Organizations (HROs) – those operating in complex, high-risk environments (like aviation, nuclear power, hospitals) that nonetheless have excellent safety records. HRO theory identifies key habits and principles that help these organizations avoid catastrophes despite the potential. Adopting some of these high-reliability habits on a construction site can significantly improve safety performance and consistency.

## The five classic HRO principles are:

Preoccupation with Failure: HROs are always alert to the possibility of error or failure, even small ones. They treat any lapse as a symptom of a system weakness, not a one-off.

1. On site, this means not becoming complacent when things are going well. Near-misses and minor incidents are treated as valuable warnings. For example, if a tool is dropped from height and nobody is hit, a high-reliability mindset treats that as a serious event to investigate: Why did it happen? How do we ensure nothing ever falls again? Instead of shrugging it off ("no harm done"), they might implement tool lanyards universally and improve housekeeping on scaffolds, seeing that drop as a potential precursor to a fatal accident.

2. Teams might start meetings by asking "What could go wrong today?" or reviewing recent close calls. This constant sense of "chronic unease" (not paranoia, but healthy concern) keeps them proactive.

3. Reluctance to Simplify: HROs resist oversimplifying interpretations; they dig into details and don't assume away complexity.

4. In construction, that means recognizing that accidents usually have multiple causes. So when identifying hazards or investigating incidents, avoid "simple" explanations like "worker didn't follow rules." Instead, consider context: Was the rule clear? Was time pressure a factor? Was training adequate? Was the environment unusual? By exploring the nuances, you implement more effective fixes.

5. Also, be cautious with generic statements like "be careful" – that's oversimplified guidance. Instead, give specific caution ("use three-point contact on that ladder because we had rain and rungs are slippery" – acknowledging context).

6. Different trades and tasks carry unique risks; a one-size-fits-all approach to safety might miss subtle differences. High-reliability construction managers consult trade experts to understand specific complexities (e.g., how formwork pressure evolves during a concrete pour – to manage that risk, you need the detailed understanding, not just "don't overload" platitude).

7. Sensitivity to Operations: HROs keep a keen eye on the front line; they maintain situational awareness at all levels of the organization.
8. For construction, this translates to managers spending time in the field (the "gemba walk" concept from lean manufacturing). A project manager who regularly walks the site, talks to crews, observes how work is actually happening, is more likely to catch issues (like a crew taking an improvised approach that might be unsafe) and also more likely to hear concerns employees voice only informally.
9. It also means communication flow: workers should report small anomalies (equipment making odd noise, slight movement in a trench wall) without fear – because those might be early signals of bigger problems. The team is trained to pay attention to gut feelings and subtle cues and act on them.
10. Safety and construction progress meetings include discussing ongoing operations and any changes. If, say, a crew is behind and rushing, a sensitive organization senses increased risk and might add a safety observer or slow down rather than plow ahead blindly.
11. Using real-time data: some HRO-inspired sites use technology like sensors (for dust, noise, structural movement) to get immediate feedback on operational safety conditions.
12. Deference to Expertise: In HROs, decisions migrate to the people with the most relevant knowledge, regardless of rank.
13. Construction application: If a crane operator or an ironworker rigging a load says "I'm not comfortable with this lift plan," their expertise is heeded and the plan is reassessed, even if schedule suffers. The hierarchy is flattened when it comes to safety critical moments – the person on the ground with intimate knowledge gets the say.
14. Build a culture where workers are empowered to halt work if they see danger (Stop Work Authority) and where that action is praised, not penalized, because those workers often have the closest view of evolving hazards.
15. Example: During a complex concrete pour, if a vibrating crew member notices formwork deflection, they should feel completely authorized to call a halt – their on-the-spot expertise in noticing that is critical, outranking the plan or the project timeline in that moment.
16. It also means bringing in specialists when needed (e.g., defer to a geotechnical engineer's advice if unusual soil condition arises, rather than super or PM guessing).
17. Commitment to Resilience: HROs anticipate that not everything will go as planned, so they develop capacities to respond and contain errors.
18. For a construction site, resilience can mean having contingency plans: e.g., an extra pump on standby for water ingress, a standby crew to relieve workers in extreme heat, or drills for emergency response (so if something does happen, team reacts effectively, minimizing harm).
19. Training is key: cross-train workers in basic first aid, extraction from confined space, etc., so the team can self-stabilize a situation while waiting for professional responders if needed.
20. Learn and adapt: after any incident or near-miss, implement changes swiftly across the board (and share that knowledge company-wide or even industry-wide). This learning agility is part of resilience – the organization bounces back stronger because it genuinely internalizes lessons.

21. On a daily level, resilience is also about housekeeping and readiness: keeping pathways clear (resilient to allow evacuation), keeping equipment well-maintained (resilient against failures). It's defensive design in safety operations.

**Implementing HRO Habits Practically:** - Start meetings (even daily toolbox talks) with a quick scenario: "If X went wrong, what would we do?" – This fosters preoccupation with failure and resilience thinking. - Encourage questions and even healthy skepticism: If someone says "I think our fall protection on that edge isn't adequate," don't brush it off if it technically meets OSHA; discuss and perhaps take extra measure. That's refusing to simplify and deferring to that person's perspective. - Track small failures (like minor equipment damage, human errors caught in time) and treat them as systemic improvement opportunities. Maybe do a monthly "error review" where these are anonymized and discussed. - Leadership modelling: project leaders admitting their own near-misses or mistakes openly helps set an HRO tone (no complacency, we can all improve). - Celebrate catches: if a worker stops a pour because they see something odd, reward that behavior publicly. It reinforces attention to detail and active engagement. - Use advanced tools: some HRO-minded sites use "safety observation" programs where all staff are tasked to observe and report a number of safe or unsafe acts per week, fueling positive feedback loops and constant vigilance.

The beauty of HRO habits is that they create a culture where safety is truly ingrained – not because of rules, but because everyone from management to new hires shares the mindset that we can't take safety for granted for even a second and we trust the team's expertise and instincts to keep us safe. Construction may never eliminate risk, but by borrowing these high-reliability habits, it can dramatically reduce the frequency and severity of incidents, even under the pressures of schedules and complexity.

## The Best Practice Ladder

Safety improvement is a journey with successive rungs of performance – one can think of it as climbing the best practice ladder. At the bottom is legal compliance and reactive measures; at the top is a deeply ingrained safety excellence where injuries are virtually non-existent. It's useful to understand these rungs, so a project or company can assess where they are and chart a course upward.

Ladder Rungs (Maturity Levels): Various models exist (e.g., the Bradley Curve by DuPont, or the cultural maturity model often given as five stages: Pathological, Reactive, Calculative, Proactive, Generative). To simplify, consider: 1. Pathological (Chaotic) – "Who cares about safety? We just care about getting the job done." At this bottom rung, there's ignorance or even defiance of safety. Accidents are high, but management treats them as normal cost of doing business or blames workers entirely. (Hopefully no modern professional projects are here, but some small outfits or past eras were like this). 2. Reactive (Compliance-driven) – "We take safety seriously, especially after something happens." This rung follows rules mainly after incidents or when forced by law. There is some structure (maybe basic safety program) but the culture is to fix problems as they arise. Accident rates may fluctuate. Safety is viewed somewhat as an external requirement ("OSHA says...") rather than internal value. 3. Calculative (Systematic) – "We have systems and checklists for safety." At this rung, organizations implement formal safety management (hazard reporting systems, regular training, metrics tracking). It's largely management-driven and bureaucratic – lots of procedures, perhaps focused on lagging indicators and compliance. Workers participate by following procedures. Incident rates usually drop here compared to reactive stage, but there may still be a feeling of "safety is management's job" among workforce. 4. Proactive (Integrated) – "We work on preventing all injuries – safety is a key part of planning and operations." Here, the culture starts actively involving everyone; near-misses are eagerly reported and addressed, leadership is visible in safety efforts, and innovation in safety is supported. The organization looks outside itself for best practices and tries new approaches (like wellness programs, human performance training, etc.). Safety and productivity are seen as complementary, not competing. 5. Generative (Excellence) – "Safety is how we do business – it's in our DNA." At the top rung, safety is truly owned at all levels. Workers will self-organize to fix hazards without waiting, and will speak up if anything seems off, with full trust and backing. The organization benchmarks itself against perfection, not peers. They share best practices industry-wide because the goal is zero harm everywhere. Incident occurrence is very low (and when something minor occurs, it's treated with utmost seriousness to learn from it). In a generative culture, it's not unusual to see workers stop work over a concern even in absence of management – because the culture supports that fully.

**Climbing the Ladder:** - From Pathological to Reactive: Often takes a big event or regulatory enforcement to jolt an organization into caring about safety. Education about human cost and legal consequences also helps. Once leadership acknowledges safety matters, you see policy statements and basic compliance efforts. - From Reactive to Calculative: This typically involves putting structured programs in place: hiring safety professionals, establishing rules and PPE requirements, doing incident investigations and some training. It's climbing by building a scaffold of procedures. - From Calculative to Proactive: The shift here is cultural – engaging employees, moving from lagging metrics (e.g., after injuries) to leading metrics (hazard hunts, safety observations), and making safety part of performance evaluations. Leadership starts to inspire rather than just enforce. You might implement behavior-based safety programs or safety incentive programs carefully balanced to encourage reporting. - From Proactive to Generative: This is the hardest jump – it requires sustaining a high level of vigilance and caring even when things are going well. It's where safety becomes truly values-driven. Techniques include more advanced HRO practices (as discussed in 11.5.3), full integration of safety into all business processes (like procurement considering safety of equipment, design for safety up front, etc.), and often external audits or even pursuing ISO 45001 certification to validate systems. - Each rung builds on the last. You can't skip directly from Reactive to Generative; you need the systems (calculative) and the engagement (proactive) steps to get there.

**Project Scale Application: -** Within a single project, you can have mini-versions of this ladder. Early on (especially if new crew or subcontractors not aligned), you might start at a lower rung and deliberately cultivate maturity. Many big projects do a "safety culture survey" at the start, then mid-project, then end to measure progress up the ladder. - Example: At project start, you find workers only wear PPE when safety officer is around (reactive compliance mindset). Through constant coaching and involving them in safety planning, by mid-project they're reminding each other to tie off (proactive/team mindset). By project end, maybe they spontaneously developed a new best practice (like a creative tool tethering system they rigged up) – that shows generative behavior (innovating for safety without being told).

**Benefits at Top Rungs: -** Organizations at generative level often achieve zero lost-time injuries for long periods. Workers feel safer and valued, which improves morale and productivity. - It also means if an unforeseen crisis hits (say a natural disaster during the project), the team's resilience (HRO style) likely handles it better, minimizing harm, because they are adaptive and deeply committed.

**Sharing Best Practices: -** Part of being at the top is helping others up the ladder too. Generative organizations often share their methods at conferences, within industry groups, etc. On a multi-contractor site, a generative GC might coach subs who are weaker on safety, effectively pulling them up a rung by mentorship.

**Continual Climb: -** The ladder isn't one-way; without diligence, organizations can slip down (especially under new management or extreme pressure). So continuous assessment and reinforcement are needed. This is where tools like regular culture surveys, third-party audits, or corporate oversight ensure you don't fall from proactive back to calculative due to complacency. - Celebrate each rung's achievements – e.g., when near-miss reporting doubles (showing proactivity), recognize that milestone. But also keep the vision clear that we aim for the next rung: "Good, we have a reporting culture now (proactive), next we want to see crews solving issues on their own and truly believing zero incidents is possible (generative)."

**Applying Ladder Concept on Projects: -** At project kickoff, articulate the safety culture goal (aim for at least proactive). Provide training not just in rules but in mindset – e.g., hold workshops on why we report near misses or how to challenge each other constructively on safety. - Use visual aids – some companies literally post a ladder graphic in the break room, marking where they see the current culture and aiming upward, with specifics ("At level 3 calculative now: we do regular inspections and meetings. To reach level 4: we need more active worker interventions and hazard suggestions."). - Engage subcontractors by including them in cultural development activities – if a sub is stuck at a compliance mentality, pair them with a mentor or put their foreman through additional leadership safety training.

Viewing safety improvement as a ladder or maturity model helps teams stay motivated – there's always a higher level to reach. By understanding what each rung looks like, a project can set concrete targets to climb higher. Reaching the top "generative" rung is challenging and rare, but every move up yields significant safety performance gains and is worth the effort. The ladder metaphor also reinforces that safety is a progression, not a one-and-done task – an ongoing journey of betterment that continues even when legal compliance is long surpassed.

This chapter has journeyed from understanding U.S. regulatory standards to embracing global best practices, all with the aim of equipping construction leaders and teams to manage projects safely, ethically, and efficiently. The consistent thread is that knowledge and proactive management are key – whether it's knowing OSHA rules, applying building codes meticulously, or pushing beyond compliance toward a culture of excellence. A construction manager who masters these topics becomes not just a builder of structures, but a builder of a robust safety environment where workers thrive and projects succeed even under the toughest conditions.

**Operational Takeaways for Today's Job:**

- Treat OSHA guidelines as a daily to-do list. For instance, if you're running a trenching operation, ensure you have shoring or a proper slope per OSHA regs before anyone enters. Integrate these rules into your normal work routine so compliance is just part of "how we do things here."
- Stay up-to-date on required safety gear and practices. If a task requires fall protection above 6 feet, no exceptions – check that all workers at height have full fall-arrest setups and are using them correctly. Being lax on regulatory requirements even once can lead to accidents and fines.
- Use toolbox talks to remind the crew of specific regulations relevant to the day's work. For example, before using a lift, remind everyone of the OSHA rule about staying clear of the swing radius. These mini refreshers help everyone remember that rules are there to keep them safe.
- Keep records proactively. Don't wait for an inspector to ask – do your own audits of paperwork like equipment inspections, training certifications, and safety meeting logs. Making sure these are current and accessible means you're always inspection-ready and, more importantly, that you haven't missed a safety step.
- If you encounter a code or standard you're not sure about, get clarification before proceeding. It could be as simple as calling the safety manager or looking up an ANSI standard for scaffold tags – better to take a short pause to ensure compliance than to assume and risk a violation or hazard.

# Chapter 12

# Professional Licenses and Certifications

## Professional Engineer (PE), Structural Engineer (SE), Architect (RA) Licenses
## What PE and SE Licensure Actually Entail

Becoming a **Professional Engineer (PE)** in the United States is a multistep journey. First, an individual typically earns an accredited engineering degree and passes the Fundamentals of Engineering (FE) exam to become an Engineer Intern. After accruing a required period of experience under existing PEs (usually four years), the candidate must pass the Principles and Practice of Engineering exam in their discipline. Only then can they be licensed as a PE in their state. Once licensed, a PE is legally permitted to "stamp" and sign engineering documents, taking responsibility for their adequacy. In concrete terms, a PE's seal on a structural beam calculation or a stormwater drainage plan signifies: "*I attest that this has been prepared with appropriate care and meets the applicable codes and standards*."

Critically, a PE's obligation is to hold public safety paramount—a creed codified in engineering ethics. Licensure isn't merely a title; it's a commitment to practice only in areas of one's competence and to speak up if unsafe conditions are observed. Many engineering projects involve teams of people (drafters, junior engineers, specialists), but the licensed PE in *responsible charge* must supervise and review the work. That PE is expected to catch errors or omissions because legally and ethically, they are on the hook. In many states, if an engineering failure occurs, the question becomes: Who was the PE of record for that project?

For **Structural Engineers (SE),** a few states require an additional licensure step beyond the PE. Structural engineering often carries *inherently high risk*, so states like Illinois and Hawaii mandate a dedicated SE license for anyone practicing structural design. Other states (California, Washington, Oregon, among others) impose SE requirements for specific "significant structures" such as high-rises, hospitals, large public assembly buildings, or essential facilities. This typically means passing a rigorous 16-hour NCEES Structural Engineering exam (on top of the PE exams) to demonstrate advanced competence in designing buildings for earthquakes, wind, and other extreme loads. The additional SE designation signals a higher benchmark of expertise. In practical terms, if you're constructing a 50-story skyscraper in San Francisco or a long-span bridge, the law may require an SE to be

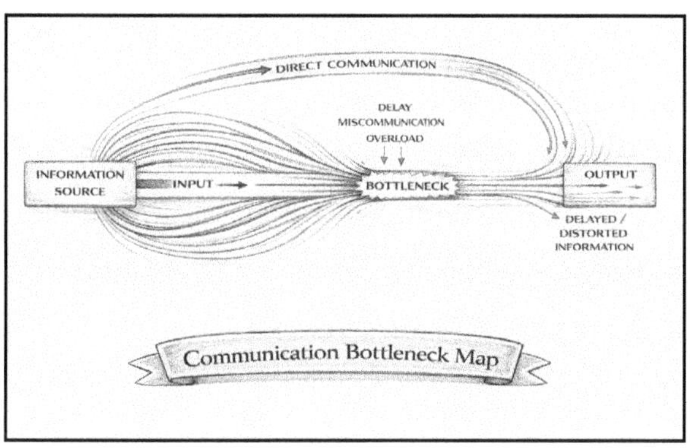

the one stamping those structural plans. Even in jurisdictions that don't *require* a separate SE license, many structural specialists pursue it voluntarily as a mark of elite qualification.

What does licensure entail day-to-day? For one, licensed engineers must maintain familiarity with ever-evolving building codes, standards, and best practices. They also take on personal liability – both legally and in terms of reputation. Being a PE or SE means when you sign drawings or reports, you are effectively saying "the buck stops here." It entails readiness to justify your design decisions if challenged and to ensure unlicensed team members are properly guided. It also means navigating the limits: a PE must not seal work outside their area of competence (for instance, a licensed electrical engineer shouldn't stamp structural plans). In summary, PE/SE licensure is a mix of **credential and responsibility**: it grants authority to make pivotal design judgments, coupled with accountability for those judgments.

## The Role of the RA (Registered Architect)

Just as engineers are licensed, so too are architects – commonly termed *Registered Architects (RA)* or simply licensed architects. Architecture licensure follows a parallel structure: typically a professional degree in architecture, a multi-year internship, and a comprehensive exam series (the ARE – Architect Registration Examination). Once licensed, an architect can legally sign and seal architectural drawings and is recognized as the design professional of record for building projects.

The role of an RA on a project is often that of the design lead and coordinator. Architects are responsible for the overall vision of a building – its functionality, aesthetics, and how it meets user needs – but they also must ensure the design conforms to building codes, zoning laws, ADA regulations, fire safety requirements, and so on. In most jurisdictions, any commercial or public building (and many residential structures above a certain size) must have an architect's seal on the plans submitted for permit. This requirement acts as a quality gate: it ensures that someone with appropriate qualifications has reviewed the life-safety aspects of the design (egress routes, fire separations, structural layout coordination, etc.).

One could say the architect is the master integrator of the design team. They collaborate with structural and MEP (mechanical, electrical, plumbing) engineers, interior designers, landscape architects, and others. While each discipline engineer will stamp their own technical drawings, the architect often stamps the overall set and carries broad responsibility for the building's design being coherent and safe. For instance, if there's a conflict where ductwork is shown running through a beam, the architect (as coordinator) should catch and resolve that with the engineer.

Legally, RAs are bound by similar public protection duties as PEs. They must not practice beyond their expertise (an architect wouldn't design an exit stair that violates code, for example, and then try to hide behind ignorance). They are subject to state board rules and can face discipline for gross negligence or misconduct. A notable distinction: some states allow certain small projects or single-family homes to be designed without an architect's seal (often an exemption for simple buildings), but any sizable or risky structure will demand an RA's involvement.

In construction management contexts, the presence of a licensed architect on the team brings a layer of quality control and accountability. Architects will often perform site observations during construction to ensure the work matches the design intent and code requirements. They issue reports or certifications at milestones (such as a "substantial completion" certificate) which a licensed professional must sign. In essence, the RA's role is both creative and regulatory – they are the creative force shaping the project and the licensed gatekeeper ensuring that the creative vision doesn't compromise safety or legal standards.

## Licensure and the Chain of Responsibility

Professional licensure establishes a clear *chain of responsibility* on projects. At each major juncture of design and construction, a licensed individual is expected to be the final authority reviewing and approving the work of others. This chain typically works as follows: junior engineers or architects (who might be unlicensed or "in training") perform much of the analysis, drafting, and detail work. They operate under the guidance of a licensed mentor or supervisor. Before documents leave the office, the licensed PE or RA in charge must review, correct, and ultimately stamp them. By stamping those documents, the license-holder takes ownership. If there's a serious error in the plans, it's not an excuse to say "my intern made a mistake" – the state board will point to the name on the seal.

In design firms, this chain of responsibility is often formalized. For example, a structural engineering firm might designate a particular PE as the Engineer of Record (EOR) for a project. Others contribute, but the EOR's license is on the line for the final product. On the architecture side, the Architect of Record has a similar role. This hierarchy ensures accountability flows upward: tasks can be delegated, but responsibility cannot be abdicated. It also empowers the licensed professional to insist on quality – since their neck is on the line, they have authority to mandate necessary checks, peer reviews, or corrections before work proceeds.

This chain continues into the field during construction. Many jurisdictions require a licensed professional's involvement during key stages of building. For instance, special inspections (like inspecting structural steel welds or concrete strength) often must be overseen by a licensed engineer or their certified representative, with reports ultimately signed by a PE. If a contractor proposes a change that could affect structural integrity (say, substituting a different beam size), typically that change must be reviewed and approved by the licensed engineer and sometimes the architect as well. Even shop drawings for critical items (like trusses or curtain walls) often require a delegated design stamp – meaning the subcontractor's engineer, who must be licensed, takes responsibility for that component under the overall design team's eyes.

Importantly, licensure also delineates where contractor responsibility ends and design responsibility begins. A general contractor is usually not required to hold an engineering or architecture license (though they often have a contractor's license, which is a different kind of credential focusing on business competency and trade knowledge). Contractors build according to the plans and specifications provided by the licensed design professionals. If they spot a discrepancy or a likely problem, they should flag it, but they are not authorized to unilaterally redesign on the fly – doing so veers into the unlicensed practice of engineering/architecture. The proper chain is that any necessary design change is routed back to the PE or RA to evaluate and approve. This process has been written in the blood of past failures: when unqualified persons make ad-hoc design calls, the results can be disastrous.

## Maintaining Licensure and Disciplinary Risks

Earning a PE or RA license is not a one-and-done deal; maintaining that license is an ongoing commitment. States require licensees to renew periodically (often every 1 to 3 years), and most mandate Continuing Education (CE). For example, a state might require 15 Professional Development Hours (PDH) of coursework or training per year, including a subset on ethics or building codes. Failing to complete these hours or to renew on time can lead to an inactive or expired license – which, if not caught, could halt a project or even lead to practicing illegally. A cautionary tale: an engineer in one state forgot to renew and kept stamping plans; when it came to light, the board not only voided those plans until re-stamped by a valid licensee, but also issued fines for unlicensed practice (even though it was unintentional). The administrative side of licensure (tracking renewals, CE credits, address changes, etc.) is thus a non-trivial part of a professional's responsibility.

Disciplinary risk is the counterpart to the privileges of licensure. State boards can investigate complaints and impose penalties ranging from reprimands and mandated training up to suspension or revocation of the license. Common causes for discipline include negligence, incompetence, ethical violations (like conflict of interest or plan-stamping without oversight), and criminal convictions. Often these cases arise from seemingly small lapses that snowball. For instance, an engineer might sign and seal a set of plans that they didn't thoroughly check (perhaps under pressure of deadline). If those plans turn out to have serious errors, that's grounds for negligence. In Florida, for example, recent disciplinary cases show engineers cited for signing off on designs with obvious safety flaws or code violations – like a fire protection system that wouldn't actually work in an emergency. Penalties in such cases have included fines, required ethics courses, probation, or license suspension. In egregious situations (falsifying documents, practicing drunk, abandoning projects), boards will revoke a license entirely, effectively ending one's professional career.

A licensed professional must also be mindful of where they are licensed. Licenses are state-specific; practicing in another state requires obtaining a license there (usually via reciprocity if one has a good record elsewhere). Working outside one's jurisdiction without a license is considered unlicensed practice. Boards do communicate with each other: a violation in one state can trigger action in others. It's not unheard of for an engineer to lose their license in their home state and then see their licenses in three other states voluntarily surrendered or revoked as a chain reaction.

Then there's the gray area of "**responsible charge**". A PE can delegate work but must maintain oversight. If an engineer rubber-stamps work done by someone else without truly reviewing it (a violation known as plan stamping), that's a disciplinary offense. Boards enforce that the intent of licensure – that the licensee's judgment was applied – is not circumvented. This is especially relevant with digital work-sharing and global outsourcing of drafting; the local PE must still ensure they exercised control and review over the final documents.

Maintaining licensure is about diligence: staying educated, following rules, and self-policing one's practice. The *disciplinary* risks are real, but they underscore a key point: licensure has teeth. It's a system to remove bad actors or remediate poor practices for the greater good. Ethical professionals actually embrace this fact – knowing that the threat of discipline helps keep the profession's standards high. It reinforces that being a PE, SE, or RA is not just a status symbol; it's a live responsibility that one must continually earn.

## Construction Manager Certifications (CMAA, CM-BICSI)
## The Rise of the Construction Management Professional

In the not-so-distant past, the role of "construction manager" was often performed informally by seasoned superintendents or engineers on an ad-hoc basis. The idea of **Construction Management (CM)** as a distinct profession has truly come into its own over the last few decades. As projects grew in complexity, cost, and speed, it became clear that dedicated expertise was needed to plan and coordinate the construction process itself – not just to design the project (architect/engineer) or physically build it (contractors). This gave rise to the construction management professional: individuals whose primary training is in orchestrating all the moving parts of construction, from scheduling and cost control to safety and contract administration.

The formation of the Construction Management Association of America (CMAA) in 1982 is a landmark in this evolution. It signaled that CM had matured into a profession deserving its own body of knowledge and standards of practice. Over time, universities began offering degrees in Construction Management, and large owners (especially in the public sector) started hiring third-party CMs to represent their interests on major projects. We saw the emergence of project delivery methods like **CM-at-Risk** and **CM-as-Agent**, where the construction manager plays a lead role either as a contractor or as an owner's representative, respectively. In both cases, having a skilled CM greatly improves communication and outcomes by providing leadership focused on how the project gets built (not just what gets built).

By the 2000s, the construction industry began to widely recognize that effective construction management is pivotal to project success. High-profile project failures or delays often traced back to poor coordination rather than technical design issues. For example, when Boston's "Big Dig" megaproject faced cost overruns and schedule slips in the 1990s, analyses pointed to management and coordination challenges on a massive, multi-contractor job – underscoring the need for strong CM practices. Conversely, projects like the Olympics or large airport expansions that finished on time often credited robust construction management teams employing systematic techniques.

Today, we have what could be called a **professionalization of construction management**. Many CMs no longer come solely from the trades; they are often college-educated, tech-savvy in project management software, and versed in formal PM techniques (critical path method scheduling, risk management plans, lean construction, etc.). Organizations such as CMAA have published *Standards of Practice* that codify the CM body of knowledge, similar to how engineering has design codes. With this professionalization came the push for certifications to validate a CM's expertise – a way of saying "this person meets a national standard for construction management proficiency."

In short, the rise of the construction management professional parallels the growing complexity of construction itself. Modern projects demand a conductor for the construction orchestra, not just skilled musicians. The industry's response has been to recognize construction management as a profession in its own right, with career paths, certifications, and associations to support it.

## The CMAA CM and CCM Tracks

The **Construction Management Association of America (CMAA)** developed a certification program to identify truly qualified construction managers. The flagship designation is the **Certified Construction Manager (CCM)**, which has become increasingly known as the gold standard credential for CMs. Earning the CCM isn't a trivial process. Candidates must meet a combination of education and experience requirements (typically, a construction-related degree plus at least 48 months in CM roles, or more years of experience if coming from a non-degree background). They also need letters of reference attesting to their capabilities, and must pass a comprehensive exam. In 2006, the CCM became accredited by ANSI under ISO 17024, meaning it meets international standards for personnel certification programs – a sign of its rigor.

For a construction manager, obtaining the CCM is somewhat akin to an engineer getting a PE license: it formally validates your expertise. While not legally required (CMs are not licensed by states in the way engineers/architects are), it carries weight in the industry. Many owners and agencies now *prefer or require* a CCM-certified individual to lead their projects. The U.S. Army Corps of Engineers, for instance, has embraced the CCM and encourages its staff to earn it, considering it a measure of excellence in project delivery. Owners find reassurance that a CCM has demonstrated knowledge of CMAA's **Construction Management Body of Knowledge (CMBOK)**, covering areas like cost management, time management, safety, contract administration, and quality – the full spectrum of what a CM should handle.

CMAA actually offers more than one track. In addition to the full CCM for experienced professionals, there's a stepping-stone called the **Construction Manager-in-Training (CMIT)**. The CMIT program is aimed at early-career individuals, even fresh graduates, to signal their commitment to the profession. It involves an entry-level exam on CM fundamentals and is often structured in levels (demonstrating progressive knowledge). Someone might become a CMIT, gain a few years of experience, then advance to CCM when eligible. This stackable credential approach helps young CMs map out a path: first learn the ropes and terminology (CMIT), then prove you can run the show (CCM).

Within the CCM universe, there aren't multiple "levels" of certification beyond CMIT/CCM, but CMAA does maintain specialty practices through its College of Fellows and such. The key distinction to know is that **CCM is a broad certification for general construction management practice**. It's intentionally not segmented by project type; a CCM is expected to know how to manage a wastewater plant build or a high-rise construction using the same fundamental principles. That said, each CCM will have their domain of experience.

What sets a CCM apart on a job site or in the boardroom? Often, it's the systematic approach and familiarity with industry best practices. For example, a CCM is trained to set up a project execution plan from day one, to use risk registers, to enforce a safety program, and to communicate in a standard, documented way. They also pledge to uphold a code of ethics, similar to licensed engineers. For a company hiring a project manager, seeing "CCM" after a candidate's name is a strong indicator that this person has been vetted and tested against the industry's benchmark of competence. It's not just theoretical either – the eligibility requirements ensure they have real project experience, not just book knowledge.

**To give a sense of adoption:** as of late 2010s, the number of CCMs was growing rapidly. CMAA celebrated surpassing 4,000 CCM holders in 2018, noting a 132% increase over five years prior. That growth has continued, fueled by owners' preference for certified CMs. In fact, CMAA's CEO was quoted saying, "*Owners... increasingly prefer to have a CCM lead their project teams,*" highlighting that the market is driving this credential into becoming a default expectation.

**The CMAA offers a clear track:** start as a CMIT (optional but beneficial for newcomers), gain experience, and then earn the **CCM** to distinguish yourself as a top-tier professional. It's a career investment. Many employers will even encourage or pay for their rising managers to get certified. The presence of a CCM can influence winning a job; it gives clients confidence that the person running their multi-million dollar construction project truly knows what they're doing according to an industry-recognized standard.

(Figure 12.2: Construction Manager Certification Pathways – from CMIT to CCM and beyond)

## CM-BICSI for Technology Infrastructure

As construction projects become more technologically complex, a niche area of construction management has emerged at the intersection of building and IT infrastructure. Consider a modern hospital or a high-tech campus – not only are we dealing with concrete and steel, but also with extensive **information and communications technology (ICT)** systems: data centers, networking cabling, wireless access point layouts, security systems, audiovisual setups, and more. Managing the implementation of these systems requires specialized knowledge that goes beyond traditional construction trades. This is where BICSI comes into play.

**BICSI** stands for Building Industry Consulting Service International, a professional association that focuses on ICT infrastructure. BICSI has developed its own set of certifications to train and validate professionals in designing and installing communications systems (things like structured cabling, fiber optics, network infrastructure). For construction managers venturing into projects heavy on tech (such as data centers, smart buildings, large office networks), a general CM background might not be enough – you benefit from having ICT-specific project management skills.

One notable certification is the **Registered Telecommunications Project Manager (RTPM)** offered by BICSI. This credential is tailored for project managers in the telecommunications/IT systems arena, effectively a construction manager for technology infrastructure. The RTPM course and exam cover how to lead ICT projects successfully – including budgeting, stakeholder management, developing deliverables, scheduling, and quality control for cabling and telecom installations. In other words, it teaches a PM everything from coordinating with designers of a data network to overseeing the installers pulling cables and terminating fiber, all while ensuring the result meets industry standards.

You might see "CM-BICSI" referenced in some job postings – essentially meaning a **Construction Manager with BICSI certification** (like RTPM or the more design-oriented RCDD). The **RCDD (Registered Communications Distribution Designer)** is another well-known BICSI credential, focused on the design aspect of telecom systems. While RCDD is for the designer/engineer role, a CM who has an RCDD on their team or also holds an RTPM is well-positioned to bridge the typical gap between construction and IT. For example, integrating the installation of miles of CAT6 cables, servers, and wireless routers into a construction schedule requires understanding both the construction process and the technical requirements (testing, certifying cables, ensuring server rooms are built to spec with cooling and power). A BICSI-trained manager brings that blend of knowledge.

Technology infrastructure has sometimes been an afterthought in construction, but not anymore. Owners expect buildings to be smart and fully functional on day one. A delay in IT systems can delay occupancy just as much as a delay in the elevator installation. Thus, having *CMs who speak the language of technology* is increasingly valuable. BICSI certifications effectively credentialize that capability. They provide **professional legitimacy** in ICT management – demonstrating to clients that the CM understands the latest standards and best practices for IT systems integration.

A construction firm bidding on a new corporate campus might highlight that their project manager is RTPM-certified, meaning the client can trust them to handle the extensive communications network build-out. Or a contractor doing a large stadium (with complex AV and security systems) might hire a BICSI-certified specialist to manage those subcontractors and coordinate with the base building team.

**CM-BICSI** refers to the convergence of construction management and ICT expertise. BICSI's project management and design certifications ensure that those managing technology infrastructure deployment have proven skills. As our industry leans more into integrated systems (think smart buildings with IoT devices, advanced security and data needs), this kind of credential will only become more pertinent. A construction manager who can oversee pouring concrete and also understand fiber optic cable testing is a potent combination – and exactly what many complex projects need today.

## Why CM Certification Is Becoming the Default Expectation

Just as the Professional Engineer license became a baseline expectation for engineering consultants, **Certified Construction Manager (CCM)** and related CM certifications are swiftly moving from "nice-to-have" to "must-have" in many sectors. There are a few reasons driving this trend, all rooted in the increasing demands for accountability, consistency, and professionalism in project delivery.

**First**, large owners and agencies have signaled their confidence in certification. We mentioned the Army Corps of Engineers preferring CCMs; they're not alone. Transit authorities, hospital systems, universities, and Fortune 500 companies with big capital programs often either require or give selection preference to firms staffed with certified CMs. When an RFQ (Request for Qualifications) for a project management services contract lists "CCM required" for the proposed project manager, that's a strong message to the industry. It essentially says: *if you want this job, you better have people who took the time to get certified.* Owners have been key to the growth of CCM adoption, because they see value in it – it serves as a shorthand for quality. In competitive procurement scoring, having a CCM might be the differentiator that scores those extra points.

Second, the complexity of managing construction has ballooned, and with it the risk. Budgets are tight, schedules are aggressive, and there's less room for error. A certified CM is presumed to be up-to-date on best practices and ethical standards, reducing the risk of mismanagement. The Bureau of Labor Statistics itself has noted the growing movement toward certification in construction management, reflecting an industry acknowledgment that the job is specialized. The certification provides an *objective measure* of a person's knowledge, which is invaluable in a field where outcomes can be somewhat subjective. It's a quality control mechanism – much like ISO certification for processes, CCM is certification for people.

Third, unlike engineers and architects, construction managers don't have state licensure to fall back on. Anybody can, in theory, call themselves a construction manager or project manager. That means the range of competence out there is very broad – from truly expert to dangerously inept. Certification fills this vacuum by setting a baseline. It filters the pool: those who are serious and competent will pursue certification, and those who aren't will likely not bother. This dynamic is gradually making certification a **de facto requirement** even if not a de jure one. In other words, you could operate without it, but you may struggle to win clients or the trust of stakeholders in higher-level roles if you lack credentials behind your name.

Fourth, the career development perspective cannot be ignored. Up-and-coming professionals see that obtaining credentials like CCM, or specialized ones like PMP (Project Management Professional, which we'll discuss later), can accelerate their advancement. It's become part of the professional ladder in many organizations: entry level, then maybe get a CMIT; mid-level, aim for CCM when you qualify. Firms know that if they invest in their people's certifications, they are also investing in better project outcomes and marketability.

We are reaching a point where project owners might ask in an interview or proposal presentation: "Who is the construction manager on my project, and what are their credentials?" If Firm A presents *John Doe, CCM and Firm B presents Jane Smith, no certification mentioned*, the owner's confidence tilts towards John Doe, all else being equal. This is especially true in an environment where projects are under public scrutiny. If a project fails or has issues, having a certified professional in charge shows that the team was vetted. It's a bit like a safety net – "We put a certified pro at the helm" is a defensible stance.

CM certification is becoming the default expectation because it delivers value to all parties: owners feel assured, CMs gain knowledge and status, and the industry as a whole gets a more standardized skillset to rely on. The trajectory is similar to what occurred in fields like IT (where certifications like Cisco's or Microsoft's became expected for certain roles). We can foresee a time in the near future where an uncertified construction manager on a major project will be an anomaly. As one industry CEO put it during the CCM surge: "*It's a great time to be a CCM... Owners emphasize effective, accountable project delivery as key to managing risk, and they are turning to CCMs to deliver just that.*" The writing is on the wall – or rather, on the resume.

(Figure 12.4: Growing Prevalence of CM Certification – Graph of CCM adoption over time)

## PMP: The Project Manager's Toolkit

Construction is a subset of the larger world of project management, and one of the most recognized credentials in that larger world is the **PMP (Project Management Professional)**. Offered by the Project Management Institute (PMI), the PMP is not construction-specific – it's used in IT, finance, healthcare, you name it – but it provides a robust framework for managing any project. In construction firms, especially at the corporate level or for those involved in multi-industry projects, having PMPs on staff is seen as a plus because it signifies mastery of universally accepted project management practices.

To earn a PMP, one must have a certain amount of project management experience and education, and then pass a notoriously rigorous exam that covers the **PMBOK (Project Management Body of Knowledge)** areas: Integration, Scope, Schedule, Cost, Quality, Resource, Communications, Risk, Procurement, and Stakeholder Management. These knowledge areas resonate strongly with construction management as well: after all, a construction project needs scope control (design change management), schedule planning (critical path method), cost management (estimating, budget tracking), quality assurance, procurement (buyout and subcontracting strategies), and so on. The PMP formalizes the understanding of these areas and introduces standardized terminology and techniques. For example, even a seasoned construction manager might learn new formal risk management techniques through PMP training (like qualitative and quantitative risk analysis, decision trees, etc.) that they hadn't been explicitly using before.

One might ask, "If I have CCM, do I need PMP, or vice versa?" These credentials are complementary. The **CCM is construction-specific**, testing knowledge of constructability, design-phase roles, safety, etc. The **PMP is industry-agnostic** but very deep on process. A construction PM who holds both likely has a strong grasp on both the unique aspects of construction and the structured methodology of project management. Some firms require their project executives or program managers to obtain a PMP because their role involves high-level coordination and perhaps dealing with clients' internal PMOs (Project Management Offices) that speak the PMI language.

The impact of PMP in construction can often be seen in how project controls are implemented. A PMP might introduce a work breakdown structure (WBS) approach that's more detailed or systematic. They might champion a change management process that mirrors PMI's change control recommendations, ensuring that every change has an impact analysis and formal approval. They're likely to use risk registers proactively. In essence, a PMP brings in the toolkit and discipline of classic project management.

From a numbers perspective, PMP is huge globally – over 1 million PMP holders worldwide. In construction specifically, PMI launched a Construction Extension to the PMBOK and even a Construction Professional in Built Environment Projects (PMI-CP) certification recently to tailor their offerings more to construction. That shows the crossover: the project management community recognizes construction as a big user base, and the construction industry recognizes PMI as a repository of best practices.

Attaining a PMP often also correlates with career advancement and salary benefits (PMI surveys frequently show PMP holders earn higher median salaries than non-certified PMs in many industries). It's a respected credential that can open doors to roles beyond just construction, such as management consulting or owner's representation roles, because it's widely recognized. If you tell a tech company or a public agency "I'm a PMP," they immediately understand you've met a global standard for project management knowledge. It's like being multilingual – you can talk the language of construction and the language of generic project management.

The **PMP is the project manager's toolkit writ large**. For construction leaders, it provides structured approaches that can be applied to improve project outcomes. It is becoming more common to see construction managers, especially those interfacing with non-construction stakeholders, to hold a PMP. It signals that you don't just rely on intuition and experience; you also use well-defined processes and can plug into a broader professional community of project managers. In the increasingly complex tapestry of modern projects (which often involve software, hardware, public outreach, financing complexities, etc.), that broad toolkit is invaluable.

## Specialty Certifications (Crane Ops, Welding, Rigging)
## High-Risk Work Requires High-Reliability Credentials

On a construction site, certain activities carry significantly higher risk – both to the people involved and to the integrity of the project. For these **high-risk tasks**, simply having on-the-job experience isn't always enough; formal certification is demanded to ensure a worker has the necessary skill and knowledge. Three classic examples are crane operation, welding, and rigging (the attachment and hoisting of loads). In each case, the potential consequences of incompetence are so severe that industry and regulators have developed stringent certification programs.

**Crane Operations:** Modern cranes can lift hundreds of tons and reach hundreds of feet in the air; a mistake in judgment or a lapse in skill can lead to catastrophic collapse. OSHA regulations were updated in recent years to require that crane operators be certified by an accredited organization for most construction crane types. The National Commission for the Certification of Crane Operators (NCCCO) is one of the premier certifying bodies. A crane operator seeking NCCCO certification must pass both a written exam (testing knowledge on load charts, setup, safety protocols, signaling, etc.) and a practical exam demonstrating actual operating skill. Only then do they earn a certification card for, say, "Mobile Crane Operator" or "Tower Crane Operator." This is not a one-time deal; operators must recertify typically every five years. Additionally, medical clearances (to ensure they don't have impairments that could cause an accident) are often required periodically. The rationale is simple: we entrust crane operators with literal life and death responsibility every time they pick a heavy load over our heads or near buildings. Therefore, we insist on a high-reliability credential that proves their competence.

The consequences of not having certified crane operators have been illustrated too often in the past: historically, crane accidents occurred more frequently when operators were "trained" only informally. With certification becoming standard, there's been improvement in safety statistics. It's worth noting that many large contractors and cities require not just the operator but even the crane assembly/disassembly director to be certified or at least qualified through training – reflecting that every role in a critical lifting operation needs demonstrable expertise.

**Welding:** Welding is literally the glue that holds much of our structures together. A bad weld on a critical connection can mean structural failure under load. Thus, the American Welding Society (AWS) and similar bodies have long established certification processes for welders. A welder will test in a specific process (e.g., SMAW stick welding) and position (e.g., vertical, overhead) by welding test coupons that are then inspected or lab-tested (bend tests, x-rays, etc.). If the welds pass, the welder is qualified for that process/position under a given code (like AWS D1.1 for structural steel). Importantly, welding certifications must be maintained – typically the welder or their employer must submit verification every six months that the welder has been continuously performing that type of welding, or else the certification lapses. If they stop welding for a while, they need to be re-tested. This ensures that perishable skills are kept sharp. It's commonly said that welding is as much art as science, and even a certified welder needs to keep practicing to maintain the muscle memory and technique.

On critical projects, specifications often require "Welders shall be certified within the last X months for the welding to be performed, and proof shall be submitted." Inspectors might even quiz welders on site or ask to see their papers. Specialties like pipe welding for high-pressure pipelines have their own stringent certs (and sometimes each job might mandate a unique qualification test using the exact materials and positions of that job). All this is because a tiny flaw in a weld could later become the initiation point for a crack and catastrophic break. High-reliability is demanded, so high standards are set for those doing the work.

**Rigging and Signaling:** Rigging refers to the selection, attachment, and securement of loads for lifting by cranes or hoists. It may look straightforward – just chaining something up – but it's highly technical. Riggers must know how to calculate load weights, center of gravity, sling angles, capacity reductions, and ensure the lift points won't cause the load to slip or flip. A failure in rigging can result in a load dropping from height, which is often fatal to those below and destructive to the project. Thus, certifications for riggers (like NCCCO's Rigger Level I and II) have become common. A certified rigger has proven they can choose the right slings, tie the correct knots or hitches, and inspect gear for damage. Similarly, Signal Person certification (sometimes combined with rigger certs) is about knowing standard hand signals, radio protocols, and keeping persons clear of danger. OSHA requires that any signalperson used (for crane communication) be qualified, which usually means formal training and testing – often a third-party certification.

These specialty certifications ensure high-reliability human performance. When you see someone carrying a wallet card that says Certified Crane Operator or AWS Certified Welder or Certified Rigger, you have a level of assurance that they've been vetted by an objective process. It becomes part of the project's risk management. High-risk work is not the place for trial-and-error or learning on the fly; you want someone who's already proven they can do it right nearly every time.

We can draw an analogy to aviation: just as pilots are certified and periodically re-tested in simulators to handle emergencies, crane operators and others are put through their paces. And just as aircraft maintenance techs are certified to ensure planes are safe, welders are certified to ensure our steel holds together. The stakes warrant the rigor.

## Recertification and Testing Cadence

Certifications in the construction arena aren't a "get it and forget it" affair. The very nature of these high-risk skills demands that individuals stay up-to-date and continuously prove their proficiency. Thus, most specialty certifications require **recertification on a fixed cadence or proof of ongoing practice.**

Take the **NCCCO crane operator certification** as an example. As mentioned, it's generally valid for five years. Before that period lapses, an operator needs to recertify, which typically involves taking a written exam again (sometimes a shorter version focusing on safety updates and any new regulations) and in some cases a practical exam, or at least a certain number of hours of documented experience along with a clean safety record. If too much time elapses or the operator hasn't been actively working, they might have to do the full battery of tests again. This ensures that someone who got certified a decade ago isn't operating on stale knowledge – equipment evolves, standards improve, and people can develop bad habits that need correcting. Regular testing is like a calibration of the human instrument.

Welding qualifications, as also mentioned, have a semi-annual continuity requirement[45]. Some contractors employ a welding engineer or CWI (Certified Welding Inspector) who manages a database of all welder qualifications, sending reminders and ensuring welders either perform a live test or at least verify they've done similar welding in the past 6 months to keep each qualification active. On big steel construction jobs, it's common to have dozens of welders, each certified for certain weld types. The project specs might demand that the list of active qualified welders be kept on site and updated. If a welder's cert lapses mid-project, the inspector can red-tag any welds they did beyond the qualification period – meaning potentially having to re-weld or re-test those joints. No contractor wants that, so they keep close track of these expirations.

For **rigging and signaling certifications**, the cycle is often also in the 3-5 year range for renewal. These typically require refresher courses or re-examination. The idea is to reinforce fundamentals (like never exceeding sling angle limits, inspection of gear for cuts or abrasions, etc.) and to update on any new equipment or standards. As an example, if synthetic round slings became far more common since a rigger's last certification, the new test might ensure they know how to inspect and use those as opposed to only wire rope slings they might have used years ago.

Another aspect of maintaining high-reliability credentials is continuing education. Some certifications require proof of ongoing learning as part of recertification. For instance, a Certified Welding Inspector (CWI, a higher-level certification for inspectors) has to renew every 3 years and on the 9-year mark must show evidence of continuing education or take a renewal exam. Similarly, a PMP must earn professional development units to renew, and even a CCM has to show continued professional growth. So the concept isn't unique to craft skills; it's universal: *skills atrophy and knowledge evolves, so periodic renewal keeps people sharp and current.*

From the employer's perspective, supporting recertification is a no-brainer. It's far cheaper to have an experienced operator take a one-day refresher and test than to deal with an accident caused by an out-of-practice or uninformed one. Many companies schedule re-testing during off periods or have on-site testing events where a certifying agency comes and tests a bunch of employees at once.

One challenge with recertification is managing the logistics – ensuring people remember and prioritize it. That's why the better organizations have implemented training tracking systems, essentially HR databases that flag when someone's certification is nearing expiration. If a tower crane operator's cert will expire in 4 months, the company gets a flag and can schedule them for recertification before they are out of compliance. If an expired operator were to cause an incident, and it comes to light their cert had lapsed, the liability and legal implications would be severe. It could even violate OSHA regulations, leading to citations.

Recertification and periodic testing are an integral part of what makes these credentials meaningful. They're not just a paper you hang on the wall; they represent an *ongoing commitment to excellence.* The cadence varies by field – 6 months for welding continuity, a few years for heavy equipment, etc. – but the principle is consistent. Construction is dynamic: new techniques, new safety devices, even new hazards (like how to handle synthetic slings catching fire, or new welding alloys) can emerge. The recertification is the mechanism to loop back and catch everyone up to the latest standard and to verify that skills haven't rusted. High-reliability work demands that the people doing it are, themselves, kept highly reliable through these recurrent evaluations.

## The Role of the Competent Person Certification

The term "**competent person**" comes straight from OSHA regulations, which define it as someone "capable of identifying existing and predictable hazards in the surroundings or working conditions…and who has authorization to take prompt corrective measures to eliminate them". In many construction activities – such as excavation, scaffolding, confined space entry, demolition, and others – OSHA requires that a competent person be on-site or on-call to oversee those operations with an eye toward safety. Unlike a license or third-party certification, "competent person" is a status conferred by the employer, but typically the employer will only designate someone as competent after they have received appropriate competent person training in that domain.

For example, consider trenching and excavation. OSHA says that for excavations over 5 feet deep (where cave-in protection is required), a competent person must inspect the trench daily and after rainstorms or other hazard-increasing events. This competent person should be trained to recognize signs of impending cave-ins (like tension cracks on the soil surface), know the different soil types (stable rock vs. Type C soil) and protective systems (sloping, shoring, shielding), and crucially, have the authority to stop work and order evacuation if needed. If the trench seems unsafe, the competent person can tell everyone to get out – and the company must support that decision. The authority part is key: it's not enough to know the hazard; you must be empowered to fix it on the spot.

While one can't go to, say, OSHA and get a "Competent Person License," many training firms and organizations offer **Competent Person Certification courses.** These are usually 1-2 day classes focusing deeply on the specific topic's OSHA regulations and practical hazard control. Upon completion, the individual receives a certificate from the training entity. For instance, you might attend a "Competent Person for Scaffolding" course. You'd learn how to inspect scaffold components, how to calculate loads, proper assembly techniques, tie-offs to structures, etc., as well as OSHA's scaffold standard inside and out. With that knowledge, your employer could confidently say, "Yes, John is our scaffold competent person on this site," and if an OSHA inspector asks, they'll likely accept that designation, especially if John can demonstrate his knowledge (and possibly show that training certificate as evidence of training).

The role of the competent person is vital because it decentralizes safety oversight to the field level. It's not practical for an OSHA inspector to be on every site or a safety manager to watch every crew continuously. By requiring a competent person, OSHA in effect deputizes a worker or supervisor to be the on-site safety officer for that operation. This person is scanning for danger, pre-planning the work to mitigate risks, and ready to intervene. For example, a competent person for cranes (often the lift supervisor) will ensure the crane is set up on stable ground, correct mats are used, weather conditions are monitored, etc., before giving a lift the green light.

In some cases, regulations explicitly tie certain actions to a competent person. For instance, before someone enters a 20-foot-deep trench, a competent person must have determined the protection system (like a trench box) is adequate, beyond just an engineer's design, someone on site must daily verify conditions. Or take blasting operations: the "Blaster-in-Charge" functions as a competent person who must check that everyone's at a safe distance and all misfires are accounted for. Or demolition: a competent person must inspect the structure daily for stability issues.

Competent person training also dovetails with other certifications. You might have a certified crane operator who is also the competent person for crane assembly on site, or a certified scaffolder who is the competent scaffold inspector daily. The training and authority give an extra layer of assurance that not only are workers skilled, but they are also actively managing risk as the work unfolds.

One should note that being labeled competent can have legal implications. If an accident happens and OSHA investigates, they will ask: was a competent person present, and did they fulfill their duty? If a trench collapsed and a competent person hadn't been checking or wasn't properly trained, citations will follow – potentially even personal liability in some cases if negligence is extreme. Thus, employers must choose these individuals carefully and support them. A competent person's call to stop work can sometimes cause tension (schedule delays, added cost to fix an issue), but wise management will back them up because the alternative might be an injury or a failure that is far costlier.

**Competent Person Certification (training)** equips individuals to take on this critical safety role. It's an example of how construction, as an industry, creates internal checks and balances: empowering those on the front lines to be guardians of safety and structural stability. When you hear "Joe is our excavation competent person," it means Joe has both the knowledge and the clout to call the shots when it comes to trench safety. That role has saved countless lives by catching hazards in real time and correcting course – exactly what it's intended to do.

## Temporary Works and Task-Specific Credentials

Not all construction work is about the permanent parts of a structure. **Temporary works** – like scaffolding, formwork, shoring, bracing, and other provisional structures – are essential to build the permanent ones, yet they often don't get the same spotlight. However, history has shown that failures in temporary works can be just as deadly and damaging as failures in finished buildings. Because of that, there are specialized credentials and requirements tied to these tasks, ensuring those who design, erect, or inspect temporary works are qualified.

Consider **scaffolding**, which provides workers access to heights. A scaffold collapse can injure many workers at once (and pedestrians too, in urban areas). Erecting scaffolds is typically done by trained crews, and OSHA requires a competent person to supervise and inspect scaffolds. Some jurisdictions and companies add on a requirement that a "**Scaffold Erector**" or "**Scaffold Inspector**" hold a certificate from a recognized training program (e.g., a 40-hour scaffold builder course). There are even manufacturer-specific certifications – for instance, a company might require that anyone assembling system scaffold must have completed the manufacturer's training program for that system. The idea is that scaffolding is a temporary work which, if under-designed or improperly built, is *highly failure-prone*, so only those who have proven they know the ins and outs should be in charge of it.

Formwork and Shoring for concrete structures is another critical temporary work. We pour wet concrete into formwork molds supported by shores; if that setup fails, tons of concrete can come crashing down. Many might recall incidents of formwork collapse causing fatalities and massive project setbacks. For complex or high concrete pours, it's common (or required by code) that a licensed engineer design and sign off on the formwork/shoring plan, effectively treating it like a real structure. That engineer may also need to inspect it in the field prior to pour. Additionally, contractors often certify their formwork carpenters or foremen through internal programs or external courses. Organizations like the National Concrete Formwork Association (for example) provide guidelines and perhaps training modules. So while the crew assembling it may not need individual licenses, the design and the inspection often rest on credentialed shoulders. As a rule of thumb, OSHA says any *temporary support of a structure*, like falsework, that will handle loads beyond a certain size should be designed by an engineer. For trenches: if deeper than 20 feet, the protective system *must* be engineered[49]. Similarly, for excavations support (sheeting, soldier piles, etc.), an engineer's plan is typically mandated beyond simple cases. These are task-specific credentials at work – you need an *engineer's seal* for some temporary conditions, which is a credential applied to a task, not a permanent element.

**Cranes & Critical Lifts:** We touched on crane operator certification, but for particularly heavy or complex lifts (so-called critical lifts, perhaps involving multiple cranes or lifting near capacity), projects will often require a critical lift plan stamped by a PE and a specialized Lift Director (who might have a certification from say, NCCCO as a Lift Director). This person's job is purely to ensure the lift is planned and executed safely – it's a credentialed role for a specific high-risk task. They'll check the rigging calculations, crane configuration, wind conditions, etc., and have stop-work authority.

Concrete Pumping has certifications for operators, Tower Crane Erection has certifications for those crews, Explosive Demolition (blasters) are licensed separately as well. Each of these is highly task-specific. For example, an explosives handler needs a BATFE license and often state certification, given the obvious hazards.

Another example: **NDT (Non-Destructive Testing)** technicians who do ultrasonic or x-ray testing on welds are certified to Level I, II, or III per ASNT (American Society for Nondestructive Testing) standards. That's very task-specific – they aren't general inspectors, they specifically have a credential to run and interpret NDT equipment.

In the realm of safety, think of certifications like **HAZWOPER** 40-hour training for hazardous waste operations (cleanup sites, etc.), or **Confined Space Entry Supervisor** training for those who monitor confined space works. These, too are targeted at certain hazardous tasks. If a project involves entry into tanks or sewers (confined spaces with potential toxic gas or low oxygen), you'll want folks with that training and possibly additional rescue training on site.

And then there are site-specific credentials mandated by clients: e.g., to work on a petrochemical plant, you might need a site safety orientation card from that owner (like Exxon or Shell have their own) plus a Transportation Worker Identification Credential (TWIC) if it's near a port. While not a skill certification, it's another example of credentials layered onto specific scenarios.

The overarching theme is, *for temporary and task-specific works that carry unusual risks*, the industry doesn't leave it to chance. They overlay requirements – whether it's a professional license, a third-party certification, or an intensive training – to make sure that those tasks are planned and executed by qualified people. And when such systems are in place, failures are less frequent. Unfortunately, when corners are cut, we often see disasters. A study of temporary structure failures found they often occur when these elements are treated as secondary or "not engineered".

For leadership, the takeaway is clear: identify those non-permanent but critical parts of your project and **ensure the right credentials are on the job for them**. Don't say, "It's just temporary, we'll wing it." If anything, temporary works deserve extra scrutiny because they don't benefit from years of code development and oversight that permanent structures do, yet they carry real loads and real dangers. Whether through bringing in a licensed engineer for a scaffold over a street, or making sure your excavation foreman attended that trench safety class, it's about matching specialized tasks with the people holding specialized credentials to manage them. It's an investment in preventing the kind of accident that not only endangers lives but can tarnish reputations and halt projects in an instant.

## Why Licensing Matters for Accountability

## Accountability in a Multi-Stakeholder Environment

Construction projects are multifaceted endeavors involving owners, designers, contractors, subcontractors, suppliers, regulators, and sometimes the public at large. With so many stakeholders, it's easy for finger-pointing to occur when something goes wrong. One of the critical roles of professional licensing and credentialing is to create **clear lines of accountability** amid this crowd. In a multi-stakeholder environment, everyone has defined roles, but the licensed individuals carry an extra weight: a legal and ethical accountability for certain decisions and outcomes.

Think of it as an accountability **anchor**. For example, an architect and an engineer design a building – they are accountable (to the public and to the owner) for the safety and code-compliance of that design. The contractor builds it – they are accountable for following the plans and performing the work safely and to standards. However, if during construction a question arises like "this beam seems undersized" or "this foundation is on questionable soil," who ultimately must address it? The licensed structural engineer in charge of the beam design or the licensed geotechnical engineer who vetted the soil. Licensing ensures that we know who must step up and resolve certain issues; it's not left to a round-robin of "not my problem."

Furthermore, licensed professionals have a duty to each other and the project through the "standard of care." In practice, this means a PE or RA should perform their services as a reasonably prudent professional would. If multiple licensed professionals are on a project, each must still independently exercise judgment and not assume "someone else probably checked that." For instance, a structural engineer might rely on a geotechnical report for soil capacity – but if something seems off, they need to question it. Each license carries the expectation to proactively uphold safety, even if that means sticking one's neck out in a team setting. The accountability is personal. The stamp on the drawing has one person's name – you can't hide behind the company or the group if negligence occurred on your watch.

In large projects, you might have dozens of PEs (civil, structural, electrical, mechanical, etc.) and one or more RAs, plus perhaps CMs with CCMs, etc. Each one is a stakeholder with specific accountability. Licensure helps delineate boundaries: the architect is typically responsible for life-safety code aspects and overall coordination, the structural PE for structural integrity, the MEP PEs for their systems, and so on. When an issue arises in one of those domains, the corresponding license-holder is expected to take ownership. This chain-of-command by discipline is what keeps things orderly. Without it, tasks could fall through the cracks, or everyone could assume someone else was responsible.

Consider when a failure happens and litigation or investigation begins. One of the first things examined is: who was the professional of record for X aspect? If a building façade panel falls, they'll look for the structural engineer of record and the specialty engineer (if any) who designed the panel connections. Those names, by virtue of licensure, mark who had responsibility. In depositions (which we'll discuss soon), each stakeholder is asked to speak to their domain. It becomes very clear, for example, if an electrical fire happened, the PE who sealed the electrical plans will be questioned about the design and whether it met code and good practice. Licensing thus forces a clarity of who answers for what. In complex environments, that clarity is protection against diffusion of responsibility.

Moreover, in multi-employer worksites (common in construction), OSHA and safety regulations also tie into this concept. There's the idea of the controlling employer, the creating employer of a hazard, the correcting employer, etc. But if there's a serious safety issue, say a crane setup that's patently unsafe, a licensed engineer (if present in oversight capacity) might be expected to intervene or at least has the authority to do so. The crane's assembly might even have required a PE sign-off. That license injects accountability where multiple parties interact – e.g., the general contractor might say "we left it to the crane subcontractor," but if a PE stamped the assembly plan, that individual is in the spotlight for what went wrong.

**Licensure and certification anchor accountability** by saying: this person has the duty to act in this sphere. It is not meant to exonerate others from common sense or contractual obligations, but it does provide a backstop. When everyone is licensed or certified for their role, you have a web of accountability that covers the project. Each professional is like a node that holds a piece of the safety net, making sure that within their purview, things won't fall.

## Licensure as a Filter Against Negligence

Professional licensure acts not only as an enabler (granting authority to practice) but also as a **filter to keep out or penalize negligence and incompetence**. The filtering mechanism operates in a few ways: entry qualifications, ongoing requirements, and the threat of disciplinary action. Together, these create a quality-control loop that, while not perfect, significantly reduces the likelihood that someone unqualified designs or oversees something critical.

**At the entry level,** obtaining a license already filters out many who might not be up to the task. The exams are difficult and the experience requirements mean you can't just waltz in and get licensed without proving yourself. The result is that a person who achieves, say, their PE, has demonstrated a baseline of knowledge and an ability to apply it. They have been filtered through an educational gauntlet and practical mentorship. In contrast, without licensure, anyone could claim to be an "engineer" and attempt designs – the only real filter then would be lawsuits after failures, which is a much more reactive and dangerous way to cull out incompetence. By making licensure a prerequisite for independent practice, we largely prevent inexperienced or uneducated individuals from taking on work that could endanger others.

**Continuing education** requirements are another filter mechanism. They ensure that even competent professionals don't become negligent through obsolescence. A structural engineer licensed in the 1970s might have been top-notch then, but if they never learned anything new, they'd be behind on modern codes, materials, and analysis methods by now. Requiring ongoing learning (and in some states, specific ethics training periodically) nudges professionals to stay current and thereby avoid negligent practice that stems from outdated knowledge.

The **disciplinary system** is perhaps the most direct filter against negligence. When a licensed professional does commit negligence – for instance, signing off on deficient plans, or failing to properly oversee work – and it comes to the attention of the board, consequences follow. If the case is clear cut, the board can revoke or suspend the license. That immediately filters that person out from doing further harm (at least within that jurisdiction and often, by reciprocity, in others). For less severe issues, fines or mandated additional training can correct behavior before it leads to something worse. There's an educative effect too: disciplinary cases are often published (with names or at least scenarios) so that other professionals learn "don't do that." It sets a precedence and deterrence. The existence of this oversight is a mental check every time a professional might consider taking a shortcut – the little voice saying, "I could lose my license if I do this improperly."

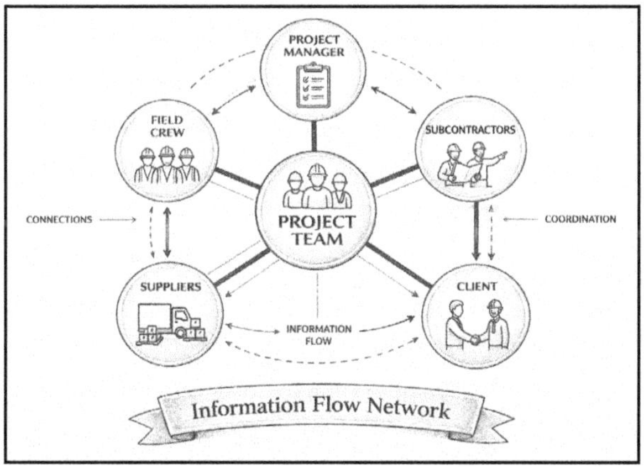


The filter metaphor is apt when you consider some states' periodic review of license holders. Certain states require a licensee to self-report any criminal convictions or any malpractice judgments when renewing. Those too can trigger board reviews. For example, if an engineer was found in court to have been grossly negligent (even if the board hadn't known yet), that can lead to license action. The goal is to filter out bad actors systematically, not just leave it to chance. It's a self-cleansing system to an extent.

One might question: Does it really work? There have been cases where engineers or builders without proper credentials have done work – but they often get caught precisely because at some point, someone asks for the stamp or the license number. When they can't produce it, the filter engages (project stops, authorities called). And in cases where licensed folks did something wrong, you often see swift actions. For instance, in Florida, the Board of Professional Engineers routinely reports on how many licenses they revoked or suspended each year, with negligence being a common cause. Knowing that, engineers in Florida are highly motivated to peer-review each other's work, follow building codes to the letter, and not sign anything they aren't sure about. The risk of losing one's livelihood is a strong filter indeed.

Moreover, licensure adds a legal lever for accountability through the concept of negligence per se. If someone violates a provision of the engineering licensure law (say practicing outside their competence) and something fails, in court it can be simpler to establish negligence because they broke a black-and-white rule intended to protect public safety. In depositions and trials, lawyers often latch onto whether a professional followed the standard of care and licensing rules. If not, it's painted as negligence inherently. Thus, professionals have internalized that following the rules is not just good practice, but essential self-protection.

It's important to note that licensure doesn't filter out all mistakes – humans are fallible and even good professionals can err. But it does filter out patterns of negligence. A one-off minor mistake might lead to a lawsuit but not loss of license; however, repeated sloppy engineering will eventually hit the board's radar, and that license can be pulled. And that person is effectively filtered out from making more mistakes under the mantle of a professional license.

In sum, licensure serves as a proactive gatekeeper to entry and a reactive mechanism to expel or reform those who fall below the accepted standard. It raises the overall reliability of the professional pool. One could argue it's like quality assurance for people: vet them, monitor them, and remove or retrain those that don't meet quality, thereby reducing the incidence of professional negligence that impacts public safety.

## What to Expect in Depositions, Hearings, and Investigations

When something goes awry on a project – be it a structural failure, a building defect, or an accident – a chain of events is set in motion to determine what happened and who (if anyone) is at fault. For licensed and certified professionals, this often means being involved in depositions, hearings, or investigative interviews. These can be daunting, but understanding what to expect can help professionals navigate them while upholding their responsibilities.

**Depositions (Civil Litigation):** A deposition is sworn testimony taken outside of court, usually in a lawsuit's discovery phase. If you were the engineer of record, the architect, the construction manager, or any credentialed expert on a project that's now in litigation, you can bet you'll be deposed. Lawyers will ask you, under oath, to explain your role, your decisions, and your knowledge of the events. Expect very detailed questions. For instance, an engineer might be asked: "*When you designed Connection X, what load did you assume? Why? Show us where that's documented. Did you consider an alternate design?* Who reviewed your calcs?" They will probe for any indication of oversight or negligence. Having a license means you're held to the professional standard of care, so attorneys often bring in expert witnesses who are also licensed to opine if you met that standard or not.

Be prepared to speak to **your process** – did you follow the codes? Did you communicate properly with other stakeholders? One common theme in depositions is exploring communication: e.g., "*Mr. Architect, when the contractor RFIs suggested a cheaper detail, why did you approve it? Did you consult with the structural engineer?*" They are looking for disconnects or lapses. As a professional, if you documented your decisions and followed standard practice, you'll have solid answers. If you cut corners, those will painfully come to light.

It's also common to be asked about your *credentials and experience*. Opposing counsel might try to undermine your credibility: "*You're not a structural engineer, correct?* Yet you made a structural judgment here..." or "How many times have you dealt with this building type before?" They might even ask if you've ever failed the PE exam or had any disciplinary actions – those are fair game if relevant to competency. It's not personal; it's strategy. Knowing this, professionals should maintain up-to-date resumes and records of continuing education, because those could become exhibits to show you were well-qualified for the task.

**Hearings (Licensing Board or Arbitration):** If the issue triggers a licensing board investigation (for example, a formal complaint was filed that an engineer was negligent or an architect violated a code of conduct), there may be an administrative hearing. This is somewhat like a mini-trial but in front of the board or an administrative law judge. Here, expect to explain technical matters in lay terms – board members are usually peers (other professionals) plus public members. They'll want to know what you did or didn't do that a prudent professional would have. If it's an ethics issue (perhaps plan-stamping or misrepresentation), they will ask you to account for your actions and possibly why they shouldn't sanction you. Unlike depositions, which are fact-finding, a hearing is decision-making: your license could be on the line. It's a serious affair where having counsel and perhaps witnesses (like a respected colleague attesting "I would have done the same in that situation") is common. The best approach is honesty and demonstrating that either a) you did everything right or b) if you erred, it was an understandable human mistake, not gross negligence or willful misconduct.

Sometimes a construction dispute goes to arbitration (especially if contracts required it). In arbitration, a panel (often including engineers or construction experts) hears the case. A licensed professional might have to testify similarly as in court. The difference is arbitrators might cut through legal posturing and ask very direct technical questions themselves. They might ask a structural engineer: "*Show us the formula or method you used. We notice it doesn't account for X; isn't that a concern?*" So one must be prepared to justify methods to experts, not just laypersons.

**Investigations (Accident or Failure Investigations):** If there's, say, a collapse or a fatal accident, investigations may be done by OSHA, NTSB (for certain structure failures or bridge collapses, etc.), or independent forensic teams. Professionals will likely be asked to provide documents (plans, emails, inspection logs) and to give interviews. In such settings, even if not under oath, treat it with the same rigor – the investigation report could lead to legal action later. Investigators often want the sequence of events and decisions. If you're the site safety competent person and a trench collapsed, OSHA will interview you: *What did you observe? When did you last inspect? Did you warn management of risks?* If you're an engineer and something failed, forensic engineers might question the choices: *Why that material? Did you consider the unusual load that occurred?* It can feel interrogative, but the aim is to learn the truth and prevent repeats. Cooperating is usually wise; stonewalling often raises suspicion.

Through all these scenarios, a few guiding principles help credentialed professionals:

- **Documentation is your ally:** If it's not in writing, it's as if it didn't happen. Keep design notes, meeting minutes, approval signatures, etc. In deposition or hearing, producing a contemporaneous email where you warned "this change could reduce factor of safety" can be a shield.

- **Know your scope and stick to it:** Don't answer beyond what you were responsible for. If you're the architect and asked a structural question, it's fine to say, "That was under the structural engineer's purview; I relied on their expertise." Conversely, don't try to shrug off something that was squarely yours.

- **Professional demeanor:** These settings test not just your work but your conduct. Staying calm, factual, and objective goes a long way. Board hearings in particular judge if you take the matter seriously and show contrition if appropriate. Arrogance or deflection can prompt harsher outcomes.

- **Legal and insurance support:** If you're licensed, you often carry professional liability insurance which may provide an attorney in a lawsuit, and some professional societies offer legal advisory for board issues. Use those resources. Walking into a deposition cold is not advisable; preparation with counsel is key.

Ultimately, being credentialed means you will be front and center when there's scrutiny. That's part of the deal. It can be stressful, but it also means you have the authority to make a difference in outcomes. Those who handle it well often help clarify issues and even defend rightful decisions. Knowing what to expect – detailed technical queries, challenges to your decisions, and examination of your qualifications – lets you prepare accordingly. And if you prepared in your actual work (by doing it diligently), then these proceedings, while serious, tend to affirm that the system of professional accountability works as intended: by examining actions under the lens of standards of care and determining if the professional acted appropriately. In other words, it's a crucial part of the accountability process that ultimately keeps our industry safe and reliable.

**Operational Takeaways for Today's Job:**

- Match tasks with qualifications. Before a specialized job starts (like electrical work, welding, or operating a crane), verify that the person assigned has the required license or certification. If not, pause and find a qualified individual – allowing unqualified work puts everyone at risk and can violate regulations.
- Use your experts. If you have a Professional Engineer (PE) or certified specialist available (even by phone), run critical questions or changes by them. For example, if you hit an unknown soil condition, a quick consult with a geotech engineer can ensure you're proceeding safely and correctly.
- Encourage ongoing training for your team. Suggest or arrange for key crew members to get their OSHA 30, first aid/CPR, or other relevant certifications. A well-trained crew is safer and more efficient – even a short refresher or course can pay off on site through fewer mistakes.
- Lead by example in professional development. If you're a superintendent or foreman, maintain your own certifications (like a CM certification or safety credentials) and let your crew see you doing it. It sends the message that learning never stops and it motivates them to follow suit.
- Keep a certification tracker. Maintain a simple chart of who on your crew is certified to do what (forklift, rigging, scaffolding competent person, etc.) along with expiry dates. Check it monthly so that nobody's qualification lapses and you always know who can legally and safely perform a given task.

Chapter 13

# Compliance and Enforcement Mechanisms

Construction projects operate under a web of regulations, contracts, and safety standards. Knowing the rules (as covered in Chapter 11) is only half the battle – the other half is ensuring those rules are followed on site and understanding what happens when they're not. This chapter examines how safety and building regulations are enforced in practice, and what EPC construction managers must do to lead their teams in compliance. From unannounced OSHA inspections to the fine print of insurance policies and bonds, enforcement mechanisms create both legal obligations and powerful incentives to maintain safe, high-quality work. We also explore how codes evolve after disasters, the special challenges of fast-tracked EPC jobs, and how experienced leaders use compliance as a strategic advantage. The goal is practical knowledge: a field leader who anticipates inspections, prevents violations, and navigates the aftermath of any issues – all while keeping projects on track and people safe.

## OSHA Inspections, Citations, and Penalties

On American jobsites, the Occupational Safety and Health Administration (OSHA) is the primary enforcer of worker safety laws. OSHA inspectors can arrive without warning, so a superintendent or EPC construction manager should always be prepared for that knock on the trailer door. Inspections may be triggered by several events: an employee complaint, a referral from another agency, a recent serious accident or fatality (which by law must be reported to OSHA within 8 or 24 hours, depending on severity), or as part of OSHA's programmed initiatives (for example, emphasis programs targeting hazards like falls or trenching). In any case, the inspection follows a well-defined process:

- **Opening Conference:** The OSHA compliance officer presents official credentials and explains the scope and reason for the visit. If it's a focused inspection (say, on fall protection), they'll outline that. If it's a response to an incident or complaint, they'll inform the employer of that as well. The manager should welcome the inspector professionally, verify their ID, and involve the site safety representative at this kickoff meeting. It sets the tone – cooperation and transparency are key from the outset.

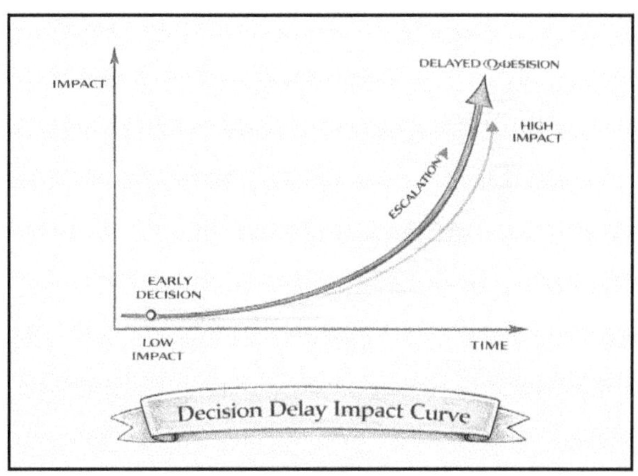

Decision Delay Impact Curve

- **Walkaround Inspection:** The inspector, often accompanied by an employer rep and sometimes an employee rep, will tour the jobsite. They observe work in progress, ask questions, take photographs or measurements, and may privately interview workers. The key here is cooperation – obstructing an inspection is never wise. If the inspector identifies an imminent danger (a hazard likely to cause death or serious harm immediately), they will ask that work be stopped in that area until it's fixed. As a manager, you should take notes on any hazards pointed out and, if possible, correct them on the spot. Prompt corrections can demonstrate good faith and sometimes even reduce potential penalties.

- **Closing Conference:** After the walkthrough (which might last hours or even span multiple days for a large site), the inspector holds a close-out discussion. They won't issue citations on the spot, but they will summarize any apparent violations found. This is an opportunity for the manager to clarify facts, provide documentation (e.g. training records, inspection logs) that might address an issue, and demonstrate a proactive attitude. If you've fixed a minor issue during the inspection, respectfully point that out. The inspector will explain that any official citations and penalties will be sent by mail after OSHA's area office reviews the findings.

Following an inspection, OSHA issues citations by mail, typically a few weeks later. Each citation document will cite the specific OSHA standard allegedly violated, classify the severity, and propose a monetary penalty along with a date by which the hazard must be abated (fixed). Citations are classified by severity and characteristics, generally as follows:

- **Other-than-Serious:** For hazards unlikely to cause serious harm or death. These often carry smaller fines (or sometimes no fine at all), but they still require correction.
- **Serious:** For hazards that could likely cause an accident or illness resulting in serious injury or death. As of 2025, an OSHA serious violation can carry a penalty of up to roughly $16,550 each (this maximum is adjusted annually for inflation).
- **Willful:** For violations where the employer knowingly failed to comply with a legal requirement or acted with plain indifference to worker safety. Willful violations are penalized most heavily – up to about $165,000 per violation. In cases of extreme willful negligence resulting in a fatality, OSHA can refer the case for criminal prosecution under federal law (though such cases are rare and usually prosecuted as misdemeanors).
- **Repeat:** For a violation that is substantially similar to one for which the company was previously cited (usually within the last 5 years). Repeat violations carry penalties up to the same level as willful (around $165k each), meaning a company that fails to learn from past mistakes will pay dearly.
- **Failure to Abate:** If an earlier violation was cited and not fixed by the prescribed abatement date, OSHA can assess additional fines (potentially over $16,000 per day beyond the deadline) until the issue is corrected. This is essentially a meter that keeps running until you resolve the problem.

To put these numbers in perspective, consider some recent enforcement cases. In mid-2024, OSHA publicized that a contractor in Guam faced over $1 million in proposed fines after inspectors found employees working in unprotected 8-foot-deep trenches – the company had a history of ignoring trench safety and received nine willful citations in one go. Around the same time, a New Jersey construction firm was cited for 32 violations across multiple sites (including nine willful and nine repeat violations) for repeatedly exposing its workers to fall hazards and silica dust, racking up about $819,000 in penalties. These headline-grabbing fines illustrate OSHA's willingness to drop the hammer on employers who don't correct dangerous practices, especially repeat offenders. The message is clear: the cost of non-compliance can far exceed the cost of prevention.

A construction manager should also know that OSHA has a Multi-Employer Worksite Policy. On a typical large site, you might have the general contractor, multiple subcontractors, staffing agency workers, and vendors all present.

OSHA can hold more than one employer responsible for the same hazardous condition under this policy. In OSHA's framework, there are four roles an employer might play on a multi-employer jobsite:

- **Creating Employer:** The one that caused a hazardous condition (e.g. a trade contractor who leaves a floor opening uncovered).
- **Exposing Employer:** An employer whose workers are exposed to the hazard (even if that employer didn't create it).
- **Correcting Employer:** An employer responsible for fixing the hazard (such as a subcontractor brought in to repair a faulty scaffold).
- **Controlling Employer:** The employer with general supervisory authority over the worksite, typically the general contractor or construction manager, who has the power to prevent or correct safety issues through their broad oversight.

If you are the general contractor or CM, OSHA usually views you as a controlling employer on site. That means OSHA expects you to exercise "reasonable care" to prevent or detect violations by any party. In practice, this doesn't mean you must watch every worker constantly – but you must implement a robust safety plan, monitor the work, and address hazards that are obvious or that you should know about through diligent supervision. You cannot turn a blind eye and say "that's the subcontractor's problem" if one of your subs is, for example, removing trench shoring to save time or disabling safety guards on a saw. Ignorance is not an acceptable defense. OSHA can cite multiple employers for the same issue – for instance, if a sub's workers are exposed to a fall hazard that the GC could have had corrected, both the sub (as the exposing employer) and the GC (as the controlling employer) might receive citations. The multi-employer doctrine essentially ensures that safety is everyone's responsibility on a jobsite, and that those with overarching control maintain active oversight. For the seasoned construction manager, the takeaway is: you must actively manage safety across all trades, not just your direct employees. If you see it, you own it. **For EPC projects with many subs and vendors, assume OSHA can treat you as the controlling employer by default. If you see a hazard and don't act, you own part of the outcome.**

What should a construction manager do to avoid OSHA citations in the first place? The best strategy is to run your project as if an OSHA inspector were on site every day. That means conducting your own frequent safety audits, encouraging workers to speak up about hazards without fear of retaliation, and fixing problems proactively before someone else finds them. We know the common trouble spots by now: fall protection (the number one killer in construction and perennially OSHA's most cited violation), proper scaffolding and ladder use, trench safety (protecting any excavation over 5 feet deep), electrical safety (locking out power, using GFCIs), hazard communication (labeling chemicals and training crews on them), machine guarding, and so on. OSHA helpfully publishes an annual "Top 10" list of violations, and construction standards dominate that list year after year – fall protection, hazard communication, scaffolding, ladders, lockout/tagout are usually all in the top ranks. A smart manager uses that knowledge as a checklist: before OSHA ever arrives, make sure guardrails and personal fall-arrest systems are in place for work at heights, that trenches over 5 feet have shoring or are properly sloped, that electrical panels are closed and live wires are locked out, that workers have and use the right PPE, etc. Many companies even implement surprise internal inspections or hire third-party safety consultants to do mock OSHA audits, catching issues before the real OSHA does. This not only reduces the chance of citations; more importantly, it prevents injuries – which is the ultimate goal.

Despite all best efforts, let's say OSHA does issue citations. How you respond is critical. The employer has the right to contest citations or penalties through the Occupational Safety and Health Review Commission, but there's a short window – 15 business days from receipt of the citation – to file that contest. Often, before resorting to litigation, employers can request an informal conference with OSHA's area director to discuss the citations. In that meeting, you might negotiate penalty reductions or extended abatement deadlines, especially if you can show you've already fixed the cited problems or if you have a good overall safety record. Remember, OSHA penalties aren't just a slap on the wrist; beyond the dollars, they become part of your company's compliance history. A serious or willful violation on record can trigger higher scrutiny on future projects and heftier fines for any repeat violation. There's also the reputational hit: OSHA citations are public records, and OSHA often issues press releases for significant cases (especially willful or repeat violations). Prospective clients and bidding authorities do check safety histories. Many owners prequalify contractors based on their OSHA Experience Modification Rate (EMR) or past citation history – a string of serious violations can knock you out of contention for elite projects that demand safe contractors.

OSHA enforcement is a key mechanism pushing contractors toward safer practices. The prospect of an unplanned inspection and the bite of a hefty fine (or even a work shut-down order in extreme cases) shouldn't induce paranoia, but rather vigilance. **On an EPC site, run the job as if an OSHA compliance officer could join your daily walk every day.** A wise construction manager fosters a culture of compliance before an inspector ever arrives – training crews on the rules, correcting hazards daily, and treating safety as a non-negotiable part of the work. **If an OSHA visit does occur, your team should follow a simple drill: stay calm, designate one spokesperson, be truthful and courteous, fix what you can immediately, and treat the visit as an external audit that makes the project better**. As one saying goes, "Compliance is not optional" – but beyond avoiding penalties, remember that it's about every worker going home in one piece. Lead with that principle in mind, and you'll find OSHA is not an adversary but rather an important checkpoint ensuring you're doing right by your team.

## Insurance, Liability, and Bonding Requirements

While OSHA inspectors and building code officials enforce the rules through laws and regulations, another powerful set of enforcement mechanisms comes from the business side: insurance companies, contract liabilities, and surety bonds. In construction, you typically can't even set foot on a major project without proper insurance and bonding in place – and those entities have a keen interest in how safely and responsibly you operate. In essence, they are silent regulators: they incentivize compliance and penalize poor practices through premiums, coverage terms, and legal/financial consequences. A seasoned construction manager recognizes that keeping insurers, sureties, and clients confident in your safety performance is just as important as passing inspections.

Start with insurance. A construction firm usually carries multiple policies: general liability (for third-party injuries or property damage), workers' compensation (for employee injuries on the job), builder's risk (for damage to the project itself during construction), professional liability (for design errors or omissions, if you have a design-build role), among others. All of these insurers care deeply about safety performance, because accidents and failures cost them money in claims. If your company develops a reputation for safety problems, expect to see it reflected in higher premiums and even difficulty renewing coverage. For example, workers' compensation insurance uses an Experience Modification Rate (EMR) system to adjust premiums. An EMR of 1.0 is considered industry average risk. If your firm has more injuries and claims than average, your EMR will rise above 1.0 (say 1.20), which might mean you pay 20% higher worker's comp premiums – a direct financial hit as a result of those incidents. Conversely, a stellar safety record (EMR below 1.0) can save you money and even become a selling point to clients. **Many large EPC owners and GCs prequalify bidders by requiring an EMR at or below a defined threshold**. Many large project owners and GCs prequalify bidders by requiring an EMR at or below some threshold (often 1.0 or even 0.9) because they equate a low EMR with a lower-risk, more reliable contractor. In short, insurance costs serve as a continuous enforcement mechanism: unsafe companies literally pay for their mistakes through higher insurance, and they often get shut out of projects that only want safe players.

Insurance carriers also directly enforce risk control measures. It's increasingly common for insurers to require evidence of strong safety programs, training, and oversight practices, especially on high-risk projects. For instance, an insurance underwriter might mandate that a contractor implement a formal hazard audit system or maintain detailed safety inspection logs as a condition for coverage on a complex demolition or deep excavation project. Insurers learn from expensive failures. When there's a disaster and they have to pay multimillion-dollar claims, they respond by tightening requirements to prevent a recurrence. A classic example: after a series of high-profile structural collapses that led to massive insurance payouts and even criminal investigations, the insurance industry upped its requirements for design and construction quality control. They know that paying a bit more upfront for prevention saves them from huge payouts later. As a result, your policy renewals may come with detailed questionnaires about your safety training programs, past incident history, and even surprise jobsite visits by insurance loss-control consultants. If you do suffer a major incident, expect your insurer to investigate and potentially raise your deductibles or premiums in the future. One clear metric: after a widely publicized construction failure caused by negligence, the entire industry's insurance premiums for that sector can rise. In effect, every major failure sends ripples through the insurance market – a stark reminder that the cost of a safety failure isn't just the OSHA fine; it's higher insurance costs, higher bid qualifications, lost work, and a damaged reputation.

Now consider contractual liability and bonding. Construction contracts are full of clauses that function as enforcement levers. Nearly all contracts explicitly require compliance with safety laws and regulations. If you violate OSHA regulations or fail to follow applicable building codes, you could be in breach of contract with the project owner. Owners also often include even stricter safety requirements in contracts – for example, mandating a full-time safety manager on large jobs, requiring all supervisors to have OSHA 30-hour training, or stipulating that certain high-hazard work (like crane lifts over occupied facilities) have an independent engineer's review. If a contractor flouts those requirements and an incident occurs, the owner may have grounds to terminate the contract for default. Even if you're not terminated, you will likely be liable for the damages. For instance, if a fire or structural collapse on site (traceable to the contractor's negligence) causes a months-long delay, the owner may pursue the contractor for liquidated damages or other costs related to that delay.

Liability can extend to individuals as well – project executives, project managers, superintendents, and engineers have, in some egregious cases of neglect, been personally sued or even criminally charged. This isn't common, but it does happen when willful disregard for safety leads to serious harm. One example: after a deadly building demolition collapse in Philadelphia in 2013 (where a wall collapsed onto an occupied store, killing six people), the contractor and the site engineer were criminally charged, and city officials were sued for lax oversight. In less extreme scenarios, personal liability might mean professional consequences like loss of one's license. Think of design professionals: after a famous walkway collapse at a hotel in the 1980s, the structural engineers of record lost their PE licenses for life due to negligence in design and review. These kinds of consequences underscore that safety lapses can end careers, not just projects. As a construction leader, you carry the weight of not only your company's livelihood but potentially your own, if you were willfully complicit in ignoring safety.

Bonding (performance bonds and payment bonds provided by surety companies) adds another layer. Performance bonds guarantee that a contractor will complete the job, and payment bonds guarantee that subs and suppliers get paid. While bonds primarily protect financial and performance aspects of the project, here's the connection: a contractor with a poor safety record is a riskier bet to actually finish a project on time and without costly interruptions. Surety underwriters quietly factor your safety history and EMR into their evaluation of your firm's reliability. If you've had a fatal accident or major loss on a job, that project likely experienced delays, investigations, maybe lawsuits – all of which cost time and money. The surety's worst nightmare is a bonded contractor defaulting and leaving them on the hook to hire completion contractors. Chronic safety problems can be a red flag for default risk, as they may indicate disorganized management or impending financial strain from claims. Thus, a bad safety record can lead to higher bond rates or even difficulty obtaining bonds for new work. In some cases, if a project under bond starts experiencing serious safety issues, the bonding company might step in behind the scenes – they might send their own representative to quietly check on the project status or press the contractor to beef up oversight. Remember, the surety isn't in business to lose money; like an insurer, it will push for whatever measures are necessary to protect the project's success.

On large projects, even lenders and financiers get involved in enforcement. A bank financing a big development might require regular safety reports as part of the loan covenants, or reserve the right to inspect the site. It's not unheard of for a lender to threaten to withhold funding draws if they believe the project is being mismanaged into a dangerous situation. Essentially, anyone who has money on the line – insurers, sureties, clients, lenders – becomes an enforcer of safe, lawful conduct. They may not issue citations like OSHA, but they can hit where it hurts by raising costs or pulling support.

Let's talk about a subtle but very real enforcement mechanism: **invalidated coverage**. Insurance policies have conditions, and if you blatantly ignore them, a claim might be denied. For example, most liability policies expect the insured to follow the law and maintain "reasonable precautions." If a contractor was contractually required to use certified welders for a critical structural weld but instead used unqualified workers who caused a collapse, the insurer might invoke policy language to deny the claim due to willful misconduct or material breach of contract terms. Or consider that many contracts stipulate key personnel must have certain qualifications (say, a certified safety professional on site). If a contractor skirts that and something goes wrong, they may not only face safety consequences but could also forfeit some insurance or bond protections. The lesson: failing to use properly trained, certified people can invalidate contracts and coverage. As a construction manager, part of your job is ensuring that project and insurance requirements (safety training, equipment certifications, hazard communication programs, etc.) are upheld so that, if an incident does occur, your company isn't left footing the entire bill alone due to a technical escape clause. The fine print matters – enforcement isn't only what OSHA or the city says, it's also what your contract and insurance policy say.

Finally, consider the aftermath of a disaster from a financial perspective. A major catastrophe can put a contractor out of business if they aren't well-insured and well-capitalized. Even if insurance covers direct costs, the indirect costs (future premium hikes, legal defense, lost client trust and opportunities) can be enormous. A poignant line often quoted is that sometimes "the price of haste is the survival of the corporation." In other words, one big safety failure can bankrupt a company either directly or through ripple effects. This is enforcement by market forces: companies that develop a bad safety reputation may simply not survive in the long run. Conversely, those with excellent safety records gain a competitive edge – they have lower insurance costs, fewer disruptive incidents, and often enjoy a reputation that helps them win more business. Increasingly, public agencies and private developers alike want to hire contractors who treat safety and compliance as fundamental. They see it as a proxy for overall quality and professionalism.

In summary, while OSHA inspectors and code officials issue the citations and notices, the financial ecosystem of construction (insurers, sureties, contracts, and courts) provides another powerful layer of compliance enforcement. As a manager, paying attention to this means working safely not just to avoid a fine today, but to keep your company insurable, bondable, and reputable for the long haul. It means documenting everything – training sessions, safety meetings, equipment inspections – because if something does happen, those records prove you were conscientious (which can mitigate liability, or even be required by your insurance). Think of every preventive measure and certification not as overhead, but as an investment in "compliance insurance." In fact, one could say that a thorough safety plan is itself a form of insurance – every accident you prevent is a claim you never have to file. Modern practice bears this out: many owners and insurers now demand to see safety plans, maintenance programs, change-management logs, and more before contracts are awarded or policies underwritten. They're essentially saying: show us you run a tight ship, or we won't take the risk on you. The takeaway for a construction leader is clear: manage your projects in such a way that any stakeholder – whether an OSHA inspector, an insurance auditor, a bond underwriter, or a client – could walk in, review your practices, and come away satisfied that you're in control of the risks. That not only keeps you compliant and insured, it also means your people are safer and your project is far more likely to finish without disaster.

## Regulatory Lag and Code Reform After Disasters

Safety regulations and building codes are often, as the saying goes, written in blood. A sad reality of our industry (and others) is that major disasters frequently expose gaps in codes or weaknesses in enforcement – and only then do the rules get updated. Regulatory lag refers to the delay between the emergence of a hazard or new construction practice and the development of standards or codes to manage it. In a fast-evolving construction world, technology and techniques can outpace the rulebooks. Alternatively, everyone might be vaguely aware of a hazard in theory, but until a high-profile failure occurs, there isn't enough impetus to memorialize the solution in the code. This section looks at how building and safety regulations adapt (sometimes sluggishly) after catastrophes, and what that means for construction managers. Ideally, we should learn from these events even before the law requires us to – because being merely "to code" is not always enough to guarantee safety.

Several historical examples illustrate the familiar pattern of tragedy leading to reform:

- **Skyline Plaza Collapse (1973):** This was a 26-story residential building under construction in Virginia that partially collapsed, killing 14 workers and injuring dozens. Investigators found that the concrete floors hadn't fully cured before the temporary shoring was removed and additional loads were placed on them. In the rush to stay on schedule, workers were "pulling supports out too fast… trying to hustle the job too fast," according to one crew member. At the time, there was no specific OSHA regulation detailing how long to leave formwork and shoring in place for multi-story concrete pours – it was left to industry practice. The Skyline disaster was a wake-up call that led to more explicit standards on formwork stripping and reshoring in the OSHA concrete construction rules. Essentially, a rule that seems obvious now ("don't remove supports until concrete is strong enough") was cemented into regulation because the absence of that clarity proved fatal. Today, every competent site team knows to follow engineered reshoring plans and verify concrete strength before stacking additional loads, but it took a tragedy to hard-code that into the rules.

- **Hurricane Andrew (1992):** Not a man-made accident, but a natural disaster that exposed huge weaknesses in building codes. Andrew's 165-mph winds devastated South Florida, flattening thousands of homes. Post-storm investigations showed many buildings weren't adequately tied together or anchored to withstand high winds; in some cases construction shortcuts and a patchwork of lax local codes contributed to the destruction. In response, Florida undertook a sweeping reform of its building codes – moving to a unified Florida Building Code (FBC) in 2002 with significantly strengthened requirements for hurricane resistance (e.g. stronger roof tie-downs, impact-resistant windows, improved inspection regimes). Essentially, decades of complacency in code requirements were erased by a single day of extreme wind. Buildings constructed to the new code have since fared much better in subsequent hurricanes. The lesson here was that sometimes it takes an outside shock to spur regulatory modernization. The lag was that older codes didn't reflect the known science of high-wind engineering; the reform was essentially catching up to what engineers already knew could be done to make buildings safer. A construction manager who lived through that era in Florida will tell you: before 1992, a lot of things were done by habit that, after 1992, were required by law to be done differently (and better).

- **Northridge Earthquake (1994):** A 6.7 magnitude quake struck the Los Angeles area. Even though the region had modern seismic codes and most buildings survived, an unexpected failure mode was observed: brittle fractures in the welded steel moment-frame connections of many modern steel buildings. Steel frames were thought to be the gold standard for ductile earthquake resistance, yet Northridge revealed cracks where beams joined columns – a hidden vulnerability. This prompted a deep investigation by FEMA, SAC, and others, and ultimately led to major code changes. Building codes (through the AISC standards and model codes) were revised to require tougher welding procedures, new connection designs (like reduced beam sections or "dogbone" cuts), and more rigorous inspection of welds in seismic zones. In this case, regulators didn't wait for buildings to collapse (thankfully none collapsed from those connection cracks during Northridge) – the near-miss was treated with the same seriousness as a collapse. It was a proactive example of reform triggered by lessons learned rather than fatalities. Still, it took the earthquake to reveal the gap. The code improvements in the late 1990s ensured that new steel structures wouldn't have that same flaw, and retrofits were recommended for existing buildings. For a construction manager, Northridge is a reminder that just because something has performed well historically doesn't mean it can't fail in new ways – and when it does, expect the rules to tighten.

**These examples underscore a recurring theme:** regulations are often reactive, not proactive. There's an inherent lag as industries and governments tend to "prove the need" for a rule via evidence – unfortunately, sometimes that evidence is a collapsed structure or a line of coffins. From a construction manager's standpoint, understanding this pattern is crucial. It means you cannot assume that just because something is "up to code" or legal today, it is actually safe or best practice. We should keep an ear to the ground on industry lessons learned, not just the letter of current codes. Often, after a near-miss or disaster, engineering societies and safety organizations will publish recommendations long before official codes change. A true safety-conscious manager will implement those best practices early, voluntarily, rather than waiting for them to become mandatory. For example, years before Florida made condo re-inspections mandatory statewide, a diligent property manager could look at the Surfside tragedy and say: "We're going to start voluntary structural health monitoring of our older buildings right now." Similarly, if you're building something unique or using a novel method that isn't fully addressed by existing code, you should seek extra peer reviews and apply conservative engineering judgment – don't hide behind "well, no law explicitly forbids this." Our ethical duty to protect lives goes beyond the letter of the current rules.

It's also worth noting that sometimes regulatory lag isn't due to ignorance of the problem, but due to bureaucracy or politics. Code committees can take years to agree on changes, and even after model codes update, local jurisdictions might take additional years to adopt them. This adds an adoption lag on top of the development lag. For instance, a hazard might be identified in 2018, a model code change might be formulated by 2021, but a given city or state might not implement that change until 2024 or later, depending on their code cycle. In the meantime, projects go on under the older rules. This is why a conscientious manager often must go above and beyond minimum requirements. A practical tip: stay connected with professional organizations (ASCE, ACI, OSHA updates, etc.) and subscribe to industry publications that discuss failures and fixes. By keeping current on emerging risks and the "word on the street" about needed changes, you can anticipate where the codes will go, and often you can self-impose higher standards now. It may cost a bit more or require an extra step, but it's far cheaper than learning the hard way via an accident.

Regulatory lag teaches us that safety leadership often requires looking forward. As the saying goes, "Wise men learn from others' mistakes." We have an unfortunate library of "blood precedents" in construction – each with its lesson. Don't wait for the code to catch up if you already know a safer way. Being proactive not only protects your people and project, but when the rules finally do change, you'll be ahead of the curve and not scrambling to adjust. As an EPC construction operator, your standard should be what is reasonably knowable and preventable today, not just what is minimally required by the last code cycle.

## Enforcement Challenges in Fast-Track Projects

The pressure of time is a defining feature of modern construction. Owners want projects faster and cheaper, and this drive gave rise to fast-track construction, where design and construction phases overlap to compress the schedule. Fast-tracking might mean starting foundation work before the final structural design is 100% complete, or fabricating steel while drawings are still being reviewed. When done prudently, fast-track methods can shave months off a timeline. But when schedule rushes ahead of process, maintaining safety and quality compliance can become significantly more difficult. Oversight mechanisms – both internal (like QC checks) and external (like inspections) – can be strained by the pace and complexity. In a "go, go, go" environment, the temptation to cut corners increases, and enforcement often struggles to keep up.

Fast-track project environments have been factors in several notorious failures. We've already discussed some in the previous section. To avoid repeating those same examples, let's look at a few other cautionary tales:

- **CityCenter Construction, Las Vegas (2007–2008):** The $9 billion CityCenter project in Las Vegas was one of the largest private commercial developments ever, featuring multiple high-rise towers being built simultaneously on a very tight schedule. Work was running 24 hours a day in multiple shifts. Under that frenzied pace, the project saw an unusual spike in fatalities – nine construction worker deaths in 16 months. This alarmingly high toll led to investigations that found safety oversight had been falling through the cracks amid the rush. Crews were tired, coordination among trades was chaotic, and Nevada's OSHA (a state-plan OSHA) was later criticized for not being tougher sooner. Things came to a head when workers, fed up and scared after a string of fatalities, staged a walk-off protest in mid-2008 to demand safer conditions. The publicity and pressure forced a safety stand-down and a comprehensive review of practices on site. Federal OSHA stepped in to assist the state in a sweeping investigation, which resulted in dozens of serious citations and a new commitment to enforcement. Remarkably, after the workers' one-day strike and the subsequent safety overhaul, no more deaths occurred on the project going forward. CityCenter's troubles illustrate how fast-track mega-projects can overwhelm the usual safety systems: with design changes happening on the fly, hundreds of subcontractors and workers on site around the clock, communication and supervision had broken down. Enforcement in this context required an external shock (media attention and federal intervention) to recalibrate the balance between speed and safety. For us as managers, the take-home message is that if you ever find yourself thinking "we don't have time to pause for safety," then you're on a dangerous road. At CityCenter, each individual incident probably had an immediate cause, but the root cause across them was the schedule-first culture. Only when that culture was reset (by essentially slowing down and retraining) did the fatalities stop.

- **Skyline Plaza & Other Concrete Frame Collapses:** Earlier we talked about the 1970s Skyline Plaza collapse caused by rushing concrete work. Unfortunately, that pattern didn't end there. Similar failures have occurred when fast-track pressure led to formwork or temporary supports being removed too soon. In 2004, a parking garage under construction at the Tropicana Casino in Atlantic City collapsed, killing 4 workers – investigators pointed to design changes and possible premature removal of supports as contributing factors, all under a rush to meet deadlines. In 2012, a 5-story building under construction in Mexico City collapsed during the concrete pour of the top floor, killing workers; again inadequate shoring and rushing were blamed. In these cases, enforcement should have come from the project's own procedures (e.g. requiring sign-off before stripping forms), but those procedures were either not in place or not followed. Code officials typically only inspect periodically and might not catch a contractor jumping the gun by a few hours or days on curing time. Thus it falls to the site management team to enforce the rules of structural safety. The recurring tragedy is that, under schedule pressure, someone decides "ah, it's probably fine" and removes bracing or shores early – and then everything comes down. The pattern is so well-known that it's now a staple of construction training: never load green concrete, never assume yesterday's pour is strong enough unless proven. Yet, when fast-track project managers treat time as more precious than safety, the old mistakes resurface.

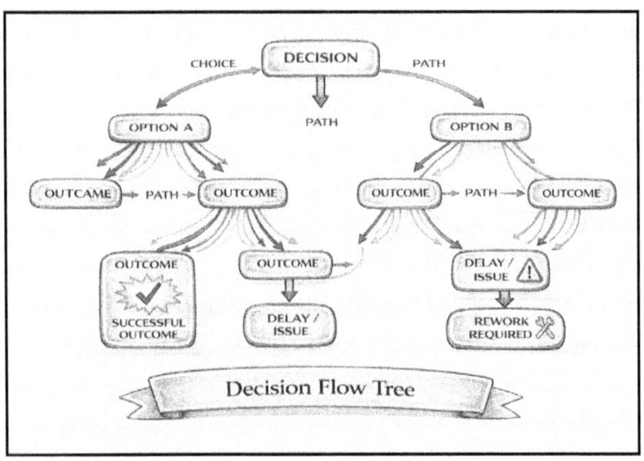


- **Fast-Track Design Changes:** Some failures highlight how rapid design changes under fast-track can circumvent normal checks. Imagine a scenario where steel connections are redesigned in the field to speed up erection, without a thorough engineering review. This contributed to the famous 1981 hotel walkway collapse (as discussed in prior chapters) – a last-minute connection detail change doubled the load on a rod, but calculations weren't redone due to time, and disaster ensued. In another case, during the construction of a large retail store, a decision was made to eliminate certain braces to open up floor space (for aesthetic reasons) while construction was ongoing; the change was sketched quickly, approved hastily, and workers removed the braces – causing a partial collapse. What went wrong? In a properly sequenced project, a significant design change would go through rigorous analysis and perhaps independent review. In a fast-track "design-as-you-build" situation, that luxury is gone. Enforcement from the city code side is challenging because inspectors may not even be aware a change has been made if it isn't formally submitted. And the internal enforcement (project managers insisting on calcs and peer review) can falter under pressure to just "get it done." The lesson for managers is to build your own safety nets when the schedule is tight. For example, even if it's not required, you might hire an outside engineer to double-check any critical design tweaks made under time pressure. Or insist on halting work for 24–48 hours to analyze an issue properly, even if higher-ups are screaming about delays. These are hard calls, but the alternative is risking a collapse that will stop the project far longer. **On EPC jobs with design and construction under one umbrella, the temptation to make field-driven design changes informally is especially high; your job as a construction leader is to insist that every structural or high-energy change goes back through an engineer of record and a documented MOC, even when it hurts the schedule.**

**Why is enforcement so much harder on fast-track projects?** A few key factors stand out:

- **Compressed Decision Cycles:** Normally, designs are completed, reviewed, and approved before construction, and inspectors have a clear final plan to check against. In fast-track, designs and field conditions evolve continuously. Inspectors might be looking at partially complete plans or a stack of design revisions issued at lightning speed. Ensuring that what gets built actually matches an approved, safe design becomes like hitting a moving target. The feedback loops are shortened or skipped. Coordination meetings that would typically flag inconsistencies might be rushed or omitted. Errors that would normally be caught in design reviews instead get built into the project. From an enforcement view, a building official might approve the foundation and structure as originally submitted, but if the design team makes later changes to, say, add a heavier rooftop unit or revise beam layouts, who catches that? On a traditional job, any design change must go back through the permit process in a revision submission. On a fast job, there's a risk that changes are made "on the fly" to save time, with formal approval to follow later (or not at all). This is obviously not legal, but it does happen. The controlling mechanism – thorough plan checking – is undermined by constant change. So unless the project imposes strict internal controls (like "no field change without written engineer approval and city sign-off"), things can slip through.

- **Multiple Parallel Activities:** Fast-track often means multiple trades and phases overlapping. You might have steel erection still ongoing on the top floors while interior buildout has started below, while also still receiving revised drawings from the architect. The site gets very congested with workers and tasks. This makes it harder for supervisors and safety personnel to keep tabs on everything. Hazards proliferate: more people in a confined space, more temporary setups, more interference. A traditional schedule might have had clear separation – e.g. structure complete, then enclosure, then finishes – allowing focus. In fast-track, all these overlap, and it's easy for critical inspections or tests to be missed amid the chaos. For example, if concrete slabs, structural steel, and MEP rough-in are all happening in adjacent areas simultaneously, the site safety officer and quality inspectors are stretched thin bouncing around. They might miss that some rebar was omitted or a weld wasn't done properly in one zone because they were busy dealing with an issue in another. Also, because of overlap, sometimes work is done out of sequence or in temporary conditions that are not well-covered by code. Enforcement agents (OSHA or city inspectors) often rely on the contractor to manage those temporary states safely, since they can't monitor full-time.

- **Human Factors – Fatigue and Stress:** Fast-track schedules often demand long hours, night shifts, and rapid decision-making. Workers and managers alike get fatigued and stressed. Tired people are more prone to mistakes and less attentive to safety. The enforcement system – whether it's a foreman checking work, or a worker following a safety rule, or an inspector doing rounds – all rely on humans being vigilant. In a fatigued, hurry-up state, the whole safety net weakens. It's not hypothetical: industry data shows that when projects are in schedule crunch mode (lots of overtime, second shifts), injury rates tend to climb. If folks are pulling 70-hour weeks to hit a deadline, you start seeing near-misses and incidents that wouldn't happen under normal hours. Some progressive projects anticipate this and, counterintuitively, impose limits on work hours or ensure rotating shifts to give rest, even if it means bringing in extra crew, because they know an exhausted workforce is an unsafe workforce. But not all projects do this – and fast-track jobs by their nature push the limits of human endurance. A construction manager must be acutely aware of this and possibly be the voice that says "we need an extra safety meeting" or "we're going to add a second safety officer for the night shift" or even "we need to slow down for a day to let people recharge," when he or she senses that fatigue is endangering compliance.

- **Cultural Temptation to Cut Corners:** When a finish date becomes almost sacred (often tied to big financial incentives or avoidance of liquidated damages), some project teams start rationalizing risk-taking. This could be as small as skipping a step ("we'll pour the concrete without waiting for the lab test, it's probably fine") or as severe as ignoring a required safety protocol ("we won't bother re-installing the guardrails for this one-day task to save time"). These are enforcement failures at the project management level – someone consciously decides to bypass a rule to save time. Unless there's a very strong safety culture in place, fast-track pressures make these lapses more likely because everyone is laser-focused on speed. A telling quote from a veteran superintendent: "When schedule or cost take precedence, safety and quality are the first to pay." In fast projects, it takes enormous discipline and leadership backbone to continually pump the brakes and say "No, we do it right," especially when the schedule dragon is breathing fire. But that is exactly what is required to avoid disaster – otherwise, the project becomes an accident waiting to happen.

Given these challenges, how can we maintain compliance and safety on fast-track projects? It requires deliberate strategies and sometimes external mandates to force the issue. One effective approach is implementing mandatory "hold points" in the schedule – enforced pauses where work cannot proceed until a certain quality or safety check is done (and done correctly). For example, many state building codes or project specifications now include a hold point after concrete placement: you cannot remove forms or shores or apply loads to the new concrete until an inspector signs off that a specified concrete strength has been reached (often verified by break tests of field-cured cylinders). This is a direct legacy of past failures. Another common hold point is before a major lift or critical structural activity: requiring a third-party engineer to review the rigging plan or method statement and certify it's okay. These hold points create buffers against rushing. They inject a bit of "slow down" into the DNA of a fast project. Some contracts even explicitly allow schedule relief (extensions) if a safety hold point causes a delay – in other words, they'd rather have you finish late than have an incident. This is a positive development in contracting: writing "safety time" into the plan. On an EPC contract, treat hold points as contract milestones, not optional courtesy checks.

On the design side, some owners combat fast-track risk by requiring independent engineering reviews for complex or novel designs, even if not strictly mandated by code. For instance, on a design-build high-rise, an owner might hire an outside structural consultant to do a peer review of the design in parallel with construction, to catch errors that the overlapping design/construction process might miss. Additionally, modern collaborative delivery methods like Integrated Project Delivery (IPD) can mitigate some of the adversarial nature of schedule pressure. In IPD, all parties (owner, designers, contractors) share the risks and rewards via a single contract, including metrics for safety and quality, not just schedule. This means nobody has an incentive to quietly cut corners – transparency is higher and the team collectively owns the outcome. On some fast projects, teams even institute something called a "Safety Timeout" whenever a critical sequence is about to start: essentially a short, all-hands pause where everyone – management, workers, inspectors – double-checks that prerequisites are complete and risks mitigated before proceeding. It's a way to enforce patience within an impatient framework.

We're also seeing insurers and financiers stepping in here. As mentioned earlier, a surety or insurance carrier might flat-out refuse to underwrite a project if they feel the schedule is unrealistic without extra safety measures. Some public clients now ask bidders, "Show us how you will maintain safety on this aggressive schedule." They expect detailed staging plans, additional supervision, maybe commitments to hire extra crews to avoid burnout, etc. If you say you'll do it in 18 months, you'd better explain how you'll do it safely in 18 months. And they will hold you to it. If a project starts slipping and the only solution management offers is "we'll work faster/longer," a savvy owner or lender might counter with "and what about additional safety resources or adjustments?" As a project leader, you may have to articulate to higher-ups: "If you want me to accelerate, here's what we need in terms of staffing, safety protocols, and maybe adjusted expectations." If those conditions aren't met, the tough but ethical stance is to say that acceleration beyond a point isn't feasible without undue risk. It can take courage (and might be unpopular with executives), but that's part of leadership – sometimes protecting people and the project means telling stakeholders what they need to hear, not what they want to hear.

A core economic truth about fast-track is worth remembering: rushing can cost far more than it saves. Any calculation of schedule vs. safety needs to consider the downside. Shaving a week off the schedule might save some overhead and please the client in the short term, but if that rush leads to a major failure, the project could be set back months or years and cost tens of millions in losses. One industry analysis compared many fast-track failures and found that the downstream costs (investigations, rebuilds, lawsuits, insurance hikes, lost reputation) vastly outweighed the gains of shaving a few weeks off the schedule. In extreme cases, a major failure can even bankrupt a contractor or permanently tarnish an owner's program. Thus, any rational cost-benefit mindset should put safety delays in perspective: a day's delay to address a serious concern might cost $50,000, but a collapse could cost $50 million. As a manager, you sometimes have to make that business case to those pressuring you – "We can invest a little time now to avert a catastrophe that would blow the budget and schedule completely later." Most reasonable owners, when faced with that logic (especially if you can cite examples), will relent.

Fortunately, many project owners and agencies have become wiser on this front. They now view schedule realism as a component of safety. Submit an unrealistically short schedule in a bid, and a savvy owner might actually mark you down or disqualify you, reasoning that you're planning to cut corners or drive people into the ground. Regulators, too, are pushing back on rushed work. Some city permitting departments require a construction logistics plan that shows, for example, how you'll separate the public from construction activities – you can't just say "we'll work double shifts and keep the sidewalk open" if that's unsafe. If someone tries to fast-track by, say, doing demolition over an active street without proper closures, they simply won't get the permit. So in some respects, the industry and regulators are codifying sanity checks against extreme fast-tracking.

Fast-track projects demand a higher level of vigilance and creativity in enforcement. The mantra should be "Fast, but safe." To achieve that, one must intentionally plan safety into the schedule from day one. That means acknowledging where you must slow down (critical structural operations, testing periods, tie-ins, etc.) and protecting those moments in the schedule. It means potentially spending more on supervision, training, and planning to keep up with the accelerated pace. It definitely means empowering people at all levels to speak up – workers need to know that if they see a problem, management will support pausing the work to address it, deadline or not. The culture on a fast job should be that anyone can call a timeout for safety, and they'll be listened to. As an industry proverb puts it, "We can make up time, but we can't undo a tragedy." If you're managing a fast-track project, you almost have to over-communicate and over-enforce standards to counteract the natural slide toward haste. Stay very close to the work, do frequent check-ins, and celebrate those times when someone says "Hold on, let's do this right" – because that person likely just saved the project.

## Using Compliance as a Strategic Advantage

Up to now, we've often discussed compliance – with OSHA regulations, building codes, contractual requirements – in the context of avoiding negative outcomes (citations, accidents, legal trouble). But experienced construction leaders know there's another side to this coin: strong compliance and safety practices can be a strategic advantage. **On EPC pursuits, your safety record, EMR, and demonstrated compliance systems often decide whether you even make the shortlist**. In other words, doing things "by the book" (and even beyond the book) isn't just about avoiding penalties; it can actively help you build a more successful project and a more competitive business. In this section, we explore how going above and beyond minimum requirements can differentiate a contractor, and how fostering a culture of compliance can directly contribute to project success and business growth.

**Go Beyond Minimum Standards:** Perhaps the most defining trait of companies that use compliance as an advantage is that they don't stop at merely obeying the rules. They look for ways to go above and beyond in sensible ways that actually help the project. **For EPC CMs, beyond compliance usually means setting clearer internal limits – wind thresholds for crane work, lower trigger heights for fall protection, and more conservative cure times for removals – than the minimums, then protecting those limits when pressure mounts.** For example, fall protection standards say guardrails or fall-arrest systems are needed at 6 feet and above. A company striving for zero harm might set an internal policy of 100% fall protection at 4 feet, anticipating that extra margin removes even the temptation for workers to take chances on ladders or low roofs. Another example: it may not be legally required to have a dedicated safety professional on a small job, but an experienced manager might assign one anyway for a complex task, because the cost is small compared to the value of focused oversight. "Beyond compliance" might also mean pursuing certifications like ISO 45001 (safety management systems) or OSHA's VPP Star status – not for bragging rights, but because the process of qualifying for those forces you to strengthen your procedures. These programs can bring fresh ideas and rigorous audits that make you better. And yes, you do get bragging rights that translate into business: a VPP Star flag flying on your project tells every passerby (and potential client) that this site meets the highest safety standards in the country.

**Foster Ownership and Pride in Compliance:** The best construction leaders make safety and quality a source of pride for the whole team. They celebrate when the project hits a milestone like "100 days injury-free" or when it passes a surprise OSHA inspection with no citations. They might throw a barbecue for the crew or give out recognition awards. This positive reinforcement turns compliance from something people feel they have to do into something they want to do. It becomes a shared goal, like hitting a production target. I recall a project that created a friendly competition between crews for the safest work area each month – the winning crew got merch or a bonus. It sparked engagement; crews would actually audit their own areas to make sure they beat the other guys. The result was not only zero accidents, but the site overall was much tidier and more productive, because in the act of competing on safety, they also organized their materials better and communicated more. In short, culture eats policy for breakfast – if you build a culture where complying with safety and quality standards is simply "how we do things here," you unlock the full strategic value of it. You're no longer pushing compliance; it's pulling you.

**To illustrate the ultimate strategic mindset:** I once visited a project run by a superintendent with 30 years' experience. He told me, "I use inspections for me, not against me." By that he meant he welcomed the city inspector or the insurance auditor on site, because he viewed their perspective as an extra set of expert eyes that might spot something his team missed. If they found an issue, he treated it like free consulting to improve his project. That is a far cry from the old-school "hide from the inspector" mentality. This superintendent had virtually no rework on his jobs and a stellar safety record. Subcontractors would vie to work with him because they knew he ran a tight but fair ship – they'd be held to high standards but also taken care of. And not surprisingly, owners of his projects often requested him by name for their next job. His personal reputation for compliance excellence became a career booster.

At a company level, many firms now strive to be "the safest contractor in X" not just for altruism, but because it correlates with being the most successful. It's part of the professionalization of construction management – treating safety and compliance not as annoyances or merely moral obligations, but as integral to doing good business. We see this in trends like Safety by Design and Prevention through Design (PtD), where safety considerations are built into the design and planning phase, or in contractors investing in training programs to get every worker OSHA-10 or OSHA-30 certified (even if not required) because it raises the baseline knowledge on site.

Using compliance as a strategic advantage boils down to a change in mindset: from seeing regulations and enforcement as external pressure to seeing them as helpful guidelines and tools that, when embraced, make our projects better. When you lead with that mindset, a few notable things happen. Your projects tend to run smoother – people know the expectations, there's less firefighting of problems that could have been prevented, and there's a sense of order. Your teams gel, because nothing builds camaraderie like a shared purpose of "we look out for each other." Clients notice the difference – a site that is clearly well-managed, with workers in proper gear, housekeeping in order, and an upbeat atmosphere, instills confidence that the job is in good hands. And perhaps most importantly, you go home at the end of the day knowing you did the right thing, and you sleep well at night instead of worrying if tomorrow's the day something will go wrong.

In the big picture, compliance is not just about avoiding negatives; it's about creating positives. A mature safety and compliance culture pays for itself many times over. It delivers measurably higher productivity (fewer stoppages, smoother workflow), lower costs (less rework, no fines, reduced insurance claims), stronger workforce loyalty (people stay where they feel safe and respected), better project outcomes (on-time, quality builds), and a sterling reputation that opens doors. In a competitive industry, those are tremendous advantages. As the construction sector continues to evolve – with pressures like labor shortages, tighter margins, and higher client expectations – the firms and leaders who thrive will be those who treat compliance not as the ceiling but as the floor, and who cultivate safety and quality as core values. In the words of an old construction adage reframed for today: "Do it right, do it safe, and the job will succeed." Compliance is doing it right; when you leverage that, you're not only building projects, you're building a foundation for long-term success.

**Operational Takeaways for Today's Job (EPC CMs)**

- Run one "mock OSHA walk" this month with your safety lead and a superintendent; fix at least three issues before an inspector ever could.
- Verify your project EMR, recent citation history, and insurance conditions; brief your team on what is at stake if safety drifts.
- Identify two safety or quality hold points on the critical path (e.g., shoring removal, first energization) and make them nonnegotiable schedule gates with written signoff.

## Part IV – *Building Teams, Systems, and Adaptability*

**Roles:** *Organizational leaders* – construction managers, superintendents, project executives, and team leads.

**Focus:** Emphasizes the human and organizational elements of project execution. Part IV explores how to train and mentor crews, foster a strong safety culture, embrace technology, and cultivate continuous improvement. The goal is to turn each jobsite into a high-performing organization that learns and adapts over time.

**Operational Decisions Informed:** How to invest in your team and processes for long-term success. After Part IV, a leader will be equipped to decide how to improve training programs and toolbox talks, conduct effective after-action reviews, implement quality audits or peer reviews, and embed a "learning culture" on the team. This section helps construction leaders build resilient teams and workflows – so that even as conditions change, new technologies emerge, or crises arise, the project can adapt and continue to thrive without sacrificing safety or quality.

# Chapter 14

# Team Training and Development

**Roles:** Organizational leaders – EPC construction managers, superintendents, project executives, and team leads.

**Focus:** Emphasizes the human and organizational elements of project execution. Part IV explores how to train and mentor crews, foster a strong safety culture, embrace technology, and cultivate continuous improvement. The goal is to turn each jobsite into a high-performing organization that learns and adapts over time. The emphasis is on building cross-functional EPC teams that can absorb design changes, supply disruptions, and schedule shocks without losing control of risk.

In the dynamic world of construction, effective team training and development is a cornerstone of project success. Unlike compliance-focused training (such as OSHA safety courses covered elsewhere in this book), the emphasis here is on cultivating a field-ready, versatile workforce that can adapt to challenges on the fly. Modern construction leaders recognize that training isn't a one-off event or box to check – it's an ongoing investment in their people, yielding safer operations, higher quality, and resilient project delivery.

This chapter reexamines team training and development with a practical, field-oriented lens. We will explore how cross-training can build operational redundancy to keep projects moving even when key personnel are unavailable. We'll discuss establishing feedback loops so that real-world performance data and crew input continuously improve training programs. Recognizing the human limits on any team, we'll address fatigue and focus management – ensuring crews stay alert and aware on the jobsite. We'll also draw lessons from high-reliability industries like nuclear submarine crews and offshore oil rigs, which achieve exceptional safety and consistency through relentless training. Finally, we'll outline practical strategies to sustain field-level readiness and skill versatility over the long haul. All examples and recommendations are geared to U.S. construction contexts and challenges, providing fresh case examples not covered in other chapters. The goal is a comprehensive, applied guide to developing construction teams who are competent, confident, and prepared – every day, every shift.

## Cross-Training for Operational Redundancy

One of the most powerful strategies for strengthening a construction team is cross-training employees to handle multiple roles. Cross-training means teaching crew members skills outside their primary trade or job function so that they can step in for each other when needed. The objective is to build operational redundancy – no critical task should be known by only one person, and no project should grind to a halt because a specialist is absent or overloaded. In the words of one roofing estimator, "No single person should have all the knowledge when it comes to productivity and functionality... If the loss of a key person happens, companies can reduce the risk by ensuring that vital information is retained within the company." In practical terms, a cross-trained crew is more resilient and can maintain momentum despite unexpected disruptions.

**Benefits of Cross-Training:** Cross-training yields immediate and long-term benefits for both projects and personnel. First, it provides insurance against labor shortages or absences. If a certified equipment operator calls in sick or a foreman is pulled away, another trained worker can step up with minimal downtime. For example, a large general contractor that implemented systematic cross-training found they could reduce idle time between tasks – crew members no longer waited around for a lone specialist, since others were qualified to carry on the work. This flexibility kept productivity up and schedules on track. Second, cross-training improves team communication and morale. When craft workers learn each other's tasks, they gain appreciation for what their teammates do and can anticipate needs across trades, leading to smoother coordination. It "breeds encouragement of knowledge sharing and leads to a more knowledgeable and versatile team," as one industry professional observed. Additionally, giving employees opportunities to learn new skills combats monotony and burnout. Crew members become more engaged when they can grow their capabilities, which in turn boosts retention. Rather than losing talent to boredom or lack of advancement, companies that invest in cross-training often see higher loyalty and job satisfaction, because workers feel valued and see a path for growth. In short, a well-executed cross-training program makes the team more adaptable, more productive, and more motivated.

**Implementing Cross-Training:** To launch cross-training on your team, start by identifying mission-critical skills and "single points of failure." These are tasks for which only one person or a single trade contractor is currently responsible. Prioritize these areas for skills redundancy. For instance, if only one crew member can program the surveying GPS or only one electrician is licensed to operate a particular piece of equipment, those are good candidates for cross-training. Next, create a plan for knowledge transfer. Many firms use a mix of on-the-job shadowing, short courses, and mentoring to spread skills. A field laborer might spend a few hours a week shadowing the grade checker to learn basic surveying, or carpenters might be sent to a one-day rigging certification class so they can assist the ironworkers during lifts. At the New Jersey Department of Transportation, managers found it effective to conduct cross-training during seasonal slow periods – in their case, July and December – when staff had downtime to learn new roles without work piling up. They paired experienced employees with learners and even had the trainees practice by redoing a completed task as if they were responsible for it, which built confidence in a no-risk setting. Construction contractors can take a similar approach: use rainy days, winter months, or brief lulls between project phases as opportunities to run internal training sessions or send workers to outside training programs. The training itself can be informal and hands-on. One crew in a small construction business rotated team members through various tasks – their carpenters learned basic plumbing hookups, and vice versa – by having seasoned workers teach brief workshops in the laydown yard. They also utilized external resources for specialized skills (for example, sending a few people to get certified in welding or first aid). The result was a more agile workforce that could take on a wider range of projects. When a new job required skills the team didn't originally have, someone on the crew had been cross-trained or certified to handle it, enabling the company to bid on and execute work they might have otherwise subcontracted out.

It's important to note that cross-training does have limits and must be managed thoughtfully. Highly technical skills (like structural engineering calculations or licensed electrical work) can't be fully taught to someone lacking the prerequisite background. The goal isn't to make every worker a master of every trade, but to cover the fundamentals so they can at least carry the load temporarily. As the NJDOT example illustrated, someone in data entry can't become an engineer overnight, and an engineer may not sign off on financials, "but they are able to learn some aspects of any job to assist when needed." Management should communicate clearly that cross-training is not about doubling workload or replacing jobs – it's about teamwork and coverage. Workers may initially worry they're "training their replacement" or being asked to do more without compensation. Overcome this by involving the team in the planning, making participation voluntary, and rewarding employees who expand their skill sets (through recognition, small pay bumps for new qualifications, or preferential consideration for promotions). Protect each person's primary role – assure them that helping in another area is a temporary backup plan, not a permanent shift of duties. By addressing these concerns and highlighting the personal benefits (like increased advancement opportunities and higher civil service exam scores reported among cross-trained staff), you can foster buy-in.

In practice, establishing a successful cross-training program may involve creating a skills matrix for your crew – a simple table listing crew members down one side and key skill areas across the top. Mark who is proficient in what, and target the gaps where only one name (or none) appears under a critical skill. Then systematically arrange training to fill those gaps. As with any training, follow-up is key: refresh skills periodically if they aren't used often. For example, if you cross-trained a concrete finisher to operate the telehandler, but months go by without doing it, schedule a quick refresher or have them operate it during a less critical task to keep the skill alive. Cross-training is not a one-and-done effort – it should be an ongoing part of your team culture, with regular opportunities for people to learn new things. Ultimately, by embracing cross-training, construction companies large and small can cultivate a multi-skilled crew ready to tackle whatever the day throws at them. As one industry article concluded, firms that invest in cross-training "witnessed significant improvements in collaboration and project outcomes" – from reduced downtime to more seamless teamwork – positioning themselves for success in an ever-changing industry. For EPC construction operators, target cross-training at failure points: have at least two people able to run each critical pre-task briefing, complete each daily equipment inspection, and manage each high-energy permit so the system doesn't depend on one hero staying late.

## Feedback Loops for Continuous Improvement

Training shouldn't be static or delivered in a vacuum – the best construction companies treat training as a closed-loop system that evolves with input from the field. In simple terms, this means establishing feedback loops where lessons from actual project performance feed back into refining the training content and methods. A feedback loop in training is essentially a continual conversation: workers provide feedback on what they encounter and how training helped or fell short, and managers use that information to adjust future training accordingly. Rather than blindly repeating the same orientation slide deck or skills course each year, companies with feedback-driven training programs ensure that what they teach remains highly relevant to real jobsite challenges. This approach keeps training effective and targeted – a crucial factor in an industry where conditions and techniques are always changing.

**Performance-Based Training Adjustments:** One practical way to implement a feedback loop is through performance data and incident reviews. Whenever there's a notable event – say a quality issue like recurring concrete formwork errors, or a safety incident like a near-miss with a backing vehicle – treat it as valuable feedback. Analyze why it happened and ask: did our training prepare the crew for this? If not, what can we change? Perhaps your analysis finds that several crew members weren't clear on lockout/tagout procedures, contributing to an equipment near-miss. That's a signal to update your training curriculum and emphasize that topic more, or even hold a special session on it immediately. In fact, industry best practices endorse this responsive approach. The Associated General Contractors (AGC) recommend holding targeted safety trainings to address all safety incidents: identify the incident's details, then "quickly follow up by communicating targeted messages designed to address specific safety hazards involved to avoid similar future incidents." This message can be delivered in many forms – a bulletin, an extra toolbox talk, a formal training module – as long as it reaches the team promptly. For example, if a crew had a scaffold plank fail due to improper inspection, the company might issue a safety bulletin that same week reviewing how to inspect planks, and then incorporate a hands-on scaffold inspection drill into the next training meeting. The feedback (an incident) directly triggers a training improvement (targeted learning to prevent recurrence). Over time, this creates a culture of learning from mistakes rather than blaming – every slip or lapse becomes an opportunity to strengthen skills and procedures.

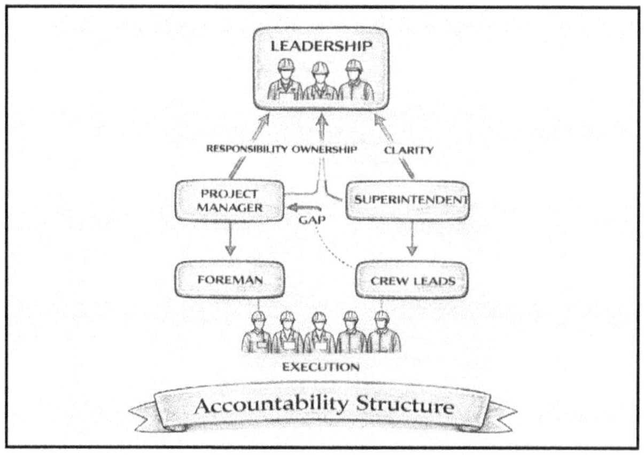


Feedback loops also apply to positive performance. Pay attention to areas where crews excel and reinforce those in training as best practices. If, for instance, your concrete crew consistently achieves outstanding slab finishes, invite the lead finisher to demonstrate his techniques in the next training session or field forum. Capture what's working well and make it part of the standard training so new hires learn the same tricks of the trade. In high-reliability industries, this is known as "preoccupation with failure" (always watching for what could go wrong) paired with a commitment to resilience, but it also means celebrating and replicating success. The goal is to create a virtuous cycle: training drives better performance, and observations of performance drive better training.

Gathering Crew Feedback: Another critical source of insight is the workers themselves. The people on the front lines often know better than anyone where skill gaps lie or which procedures are confusing. Construction leaders should actively solicit feedback from their teams about training needs and training quality. This can be done informally or formally. On the informal side, encourage supervisors to ask their crews during safety meetings or after a task: "What could we learn from this? Is there anything we should train on more?" Create a culture where it's normal for a crew member to speak up and say, for example, "I'm not totally comfortable with that new laser screed – can we get a refresher on it before the next slab?" When workers raise their hands with a concern or suggestion, act on it by scheduling that refresher or providing additional coaching. On the formal side, consider short surveys or debriefs at key points. Some companies do end-of-project "lessons learned" sessions, which are perfect for identifying training improvements: maybe the team found the need for better soil compaction training after struggling with subgrade on this job, so you plan a new compaction workshop before the next project starts. Even more routinely, a quick eval at the end of a training class ("Was this useful? What topics need more clarity?") can give instructors valuable input to tweak the program. Modern training practice often uses post-training surveys or even digital tools (Learning Management Systems with feedback forms) to capture participant input. The key is to close the loop – analyze the feedback and act on it. If several apprentices report that the steel erection training didn't cover certain bolt-up techniques they later needed in the field, don't let that critique gather dust. Revise the training materials and methods to fill that gap. And importantly, let the crew know you've made the change (e.g. "We heard your feedback, and next time we're including a bolting demo"). This not only improves the program but also builds trust that management listens and continuously improves the training experience.

Regular Coaching and Check-Ins: In addition to structural feedback loops, day-to-day coaching interactions serve as mini feedback cycles. Rather than waiting for an annual review, many construction supervisors now engage in frequent, focused feedback sessions with their crew members. A 15-minute weekly check-in can cover what the worker is doing well, where they can improve, and what training or support they might need. These bite-sized conversations make feedback a normal part of work, not a dreaded event. For example, a foreman might schedule a biweekly chat with each apprentice to review their recent work (perhaps using a simple template to stay focused). In one meeting, the foreman might commend the apprentice's safe rigging practices (positive feedback) and suggest they practice using a laser level because some readings were off (constructive feedback). Together they could agree that the apprentice will shadow the layout engineer next week to get more practice with the laser level (closing the loop by prescribing training). Such continual coaching ensures that training is responsive to individual performance – those who need extra help in an area get it promptly, and those who are ready to advance to more complex tasks are identified and mentored. It also reinforces a growth mindset: feedback isn't punishment, it's about development. When crews see that every issue raised is met with a plan to learn and improve (instead of blame), they become more open about admitting mistakes or skill gaps early, before they turn into bigger problems.

Ultimately, integrating feedback loops in team training transforms a construction organization into a learning organization. It prevents stagnation of skills and helps the team adapt to new techniques, technologies, and hazards proactively. As one construction training publication put it, "When learning becomes an ongoing habit rather than a one-time event, your team becomes more resilient, more innovative, and better prepared for the challenges ahead." The loop may start anywhere – a project post-mortem, a near-miss report, a worker's suggestion – but it should always end with concrete action to enhance the training program. To make feedback loops real, pick two mechanisms and standardize them: a 10-minute weekly "learning from last week" huddle, and a simple near-miss log that is reviewed at the EPC coordination meeting, not buried in a safety folder. By continually aligning training with on-the-ground reality, construction leaders ensure their teams are always sharpening their edge, with up-to-date skills and knowledge that drive safer and better performance.

## Fatigue and Focus Management for Jobsite Awareness

Construction is tough, physical work often performed under tight deadlines – factors that can unfortunately lead to fatigued workers and lapses in focus. Managing fatigue and maintaining focus on the jobsite is a critical aspect of team development, because even the best-trained crew can be compromised if they are overtired or distracted.

Fatigue has been called a "silent and pervasive threat" in construction, one that "might seem trivial compared to obvious dangers, but its consequences can be catastrophic." This section examines why alertness matters so much in our industry and outlines practical strategies to keep crews fresh, attentive, and safe.

The Impact of Fatigue: Simply put, a fatigued worker is an unsafe and inefficient worker. Physical and mental fatigue dulls concentration, slows reaction times, and impairs decision-making. Studies have equated fatigue with impairment similar to alcohol intoxication. For example, losing just two hours of sleep from the recommended eight can produce psychomotor slowdowns comparable to having a 0.05 blood alcohol content – about 2 to 3 beers for an average person. If someone gets no sleep at all, the effect is akin to a staggering 0.20 BAC – far beyond legal intoxication. Now consider that in construction, we routinely have workers driving heavy trucks early in the morning or operating cranes and forklifts. Doing so at a fatigue level that mimics a few drinks is obviously a recipe for disaster. It's no surprise then that an estimated 13% of workplace injuries are related to fatigue, and workers who sleep less than 5 hours in a 24-hour period are more than three times as likely to be injured on the job compared to those getting a healthy 7+ hours of sleep. These statistics are eye-opening – literally – and they underscore that fatigue is not a minor issue but a major safety hazard.

Construction work schedules often exacerbate fatigue. Long shifts (10, 12 hours or more), night or early-morning work, and stringing together too many days without rest all increase fatigue risk. In one National Safety Council survey of safety-critical industries, 100% of construction workers surveyed had at least one risk factor for on-the-job fatigue, and nearly half of construction employers reported that worker fatigue had led to safety incidents on their sites. Think about that: essentially every construction worker faces some fatigue risk, and in many companies nearly one out of two incidents has fatigue as a root cause. Fatigue doesn't just come from lack of sleep; it can accumulate from repetitive or strenuous tasks, harsh weather exposure, or even intense mental focus without breaks. Workers might "push through" feeling tired, but eventually cognitive lapses or microsleeps (brief involuntary dozing off) can occur. In construction environments, zoning out for even a few seconds at the wrong time – missing a load swing, forgetting a lockout, mis-stepping on a scaffold – can be fatal.

**Focus and Distraction:** Alongside fatigue, loss of focus due to distractions is another threat to jobsite awareness. Modern jobsites are full of potential distractions: smartphones in pockets, noisy equipment, multiple activities competing for attention, or personal stress that preoccupies a worker's mind. A momentary distraction can steal someone's focus from the task at hand. Consider a rigger glancing at a text message while a crane is in motion, or an equipment operator mentally replaying an argument with his spouse instead of watching his blind spots. Distraction-related accidents – from vehicle collisions to slips and trips – have become enough of a concern that many contractors now implement "no phone" policies on site. In fact, experts advise a zero-tolerance policy for personal device use on active worksites: "Even one use of a cellphone on the job can lead to dangerous distractions, so it's best to keep these things at home or off the jobsite." Some firms collect phones at the gate or use apps that lock the phone during work hours, while others revert to old-school communication like two-way radios to remove the temptation of texting. Beyond electronics, noise is another factor – constant loud noise can reduce concentration and cause "tunnel hearing," where people miss auditory cues. High noise levels also stress the body, contributing to fatigue. OSHA warns that prolonged exposure to noise over 90 dBA not only risks hearing loss but can impede communication needed for safety. Thus, managing distractions might include providing quiet break areas, rotating workers out of high-noise zones periodically, or using hearing protection with built-in radios for critical communications.

**Strategies for Fatigue Management:** So how can construction leaders manage fatigue among their teams? A starting point is education and culture. Make fatigue awareness part of your safety training – teach workers the signs of fatigue (both in themselves and co-workers) and encourage an attitude that fatigue is a legitimate safety concern, not a personal weakness. Some companies have even held “Fatigue Awareness” toolbox talks where they discuss how a person’s alertness can degrade over a long week, and encourage the team to watch out for each other. Emphasize that it’s okay to speak up if you’re too tired, just as it’s okay to refuse work if you feel it’s unsafe. A non-punitive approach is crucial: if a worker admits fatigue, the response should be support (like finding them a less hazardous task or a short rest break) rather than ridicule or punishment. This requires management to set the tone – for instance, supervisors should avoid scheduling outrageous hours or calling people back in with only a few hours of sleep. Planning and policy play a big role. Consider fatigue in project scheduling from the outset. If possible, avoid make-or-break tasks being done at 3 AM or at the tail end of an 80-hour workweek. Some best practices drawn from high-reliability sectors include: limiting night shifts to under 10 hours and limiting any shifts over 12 hours; allowing at least 9–10 hours of rest between shifts (especially after long ones); and capping the number of consecutive night shifts a worker can do (e.g. no more than 2 in a row if they’re 10+ hours). Enforce reasonable overtime limits – it might be tempting to run full throttle to meet a deadline, but if you burn workers out, an accident or costly mistake could blow that deadline entirely. Some contractors have started scheduling mid-shift power naps or extra break time during extended night shifts, noting that even a 20–30 minute nap mid-shift can significantly refresh alertness. On critical night operations (like overnight highway bridge moves), it’s not unheard of to see cots in the job trailer for this purpose. Additionally, when switching workers from day to night shifts, give them a rest day in between to adjust. And absolutely avoid back-to-back shifts – no one should finish an evening shift and then be expected to come in for an early morning pour with just a catnap in between.

Another strategy is health and wellness support. Often, fatigue stems from factors off the job: sleep disorders, poor sleep environments, or lifestyle issues. Employers can help by providing resources on proper sleep hygiene. Something as simple as a pamphlet or a 5-minute segment in orientation on tips for better sleep (dark, cool room, avoiding screens or alcohol before bed, etc.) can raise awareness. Some companies partner with wellness programs to offer screenings for sleep apnea and other disorders – since millions of Americans unknowingly suffer from conditions that severely disrupt sleep. Treating those conditions (often through medical intervention or devices like CPAP for apnea) can dramatically improve a worker’s daily alertness and health. Encouraging overall fitness and stress management indirectly helps reduce fatigue as well.

**Maintaining Focus on the Job:** To bolster focus and minimize mental mistakes, consider implementing daily or weekly rituals that center everyone’s attention. Many crews do a morning stretch-and-flex routine followed by a brief “safety and focus” huddle. During this tailgate meeting, the superintendent or foreman can remind everyone of the day’s high-risk activities and explicitly encourage the crew to stay alert. This is also a great time to ask if anyone isn’t feeling 100% (maybe someone’s fighting a cold or didn’t sleep because of a sick child at home) – if so, adjust their tasks accordingly. Regular toolbox talks on situational awareness also keep the concept fresh. Remind workers that complacency is the enemy: an old-timer who’s done a task a thousand times might start going on autopilot, precisely when an unusual hazard can sneak in. Rotate workers through tasks when feasible to prevent fatigue from tedious, repetitive work. And ensure people take their breaks. One cultural challenge in construction is the macho ethos of powering through lunch or working till you drop. Leadership must counteract that by actively insisting on breaks and modeling the behavior (e.g. the foreman actually shuts off the machine for 15 minutes at break time and has a rest, sending the message that it's not only allowed but expected). Enforcing distraction controls is equally important. As noted earlier, banning cell phone use during work (except for designated supervisors or in emergencies) removes one huge source of distraction. Make it a policy, communicate it clearly, and follow through with enforcement consistently. Some contractors have workers sign a pledge or include it in the site safety plan that phones stay in lockers or pockets except in safe zones. Supplement rules with training on the why: help workers understand that a split-second of looking at a phone or joking around can lead to a lifetime of consequences. For example, a company might present case studies at a safety meeting of incidents caused by distraction – such as a young laborer on another site who was on his phone and walked behind a backing truck, with tragic results. These real stories can hit home.

Moreover, engaging workers in conversations about distraction can surface hidden issues. Perhaps crew members will mention that they struggle to concentrate in the afternoon heat, or that a loud generator on site makes it hard to communicate and stay focused. With that feedback (tying back to our previous section on feedback loops), management can take action: maybe set up a shade tent for afternoon breaks, relocate or enclose the generator to cut noise, etc. One builder suggests holding "refocus meetings" whenever there's a noticeable dip in alertness – essentially a quick timeout to regroup and remind everyone to keep their head in the game. Even a mid-shift stretch or a quick team check can serve to reset attention spans. And don't underestimate leadership by example: if the crew sees supervisors rushing, skipping breaks, multitasking on a phone call while signaling a crane, they'll assume that's the norm. Instead, leaders should model attentiveness – put the phone away, follow safety protocols without cutting corners, and visibly take care of their own well-being.

Managing fatigue and focus is about creating conditions where workers can perform at their best mental and physical capacity. It involves scheduling sensibly to allow rest, educating the crew on risks and self-care, enforcing anti-distraction policies, and building a jobsite culture that values alertness as much as any technical skill. One safety manager framed it well: "Emergency preparedness on construction sites isn't just about having a plan – it's about ensuring every worker knows it by heart. A well-prepared site is a safer site, ready to respond effectively when seconds matter." That sentiment extends beyond just emergencies – if every worker is mentally present and not running on fumes, they can respond to any sudden hazard or change in the work effectively, when seconds (or split-seconds) matter. By treating rest and focus as essential components of training and daily work, construction leaders protect their people and projects from the costly toll of fatigue-related errors. As an EPC CM, treat fatigue as an operational hazard. Build planned breaks around high-risk windows (night pours, tie-ins, heavy picks), and give supervisors explicit authority to rotate or stop a crew when focus is clearly gone.

## Training Systems Modeled After High-Reliability Industries

Construction can draw valuable inspiration from other industries that operate in high-risk, high-consequence environments. Sectors like nuclear power, military aviation, nuclear submarine operations, and offshore oil drilling are often referred to as High-Reliability Organizations (HROs) because they have systems in place that achieve exceptional safety and consistency even under challenging conditions. While building a skyscraper or highway isn't exactly the same as steering a nuclear submarine or running an offshore platform, the underlying training principles of those fields can be applied to construction team development. This section explores a few key practices from high-reliability industries – such as immersive training, rigorous drills, and a culture of safety – and discusses how similar approaches can elevate training in construction.

**Relentless Drills and Simulations:** One hallmark of high-reliability training is constant, realistic practice for critical scenarios. Take the example of U.S. nuclear submarine crews. Every submariner goes through intensive training not only in their specialty but in damage control and emergency procedures for the entire vessel. New crew members must study and be tested on every system on the sub – even a cook or yeoman learns the basics of the reactor, navigation, weapons systems, and especially firefighting and flood control. Earning one's "dolphins" (qualification insignia) requires demonstrating knowledge well beyond one's normal duties. And the training never stops: "the crew constantly trains to handle various casualties, such as fire or flood," with drills built into the daily routine. A typical submariner's schedule might include 6 hours on watch, followed by 12 hours where a portion of that time is dedicated to training and drills – every single day. They simulate emergencies like an onboard fire, a leak, or reactor scram repeatedly so that if it ever happens for real, responses are second nature. Construction teams can adopt a similar mentality by incorporating regular drills for emergency and high-risk events. For instance, consider conducting mock drills for scenarios like a trench collapse rescue, a tower crane malfunction, or a severe injury on site. Everyone on the crew should know their role if such an event occurs (just as every submariner knows their station in a fire or flood). A commercial building contractor might hold a quarterly "jobsite emergency drill," where without advance notice they stage a scenario – say, a fire in the building – and evaluate how workers respond: do they know how to sound the alarm, use fire extinguishers, and evacuate in an orderly way? These drills reveal gaps in readiness and reinforce training in a visceral, hands-on manner. And it's not only about accidents; construction crews can drill for operational contingencies too. Imagine practicing a scenario where a critical piece of equipment breaks down: the team walks through the backup plan (perhaps rigging a temporary hoist if the crane is out, or using alternative methods to pour concrete if the pump fails). By war-gaming these "what if" situations, the crew builds collective confidence and problem-solving skills that pay off when real surprises happen.

**Simulator and Scenario-Based Learning:** Another practice from high-reliability industries is the use of simulators and scenario-based training. Commercial aviation and nuclear power both famously use simulators to train operators in normal and emergency conditions. Nuclear power plant operators, for example, spend a significant amount of time every year in a control room simulator that is an exact replica of their plant's control room. They run through everything from routine startups to worst-case emergency sequences. In fact, at Duke Energy's nuclear plants, once an operator is fully qualified, they are still required to attend training every five weeks, including simulator time, to keep skills sharp. Over a year, a nuclear plant worker might log 80–200 hours in training refreshers and drills. Construction is increasingly adopting this mode of training with technologies like VR (Virtual Reality) and equipment simulators. Large contractors and trade unions have started using crane simulators, excavator simulators, and VR modules for things like welding or fall protection training. These tools allow crews to practice in a realistic but safe environment. A VR simulation can, for instance, put a trainee in the driver's seat of a tower crane during high winds or simulate the process of performing a confined space rescue. The trainee can learn the correct responses and experience the scenario without real risk. While not every company will have high-end simulators on site, many trade training centers do, and even tabletop exercises or role-playing scenarios can be effective. A project manager might run a tabletop drill with the site supervisors: "It's midday, summer, we have a scaffold collapse on the west elevation. Let's walk through what each of you does in the first 10 minutes." This kind of scenario-based discussion can uncover weak points in your emergency planning or training that you can then address proactively (maybe the need for more people trained in first aid, or clarity on who calls 911 versus who secures the area).

**Borrowing HRO Principles:** High-reliability organizations also operate on certain core principles that can be instilled via training and leadership. Some of these principles include: preoccupation with failure (constant vigilance for what could go wrong), deference to expertise (empowering the most knowledgeable person in a situation, regardless of rank), a reluctance to oversimplify problems, sensitivity to operations (maintaining situational awareness), and commitment to resilience (being prepared to recover from setbacks). In practical construction terms, deference to expertise might mean training your teams that during a critical lift, if the crane oiler or an experienced ironworker sees a problem and yells "stop," everyone listens – even if a senior manager is giving orders. It means instilling a culture (through training scenarios and supervisor coaching) where anyone can speak up about a safety concern and be heeded. Preoccupation with failure can be encouraged by incorporating lessons from past failures (both internal and industry-wide) into training. High-reliability industries meticulously dissect near-misses and accidents to learn from them; construction teams should do the same. If there's a collapse or incident reported in ENR or OSHA bulletins, share it in safety meetings: "This happened on a job in Ohio; how can we make sure it never happens here?" Teach crews to always be asking, "What could go wrong next?" not to paralyze them, but to keep them alert. Commitment to resilience ties back to cross-training and drills – it's about having depth in your team and plans B, C, and D ready so the unexpected doesn't become unmanageable.

Examples of High-Reliability Training in Construction: Some forward-thinking construction organizations have already adapted HRO-like training practices. For example, certain deep foundation and tunneling contractors, whose work approaches the complexity of mining, conduct rigorous emergency preparedness training. Crews boring a subway tunnel under a major city might undergo specialized tunnel rescue training alongside firefighters, practicing how to respond to tunnel fires or worker entrapments as a combined team. Likewise, large industrial contractors working in petrochemical plants often coordinate with facility owners to practice plant emergency drills (like simulated chemical leaks or explosions) so that all contractors on site know how to react and interface with the plant's emergency response. Another arena where high-reliability training is evident is in prefabrication and modular construction factories – these blend manufacturing with construction. Some prefab facilities have adopted lean manufacturing training techniques, including regular "stop-the-line" drills, mistake-proofing exercises, and continuous improvement meetings, very much in line with high-reliability concepts of stopping work at any sign of trouble and iterating to improve processes.

Adopting high-reliability training might simply start with increasing the realism and rigor of your current training. If your orientation for new hires currently involves an hour of PowerPoint, consider adding a hands-on component: e.g. a new hire must perform a harness inspection and put on fall protection correctly as part of orientation, not just watch slides about it. If your weekly safety meetings cover generic topics, try occasionally running a scenario-based discussion as described above. Implement a "near-miss reporting" system if you don't have one, and train crews that reporting a near-miss (like a piece of falling debris that almost hit someone) is a positive, blameless action that will be used to strengthen safety measures. This mirrors the aviation industry's confidential incident reporting systems which have vastly improved airline safety by surfacing problems before they cause accidents. The payoff for adopting these high-reliability-inspired practices is huge: fewer accidents, better quality, and a team that can handle surprises with calm and competence. One study cited by the National Safety Council found that the hidden costs of fatigue and human error in construction (from injuries, rework, turnover, etc.) can cost a typical firm millions per year. High-reliability training directly targets those human factors.

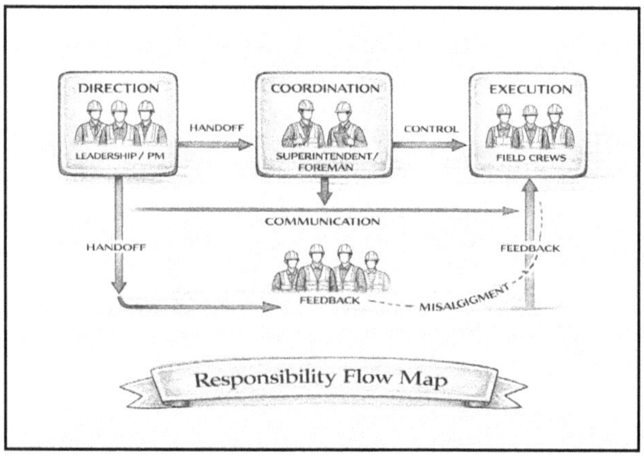


While construction will always have its unique challenges, learning from industries where failure is not an option can only improve our outcomes. Whether it's as simple as doing fire drills on site or as ambitious as building a full training simulator for crane ops, embracing these practices moves a construction organization closer to that high-reliability ideal. The third edition of this book emphasizes an applied, field-oriented tone for a reason – it's about real practices that work on real jobsites. High-reliability training concepts have proven their worth in nuclear subs sailing blind under the ocean and oil rigs braving hurricane-force winds; appropriately adapted, they can prove their worth on your projects as well, by ingraining a deeper level of preparedness and professionalism in your team.

## Strategies to Sustain Readiness and Versatility in the Field

Training and development is not a "flavor of the month" initiative – it must be sustained over the long term to truly transform a team. This final section focuses on practical strategies for maintaining a high level of field readiness and skill versatility among your workforce day in and day out, year after year. Construction crews often face ebb and flow in workload, high turnover in some trades, and the tendency to revert to old habits under pressure. To counter these challenges, leaders should aim to create a culture of continuous improvement and learning on the jobsite. The following strategies can help ensure that training gains stick, new skills are regularly exercised, and the team remains versatile enough to meet evolving project demands.

1. **Make Learning Ongoing and Habitual:** Encourage the mindset that learning doesn't only happen in formal training sessions – it's an ongoing part of work. One way to do this is by incorporating micro-learning opportunities into the regular schedule. For example, institute a weekly 10-minute skill spotlight during a toolbox talk: one week a foreman might quickly demonstrate a proper knot for rigging, another week an engineer might show how to use a new laser layout tool. These bite-sized refreshers keep skills sharp and introduce new techniques without taking much time. Some companies dub these "toolbox training bites." Over a year, that's 50+ mini lessons that cumulatively cover a lot of ground. Another idea is maintaining a jobsite reference library or digital resource hub – something as simple as a tablet or binder in the trailer with how-to guides, safety data, and equipment manuals that any crew can consult. When a worker encounters something unfamiliar, they should know that answers are readily available (and management should train them on how to

use these resources). The easier you make it for people to learn in the moment of need, the more they will do so rather than improvising or ignoring the issue.

2. Crucially, leadership must participate in ongoing learning as well. If superintendents and project managers demonstrate curiosity – say, attending new technology demos or asking craft workers to show them how a new tool works – it signals that learning is valued at all levels. As a result, crews are more likely to embrace training as a normal part of the job, not an imposition. As noted by one industry blog, "when learning becomes an ongoing habit rather than a one-time event, your team becomes more resilient, more innovative, and better prepared" for new challenges.

3. **Use the Off-Season and Downtime Wisely:** Construction often has cyclical periods (e.g. winter for road builders in cold regions, or between big projects for general contractors). Rather than viewing slow periods as unproductive, treat them as prime time for training and upskilling. Many union contractors already leverage off-season training through trade unions' skill improvement classes – non-union firms can do similarly by scheduling company-specific training or sending workers to manufacturer courses during these lulls. For instance, a paving contractor might send crews for equipment maintenance training every January when paving work pauses. A building GC might hold an annual company training day/retreat during a winter month, covering everything from new safety procedures to team-building exercises. These events not only improve skills but also reinforce team cohesion and the idea that the company invests in its people. Additionally, on projects, if there's a rain day or a waiting period for an inspection, supervisors can have a "training toolbox" ready – perhaps an exercise like practicing knots, doing a quick quiz on equipment load charts, or watching a 10-minute best-practice video – to turn idle time into learning time (instead of just sending everyone to early break).

4. **Track Skills and Plan Development Paths:** A practical tool for sustaining skill versatility is maintaining a skills matrix and individual development plans for crew members. We touched on the matrix concept in the cross-training section – listing who is skilled at what. Keep that matrix updated as people gain new proficiencies or certifications. Then use it not only to fill gaps but to plan who could advance next. For example, you might identify that an experienced carpenter has potential to become a foreman in the next year or two. With that in mind, you start giving him developmental tasks – maybe leading the crew for a day, or taking a course in reading project schedules – to prepare him for that role. Similarly, if an apprentice expresses interest in a certain specialty (say, a laborer who wants to become a certified crane rigger), create a path for them: arrange the necessary training and give them opportunities to practice under supervision. Not everyone will stay at your company forever, but those who see a career path and skill growth are much more likely to stay. It's a win-win: they gain expertise, and you gain a more capable and loyal employee. This aligns with advice on employee retention that stresses showing a future with the company through clear advancement milestones and mentorship. A versatile team is one where individuals feel their skills are always expanding; otherwise, versatility can be lost as workers stagnate or leave for growth elsewhere.

5. **Encourage Knowledge Sharing and Mentoring:** Tap into the knowledge already present in your team. Seasoned workers often have a wealth of know-how that isn't written anywhere. Create structured ways for them to pass it on. For instance, implement a buddy system for new hires – pair each new laborer or apprentice with an experienced journeyman for their first 90 days. That veteran not only teaches technical skills but also instills good work habits and safety awareness. Mentoring shouldn't be limited to craft skills; field leaders (like project engineers or young supers) can be mentored by senior project managers in the finer points of planning and leadership. Some companies start "lunch and learn" sessions where an employee with particular expertise (say, a concrete superintendent who's an expert in decorative finishes) gives an informal class to interested coworkers over lunch. These sessions can rotate topics and are a low-cost way to disseminate specialized

knowledge. Additionally, foster an environment where asking questions is welcomed. A newer employee should feel comfortable asking a veteran "Have you ever seen this situation before? How did you handle it?" – and veterans should feel proud to share, not threatened. When experienced workers teach others, it not only benefits the learners but often reinvigorates the teachers as well, giving them recognition as experts.

6. **Regularly Review and Refresh:** Construction techniques and standards evolve, and even time-tested methods can improve. Make it a practice to periodically review your procedures and training content. In team or safety meetings, take one procedure (perhaps how to set up a fall protection lifeline) and have the team discuss if there are better ways or if any new products are on the market. Solicit suggestions for improvement. This ties back to the feedback loops idea and continuous improvement. By involving the crew in updating practices, you not only get practical input but also renew their commitment to following those practices – people tend to support what they helped create. Moreover, require periodic re-certifications or refreshers for critical skills. For example, even if someone got certified as a forklift operator three years ago, you might have them do a refresher course or at least a practical evaluation each year to ensure they haven't picked up bad habits. Many safety trainings (like CPR/first aid, or OSHA 10/30) have expiration periods; keep a calendar or system to track when each person's credentials need renewing and schedule those proactively.

7. **Embrace Technology and Documentation:** In the modern construction landscape, digital tools can aid training retention. Use project management software or mobile apps to distribute updates, quizzes, or tips. Some companies send out a "Weekly Safety Quiz" via text or app – just a couple of questions that workers respond to, which keeps them thinking and may even introduce new knowledge. Others utilize learning management systems (LMS) to host short training modules that workers can complete from their phones. While field work is hands-on, even a quick mobile lesson on, say, identifying asbestos-containing materials, can be valuable before a renovation project. Another tech tool is digital skill tracking – for instance, scanning a worker's badge can bring up what equipment they're qualified on; this ensures versatility is utilized appropriately (no more wondering "can Joe drive the lull?" – the system confirms his training). On the documentation side, maintain SOPs (Standard Operating Procedures) for common tasks and ensure they are accessible. When crew members have a written reference (even one laminated sheet) on how to, for example, assemble a pipe scaffold or test a generator, it both reinforces training and allows self-directed refreshers if someone hasn't done the task in a while. SOPs also help sustain consistency as personnel change – the knowledge stays with the company, not just in individuals' heads.

8. **Celebrate and Reinforce Versatility:** Lastly, reinforce the behavior you want to see. When crews successfully adapt to a challenge because of their broad skills, acknowledge it. For example, if your cross-trained team saved the day by rapidly reassigning duties after a subcontractor no-show, share that story in the company newsletter or a meeting and praise the flexibility and readiness of the team. This positive reinforcement signals to everyone that versatility is valued. Some firms even implement incentive programs for training – like a skills milestone program where workers earn badges or bonuses for acquiring new competencies (e.g. a bonus for each additional equipment certification, or a company jacket for those who complete five advanced skills courses). This gamifies the development process a bit and keeps people interested in learning. In the field, when a crew smoothly handles an unexpected event (say sudden rain requiring quick protection of work and rescheduling), do a quick debrief after and highlight how their preparation and teamwork made the difference. These stories become part of company lore and motivate newer folks to engage in the same level of preparedness.

Sustaining field-level readiness and skill versatility ultimately comes down to building a learning culture where training is not a chore, but a continuous journey woven into everyday work. Companies that succeed in this find that their crews can take on more complex projects, respond to surprises without panicking, and require less outside help because the capability is in-house. They also tend to have better safety and quality records, since a well-trained and alert team catches issues before they escalate. As one construction management source put it, future-ready teams operate with confidence because they know what's expected, how to execute, and where to turn for information. By implementing the strategies above – ongoing upskilling, smart use of downtime, individual development planning, knowledge sharing, periodic refreshers, tech integration, and positive reinforcement – any construction leader can cultivate a crew that exemplifies readiness and versatility on the jobsite. In the long run, that translates to projects delivered with fewer hiccups and a reputation for excellence and reliability in the field.

Team training and development in construction is so much more than an orientation or a compliance checklist – it is an ever-evolving, active process that directly impacts a company's success. In this chapter, we reframed training as a field-oriented, continuous endeavor. We saw how cross-training builds a multi-skilled crew that can maintain operations through surprises and staff changes, leading to a more resilient and collaborative workforce. We emphasized the importance of feedback loops: using real project experiences and worker input to constantly refine training programs so they remain effective and relevant. We addressed the human factors of fatigue and focus, recognizing that keeping workers rested and attentive is a form of training in itself – one that can drastically reduce injuries and errors. We drew lessons from high-reliability industries like nuclear submarines and offshore drilling, illustrating that rigorous practice, simulations, and a culture of safety can elevate construction team performance to new heights. Finally, we outlined pragmatic strategies to sustain readiness and versatility, from integrating training into daily routines to leveraging slow periods for skill development, all with the aim of making learning a continuous habit on site.

The message is clear: investing in your team's development is not an optional cost – it's an essential driver of productivity, safety, and quality. A well-trained team that keeps learning will be able to adapt to new technologies, enforce safety proactively, and deliver high-quality work efficiently, even as conditions change around them. In contrast, a team that stops learning becomes brittle, prone to accidents, and unable to innovate or improve. Thus, the principles discussed in this chapter – cross-training, continuous improvement, fatigue management, high-reliability practices, and sustained skill-building – form a blueprint for creating a strong training culture. It's a culture where every crew member, from rookie to veteran, is continually growing and contributing at a higher level.

As you implement these ideas, start small if needed, but start somewhere: perhaps initiate a new feedback process after your next project, or cross-train two roles that have been pain points in the past. Maybe run an unannounced drill next month to test your emergency prep. Over time, these efforts compound into a formidable capability within your organization. Your teams will not only execute the work in front of them but will also be ready for the unexpected and prepared for the future. In the end, construction is a team endeavor – buildings and infrastructure are brought to life by people working together. By developing those people to be skilled, versatile, and ever ready, you set the stage for success on every project, big or small. This chapter underscores that the best construction companies build not just projects but people. And those well-built people, in turn, build the best projects.

**Operational Takeaways for Today's Job:**

- Integrate training into daily work. Assign a seasoned worker to a newer crew member as a buddy for the day on a complex task – on-the-job mentorship builds skills far better than leaving someone to figure it out alone.
- Use your toolbox talk as a mini-training session. Pick one practical skill or safety topic each week (like proper ladder setup or using a fire extinguisher) and have a 5-minute demo or discussion. These bite-sized lessons keep the team's knowledge fresh.
- Encourage questions and knowledge-sharing. Make it clear that anyone can ask "why are we doing it this way?" or suggest a different approach. When a laborer or field engineer proposes a good idea, acknowledge it and consider implementing it – this empowers everyone to learn and improve methods.
- Rotate tasks when feasible. Give team members opportunities to try new roles (under supervision) for a day, such as having a foreman teach a crew member how to do a simple site inspection. Cross-training not only prepares you for coverage if someone's out, but also increases overall team understanding of the project.
- Recognize and reward learning. If someone took a course over the weekend or got a certification, mention it to the team and congratulate them. Likewise, if a crew member applies a new skill effectively on the job, highlight that. A culture that celebrates development will motivate everyone to keep growing.
- Identify one critical role on your project that is single threaded and start a cross training plan this week.
- Add a standing agenda item to your weekly coordination meeting: "What did we learn last week that we must not repeat?"
- Schedule a short fatigue/focus check before any planned night or extended shift work and adjust staffing accordingly.

Chapter 15

# Problem-Solving and Adaptability

This chapter treats problem solving as a field skill, not an office workshop exercise. The focus is on how EPC construction operators think and act when drawings, weather, supply chains, or client demands change faster than the plan.

In the field, construction leaders need practical problem-solving frameworks that can be applied on the fly. Unlike formal office analyses, field solutions must be quick, adaptive, and involve the whole crew. This section explores how supers and foremen can recognize issues early, apply root-cause tools in real time, foster a no-blame culture, and turn daily problems into process improvements.

Even the best plans can break down in the field, so the first skill is early deviation detection. Frontline supervisors should continuously compare actual progress with the plan to catch slips or errors as soon as they occur. For example, a proactive manager will notice if drywall installation is slowing down before it derails successor tasks. The key is creating a culture where crew members feel safe flagging issues immediately, without fear. An issue ignored at 9 AM can snowball by noon; thus, daily huddles and site walks should explicitly ask: "What looks off-track today?" Rapid identification allows rapid correction (or escalation up the chain) before minor deviations become major rework or delays.

## Applying "5 Whys" and Fishbone in Real-Time (Without a Whiteboard)

Formal Root Cause Analysis (RCA) tools like 5 Whys and Fishbone (Ishikawa) diagrams are traditionally done in conference rooms, but field leaders can adapt them on the spot. Instead of a whiteboard session, a foreman can lead a quick 5 Whys exercise orally during a break. For instance, if a concrete pour was delayed, ask "Why?" repeatedly: why was the truck late? (Traffic.) Why was there no buffer for traffic? (No alternate route.) Why was no alternate route planned? (Lack of communication about road closure), and so on. By the fifth "why," deeper issues (like communication gaps or flawed scheduling) emerge. Similarly, a Fishbone diagram can be sketched on scrap paper or even verbally outlined. The crew brainstorms potential causes in categories (equipment, materials, methods, manpower) for a problem, then narrows down likely root causes. The goal isn't a pretty diagram – it's using the Fishbone mindset to ensure no cause is overlooked, even in a quick field discussion. These fast, simple tools help teams solve the real problem, not just the symptom, reducing the chance of repeat issues. Field use of 5 Whys and Fishbone should be practical and evidence-driven – encourage workers to share observations ("The generator failed because we skipped last maintenance"), then verify on the spot (check maintenance logs or the equipment). By integrating these techniques into daily problem-solving, crews develop a habit of attacking causes, not blaming symptoms.

Daily stand-up meetings set the tone for how problems are viewed. An adaptive leader steers the conversation from blame to systems thinking. Instead of asking "Who caused yesterday's delay?", they ask "What in our process allowed the delay to happen?". This subtle shift promotes a blame-free, problem-solving culture. Blame-based cultures kill improvement, as employees become afraid to speak up or try new approaches. Conversely, when leaders focus on systemic causes, the team feels safe to surface issues and fix process flaws. For example, if a task was done wrong, rather than scolding an individual, discuss how instructions or training might have been unclear – and how to prevent that in the future. In practice, a superintendent might say during the stand-up: "We welded the wrong beam section yesterday. Let's all learn from it. Were the shop drawings confusing? Do we need a checklist before welding?" This invites the crew to improve the system. Over time, daily stand-ups become quick RCA sessions where the team tackles root causes and shares lessons, instead of witch hunts. Field leaders also model this behavior by openly owning up to their mistakes or uncertainties – showing that the goal is learning, not punishment. The result is higher psychological safety, which research shows is critical for innovation and speaking up about problems. A no-blame stand-up thus turns daily setbacks into fuel for continuous improvement.

Every field problem is an opportunity to improve the overall process. Adaptive construction teams don't just put out fires; they capture the lesson and change how they work to prevent repeats. A practical method is keeping a running "issues log" or lessons learned journal on site. When a problem occurs – say a misaligned anchor bolt – the fix is made immediately, but then the foreman notes it down. Later, the team asks: Could a checklist or a smarter tool have prevented this? If yes, they implement that change for future jobs. This is essentially applying continuous improvement (Kaizen) at the jobsite level. For example, one company compiled all the issues from a project into a spreadsheet, identified root causes, and documented the solution and lesson learned for each. This "field improvement log" became a reference for new projects, so recurring issues (like missing parts or design clashes) were anticipated and avoided. Importantly, converting problems to improvements requires ownership: leaders must encourage crews to suggest better methods and then empower them to implement changes. It also means sharing improvements – what's learned on one site should be communicated to other superintendents and teams. For instance, if a certain make of tool is consistently failing in cold weather, a process change (different model or maintenance routine) can be standardized company-wide. By treating each field issue as a nugget of insight, frontline leaders build a culture of continuous improvement where processes evolve daily to get safer, leaner, and more efficient. In sum, today's mistake becomes tomorrow's new standard or training point, closing the loop between problem and progress.

## Resilience Under Unexpected Conditions

Construction is an arena of the unexpected – from sudden weather extremes to surprise regulatory changes – so project resilience is a critical trait. Resilience in construction means the ability of a project (and its people) to adapt to and recover from adverse conditions while maintaining safety and productivity. This section offers tactical leadership responses to stress events, delivered in the tone of a seasoned EPC construction manager coaching a team through live conditions. The focus is on keeping crews flexible and focused rather than panicked, and building buffers into plans without sacrificing standards or morale.

Project resilience is not abstract – it's the concrete ability to withstand shocks and keep going. A resilient project can absorb a delay or a disruption and find a way to still deliver. It involves contingency planning, risk management, and a mindset that setbacks are manageable, not fatal. One definition puts it plainly: resilience is the capacity to manage risks, respond to disruptions, and maintain project performance despite uncertainties. If an unexpected flood hits a site, a resilient project has systems in place (like pumps, schedule float, and alternate work ready) to recover quickly. Field leaders build resilience by anticipating "what if" scenarios ahead of time – effectively mental rehearsals of things going wrong – so when they do happen, the response is almost second nature. Resilience also means quick recovery: it's not that problems won't happen, but that the team bounces back with minimal impact. In practical terms, this could mean having backup suppliers for critical materials, cross-trained crew members who can step into different roles, and open lines of communication to resolve issues fast. In short, project resilience is adaptive capacity – the project's and team's ability to adapt, survive, and even thrive under conditions that were not part of the original plan.

## Psychological Flexibility: Teaching Crews to Adapt Without Panic

Resilience isn't just about plans and equipment – it's also a psychological game. In stressful, fast-changing situations, teams need psychological flexibility: an openness to accept what's happening and adjust behavior without falling into panic or paralysis. Leaders can coach crews in this by normalizing unexpected changes as a part of the job, rather than a failure. Psychological flexibility means people don't deny or avoid the stress; they acknowledge it and then pivot to problem-solving mode. For example, if a critical crane goes down, the crew acknowledges it's a big problem but immediately starts focusing on what can be done (Can we rent a replacement? Reassign crew to another task meanwhile?). To cultivate this, effective managers train teams with small drills: "What would we do if… (this machine failed, or that delivery didn't come)?" By brainstorming contingency actions regularly, crews build a mental muscle to handle surprises calmly. Leaders also model calm adaptability – if the superintendent stays composed and says, "Alright, Plan A fell through, let's move to Plan B, here's what we'll do…", the crew sees that reaction and mirrors it. It's important to address fear and uncertainty directly: field leaders can remind teams that it's okay to feel anxious when the unexpected strikes, but then refocus everyone on actionable steps. According to experts, highly resilient teams have a culture of candor and humility – they talk openly about problems and ask for help when needed. This psychological safety net prevents panic. A crew that trusts each other and communicates will adapt far faster to surprises, because they're solving it together rather than freezing up individually. In summary, teaching psychological flexibility means encouraging a mindset that change is not a crisis, but a challenge to be met with clear heads and collective effort.

For daily use in the field, simplify these tools: one sheet, one problem, five Why questions, and one concrete change you will test tomorrow. The goal is not a perfect diagram; it is to move one level deeper than 'because they made a mistake.'

## Creating Space for Improvisation Without Abandoning Standards

Improvisation – thinking on your feet and trying something unplanned – is often how field teams overcome sudden challenges. However, unbridled improvisation can introduce safety or quality risks, so leaders must create space for improvisation within boundaries. This involves empowering crews to make on-the-spot decisions when needed, but also establishing that certain lines (safety rules, engineering tolerances) are not to be crossed. An effective approach is: "Be creative, but check if it's safe and inform the team." For instance, suppose a particular scaffold part is missing; a crew might propose a workaround using available materials. A resilient leader will listen and evaluate: if the improvised solution meets safety standards and achieves the goal, they give the go-ahead (and perhaps inform the safety officer or engineer as required). If it violates a critical standard, they'll steer the team to another solution. Improvisation thrives when teams know the priorities. In a crunch, they should know that safety and quality trump schedule. A good mantra is "find a way, but do it safe and sound." By reinforcing that, leaders ensure nobody cuts a life-critical corner in the name of improvisation. Notably, resilient organizations practice improvisation in low-stakes scenarios (like tabletop exercises or "fire drills"). That way, when a real unstable situation hits, the habit of creative thinking is already there. Research on resilient people shows they have an "uncanny ability to improvise" – a creativity that lets them devise new solutions under pressure. Field leadership can cultivate this by celebrating crew members who come up with clever fixes that do uphold standards. For example, a foreman might praise a worker who suggested using a readily available beam as a temporary brace when a formwork element failed – as long as it was properly secured and engineered. The message is: we value your ideas and initiative, as long as they keep everyone safe and the work sound. This balance keeps the project adaptable but never reckless.

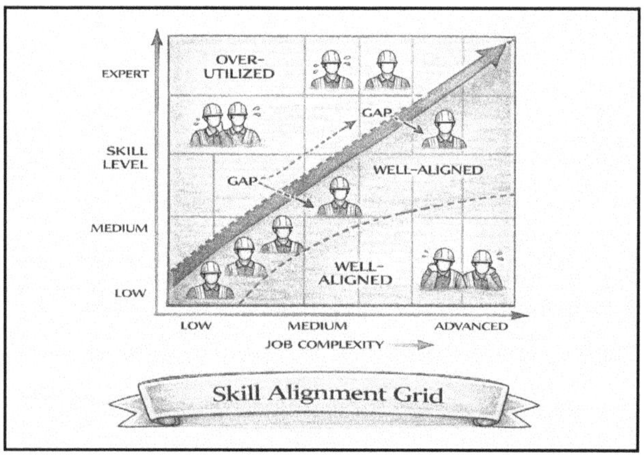


## Equipment Downtime, Weather Delays, and "No Plan B" Scenarios

Some of the most stressful unexpected conditions include critical equipment downtime, sudden weather delays, or any scenario where the original plan has no obvious Plan B. A crane breaks down with a big lift scheduled, or a flash flood washes out your excavation – now what? The worst time to figure out Plan B is in the moment of crisis, which is why good field leaders invest in contingency planning ahead of time. In fact, seasoned EPC project managers know that if you only have Plan A, you effectively have "no plan B in place for how to deal with crises". So practically, for each major project risk (key equipment, weather-sensitive work, one-shot operations), the team should brainstorm at least a rudimentary backup. For example, "If the main generator fails, can we rent one within 2 hours or bring a backup on-site?" or "If heavy rain hits during our concrete pour, do we have tarps and pumps ready?". By considering these in advance, the project avoids complete paralysis when something goes wrong. However, what about truly unforeseen scenarios? This is where prioritization under pressure comes in. If there's no Plan B on the shelf, the leader must quickly triage: What absolutely must happen now, and what can be deferred? Let's say an unexpected regulatory stop-work order lands (perhaps a permit issue). The project manager might immediately shift crews to other productive tasks away from the affected area, preserving progress elsewhere while management resolves the permit. The ability to "flex the schedule" on the fly is a hallmark of resilient field leadership – you rearrange sequence to keep people busy and project moving in some fashion, rather than standing idle. Communication is also crucial in no-Plan-B moments: the team and stakeholders should be informed honestly but confidently. For instance, "A freak equipment failure has occurred. Here's what we're doing in the interim, and here's our plan to get back on track." By communicating early and with a solution-oriented tone, leaders prevent rumor mills and maintain the crew's confidence that management is on top of it. In summary, avoiding "no Plan B" scenarios is ideal (through foresight and contingency planning), but when they do happen, improvise methodically: pause, prioritize critical needs, deploy resources to Plan B or C (even if just devised), and keep everyone informed.

On EPC jobs, define a simple rule set for improvisation: crews may improvise within defined boundaries (means, sequence, access) but may not improvise around engineered elements, design intent, or high energy controls without written approval. Make that rule explicit in pre task briefs.

Scenario Planning for Supply Chain Disruptions

Supply chain disruptions have become a frequent source of turmoil in construction projects – materials stuck on a ship, a sudden shortage of a critical component, or a vendor going bust. Field leaders can't just leave this to procurement departments; forward-thinking field leadership plays a big role in anticipating and mitigating these issues on the jobsite. Scenario planning means proactively asking "What if material X doesn't arrive on time?" and developing fallback plans. This section covers identifying which materials are high-risk, understanding the tiers of your suppliers, planning substitutions, communicating delays honestly but tactfully, and leveraging design-phase strategies like design-assist to avoid crises. It's not just about reacting to delays; it's about building buffer strategies and transparency so that the project can absorb supply chain shocks without eroding team or owner confidence.

## Identifying Vulnerable Materials and Long-Lead Items

Not all materials are equal in terms of risk. Early in a project, the team should identify long-lead items and any materials with complex supply chains or limited suppliers. Long-lead items are those that take many weeks or months from order to delivery (think custom curtain wall panels, specialty HVAC equipment, generators, etc.). During the planning phase, flag these items clearly in the schedule and procurement plan. Work closely with suppliers to get historical lead times and variability – if a certain elevator usually takes 20 weeks, assume worst-case 25–30 weeks in your plan. Once identified, treat these items with special care: initiate their procurement early (perhaps even before final design is 100% if possible), track their progress diligently, and have contingency actions ready. Modern projects often maintain a risk register where long-lead items are listed with their risk level and mitigation plans. For example, a project might note "Custom steel trusses – 5 month lead time, Mitigation: finalize design early Q1, place order by Feb 1, expedite option with fee available, alternative supplier in country X available if needed." By surfacing these vulnerabilities early, you can allocate float in the schedule (buffers) specifically for them. A good practice is also to tag these items in the schedule software so that any shift in their delivery immediately shows an impact. In short, scenario planning starts with knowing what could hurt you the most if delayed – once you know the weak links, you can fortify them.

## Tier 1, 2, and 3 Supplier Mapping and Why It Matters on the Jobsite

Supply chains have layers: Tier 1 suppliers (who sell directly to you, e.g. a local distributor), Tier 2 (who supply your supplier, e.g. a factory), maybe even Tier 3 (raw material sources or sub-components). On the jobsite, these tiers matter because a problem at a far-away tier can bite you unexpectedly. For instance, you may have a contract with a local concrete supplier (Tier 1), but if their cement comes from a single mill (Tier 2) that has a breakdown, your project is affected. Forward-looking teams do some supplier mapping: understand where critical materials are coming from in the chain. Ask vendors about their supply chain – do they have multiple sources? If your project relies on a single Tier 2 manufacturer for something like custom glass, that's a vulnerability. Why it matters on the jobsite is that it influences contingency plans. If you know your HVAC units come from overseas from one factory (Tier 2), you might plan extra shipping time or even have a backup unit from another brand spec'd just in case. Also, if there's industry news (say, a sudden aluminum shortage or a strike at a port), you can quickly assess if your tiers are impacted. Practically, project engineers or procurement leads can create a simple map or list of key materials showing Tier 1 (immediate supplier) and notable Tier 2 (manufacturer, origin, etc.). This mapping matters to the field because when delays happen, field teams often coordinate the workaround. For example, if a Tier 2 delay hits, the field might resequence work to wait for that item last or install temporary alternatives. Knowing the tiers also helps communication: you can explain to the client why something is delayed ("The factory in Germany had an outage, our hands are tied for two weeks, but here's what we're doing meanwhile..."). In essence, tier mapping = no surprises – it's part of scenario planning that extends your vision beyond just your immediate vendors to the whole supply pipeline.

## Substitution Planning: What Can You Swap Without Starting Over?

A hallmark of adaptive construction teams is having a Plan B for materials – i.e., pre-thought-out substitutions. Substitution planning asks: If Material A becomes unavailable or heavily delayed, what's Material B (or C) that could be used instead with minimal rework or re-design? It's best to identify these before you need them. During design or pre-construction, for each critical material or system, ask the design team: "Is there an alternate we can accept? How would it affect cost/schedule/quality?" Many specifications already list "or equal" products – make sure those are viable backups, not just nominal. For instance, if a certain type of fireproofing material is specified, ensure another brand could be used if supply runs short, and get that approved in advance. Effective substitution planning can save weeks when disruption hits. A real-world example: a project's custom light fixtures were delayed overseas, but the team had pre-checked a domestic supplier's similar fixture. They got quick approval from the owner to swap, avoiding a long wait. Leaders should evaluate substitution impacts carefully: swapping should not mean starting over on design or ripping out installed work. The best substitutes are drop-in replacements (same dimensions, performance, etc.) or those requiring only minor adjustments. Scenario planning here might involve mock runs: "What if the specified tile from Italy is two months late? Could we use a local tile temporarily or in lesser areas and install the Italian tile later?". By thinking these through, you may even do things like pre-order a small quantity of backup material (if affordable) or at least have supplier quotes and samples ready to go. Another tactic is writing construction contracts to allow minor material changes with streamlined approval, so you're not legally stuck. Keep the client and designers in loop – their buy-in on potential swaps ahead of time is crucial. The crew on site should also know that if they run out of Material A, there is a plan to shift to Material B (maybe in a different area or temporary use). In sum, substitution planning is about agility – finding what can be changed in mid-stream with minimal pain. It's far better to swap a component than to sit idle for weeks; a slight spec adjustment is often an acceptable price for keeping a project on track, as long as you maintain the required standards and owner's trust.

## Communicating Delays Without Eroding Confidence

Bad news is inevitable – what matters is how you communicate delays to your team and stakeholders. Field leaders must walk a line: be transparent about issues, but also project confidence that there's a plan to handle them. Hiding delays until the last minute erodes trust; but constantly crying wolf can erode confidence too. The best approach is a proactive, solution-oriented communication style. As soon as a significant supply chain disruption is known (and verified), call it out: "Team, the steel delivery from Mill X is delayed by 3 weeks due to a production issue." Then immediately follow with how you're addressing it: "We've already contacted an alternate supplier and are resequencing the decking work to keep other trades moving." By framing the problem and the mitigation together, the workforce and client hear that yes, there's a challenge, but leadership is on top of it. Regular updates are key. If a delay is ongoing (e.g., a supplier strike with unknown end), don't go silent – give weekly or even daily briefings on what's being done (even if the update is "no change, still waiting, Plan B is ready if it goes on another week"). This openness actually builds credibility. It's also important to empathize: acknowledge the impact on crews ("I know this puts pressure on your tasks, and I appreciate everyone's flexibility while we sort it out"). Internally, encourage the team to voice concerns – sometimes a labor crew might say "if material X won't come, we can't do Y," which you need to hear and address. Honesty is crucial with owners too, but avoid alarmist language. For example, rather than "We have a major delay, everything's messed up," say "We are experiencing a delay in receiving the generator. We're adjusting the schedule and exploring expedited shipping. We'll keep you posted and aim to make up time in other areas." The owner's confidence in the project can actually increase if they see competent handling of adversity. Another tip: use data and visuals in communication. Show the updated schedule with impact and recovery plans, or a dashboard of material status. When people see a clear plan, they remain confident in the leadership despite the delay. Finally, maintain a tone of earned optimism – not sugarcoating the issue, but expressing confidence in the team's ability to overcome it. Something like, "This is a challenge, but we've tackled challenges before. Our plan is sound and we'll get through this together," can keep morale steady. In short, communicate early, often, honestly, and with a plan – that's how you keep confidence intact even as you deliver bad news.

## Design-Assist and Early Procurement: Preemptive Problem Solving

One of the best ways to solve supply chain problems is to prevent them from happening. Two strategies for this are Design-Assist and early procurement, which often go hand-in-hand. In a Design-Assist approach, key subcontractors or suppliers (especially for complex systems like mechanical, electrical, facade, etc.) are brought into the project team early during design development. Their expertise helps shape the design in a way that's easier and faster to build. Critically, they can also identify long-lead items early and initiate their procurement before the final design is 100% complete. This means that while the architects and engineers are finalizing drawings, the specialty contractor is already locking in the order for, say, the air handlers or switchgear that have huge lead times. By the time the project breaks ground, those items might already be in manufacture. Early procurement through design-assist can shave months off waiting for critical components and provides cost certainty sooner. For example, if a project requires a custom generator, under traditional process you'd bid it out and order perhaps midway through construction – and then be at the mercy of a 6-month lead time. Under design-assist, the electrical subcontractor is hired during design and can place the generator order as soon as the capacity and spec are roughly known, overlapping that procurement with ongoing design and other work. This proactive approach mitigates risk from volatile pricing and availability – you're ahead of the crowd by ordering early, and if prices are rising, you've secured a slot. Another benefit: the subcontractor being involved in design means the product selected is one they know they can get or have relationships with the supplier, reducing surprise unavailability. Design-Assist also improves resilience because it's a collaborative delivery method – issues that do arise can be solved jointly by the design and construction team rather than thrown over the fence. For field leaders, leveraging design-assist means you likely have fewer "firefighting" episodes down the road, because many problems were anticipated and resolved in the preconstruction phase. If you have influence in project setup, advocate to bring key trades on board early for this reason. Owners are often receptive because it de-risks the project. Early procurement isn't limited to design-assist scenarios either; even in traditional projects, if you identify a critical item, you might ask the owner to allow an advance purchase before the rest of the project contracts – a tactic sometimes called "early release packages". In summary, preemptive problem solving via design-assist and early procurement is about tackling the supply chain challenge at its source: get the right people in early, order critical stuff early, and integrate procurement thinking into design. This way, by the time you're in the field, you're not just crossing fingers that items show up – you've actively managed those outcomes long before they could become an issue.

For EPC construction operators, the practical step is to keep a live list of long lead, single source, and safety critical items and review it weekly. Any change to those items should trigger a formal MOC, not an informal swap on the loading dock.

## Case Applications of Adaptive Leadership

Theory is important, but nothing drives lessons home like real field examples of adaptive leadership in action. This section showcases scenarios where construction leaders had to lead through ambiguity and change, and the tactics they used to keep their projects on track. Each case illustrates a different facet of problem-solving and adaptability on the job.

(The case studies and narratives would be presented here, demonstrating the principles in real situations.)

Have a backup plan ready for critical activities. Discuss "Plan B" and even "Plan C" before you start something important. For example, if a delivery is late, can you shift to another task? If a tool breaks, is there a spare? Thinking through alternatives ahead of time means you can pivot smoothly instead of stopping work.

Keep your cool and set the tone when things change unexpectedly. Whether it's a sudden storm or a design flaw discovered, your crew will follow your lead. Stay calm, communicate the new plan clearly, and roll with the punches – a composed response turns a potential crisis into a manageable detour.

After overcoming a challenge, debrief and adjust. Take a few minutes at day's end to discuss with the team: What did we learn from that surprise? Do we need to change anything going forward (like stocking an extra tool or altering a procedure)? Lock in the lessons from each problem so you're stronger for the next one.

**Operational Takeaways for Today's Job (EPC CMs):**

- Use a quick 5 Whys on one recurring field problem this week and commit to one system change instead of another reminder.
- Before approving an improvised method, ask: "Does this change touch structure, energy, or life safety?" If yes, pause for written review.
- Build and maintain a simple "red list" of long lead and safety critical materials; require MOC before any substitution.

Chapter 16

# Technology and Tools

Modern construction sites are awash in digital tools and platforms. From 3D models to smart sensors, technology promises to make field operations safer, faster, and more efficient. In integrated Engineering, Procurement, and Construction (EPC) projects especially, digital workflows tie together design, procurement, and field execution – boosting productivity but also introducing new risks around cybersecurity and digital reliability that must be managed. Yet every new tool brings both opportunities and challenges for the crews and managers on the ground. This chapter explores the key technologies shaping field work today – and weighs their real-world "blessings and curses" as experienced by field leaders. We maintain a critical eye on how these tools are actually used (or not used) in the field, distinguishing between office-centric systems and those that truly serve front-line teams.

## BIM, ERP, Scheduling, and Cost Control Platforms

Building Information Modeling (BIM): Digital 3D/4D modeling has become a cornerstone for planning and coordination. In the office, BIM is used by design and VDC teams to detect clashes and sequence complex builds. In the field, its value is more practical: superintendents and foremen pull up BIM models on tablets to visualize installations and confirm dimensions before work. For example, a foreman might check a BIM model to ensure a duct will clear a beam before crews start cutting. The advantage is clearer understanding – crews can see exactly how pieces fit, enabling more confident decisions and reducing errors. The challenge is making BIM accessible: field leaders often rely on simplified viewer apps or printed 3D snapshots, since they aren't BIM experts themselves. A model is only useful if it's up-to-date and easily viewable on site. Field teams have embraced BIM when it helps solve immediate problems (like avoiding rebar clashes), but they have little patience for clunky software or models that don't reflect on-site reality. In short, BIM's blessing is better insight into the "invisible" details of construction, while its curse can be information overload or impractical detail if not distilled for field use.

**Enterprise vs. Field Management Systems:** Large contractors often use enterprise resource planning (ERP) platforms to integrate scheduling, cost control, procurement, and more. Examples include heavy office-driven systems like SAP or Oracle for finance and materials management. However, from a field leader's perspective, traditional ERP tools offer little day-to-day utility. Site supervisors rarely log directly into an ERP to find out what to do next – those tools are geared toward accounting and high-level oversight. In practical terms, ERPs might feed the field a budget or a labor report, but they're not where a superintendent turns to solve an immediate crew or equipment issue. Because of this, many companies are heavily de-emphasizing ERP at the crew level, or interfacing it through friendlier apps. For instance, a foreman might submit time cards or receive purchase order approvals through a simple mobile interface without ever seeing the complex ERP behind it. The trend is to hide the "back-end" complexity of ERP from field users, because it's not relevant to pouring concrete or erecting steel on a given day. Instead, construction teams rely on more field-focused software for operational needs. Put simply, when a superintendent has to make a call on site, they're looking at conditions on the ground or a user-friendly app – not digging through an ERP screen.

**CMMS and Specialized Field Platforms:** In place of ERP, field operations lean on Computerized Maintenance Management Systems (CMMS) and similar tools that address on-site needs. A CMMS focuses on equipment and maintenance – for example, tracking service schedules for cranes and generators, logging repair tickets, and alerting the team to preventive maintenance. Field equipment managers use CMMS platforms to ensure machines stay safe and operational (e.g. scheduling the required 250-hour service for an excavator). Unlike a corporate ERP module, a CMMS is used directly by those maintaining and dispatching gear in the field, making it far more relevant on a daily basis. These systems help avoid downtime and accidents by keeping the nuts-and-bolts assets in shape. Similarly, many projects use specialized project management suites (like Procore, Autodesk Build, or Trimble ProjectSight) which include field-friendly modules for daily logs, RFIs, and cost tracking. The key is that these platforms are designed with the jobsite workflow in mind – often mobile-friendly and streamlined – rather than the dense data entry screens of an ERP. Field leaders appreciate tools that let them get information in or out quickly, without navigating a corporate database. The blessing of such systems is a clear, purpose-built process (submit a work order, update a log, etc.), but the curse is that they must still interface back with the central systems. If not set up well, they can lead to double-entry or data gaps between field and office. Successful deployment means picking software that speaks the language of both camps (field simplicity with office integration).

**Scheduling and Planning Tools:** Scheduling software bridges the office–field divide in a unique way. At the top level, planners use powerful scheduling programs (Primavera P6, Oracle P6, or similar) to develop the master schedule and track overall progress. These tools integrate with cost control to give management a big-picture view of timelines and budgets. On site, however, no one pulls out a Primavera Gantt chart when deciding tomorrow's crew assignments. Field supervisors instead rely on lookahead schedules (e.g. a two-week plan on a whiteboard or spreadsheet) that translate the master schedule into actionable chunks. Often, superintendents use simpler tools – even Excel or printed bar charts – to communicate the plan to foremen. The reality is that while an office scheduler maintains the fully linked critical path schedule, the field often uses a parallel system of short-interval schedules and daily huddles to actually get work done. This doesn't mean digital tools aren't used: many teams share lookahead schedules via Smartsheet or similar collaborative apps to keep everyone updated in real time. Smartsheet, for instance, has gained popularity for field scheduling because it's basically a spreadsheet with live updates – something crew leads find intuitive. The blessing of modern scheduling tech is that changes made in the office (e.g. a delivery delay) can be instantly communicated to the field, avoiding surprises. The curse is that if the tools are too complex (as MS Project often was, which is why many teams are phasing it out), field adoption drops. The lesson learned in many organizations is to simplify the user experience for field planning: give field leaders just the information they need, in a format they can tweak on site, without bogging them down in software training. In practice, that might mean the master schedule lives in a complex system maintained by a project engineer, but the day-to-day schedule is managed via a shared calendar or a lightweight app accessible on a smartphone.

**Modern Field Workflow Apps:** In recent years, a new class of dispatch and workflow tools has emerged, tailored specifically for field operations. These are lightweight, mobile-centric platforms that help manage forms, assignments, and data collection on site. They fill the gaps left by big enterprise systems, focusing on ease of use in the field. Notable examples include:

- *FastField (Quickbase)* – A mobile forms app integrated with Quickbase (a no-code database platform). FastField allows creation of custom digital forms (for inspections, work orders, safety checklists, etc.) that field staff complete on their phones or tablets. Through Quickbase, forms can be dispatched to specific people (e.g. "Fill out this excavation checklist today") and once submitted, the data syncs to the central database instantly. This combination is popular for replacing paper forms with real-time digital reports, and for assigning tasks to crews in the field with accountability. FastField's dispatch feature lets an office scheduler send a form to a foreman's mobile device, pre-loaded with job details – a far cry from the old method of faxing or calling in assignments.

- *GoCanvas* – A cloud-based mobile form and workflow platform used widely in construction. GoCanvas enables field teams to capture rich data (photos, GPS locations, signatures) on the go. A superintendent can open a GoCanvas app to complete a daily report or toolbox talk form, even offline, and submit it as soon as connectivity is available. The advantage is speed and accuracy: standardized templates reduce errors, and everything is timestamped and centrally stored. Crews appreciate avoiding end-of-day paperwork – instead, they fill out forms as they work, often with prompts to ensure nothing is missed. The challenge is managing yet another app; successful use depends on training crews and making the forms truly helpful (e.g. autofilling known info) so that using GoCanvas saves time over paper. When implemented well, GoCanvas streamlines workflows and provides managers with immediate visibility into field activities.

- *Smartsheet* – Originally a collaborative spreadsheet tool, Smartsheet has evolved into a flexible work-management platform that many construction teams use for tracking tasks and schedules. Field leaders might use Smartsheet to log progress or assign crew tasks in a simple grid that everyone can see. It's less specialized than the dedicated construction apps, but its familiarity (rows and columns) makes adoption easy. A project engineer could set up a Smartsheet for daily crew assignments or material delivery tracking; superintendents and foremen can update it from their phones as tasks are completed or issues arise. The blessing here is transparency – everyone with access sees updates in real time, preventing miscommunication. A possible curse is that without the guardrails of a purpose-built app, Smartsheet relies on disciplined user input; if people don't update their line items or if the sheet gets overly complex, it can become messy. Still, it offers a nice middle ground between complex project management software and informal texting/emailing of updates.

These modern field apps highlight a key shift: tools intended for field leaders must be straightforward, focused, and mobile-friendly. They are in many ways the opposite of monolithic ERP systems – and that is precisely why they've been enthusiastically adopted on jobsites. A general superintendent may never log into the company's ERP, but they might use a combination of a BIM viewer, a scheduling spreadsheet, and a mobile forms app every single day. The blessing of having these targeted tools is improved efficiency and data accuracy in the field. Forms get turned in on time, information flows faster, and less gets "lost in translation" because everyone's using the same platforms. The curse is the proliferation of apps and logins – the construction tech ecosystem can become fragmented if not managed well. As one industry expert observed, construction teams face "so many point solutions, so many workflows, and you have to do so many different logins," which can lead to tech fatigue. Field users don't want to juggle a dozen apps to do their job. Thus, a critical consideration when deploying these tools is integration: connecting field apps with each other and with office systems so that data only needs to be entered once and flows where it needs to. When done right, the mix of BIM, scheduling tools, cost control systems, and field apps can create a powerful digital ecosystem supporting field operations. When done poorly, it can feel like an overwhelming tangle, where technology intended to help ends up hindering. The remainder of this chapter delves into real-time connectivity, AI, and security – areas that further illustrate the balance between tech's promise and the practical reality on site.

## Real-Time Updates via Cloud & IoT Devices

**Cloud Collaboration and Instant Communication:** Modern projects use cloud-based platforms so that everyone – field and office alike – works off the same, latest information. For example, a cloud-based document system ensures that the drawings on a foreman's tablet are the current set; if a design change is approved in the office, the revised sheet is available to the crew immediately (with a notification to boot). Daily reports, RFIs, and safety observations entered in the field are uploaded in real time, allowing project managers and off-site stakeholders to see issues as they happen, not days later. This immediacy can prevent small problems from festering. Imagine a field engineer noting a conflict between two trades – she can snap a photo, mark it up on a cloud app, and tag the designers, all within minutes. Contrast that with traditional methods where the issue might not be formally communicated until a meeting the next day. Real-time cloud updates bless the project with agility: quick decisions, quick course-corrections.

However, constant connectivity has its curses. Field leaders sometimes feel tethered to their screens – an endless stream of emails, app notifications, and cloud alerts vying for attention. It's not unusual for a superintendent to receive dozens of updates a day from various apps (schedule changes, delivery notices, safety alerts, etc.). Information overload is a real risk. The crew still has to pour concrete or set steel, and too many digital distractions can actually hinder productivity or, worse, cause someone to miss a crucial warning amid the noise. Effective use of real-time cloud tools requires filtering and discipline: teams establish what "critical" notifications are (e.g. a structural change) versus those that can be checked periodically (e.g. an end-of-shift report). Field leaders often designate a person (say, a field engineer or project coordinator) to monitor the cloud feeds and flag what the superintendent really needs to know in the moment, allowing the super to stay focused on overseeing work. In sum, instant communication is powerful but must be managed, or it can devolve into digital chaos on site.

IoT Devices and Sensor Data: Perhaps the most transformative aspect of real-time tech in construction is the rise of IoT sensors and smart devices on the jobsite. These range from environmental sensors (monitoring temperature, humidity, dust, noise) to equipment telematics (reporting machine usage and health) to safety wearables (tracking worker locations or detecting falls). For example, concrete maturity sensors embedded in a pour can continuously send temperature and strength data to the cloud; instead of relying on guesswork or waiting for lab tests, the team knows exactly when the concrete reaches the desired strength to strip forms. Likewise, a vibration sensor on a temporary support structure might alert the team if unusual movement is detected, potentially averting a collapse by prompting immediate reinforcement. IoT devices essentially act as extra eyes and ears, monitoring conditions that humans can't easily watch 24/7. The blessing is obvious: better insight and often early warning of trouble. In one scenario, a site crane's telematics might report an overheating engine; the system could flag this on the supervisor's dashboard and automatically schedule a maintenance check for that evening – preventing a breakdown the next day during a critical lift. Or consider wearable proximity sensors that alert workers if they come too close to a hazardous zone (like the swing radius of a crane) – a buzzer or vibration can literally pull someone back to awareness before an accident occurs.

Yet for all these advantages, IoT in the field brings its own curses. One is alert fatigue: if sensors are too sensitive or there are too many of them, crews might start tuning them out. A classic example is an environmental monitoring system that issues constant alarms for minor threshold breaches – eventually, people may ignore the alarm, assuming it's "always like that," and miss the one time it's truly critical. There's also the issue of trust and adoption. Field personnel may doubt a sensor reading if it conflicts with their experience. We've seen cases where strain gauge or inclinometer warnings were dismissed because "we've done it this way for years and it's fine." In fact, history shows that having the data is not enough – someone must act on it. Before the 2018 FIU bridge collapse and the 2011 Indiana State Fair stage collapse, there were clear warning signs (cracking and high winds, respectively) detected and communicated, but they were effectively ignored on site. The decisions to proceed were not due to a lack of technology; they stemmed from human overconfidence and misjudgment. This underscores a crucial point: real-time IoT systems are only as good as the field culture that listens to them. A sensor that everyone ignores is as useless as no sensor at all.

Another challenge is connectivity and digital reliability. IoT devices often depend on stable networks (cellular, Wi-Fi, or mesh) to deliver data. Many construction sites are in remote or RF-unfriendly locations – think deep concrete basements, rural highways, or large steel structures – where signals can be weak or intermittent. Field teams quickly learn which tools have offline capability. For instance, a tablet-based drawing app that caches files for offline use is a lifesaver when you're working out of range; you can still pull up plans in the basement. Conversely, an IoT sensor that stops transmitting when the site internet blips can create false uncertainty (is the absence of data an "all good" or just a lost connection?). Smart implementation includes redundancy like local alarms (so an evacuation siren will sound on site even if the cloud link is down) and fallback procedures (if the digital inspection form won't load, have paper forms ready). Field leaders often develop a healthy skepticism: "Trust the tech, but have a backup." In practical terms, that might mean downloading critical information each morning and not relying solely on real-time streams when life safety is at stake.

When cloud and IoT tech work harmoniously, the result is a more transparent and proactive field operation. Problems are caught earlier, and everyone – from the craft worker to the CEO – can have a near-real-time window into the project's status. For example, many firms now use live dashboards that show key metrics like hours worked, incidents, percent of schedule complete, etc., all fed automatically from field inputs. A project executive can glance at the dashboard each afternoon and immediately call the project manager if something looks off, rather than finding out at the monthly meeting. The blessing here is better oversight and support: issues get resources thrown at them faster, successes get recognized sooner. The curse is that the field may feel constantly watched or under pressure to input data nonstop. There's a risk of turning construction into a fishbowl, where every minor hiccup is visible and stress-inducing for field staff who know management is watching the live feed. Maintaining trust between field and office is key – the data should be used to help, not to play "gotcha." Many organizations address this by focusing on using real-time data for coaching and support (e.g. "We noticed deliveries were late, do you need help expediting?") rather than blame.

Real-time cloud updates and IoT devices have undeniably improved how construction is managed in the field. They enable a level of situational awareness that previous generations of builders could only dream of. A foreman today can know the exact location of every piece of equipment, the status of every permit, and even the heart rate of workers (via biometric safety gear) at the push of a button. But with great power comes great responsibility – and complexity. Field teams must learn to navigate the sea of information, distill what matters, and maintain the human judgment to react appropriately. Technology can augment decision-making on site, but it does not replace the need for skilled leadership. The best field operations seem to strike a balance: they leverage real-time tools to enhance safety and efficiency, but they also set boundaries so that technology remains a servant to the project, not the master. The next section turns to one of the most hyped technologies of all – AI – examining how it is being cautiously introduced to enhance safety, and how field crews are responding to its promises and pitfalls.

## AI and Predictive Analytics in Safety Management

Artificial Intelligence (AI) and predictive analytics represent the cutting edge of construction technology – and they come with outsized expectations. In theory, AI could analyze vast amounts of project data and predict accidents or quality issues before they happen. It's an alluring prospect: imagine an AI system that, by crunching patterns of past incidents, weather forecasts, crew schedules, and countless other factors, could give a site a "safety risk score" each day, or warn that "scaffold collapse risk is high this week." Such capabilities are no longer science fiction; pilot programs and startups have been exploring them in recent years. But how does this translate to the messy, human reality of a jobsite? Field leaders are justifiably skeptical even as they're hopeful, viewing AI as a potential assistant on site – not an all-knowing oracle. Here we explore the blessings AI can bring – and the curses that have slowed its adoption on the ground.

**Current and Potential AI Uses:** Some real-world applications of AI in construction safety are already being tested. One example is computer vision systems using site cameras to automatically detect unsafe situations – say, identifying if a worker is missing required PPE or if a crane and personnel are in dangerous proximity. Another is predictive modeling: by feeding an AI with years of project data (accident reports, near-misses, productivity figures, etc.), companies attempt to have it flag when current conditions resemble "high risk" scenarios from the past. For instance, an AI might learn that whenever a project is 70% complete and simultaneously 2 weeks behind schedule with more than 10% crew turnover, the likelihood of a recordable safety incident jumps significantly. It could then alert management to heighten safety measures under those conditions. Predictive analytics are also used for equipment safety – telematics data combined with AI might predict that a particular bulldozer is likely to overheat and fail within the next 10 hours of use, prompting preemptive maintenance to avoid a fire or sudden breakdown.

The blessing of AI is its potential to catch subtle correlations humans miss. It's like having a tireless analyst combing through every data point from every project to tell you where lightning might strike next. Some project leaders report that AI-based safety recommendations have prompted them to intervene in ways they wouldn't have otherwise – for example, shifting a work crew's schedule because the algorithm spotted a pattern where working past 6 PM correlated with higher injury rates. In these cases, AI can act as a second set of eyes, an advisor that isn't swayed by "this is how we always do it." It brings a level of objectivity and breadth of analysis that humans alone can't achieve, especially as construction sites generate more and more data (from IoT, daily logs, etc., as discussed in 16.2).

**Now for the curse:** adoption of AI in the field has been slow, and not just because of technical hurdles – the human factor looms large. Many contractors express interest in AI, but few have fully deployed it. In fact, a 2025 industry survey found that roughly 45% of construction firms reported no AI implementation at all, and only about 1.5% were using AI on multiple projects or at an organization-wide scale. In other words, despite all the buzz, almost half the industry hasn't touched AI yet, and virtually nobody is using it everywhere. This gap between aspiration and reality speaks volumes. Field crews are often wary of "black box" recommendations. A superintendent might ask: "The computer says this scaffold is a high risk today – but why? I've got decades of experience telling me it's fine." If the AI can't explain itself in clear terms, it faces an uphill battle convincing seasoned professionals. Trust is essential: crews need to see that an AI system has credible insight, not just generic alarms.

There's also the question of data quality and relevance. AI is only as good as the data feeding it. If incident reports are not consistently reported or if they lack detail, the predictions will be off. Many jobsite variables are hard to quantify (e.g. crew morale or subtle site-specific hazards), so an AI might miss critical context that a human would catch by simply walking the site and sensing "something's not right here." Field leaders emphasize that AI is a tool, not a crystal ball. It can highlight patterns, but it can't capture everything happening on site – at least not yet. An experienced safety manager might use an AI-generated risk score as one input, akin to a weather forecast, but still rely on toolbox talks and on-site observations to truly gauge the day's safety climate.

Another barrier is integration into workflow. If AI insights come through yet another dashboard or app, busy field personnel may not consistently check it. One approach to mitigate this has been embedding AI suggestions into existing processes. For example, if a superintendent is filling out the daily plan in a scheduling app, an AI module might pop up a warning: "Tomorrow's forecast is 95°F and you have 4 roofers scheduled at 3 PM – heat stress risk is high, consider adding a water break." That kind of contextual nudge might be more effective than a separate AI report that no one reads. Even so, making sure the right people see and act on AI outputs is tricky. Some companies assign a safety engineer to interface with the AI system, translating its findings to the rest of the team – essentially a new role acting as the bridge between algorithm and field.

The culture of construction is another factor. This is an industry that prides itself on practical experience and intuition built on years of hard lessons. Introducing AI can be seen as second-guessing the experts or, conversely, as a way for higher-ups to monitor the field. There can be a perception (or misperception) that AI is a step toward replacing jobs or automating decisions that people used to make. To address this, leaders have to frame AI as a helper, not a critic. For example, some firms rolled out predictive safety analytics by sharing success stories – "Look, the AI helped prevent a potential accident by catching X – isn't that great?" – rather than using it to reprimand teams ("The AI says your site is high risk, what are you doing wrong?"). Field teams respond far better when the technology is positioned as an aid to them, not just a watchdog for corporate.

It's also worth noting that AI isn't infallible. There have been instances where predictive tools flagged risks that turned out to be non-issues, and missed others that did result in incidents. If an AI cries wolf too often, field personnel will tune it out just as they might an irritating alarm. The curse of false positives is real – it can erode confidence in the system. Conversely, if a major incident happens that the AI failed to predict, people may dismiss the whole concept ("what's the point of this thing if it didn't see that coming?"). So, managing expectations is key. Companies are learning to introduce AI gradually: pilot it, validate its recommendations against real-world outcomes, and adjust. In the interim, they keep the human in the loop for critical decisions. For instance, an AI might identify a crane lift as high risk, but it's ultimately up to the lift director and site supervisor to decide whether to postpone the lift. The AI can inform and suggest, but not command – at least at the current stage of maturity.

Despite these challenges, momentum is building. Contractors who have started with small AI applications (like optimizing equipment maintenance schedules or flagging contract risks) often report moderate success and then expand into safety or planning AI. The industry may be near a tipping point. The Royal Institution of Chartered Surveyors noted that while current implementation is low, most contractors do believe AI will significantly improve areas like progress monitoring and safety in the near future. The interest is there; it's the implementation that lags. We may soon see a shift from experimental to commonplace, especially as younger tech-savvy managers move into more senior roles.

For field leaders today, the takeaway is: stay open-minded but realistic about AI. If your project is piloting a predictive safety program, engage with it – provide feedback, help train it with better data, and see if it identifies something useful you hadn't considered. Use it as a second opinion or a conversation starter with your team ("Our risk score is high for falls this week; does that make sense given what we're doing? What can we do about it?"). At the same time, don't let the allure of AI dull your own situational awareness. A fancy dashboard is not a substitute for walking the site and talking to your crew. The blessing of AI is that it can crunch numbers and spot trends far beyond human capability; the curse is that it might lull us into thinking the computer is covering things we still need to check ourselves.

In conclusion, AI and predictive analytics hold great promise for making construction safer and more efficient. They are tools that, in the best case, can augment human decision-making by providing an early warning system derived from big data. However, in practice they are still in the early stages of adoption. Many field teams have yet to see any AI-driven input on their projects, and those that have are learning how to fit it into the traditional safety and planning processes. The successful deployments treat AI as a collaborative partner – one that learns from humans as much as humans learn from it. The less successful ones either shove AI outputs at crews without context, or ignore valuable AI insights due to mistrust or poor integration. The coming years will likely see AI move from novel pilot to standard practice, but its acceptance on the front lines will depend on proving its worth and respecting the judgment of the people it aims to assist. In the next section, we pivot to a related frontier: as we embrace more technology on site, how do we protect these digital systems and ensure they're reliable? We'll examine cybersecurity and digital reliability – topics that might seem "IT-centric" but are increasingly critical for field operations in a connected world.

## Cybersecurity & Digital Reliability

With construction processes becoming digitized and connected, the industry faces a new kind of risk that field leaders of the past never had to consider: cybersecurity and the reliability of digital systems. It's often said that "safety" in construction primarily meant guarding against physical hazards – falls, strikes, collapses. Now, there's a parallel need to guard against invisible threats like data breaches, ransomware, and system failures. This topic might sound abstract to the hands-on builder, but its impacts are very real. Imagine your project management software or BIM model suddenly becoming inaccessible in the middle of a critical operation, or a malicious actor causing your drone surveys to go haywire. Such scenarios have moved from theoretical to actual in recent years, as construction has become a target for cyber attacks. For EPC firms delivering turnkey projects, a cyber incident can ripple through engineering, procurement, and on-site operations – so robust digital risk management has become an essential part of project planning. In this section, we dissect the blessings of our digital connectivity (which we've lauded in earlier sections) and the curses in terms of vulnerabilities and reliability concerns – all from the perspective of keeping field operations running safely and smoothly.

Why Cybersecurity Matters on the Jobsite: Construction might not seem like an obvious bullseye for hackers – we're building bridges and hospitals, not running online banking sites, right? But the reality is that construction companies possess valuable assets: large financial transactions, confidential design data, personal information of employees, and project timelines that can be disrupted for leverage. Moreover, the industry's rapid digitization combined with historically lax cyber defenses has made it an appealing target. Recent surveys indicate that more than one-third of construction companies have seen an increase in cyber incidents like phishing attacks, data breaches, and ransomware in the past year. Field operations are directly affected when these incidents occur. Consider ransomware: if attackers encrypt your project files and demand payment, work can grind to a halt for days or weeks. In one documented case, a prominent construction firm's servers were hit by ransomware that crippled IT systems, delaying projects and forcing field teams to revert to manual processes while data was restored. It's not just a headache for the IT department; it's lost productivity, potentially missed deadlines, and even safety issues if, say, the latest safety inspection reports or equipment certifications become inaccessible.

Phishing and business email compromise schemes have also specifically targeted construction – for example, tricking someone into wiring a payment to a fraudulent account by impersonating a subcontractor. If a project engineer in the field office falls for such a scam, the project could lose a huge sum of money, impacting budgets and straining subcontractor relationships. These threats underscore that cybersecurity isn't just an "office problem." The field office trailer often has several computers, maybe a local server or network-attached storage, plus all the mobile devices and tablets used on site. Each of these is a potential entry point for attackers if not properly secured.

Vulnerabilities in Field Tech: The very tools that empower us – cloud apps, IoT sensors, tablets, and even heavy equipment control systems – can be avenues for cyber mischief or failure if not managed. Unlike a corporate HQ, a jobsite network might be a bit ad-hoc: maybe a 4G router in a trailer, a Wi-Fi network that all contractors use, and numerous personal devices connecting. This can be a curse for security. Shared passwords, lack of encryption, and personal devices plugging in can mean weak points. For instance, if a subcontractor's malware-infected laptop connects to the site's network, it could spread a virus that knocks out the digital blueprint station. Or think about drones and robotic total stations: these are often controlled via software – if someone maliciously interfered with them (through hacking their Wi-Fi link or the controller software), they could disrupt survey data or even cause a safety hazard (imagine a drone dropping out of the sky over a crowd).

Even without malicious intent, digital reliability is a concern. Construction happens in all sorts of environments – extreme heat, cold, dust, rain – and our devices sometimes fail us. Tablets can overheat in the sun or run out of battery. Network connections can drop unexpectedly (especially if everyone on site decides to stream a video during lunch on the same network!). Software updates, while meant to improve, can sometimes introduce bugs or compatibility issues that render a critical application unusable until patched. A field leader must have a contingency mindset for tech failures just as for any other risk. We often ask "what if the crane fails?" but now we also ask "what if our PlanGrid app fails?" For example, if the digital plan room goes down, do we have recent hard copies of drawings as backup? If the digital RFI system is offline, is there a way to get urgent information through (even if it's via a phone call or text as a stopgap)?

Practices to Mitigate Risks: Cybersecurity in the field starts with basic hygiene and training – which historically hasn't been a focus on jobsites. Field crews are now being trained not just in fall protection, but also in things like spotting suspicious emails or securing devices. A common blessing is that companies are extending their IT protections to field operations: laptops in the trailer will have antivirus and disk encryption, the site Wi-Fi is password-protected and segmented from the corporate network, and sensitive systems require two-factor authentication. These measures do add a bit of friction (and we've all heard grumbling: "I just want to open the darn file, why do I need a code from my phone?"). But they are the seatbelts of the digital world – a minor hassle that can prevent catastrophe. Field leaders should foster a culture where cyber safety is viewed as part of overall safety. One approach is to use analogies: just like you wouldn't leave the jobsite gate wide open at night for anyone to walk in, don't leave your work laptop unlocked and accessible or your passwords on a sticky note. Just as you check a scaffold's integrity, check that the firmware on the site's IoT sensors is updated (outdated software can be exploited by hackers).

Another important practice is regular backups and failsafes for digital data. Jobsite data should be automatically backed up to the cloud or an offsite server. If a tablet is lost or broken, no data should be permanently lost – at most, you replace the device and sync the data back. Many field apps now cache data locally and sync when they can; if they don't, procedures should ensure critical info (like that day's inspection results) are also recorded in a secondary way if needed. For crucial systems, having redundancy is wise – e.g. two internet sources (maybe a primary wired/cellular router and a backup hotspot) in case one goes down. During important operations (like a heavy lift coordinated through a software system), consider what you'd do if the system fails at a key moment – perhaps have a manual override or a designated "pause work" signal while the system reboots.

**Cyber-Physical Intersection:** As construction tech advances, the line between cyber and physical blurs. Consider automated equipment or site access control systems. A savvy attacker or even a virus could theoretically cause chaos – like disabling all your access badge readers (locking workers out or in), or feeding incorrect GPS data to machine control equipment (imagine an automated bulldozer getting bad coordinates). While such targeted attacks are rare, they highlight why digital reliability is a safety issue. It's not just about stolen data; it's about ensuring the systems we rely on don't lead us astray or fail at a critical moment. Field teams should validate important data coming from digital sources. For example, if your laser scanning equipment shows a layout that seems off, double-check with a manual measurement – don't assume the digital output is gospel. In one real case, a GPS unit's calibration error caused a crew to install anchor bolts several inches off; catching it required an old-fashioned tape measure verification. The motto "trust but verify" applies here: trust your tech tools to a point, but have quality control checks.

**Real Incidents and Trends:** Construction companies have already felt the sting of cyber incidents. A notable example is the ransomware attack on a large contractor that "temporarily crippled its IT infrastructure," as reported in industry publications. Projects under that company experienced communication breakdowns – emails and digital plan systems were down – forcing them to scramble with alternate methods. In another case, project data was stolen (breached) and the attackers threatened to leak sensitive project details if not paid, causing significant distraction to management. These are akin to unforeseen "digital accidents" that can disrupt operations much like a physical accident would. The trend is clear: as the sector digitizes, cyber incidents are increasing, and construction is no longer off the grid in hackers' minds. On the flip side, awareness is increasing too. Insurers and clients are now asking contractors about their cyber safeguards as part of project risk management. It's becoming part of doing business. For example, many EPC contracts now require contractors to present a detailed cybersecurity and data backup plan before work begins – digital safeguards are now written into the project playbook.

For a field leader, the practical steps to enhance cybersecurity and reliability are often straightforward: keep your devices updated (yes, those annoying update prompts matter – they often fix security holes), use strong passwords or passphrases and change them if there's any hint they're compromised, be cautious with unknown USB drives or devices (a free thumb drive found on site could be loaded with malware; better to hand it to IT to check), and report anything odd – like strange emails, ransomware messages, or systems acting weird – immediately to your IT support. Just as we encourage reporting a near-miss safety incident, reporting a near-miss cyber incident (e.g. "I almost clicked a suspicious link but realized in time") helps the organization improve its defenses. It's also wise to include contingency for digital failures in emergency planning: for instance, if an evacuation happens, have an offline list of all personnel on site that day (in case the digital system is unreachable) so everyone can be accounted for.

**Maintaining Balance – The Human Element:** A theme throughout this chapter has been that technology's effectiveness depends on the people using it. Cybersecurity is no different. The best firewall in the world won't stop an attack if an employee inadvertently gives away the keys (like logging in on a phishing site). So a big part of this is creating a mindset in field teams that cyber vigilance is part of their job. One superintendent started including a "cyber check" in the weekly site meetings – just two minutes to remind folks of things like "don't use public Wi-Fi without a VPN" or "there's a fake invoice email going around, be on the lookout." Little reminders go a long way in making it routine. It's similar to how safety culture was built up: constant, gentle reminders until it becomes second nature.

At the end of the day, the blessing of our connected construction world is tremendous efficiency and capability – we can do things faster and with more insight than ever. But the curse is a dependency on systems that must be safeguarded and kept running. A construction site now has to be defended not just against rain and theft, but against malware and network outages. Field leaders need not become cybersecurity experts, but they should be conversant in the basics and demand the same level of rigor in digital safety as they do in physical safety. This means planning for worst-case digital scenarios just as we plan for worst-case accidents. The two realms are intertwined – a cyber incident can quickly lead to physical consequences on a job, and vice versa (consider how a major accident could knock out communications systems, etc.).

Cybersecurity and digital reliability in construction operations is about ensuring that the high-tech tools we deploy remain available and trustworthy. It's a new layer of risk management that must be integrated into our operational mindset. The best field leaders embrace it: they champion secure practices, push for robust support from IT, and always have a Plan B if the tech goes down. By doing so, they turn the potential curse of cyber risk into just another manageable aspect of the job – maintaining the smooth running of the project in an age where the toolbox contains not just hammers and hardhats, but also iPads and IoT sensors. As we move forward, those who can blend construction savvy with tech savvy will be the ones to truly capitalize on the tools at hand, securely and reliably building the future.

**Operational Takeaways for Today's Job:**

- Start each day by updating and sharing critical digital information. Make sure everyone on the team is working from the latest plans, models, and schedules – a quick morning sync of tablets and cloud files prevents miscommunication later.
- Check your tech the way you check your equipment. Ensure devices are charged, data is synced or downloaded for offline use, and apps are up-to-date before work begins. A few minutes of tech prep each day can save hours of downtime if something crashes or loses connection.
- Trust but verify digital data. If a BIM model or sensor reading says everything is fine, double-check any critical details with your own eyes on site. Use digital insights to guide decisions, but rely on field experience and common sense to confirm them.
- Make cybersecurity part of your safety routine. Remind your crew to practice basic cyber hygiene – for example, not clicking suspicious links or plugging in unknown devices – just as you remind them to wear PPE or lock out equipment. Protecting your digital systems is part of keeping the site safe.
- Always have a backup plan for key digital tools. Assume a critical app or device will eventually go down at a bad time, and prepare accordingly. Keep important information (such as drawings, contact lists, and permits) accessible in an alternate form – whether as printed copies or a secondary system – so work can continue if technology fails.
- Use AI and analytics as a second opinion, not the final say. If an AI system flags a safety risk or scheduling issue, take it into account and discuss it with the team – but don't ignore your own observations and instincts. Let AI augment your decision-making on site, not override it.
- Streamline the digital tools your crew uses. Too many apps and platforms can overwhelm people and create confusion. Focus on the ones that add real value to field work, and integrate them so data only needs to be entered once. By keeping the tech stack lean and coherent, your team can spend more time building and less time battling software.

Chapter 17

# Quality and Safety as Cornerstones

Safety and quality are not optional add-ons in construction – they are foundational elements that shape every project outcome. This chapter explores how embedding formal compliance systems and a strong safety culture reinforces productivity and protects value. It emphasizes procedural compliance, the role of inspections and audits, and the necessity of independent verification. We will also examine why workers sometimes deviate from procedures and how leaders can enforce compliance effectively. In addition, we discuss how to focus quality control efforts on critical-path activities, where defects can cascade through the project. Throughout, we weave in practical examples and case studies to illustrate these principles in action.

## Proactive vs. Reactive Safety Approaches

Safety management can be broadly proactive or reactive. A proactive (or preventive) approach anticipates risks and builds safeguards into the work process before an incident occurs. In contrast, a reactive approach addresses problems only after they have happened. Proactive strategies include thorough planning, training, checklists, and audits designed to catch hazards early. Reactive strategies might involve investigations and corrective actions after an accident.

**The industrial record is clear:** organizations that invest in proactive safety measures – such as rigorous hazard analysis, stop-work policies, and regular training – suffer far fewer losses. By contrast, reactive-only cultures tolerate surprises and near-misses that eventually escalate into disasters. Every construction professional should strive to shift from "fixing problems after they occur" to preventing them in the first place. As one expert summary puts it: proactive safety "prevents exposure" to hazards before they materialize.

For many projects, safety still operates in a reactive mode: rules are revisited after an incident, procedures are rewritten after an investigation, and budgets expand only after a close call. Yet the pattern across failures is clear—warning signs almost always appeared long before the catastrophic event. Proactive safety is the deliberate choice to act on those warning signs early, to design risk out where possible, and to build checks into normal planning instead of waiting for the next accident to force change.

On a practical level, proactive safety shows up in pre task planning, before event critiques, and independent verification. Before high risk work, crews walk the plan, ask "What could go wrong?" and adjust sequencing, temporary works, or monitoring before starting. In design and planning, engineers and managers look for single points of failure, temporary conditions, and public interfaces, choosing more robust details or added redundancy rather than hoping that everything will go perfectly. Proactive teams treat near misses, minor defects, and equipment alarms as valuable information, not noise to be ignored.

Building a proactive culture means engaging the entire team in safety. For example, daily worksite inspections and pre-task briefings (often called toolbox talks) help uncover hidden risks before work begins. When these steps become routine, workers begin to welcome questions instead of resenting oversight, because everyone sees verification as a partnership toward safety, not an impediment to productivity.

Reactive safety still has its place—incident investigations, after action reviews, and regulatory reforms have all driven improvements after tragic failures like Willow Island, LAmbiance Plaza, and FIU. But a system that relies on learning only after people are hurt will always be one disaster behind. A mature project culture uses reactive learning to strengthen proactive practices: findings from incidents feed back into better pre task planning, more targeted audits, and clearer stop work criteria on the next job.

**One classic trap is complacency:** assuming "nothing will go wrong" because technically procedures exist. History shows that when people work under time pressure, are fatigued, or have incomplete information, errors and rule-breaking multiply. In such environments, workers may treat written procedures as mere guidelines, not requirements, and think "someone else will catch it." This mindset leads to dangerous drift away from safe practices. Proactive safety cultures, by contrast, institutionalize audits, hold-points, and training so that following procedures becomes second-nature.

Practical Takeaway: Cultivate a no-blame culture of verification. Encourage workers to speak up and pause work when anything seems unsafe. Use pre-task meetings, checklists, and independent spot-checks as opportunities to reinforce doing things right the first time.

## Embedding Audits into Culture

Audits and inspections – both formal and informal – are the tools we use to verify that safety and quality systems are actually working. An audit typically means a structured review (often by an outsider or management team) of safety performance and compliance, while an inspection is usually a more frequent check of the worksite or process (often by a supervisor or qualified safety person). Embedding these practices into daily routines keeps safety "top-of-mind."

For many teams, audits carry a reputation as events that happen to the job, usually when someone from the outside shows up. In high reliability work, auditing is not a rare event—it is a daily habit. The purpose is not merely to document compliance, but to confirm that the systems described in procedures, permits, and drawings are actually working in the field.

A robust audit culture uses several layers. Planned audits occur on a regular cadence—weekly or bi weekly—where supervisors and safety staff walk key areas against checklists tied to specific SOPs, temporary works, or high energy tasks. Spot checks by superintendents, foremen, or engineers reinforce that verification is everyone's responsibility, not just the safety department's. Peer audits, where one crew or trade reviews another's workface setup or documentation, bring fresh eyes and normalize constructive questioning.

However, an audit or inspection is only effective if it is taken seriously. Experience shows that when supervisors treat inspections as a checkbox exercise—"just paperwork" to satisfy regulations—hazards slip through. In many cases, companies have fallen into a pattern where inspectors "focus on getting through the checklist" rather than observing and verifying actual conditions. One industry observer notes, "When supervisors focus on 'getting through the checklist' rather than observation and verification, the system becomes blind to deviations." When workers believe "someone else has already checked the work," a dangerous vacuum of responsibility can form.

Instead, organizations should treat every inspection as an analytical process. Supervisors should walk the job with purpose: looking for non-obvious risks, asking questions, and confirming that procedures are truly followed – not just presumed done.

## Formal vs. Informal Inspections

- **Informal Inspections:** These are frequent, often daily or ongoing, checks usually done by supervisors, foremen, or the workers themselves as they perform tasks. They are quick and practical, aiming to catch obvious hazards early. For example, a crane operator might perform a brief check of his equipment and immediate work area before each lift. These informal checks are not necessarily documented unless an issue is found, but they serve to raise awareness and spot glaring problems before they grow.
- **Formal Inspections:** In contrast, formal inspections are planned, structured, and documented. They may be required by codes, done by certified inspectors or safety professionals, and follow defined checklists. Formal inspections often occur on a regular schedule or at required hold-points and generate official reports that must be addressed.

The key difference is that informal inspections keep safety front-of-mind day-to-day, while formal inspections provide documented assurance and compliance records. Both are needed. Informal inspections cultivate a vigilant mindset on site, whereas formal inspections provide independent verification and accountability. On many projects, best practice is for supervisors to conduct informal walkthroughs continuously, and for formal, documented inspections to occur at least weekly or per code requirements. Documenting findings – whether from formal or informal checks – is critical for learning. Anomalies caught on site should be logged as issues or non-conformances and tracked to completion.

**Practical Takeaway:** Establish a two-pronged inspection strategy. Require that every worker or supervisor does a quick pre-task safety check before starting work, and schedule formal inspections at key milestones.

## Enforcing Procedural Compliance

Having procedures and audits on paper is not enough – they must be actively enforced. Enforcement means ensuring that procedures are actually followed in practice, and that deviations trigger corrective action. In construction, deviations from procedure are a leading root cause of accidents. Yet workers often break rules for understandable reasons: schedule pressures, fatigue, incomplete understanding of the procedure, or a culture that downplays rules. For instance, one study found that people under time pressure, lacking full information or supervision, are much more likely to stray from prescribed steps. When staff are fatigued or overconfident, they may take "shortcuts" they justify as temporary, but which erode safety. Likewise, workers may follow their instincts if they have not been fully trained on a procedure; technology or automation can even create an illusion that a system is foolproof when it is not. In practice, deviations often arise not from malice but from gaps in training and communication.

Management must tackle these causes head-on. First, leadership must model compliance and make clear that procedures are mandatory, not optional. Tools like Stop Work Authority empower any crew member to halt work if a required procedure is being violated or is unclear. Regular supervisory reviews or audits should specifically look for common deviations (e.g. unscheduled concrete pours, lifting without proper bracing, skipping rebar cover checks) and address them immediately. Training is crucial: as one safety expert notes, “Systems fail not because workers choose not to follow procedures, but because they never fully understood them”. Therefore, ensure that training goes beyond “read the manual” – use hands-on practice, walkthroughs, and assessments so workers internalize each step. The industry is increasingly moving toward formal qualifications and re-certifications (for crane operators, welders, etc.) to prevent ignorance from leading to non-compliance.

Second, there must be consequences for willful or reckless violations. OSHA and other agencies now classify blatant disregard for safety procedures as serious offenses. Fines for willful violations can reach millions of dollars, and criminal charges are possible. For example, OSHA mandated national certification for crane operators after repeated lift accidents. What used to be a mere “good idea” in manufacturer manuals became an enforceable regulation, with legal penalties for non-compliance. Internally, construction firms need similar teeth: if a worker or supervisor knowingly skips a procedure, it should lead to disciplinary action, review of training, or removal from the task. It’s about creating accountability: workers must know they are responsible not just for output, but for following the process that ensures safety and quality.

Finally, use metrics and recognition to reinforce compliance. Track leading indicators (e.g. number of hold-point checks passed, drills conducted, near-miss reports submitted) and incorporate them into performance evaluations. Conversely, deviations found in audits should be analyzed, root-causes identified, and fixes implemented. Some companies have found success with a “just culture” that treats non-compliance as a learning opportunity – provided it was unintentional – but still enforces rules when necessary. Leaders should publicly acknowledge crews or supervisors who consistently follow procedures or catch deviations themselves. This shifts the culture from blame to shared responsibility.

Illustrative Example: The Big Blue crane collapse in Milwaukee (1999) shows what happens when procedures are ignored. The crane operator repeatedly lifted a 450-ton roof panel despite wind exceeding the crane’s safe limit. Supervisors had warned him, but schedule pressure overrode the rules. The crane tipped, killing three workers. In another incident, a Philadelphia demolition contractor directed workers to remove a load-bearing wall without an engineering review or building permit. Six people inside a store were killed when the wall collapsed. In both cases, the procedures (wind charts, engineering clearance, permits) existed on paper but were treated as obstacles to speed. The consequences were catastrophic loss of life and heavy penalties. These tragedies underscore that procedures matter – the “rules” are there because they prevent precisely the failures that occurred.

## Practical Takeaways:

- **Communicate Zero Tolerance for Deviations:** Make clear in meetings and on posters that procedures exist for safety; unsafe shortcuts will not be tolerated.

- **Strengthen Supervision:** Empower foremen and safety officers to enforce hold points. Consider "second person checks" where two qualified workers must sign off before critical steps (two-piece confirmation).
- **Reward Compliance:** Recognize teams that reach milestones with perfect safety/QC audits. For instance, hold a brief ceremony or give bonus points in quality incentives.
- P**romptly Correct and Document Deviations:** When a deviation occurs, document it as an incident (even if no one is hurt), fix it, and update procedures or training to prevent recurrence.

By combining strict enforcement (leadership, audits, consequences) with understanding (training, communication, and culture change), a construction project can greatly reduce willful or inadvertent non-compliance.

## Linking Safety to Productivity

Contrary to a common myth, rigorous safety and quality management enhance productivity rather than hinder it. Safety incidents, defects, and rework are among the largest sources of delays and extra costs on projects. Reliable, defect-free work on schedule increases worker confidence and pace. This is especially true on critical-path activities – those tasks that directly determine the project's completion date. On the critical path, a quality failure has two-fold impact: you incur rework cost and threaten the entire schedule. Thus, critical-path tasks warrant the highest level of quality control.

When safety systems fail—whether through rushed execution, unverified conditions, or ignored warning signs—the result is not a slightly faster project; it is total shutdown, massive rework, litigation, and, in some cases, the end of companies. The most expensive schedule delay is the one caused by an avoidable failure.

Linking safety to productivity begins with planning. Look-ahead schedules must treat critical verifications, inspections, and temporary works installations as real activities, not invisible tasks squeezed in around "production." Time for independent verification, equipment checks, and pre-task briefs is built into the plan and protected when pressure rises. Crews are rewarded not only for hitting dates, but for doing so without rework, incidents, or near misses that indicate corners were cut.

## Critical Path Quality Control Integration

Critical path activities need extra layers of assurance. Team planning meetings before a critical task should be more detailed. For example, before pouring a major foundation or hoisting a large structural component, assemble a focused preparatory meeting. Go through the drawings, tolerances, acceptance criteria, and sequencing in detail so everyone understands exactly what "done right" looks like. Develop a task-specific checklist or Inspection & Test Plan (ITP) that identifies all hold-points. For instance, if a large concrete pour is critical, discuss formwork setup, reinforcement placement, and concrete mix design in advance, and agree on specific surveys or concrete tests to perform before proceeding.

Similarly, assign your "A-Team" to critical activities. Do not treat these as training opportunities. Ensure the most experienced foremen and crew members are on critical tasks, and bring in specialist inspectors if needed. In one strategy, a weld on a key steel connection might be inspected by both the contractor's certified welding inspector and an independent inspector from the owner. Yes, this "belt-and-suspenders" approach costs more up front, but it avoids the huge expense and delay of repairing a hidden defect later.

Formal hold points are vital. Insert mandatory pauses in the work flow: do not allow a critical phase to continue until inspection has confirmed the previous step is correct. For example, don't start erecting structural steel until the foundation inspection is fully signed off, because fixing mislocated anchor bolts after steel erection could cost weeks of rework. In scheduling terms, these hold points can be represented as "quality hold" tasks: subsequent work cannot proceed until the hold is released. One practice is to schedule a quality check (e.g. an on-site test or sign-off) immediately after a critical installation and before its successor. Consider the case of a waterproofing membrane on a foundation wall that is critical before backfill: schedule a flood-test inspection hold. If the membrane leaks, it's far cheaper to find and repair it before backfill than after.

During execution, increase the frequency of checks on critical tasks. Instead of inspecting only at the end, verify partial progress. For example, if 1,000 feet of pipeline on the critical path is being laid, don't wait until all pipes are down to test. Instead, perform a hydrostatic test every 200 feet. This localizes any problem and limits rework. Daily walkthroughs by the superintendent or QA manager focused on the critical operation are also effective. They catch issues early when they are easy to fix.

Take advantage of technology to supplement manual inspections. Drones can survey tall or hard-to-reach work (e.g. tower crane assembly or crane lifts) much faster. Laser scanning or photogrammetry can verify critical dimensions in real time for components like precast panels or complex alignments. Even project software can be configured so that if any quality issue is flagged on a critical task, it auto-flags it as high priority. The goal is not to replace human oversight, but to give field teams better tools for immediate feedback.

Once a critical-path activity is verified, communicate that success to the team. Issue a short report or email (for example, "All 50 caissons drilled, inspected, and passed – ready for concrete placement"). This transparency shows that checks did not hold up the schedule, or documents if there was a delay. If a defect is found, escalate it immediately and log it in the project's quality system. Early warning means more options to recover – you might resequence other work or add resources around the issue.

- **Critical Path Quality Control Checklist:** Key steps include advanced planning meetings, assigning top crews and independent inspectors, instituting hold points for key milestones, increasing inspections during work, using tech aids (drones, scanners), and promptly communicating results.

- **Case Example:** On a fast-track plant project, installing large underground pipes was on the critical path. The team developed a strict verification plan: at each pipe segment, the crew checked gasket alignment and bolt torques as they worked, then an independent QA inspector observed a pressure test on that segment before it was buried. Midway, one segment failed the test (a flange was loose). Because the test was segment-by-segment, only that half-hour of work had to be redone. If they had waited to test the entire system at the end, crews would have had to uncover miles of pipeline to find the leak, potentially delaying the project by weeks. This example shows that breaking critical tasks into bite-size checks dramatically contains risk and keeps the schedule on track.

Not every task can have this level of scrutiny – it would be inefficient. But critical tasks warrant a higher "dosage" of quality management. Think of it like a surgeon taking extra precautions in a high-risk operation. Allocate your best quality resources (people, tests, time) where failures would hurt most. In practice, a risk-based mindset works: identify what failure modes would destroy the schedule or safety, and verify those specifically. Also, watch for "near-critical" tasks. An activity that currently has float could become critical if something slips; treat it with similar care if any single defect could push it onto the critical path.

By directing extra quality control effort to the project's heartbeat – its critical path – you amplify both safety and productivity. Teams learn to "do it right the first time" on the most important work, which pays off in fewer delays and less costly rework.

## Measuring Quality Beyond Code Minimums

Safety and quality management should never rely on just minimum compliance. Building codes and OSHA standards set only the floor, not the ceiling. Designing a project to just meet code means you may be vulnerable to durability issues or performance failures that the code never imagined. Truly quality-conscious organizations define clear performance expectations, workmanship standards, and lifecycle objectives above the code. Specifications must spell these out, and procurement must value quality (for instance, requiring higher-grade materials or experienced subcontractors) as much as cost.

Failures across the industry show that minimum compliance is not always enough to prevent disaster. Codes can lag behind emerging risks, and aging assets can drift far from their original design intent if maintenance and oversight do not keep pace.

Measuring quality beyond code means defining performance in terms that matter to how the asset will actually be used over its life. On a project, that includes tracking defect and rework rates, first time pass rates on inspections, and the frequency of punch list items tied to the same detail or trade. These metrics connect day to day construction decisions to long term reliability and cost, not just to turnover day.

Quality metrics should focus on outcomes, not just ticking boxes. Track indicators like first-pass yield (the percentage of work done correctly the first time), rate of defect recurrence, and commissioning success rates. These give insight into how well processes are working. Independent verification – such as third-party reviews or peer audits – strengthens accountability: it shows that the team is willing to have its work checked by experts to ensure standards are met. For example, if cover tests reveal a recurring issue in concrete strength, that is a warning to tighten the mix design or placement procedures, even if strength codes were technically met.

Documentation is the backbone of continuous improvement. Keep accurate records of all inspections, approvals, deviations, and corrective actions. These become the project's institutional memory. In future projects, lessons learned from one site (e.g. "we saw repeated failures at condition X until we added procedure Y") can guide better planning. A robust filing and reporting system – whether paper or digital – ensures the team is learning, not repeating mistakes.

Organizations committed to excellence often adopt higher standards proactively. For example, after a tragic crane accident, regulators may only raise wind-speed restrictions years later. But a quality leader would install real-time anemometers on site and enforce a stricter cutoff wind speed immediately, rather than waiting for an accident to drive changes. In summary, quality control beyond the minimum is about striving for robustness and resilience. When teams embed these higher standards into daily practice, projects gain trust and long-term performance.

A simple way to reinforce this mindset is to contrast "code minimum" with "high reliability practice" for a few key areas, then choose where the project will intentionally exceed the baseline.

| Area | Code / minimum expectation | High reliability practice |
|---|---|---|
| Structural redundancy | Meets basic strength checks for specified loads. | Provides alternate load paths; checks robustness under damage. |
| Temporary works | Designed to minimum factors; limited formal review. | Independently engineered, verified, and monitored in service. |
| Inspections | Pass required inspections at defined milestones. | Use additional internal checks; track trends in findings. |
| Maintenance planning | Basic O&M manuals turned over at completion. | Asset management plan with inspection intervals and reserves. |

When project teams and owners agree on these elevated targets at the outset, quality becomes more than "passing inspection." It becomes a shared commitment to deliver structures that will remain safe and serviceable long after the jobsite trailers are gone.

## Independent Verification as a Cultural Practice

Independent verification is a critical complement to inspections and audits. It means bringing a fresh, impartial pair of eyes to review important work steps. This is not about distrust – in fact, it's an acknowledgement of human limitations. By having a qualified outsider check your work, you catch assumptions or errors that familiarity can obscure.

Independent verification should be planned, not ad hoc. Identify the work that carries high energy or irreversible consequences, and schedule a second check by someone with no stake in the result. For example, before a critical concrete pour, an engineer or peer reviewer could verify the formwork and rebar setup. For a complicated lift, an independent lift plan reviewer (often required by regulation) should confirm that all calculations and plans are correct. The reviewer should not be a subcontractor's foreman (who might be too close to the work) but an engineer or QA professional who can question assumptions.

Verification should also occur on the jobsite. Team activities like walkthroughs and pre-task briefings inherently provide a moment of verification. During a walkdown, any team member should feel free to point out a concern.

In practice, we should schedule verification points into the project. For instance, hold points often include an independent review step: formwork approval before pour, welder qualification checks before welding, etc. Treat these as positive – they are safeguards, not delays.

A culture of independent verification pays off. History's biggest collapses often had "missed verification" built into them. The Hyatt Regency walkway collapse, the Willow Island concrete tower collapse, and the L'Ambiance Plaza collapse each had warning signs that went unverified by a second party. In many cases, a supervisor or peer had doubts but lacked the authority or process to get a full review. Each of those tragedies had a point where a second check – if it had been done – could have broken the chain of failure.

## Best Practices for Independent Verification

- **Plan Ahead:** Include verification tasks in the project schedule and scopes of work. For any high-risk activity (structural lifts, formwork design, tensioning, crane moves), assign a qualified person outside the contractor's direct team to verify critical details.
- **Level of Independence:** The verifier should not report to the same foreman who performed the work. Often this means an engineering manager, owner's inspector, or an external consultant. The goal is unbiased review of assumptions.
- **Regular Opportunities:** Use daily safety huddles and pre-task meetings as opportunities for everyone to verify each other's work. Encourage questions like "Can someone else check this before we proceed?"
- **Feedback and Closure:** When an independent reviewer raises an issue, document it and ensure it's resolved before moving on. See verification as a feedback loop – a chance to improve the system.

The more normalized this process becomes, the more efficient it is. When projects routinely practice verification, workers come to see "checking each other's work" not as a hindrance but as protection for the team. Leaders play a big role: if supervisors dismiss verification as "overkill," so will crews. But if managers reward discoveries of errors and treat them as learning moments, independent checks will be embraced. Over time, this becomes a reflex – a habit ingrained in the project's culture. The ultimate goal is that on every project, it is second nature to "look again, ask another question, and challenge certainty" before proceeding.

**Practical Takeaway:** Embed independent checks into critical workflows. For example, require an owner's engineer to inspect safety-critical welds, or have a peer engineer review temporary works drawings. Celebrate instances where a verification catch prevented an issue. Remember: independent verification is not a sign of weakness; it is the unseen rebar of reliability that strengthens the foundation of any safe, high-quality project.

## Quality Control Focus for Critical Path Activities

Quality control on critical-path tasks is where the safety–productivity link becomes most visible. As noted earlier, defects in critical tasks cause disproportionate delays and costs. Therefore, we must focus our QC efforts where they matter most. The key strategies include:

- **Front-Load Planning and Checks:** Before starting a critical activity, hold a thorough preparatory meeting with all stakeholders (foremen, superintendents, QA, subcontractor managers, and engineers). In this meeting, review plans, specs, and acceptance criteria in detail. For example, before a major concrete shear-wall pour, review the rebar layout, embed plates, and concrete mix in advance. Generate a specific checklist or Inspection & Test Plan (ITP) for that task. This preparatory phase helps prevent mistakes before work starts.

- **Assign Your A-Team:** Put your most experienced crews and supervisors on the job. Critical-path work is no time for inexperienced operators or minimal oversight. If needed, bring in extra specialists or inspectors.
- **Institute Formal Hold Points:** Do not allow work to proceed past certain milestones without verification. Insert "Quality Hold" tasks where needed.
- **Increase Inspection Frequency:** Inspect continuously for segmented work.
- **Leverage Technology:** Use drones, scanners, and digital tools to support verification.
- **Document and Communicate Verifications:** Make successful checks visible and escalate issues immediately.

**Case Example:** On a fast-track infrastructure project, large-diameter underground piping was on the critical path to commissioning. The team devised a rigorous verification scheme: for each pipe segment, the crew did an in-process check and then an independent QA inspector observed a hydrostatic test. Midway through, one segment failed the test. Because it was isolated, only that section required rework. If testing had been delayed, weeks of delay could have occurred.

In general, critical tasks should get a higher level of QA/QC effort. Use risk-based thinking: focus on what would hurt the project most if it failed, and verify those elements thoroughly.

**Operational Takeaways for Today's Job:**

- Start every task with a safety and quality check. Before work begins, confirm the environment is safe (proper barricades, PPE on, equipment inspected) and you have the correct drawings/specs on hand. Setting the stage correctly ensures you're not playing catch-up on safety or quality mid-task.
- Treat near-misses and small defects as big deals to fix. If a beam placement was off by an inch and had to be adjusted, or someone slipped but caught themselves, address it immediately. Find out why it happened and correct the process or retrain – proactive correction of minor issues prevents major ones.
- Retain and enforce procedures rigorously; conduct both informal and formal inspections routinely; plan independent checks into critical tasks; and allocate extra quality control on the critical path. These practices turn safety and quality from afterthoughts into cornerstones of project success.
- Use quality benchmarks as daily goals. For example, "zero rework today" or "every weld passes first inspection." Share these with the crew in the morning. Hitting those targets not only keeps the project on track but also emphasizes that quality and safety are how we measure success, not just speed.
- Conduct mini-audits on the fly. Pick one aspect of work each day – like electrical cord condition or scaffold levelness – and check it against standards. Involve a crew member in doing this audit with you. This keeps everyone sharp and shows that maintaining standards is an ongoing effort, not a one-time task.
- Link safety with productivity in your conversations. Reinforce messages like "an injury-free crew is an efficient crew" or "doing it right the first time avoids double work." Help the team see that rushing and cutting corners actually slow us down overall. A crew that believes this will prioritize working correctly over just working quickly.

## Chapter 18

# Performance Evaluation and Accountability

## Key Metrics (Safety, Cost, Schedule Adherence)

Performance evaluation only matters if people can see what "good" looks like in the work they do every day. On a construction project, that means moving beyond vague references to safety, cost, and schedule, and instead defining a short, visible set of indicators for each role: superintendent, project manager, and foreman. When those indicators are consistent, transparent, and reviewed regularly, they become a common language for how the project is really doing.

On a real project, performance evaluation only gains traction when the metrics are specific enough that each role can see what "good" looks like in their daily work. Instead of talking about safety, cost, and schedule in the abstract, project leaders should define a small, visible set of indicators for superintendents, project managers, and foremen that are reviewed regularly and tied to decisions.

For a superintendent, core metrics might include:

- Total recordable incident rate and near-miss reports per month
- Percentage of planned pre-task plans or JSAs completed before work starts
- Rework rate in the field (defects caught during QA/QC or punch-list items per work area)

For a project manager, practical measures include forecast accuracy, RFIs closed within target time, and the cycle time for change orders from identification to execution. For foremen, crew productivity versus plan, toolbox talk completion, and punch-list hits per work package create a clear, fair standard.

The Role of Metrics: In construction, you manage what you measure. Key performance metrics like safety statistics, cost variances, and schedule adherence provide an objective pulse on project health. Effective metrics serve as early warning signals when performance deviates from plan, enable accurate forecasting of future conditions, support resource allocation decisions, and create accountability throughout project teams. In fact, research by the Construction Industry Institute found that projects with systematic performance measurement achieve 12–18% better cost performance and 10–15% better schedule performance compared to those relying on informal tracking. Clearly, choosing the right metrics and tracking them diligently can be the difference between a project that catches issues early versus one that is blindsided by problems.

Key Construction Metrics: The foundational metrics in construction fall into a few critical categories, each gauging a vital aspect of project performance:

1. **Safety Metrics:** Safety is the paramount concern on any jobsite. Common metrics include the Total Recordable Incident Rate (TRIR) – the number of OSHA-recordable injuries per 200,000 work hours – and the Days Away, Restricted, or Transferred (DART) rate for more serious injuries. A lower TRIR indicates a safer project; for instance, the construction industry average TRIR was ~2.3 in recent years, while "world-class" safety performers target below 1.0. In addition to injury rates, leading indicators like near-miss reports and safety observation counts are tracked to gauge proactive safety engagement. High near-miss reporting rates can signify an open, aware safety culture (where workers report hazards before they cause harm).

2. **Cost & Financial Metrics:** Cost performance is typically measured with metrics like Budget Variance (actual cost vs. budget) and the Cost Performance Index (CPI) from Earned Value Management. A CPI = Earned Value / Actual Cost, where 1.0 means on budget; >1.0 indicates cost underrun (efficient spending), and <1.0 signals overruns. Project managers also watch cash flow closely – ensuring that the timing of expenditures vs. client payments doesn't put the project in a cash crunch. Profitability metrics, such as gross margin on the project, help contractors see if the job is financially successful after all direct costs. These cost metrics, when monitored in real time, let teams take corrective action (e.g. curbing an expense trend) before a budget blowout occurs.

3. Schedule Adherence Metrics: Time is money in construction. Schedule adherence can be tracked with Schedule Variance (planned vs. actual progress) and the Schedule Performance Index (SPI), which is similar to CPI but for time (SPI = Earned Value / Planned Value). An SPI of 1.0 means on schedule; <1.0 means behind schedule. More simply, teams track the percentage of tasks completed on time each week or whether key milestones are met by their due dates. For example, if only 80% of planned tasks finished in a given week, that trend warns of a likely slip in major milestones unless productivity is boosted. Schedule metrics often get visualized in dashboards as progress bars or Gantt charts so that everyone can see at a glance if the project is on track.

**Beyond the Big Three:** While safety, cost, and schedule are the headline metrics, a comprehensive evaluation includes quality and productivity as well. Quality metrics (e.g. rework rate or defect frequency) show if work meets standards the first time. Productivity metrics (e.g. labor hours per unit installed) help identify inefficiencies. For instance, tracking inspection pass rates (what percentage of inspections pass on the first try) can expose whether there are recurring quality issues in workmanship. In one case, a project team noticed their first-pass inspection rate fell below 80% in one month – a red flag that prompted immediate training and process checks to improve quality. The key lesson is that focusing on only one metric is dangerous; project success is a balancing act between safety, cost, time, quality, and other factors. A project that's on time but achieved that by cutting corners on safety or quality is not truly successful in the long run.

It is also helpful to distinguish "leading" from "lagging" indicators. Lagging indicators—injuries, schedule slippage, budget overruns, major change orders—tell you how you performed after the fact. Leading indicators—completed JSAs, field walkdowns, audits, and training hours—track behaviors that reduce the chance of those failures before they occur. A balanced dashboard for each role should contain both, so teams are not only judged on outcomes but are also rewarded for the preventative work that keeps the project inside the envelope.

## Example role-level metrics

| Role | Leading indicators (weekly) | Lagging indicators (monthly/phase) |
|---|---|---|
| Superintendent | % JSAs completed before work; safety walks completed; near-misses logged | TRIR; recordable incidents; rework tickets issued |
| Project Manager | RFIs closed on time; updated cost/ schedule forecast issued | Budget variance; number/size of change orders |
| Foreman | Toolbox talks completed; field checks of critical tasks | Punch-list items per work area; unplanned work stoppages |

What Makes a Good Metric: Not all metrics are created equal. A good metric is one that drives the right behavior and is tied to outcomes that matter. To evaluate your project metrics, consider these guidelines:

- Relevant and Aligned: Metrics should tie directly to project objectives and stakeholder priorities. If on-time completion is critical, measure schedule adherence; if client satisfaction is key, include metrics like punchlist issues found at handover. Avoid metrics that look impressive but don't influence success (sometimes called "vanity metrics"). For example, measuring hours worked is less useful than measuring work accomplished per hour. A classic mistake is using feel-good metrics that always show high scores – these can mask real problems. "Everyone wants a metric they score 95% on," but if it's that easy, it may be meaningless. Always ask: does this metric tell us something actionable about improving performance?

- Objective and Consistently Measurable: A metric must be defined in a clear, unambiguous way and be consistently tracked. If different supervisors interpret a metric differently, the data will be unreliable. Metrics like "team morale" or "work ethic" are too subjective to quantify fairly. Focus on observable, countable things (e.g. number of safety audits completed vs. planned). Consistency also means measuring at regular intervals and under the same criteria each time.

- Behavior-Driving (Without Causing Unintended Consequences): Perhaps most importantly, metrics should promote the right behaviors on site. This is where careful thought is needed – because people respond to what is being measured and rewarded, sometimes in unintended ways. It's often said that "when a measure becomes a target, it ceases to be a good measure." For example, if a team is only measured (and bonused) on schedule adherence, they may be tempted to ignore safety or quality to hit the date. In the 2005 BP Texas City refinery disaster, management had focused heavily on personal injury rates (which were low) and celebrated a "perfect safety record" in terms of OSHA injury metrics – yet they were blind to looming process safety risks in the plant. The facility lacked metrics for critical process safety conditions, so warning signs were missed. The result was a catastrophic explosion. The lesson is stark: measuring the wrong thing can give a false sense of security. A robust metric system mixes lagging indicators (outcomes like incidents or cost variance) with leading indicators (process measures like safety audit completion, preventive maintenance tasks done, percent of employee training completed). Leading indicators gauge whether you're doing the right things that prevent bad outcomes. For instance, instead of only tracking "how many accidents occurred", a leading safety metric might track "number of hazard reports filed by staff" – encouraging people to speak up about dangers. Metrics that encourage reporting and improvement (rather than only punishing failure) tend to drive right behaviors. As one industry article notes, "encourage behavior-based accountability rather than measuring performance solely by output" – in other words, track how work is done, not just the final result. When metrics are used this way, they reinforce the process and culture that lead to success, not just the end numbers.

- Hard to Game or Cheat: A well-designed metric resists manipulation. If team members can easily "game" the system to hit the metric without actually improving performance, the metric will backfire. For example, a delivery company that measured "on-time deliveries" found that one branch was rejecting orders they feared might be late – thereby improving their metric artificially. They hit 99% on-time, but only by turning away difficult jobs, harming business and customer relationships. In construction, consider a metric like "percent of inspections passed" – a team could game this by doing fewer inspections or only calling for inspection when absolutely sure of passing, but that might mean they delay necessary checks. To avoid this, pair metrics or set boundaries. The goal is to design metrics such that the easiest way to improve the numbers is to actually improve the work, not cheat the system. In short, build integrity into the metrics.

Metrics Driving the Right Behavior: When crafted and used properly, metrics can have a powerful motivating effect on the team. They create a shared scoreboard that everyone can reference. Many construction leaders institute daily or weekly metrics dashboards visible to all – for example, a safety dashboard might display the current injury-free streak, near-misses reported this week, and toolbox talks conducted. Publicly sharing this can instill pride and positive peer pressure to keep the streak going. Likewise, a project might use a labor productivity heatmap in the site office that shows which trades are hitting their productivity targets. Modern project management software makes these visual dashboards easy to set up; some companies even use mobile apps and Power BI dashboards that let supervisors check key metrics on their phone in real time. Visual tools like heatmaps are especially popular for quickly highlighting trouble spots. This visual management helps direct attention immediately to where it's needed.

In one example, a contractor implemented a real-time safety dashboard that tracked incidents by site and leading indicators like hazard observations. When management noticed one site falling behind on submitting safety observations, they intervened that week with extra safety walks – potentially preventing an accident. Fast feedback loops like this are only possible when metrics are timely and visible. As a rule, the shorter the interval between measurement and action, the more effective the metric will be at guiding performance.

A useful way to keep metrics manageable is to distinguish between "leading" and "lagging" indicators. Lagging indicators, such as injuries, delays, and major change orders, tell you how you performed after the fact. Leading indicators, such as completed JSAs, walkdowns, audits, and training hours, are behaviors that reduce the chance of those failures before they occur. A balanced dashboard for each role should include both types.

Key Metrics in safety, cost, and schedule are the backbone of performance evaluation on a project. The keys to strengthening this backbone are: choose meaningful metrics aligned with project success; include a balanced mix of safety, cost, schedule (and quality) indicators; use both outcome and process measures; and be vigilant about how metrics influence behavior. A good metric system, coupled with intuitive dashboards or reports, allows construction leaders to spot issues early and course-correct. It also gives the team a clear target to rally around – "what gets measured gets improved." By contrast, poor metrics (or no metrics at all) leave the team flying blind or, worse, incentivize counterproductive behavior. Thus, effective use of metrics is a cornerstone of accountable, high-performance project management.

## Peer vs. Managerial Review Models Peer Review and Managerial Review

Construction work is too complex for performance evaluation to be handled by a single person looking at a spreadsheet once a year. The most reliable systems combine peer review, which captures how someone actually plans, communicates, and executes with their team, with managerial review, which connects that performance to formal decisions about assignments, pay, and promotion. When the rules for both are clear, evaluation feels less arbitrary and more like a natural extension of how the job already runs.

Peer reviews are best suited to improving day-to-day practice. A superintendent might sit with another superintendent from a different project to walk through a recent lift plan, or two project managers might review how RFIs and change orders were handled on a difficult phase. The tone is coaching and problem-solving: what worked, what did not, and what could be done differently next time. Managerial reviews, by contrast, are formal checkpoints where a project executive or department leader looks at results over a larger time window and makes decisions about raises, bonuses, development plans, or whether someone is ready to take on a bigger role.

Neither model works well in isolation. A culture that relies only on peer feedback may feel supportive but rarely addresses chronic performance problems or misaligned incentives. A culture that relies only on top-down reviews can drive fear and paperwork without improving how crews work together in the field. On healthy projects, peer reviews happen frequently and informally throughout the year, while managerial reviews occur on a predictable cadence with documented criteria, so people improve as they go and also understand how that improvement is recognized.

## Comparison of Peer vs. Managerial Review

| Aspect | Peer review (lateral) | Managerial review (hierarchical) |
|---|---|---|
| Primary purpose | Improve day-to-day practices, share lessons, identify blind spots | Make decisions on pay, promotion, development plans, and assignment of roles |
| Typical use cases | Constructability review, look-back on RFIs/COs, safety walk feedback, crew leadership | Annual reviews, bonus allocation, performance improvement plans |
| Who participates | Same-level supers, PMs, or foremen from other areas or projects | Direct manager, project executive, HR or corporate leadership as needed |
| Tone and framing | Coaching, mutual learning, emphasis on improvement | Formal, tied to expectations and accountability |

| | | |
|---|---|---|
| Documentation level | Light notes, action items | Structured evaluation forms |
| Risks if used alone | Supportive but lacks enforcement | Can become rigid and fear-driven |

## Top-Down Reviews (Managerial Model)

The traditional model of performance evaluation in construction is the managerial review – a supervisor or manager evaluates the performance of their direct report. This has the benefit of clear authority and alignment with organizational goals: the manager is responsible for the output of their team, so they provide feedback and ratings to ensure each member is contributing.

Manager reviews can consider the big picture, including how an individual's work supports project objectives and company values. They typically carry weight in promotions, compensation, and long-term development planning. For example, a construction Project Manager will formally review an Assistant PM's performance against criteria like budget management, subcontractor coordination, and initiative.

This manager-as-coach approach can drive accountability, especially when feedback is ongoing rather than annual. However, it has limitations. A manager cannot observe every aspect of an employee's daily work. On a busy jobsite, a superintendent may only see end results, not the full effort or influence of each individual.

## Peer Feedback and Reviews

Peer review models involve colleagues at the same level, or other stakeholders, providing feedback on performance. This ranges from informal daily coaching to structured 360-degree review systems.

Peers often have the clearest visibility into real performance. On a construction site, crews work side by side every day. They see how work is actually planned, how problems are handled under pressure, and how decisions are made when supervision is not present. They know who consistently follows procedures, who contributes to problem-solving, and who may be cutting corners when no one is watching.

Because of this, peer input provides a more complete and realistic view of performance than top-down evaluation alone.

Where this becomes powerful is when peer feedback is structured enough to be useful but informal enough to remain honest. If it becomes too formal, it turns into paperwork. If it stays too informal, it never gets captured or acted on. The balance is to embed peer feedback into existing workflows rather than creating separate systems.

On strong projects, peer feedback shows up in moments that already exist:

- During look-ahead planning meetings when one foreman challenges another's sequence
- During safety walks when a superintendent questions a setup outside their trade
- During post-task reviews where crews discuss what went right and what created friction
- During handoffs between trades where upstream work is evaluated by the next crew

These are not labeled as "peer reviews," but they function exactly as such.

The key is intentional use. Instead of letting these interactions stay casual, high-performing teams capture patterns:

- Are the same issues being raised across trades?
- Is one crew consistently solving problems others are not?
- Are shortcuts showing up in similar conditions?

This turns peer input from conversation into data.

Another important distinction is between peer validation and peer correction.

- Peer validation reinforces good performance. It builds consistency and confidence.
- Peer correction identifies drift early, before it becomes rework or failure.

Both are necessary. A system that only corrects becomes negative. A system that only validates becomes blind.

Peer review also strengthens accountability horizontally. Instead of accountability only flowing from management downward, it exists between crews. This changes behavior. People are far less likely to cut corners when they know the next trade will immediately see the result.

## Accountability as a Driver of Motivation

Accountability, Not Blame: Accountability in a project setting means each person owning their responsibilities and outcomes, good or bad. It's about creating a culture where people do the right thing even when no one is watching, because they feel a personal responsibility to the team and mission. This concept, when implemented in a positive way, can supercharge motivation and engagement on the job. It's critical to note that accountability is not about playing the blame game or instilling fear. In fact, modern management research has flipped the old notion of accountability as finger-pointing punishment. Instead of, "Who can we blame for this mistake?", the thinking is, "How do we ensure everyone takes ownership so we learn and improve?" A healthy accountability culture is empathetic and learning-oriented. It distinguishes between "honest mistakes" and negligence. For example, the aviation industry follows a "Just Culture" philosophy: pilots or mechanics are not punished for human errors or near-misses if they report them and work to correct them, but willful violations or gross negligence are subject to consequences. This balance encourages openness (so problems surface and can be fixed) while still maintaining high standards. Construction can benefit from the same approach – workers should feel safe admitting, "Hey, I screwed up a step on that install," without immediate fear of being fired, as long as they are forthcoming so the team can remedy it. In a "just" accountability culture, people know they will be held accountable for choices like shortcuts or safety rule violations, but they also know that if something goes wrong despite their best effort, the company will focus on problem-solving over punishment. The result is people are motivated to act responsibly, not just avoid blame.

Why Accountability Boosts Engagement: When implemented well, accountability actually makes employees feel more invested and engaged in their work. Why? Because taking ownership gives a sense of purpose and control. When team members see how their individual actions contribute to project outcomes, it instills pride and motivation to excel. A Gallup study found that employees who strongly agree that their manager holds them accountable (in a fair and helpful way) are 2.5 times more likely to be engaged in their job. Furthermore, 80% of employees who received meaningful feedback in the past week are fully engaged at work. This highlights that feedback and accountability go hand in hand. If people get frequent input on how they're doing, they feel seen and supported – which increases their commitment. Accountability also builds trust and camaraderie within teams. When everyone pulls their weight and owns up to issues, team members trust each other more. Research published in the Journal of Business Ethics showed that accountability increases trust and commitment among team members, strengthening collaboration. Think of a high-performing construction crew: each person does their part (whether it's securing every bolt or double-checking measurements) not just because the boss said so, but because they don't want to let their brothers and sisters on the crew down. That peer trust reinforces engagement – you want to be at work because you're part of a reliable, accountable team.

Creating a Culture of Accountability: Building such a culture requires intentional effort. It doesn't happen by memo; it happens by consistent leadership behavior and systems that support ownership. Key steps to cultivate accountability include:

- **Clarity in Roles and Expectations:** People can only be accountable if they know what's expected. Thus, clearly define each team member's roles, deliverables, and decision-making authority. For example, if a field engineer is responsible for daily concrete pour inspections, make that explicit and give them the authority to stop a pour if it doesn't meet criteria. Avoid overlapping responsibilities or vague ownership (which leads to "diffusion of responsibility" where everyone assumes someone else will act). "Address any gaps in responsibility or blurred lines between roles, which can undermine accountability by leading to diffused responsibility," as Deloitte advises. On a practical level, tools like a RACI matrix (Responsible, Accountable, Consulted, Informed) for project tasks can help make sure for every task, one name is in the "A" (Accountable) spot. Clarity is the foundation: an accountable culture is one where everyone knows what the plan is, who is doing what, and what the targets are.

- **Empowerment and Trust:** True accountability is not micromanagement; it thrives on empowering people. This means giving team members the autonomy to make decisions and innovate in their area, along with the training and tools to do so. When people feel trusted to act, they are more likely to take ownership. A culture of "my boss has my back" enables accountability. For instance, granting a foreman the discretion to rearrange crew tasks if they see a safety issue – without having to ask three levels of permission – tells them the company trusts their judgment. "A healthy culture of accountability relies on trust in employees to take ownership of their work. It hinges on autonomy, encouraging people to make decisions and take initiative to solve problems," notes one leadership guide. Leaders should act as coaches and remove obstacles, rather than control every move. This empowerment is often backed up with resources and training (so it's not "you're accountable, figure it out alone," but "you're accountable, and we'll enable you to succeed"). As an example, a contractor might empower site crews to conduct their own safety huddles each morning (ownership at the crew level) and provide them with a short toolbox talk guide or checklist to ensure effectiveness.

- **Real-Time Feedback and Coaching:** Accountability works best when feedback is frequent, timely, and constructive. If someone is off track, they should know it now, not six months from now in a performance review. Likewise, when someone does great work or steps up responsibly, that recognition should be immediate. Regular feedback fuels the continuous improvement loop – it's much easier to course-correct small issues in real time than big issues long after the fact. Data strongly supports this: employees are 3.6 times more likely to be motivated to do outstanding work when their manager provides daily (vs. annual) feedback. Make feedback a two-way street: encourage team members to self-assess and voice concerns, not just receive critique. A good practice is the "fast feedback" approach – after a task or meeting, quickly debrief on what went well and what could improve. These conversations, while short, keep everyone aligned and accountable for making adjustments. For feedback to be constructive: keep it specific, behavior-focused, and future-oriented. Instead of "You're bad at managing the schedule," a manager might say, "We fell behind on concrete pours last week. Let's discuss what happened – perhaps we underestimated cure times – and how we can adjust the schedule or resources going forward." This way the conversation targets the issue and solution ("how we can adjust"), not personal blame, and it points to future improvement. Also, balance correction with positive feedback when deserved – constructive accountability means praising the good as quickly as flagging the bad. If a crew hit an important milestone through extra effort, a quick recognition ("I appreciate everyone staying late to get that pour done safely. That accountability to the schedule made a big difference.") goes a long way. Such acknowledgement reinforces the behavior you want and motivates the team to repeat it.

Real-Time Feedback in Action – A Field Example: Imagine during a morning site walk, the Project Manager notices a scaffold platform that wasn't fully checked by the crew before use. Rather than simply scolding the foreman in private later, a constructive approach is to pause work briefly in the moment. The PM gathers the crew and says, "I want to point out something for our safety – the scaffold tags weren't signed off this morning. Let's all take a moment now to re-inspect it together." This is immediate feedback and coaching. It does two things: it addresses the issue on the spot (accountability for safety is enforced), and it educates the crew by involving them in the correction (they see what to do differently). No one is being written up or shamed publicly; rather the emphasis is "we're accountable to do this right, so let's fix it now." Later on, the PM might quietly tell the scaffold competent person, "I need you to own this task every morning. If there's pressure to rush, remember that safety checks come first – and if you need more manpower to do it, let's talk." This reinforces personal accountability with support rather than fear. The immediate, constructive feedback loop prevents a possible accident and shows the crew that accountability is about having high standards that we all share, not about yelling when something goes wrong.

Accountability and Motivation: When people consistently see that effort and ownership are noticed and that slipping up leads to help (and learning) rather than humiliation, it creates a self-reinforcing cycle. Team members start holding themselves accountable – and even holding each other accountable in positive ways. A laborer might remind his buddy to put on his hard hat not just because a safety officer might be watching, but because that's just how our crew operates – we look out for each other. This is the Holy Grail of safety cultures and high-performing teams: accountability at the peer level driven by mutual respect and shared goals. One study on leadership found that when leaders model accountability and create a safe environment, teams were far more likely to speak up about problems and suggest improvements. In other words, engaged teams come from accountable teams. Gallup's research aligns with this: workplaces that integrate regular feedback and hold people to clear expectations see much higher engagement and performance outcomes than those that don't.

Giving Constructive Feedback (Tips): It's worth summarizing a few practical tips for managers to deliver feedback in a way that boosts accountability rather than resentment:

- *Be Timely:* Give feedback as soon as practical after the observation. Feedback has an "expiration date" – the sooner it's given, the more relevant and useful. A good rule is to address positive or negative events within that same week, if not the same day.

- *Be Specific and Objective:* Cite clear examples and data. Replace "You're not proactive" with "Yesterday, the concrete delivery was delayed 2 hours and it caught us off guard. Let's figure out how we can get earlier warnings on supplier delays – perhaps a daily confirmation call?" Specifics turn a potentially personal critique into a solvable problem.
- *Focus on Behavior, not Person:* Frame feedback around actions that can change, not personal traits. Instead of "You're bad at planning," say "The lookahead schedule wasn't communicated to the crew, which led to confusion. In the future, let's ensure every Friday we review next week's plan with the foremen." This keeps the discussion on the behavior and process, not an attack on character, which keeps the person receptive.
- *Listen and Involve:* After giving your perspective, ask for the employee's input: "What do you think? Was there something preventing you from doing X?" When people participate in the solution, they take ownership of it. It signals respect and can reveal underlying issues (maybe the employee didn't have the right tool or information). By making it a dialogue, you turn feedback into a coaching session rather than a lecture.
- *Balance Accountability with Support:* If you're holding someone accountable for a result, make sure they have the means to achieve it. For example, telling a site engineer to improve QA/QC checks is fine, but if she is swamped with other duties, you might need to re-prioritize tasks or assign an assistant. Also, express your confidence in their ability to improve ("I know you can get this done – that's why I'm counting on you. Let's touch base tomorrow on progress."). This combination of high expectations and support is what truly drives people – they know you won't accept less than success, but also that you'll help them reach it.

Real World Case – The Power of Ownership: A case study from an EPC contractor illustrates how accountability fuels engagement. The contractor implemented a policy that any crew member could issue a "Stop Work" if they spotted a serious safety hazard – no repercussions, in fact they'd be celebrated for it. Initially, workers were hesitant, fearing backlash for causing delays. To break the ice, management themselves would occasionally call out a Stop Work in morning meetings for minor hazards, to demonstrate it was okay. Soon, laborers and foremen started doing it too when needed – and each time, the company publicly thanked them in the next safety meeting and quickly fixed the issue. Far from slowing the project, this practice prevented at least a few near-misses from becoming accidents. More importantly, it energized the workforce: crews reported feeling safer and more "heard." One foreman said it was the first company he'd been at where "speaking up is actually rewarded." That sense of ownership – that anyone on the team could make the call to ensure safety – led to a significant boost in safety suggestions and improvements on site. Safety engagement (measured by number of hazard reports filed) doubled within a few months. This is a clear example that when you trust and empower people to be accountable, they don't shirk responsibility – they rise to it. It's human nature: being entrusted with important matters makes us feel valued and motivates us to prove worthy of that trust.

Accountability = Motivation = Results: When a construction crew or project staff feels accountable for outcomes, they treat the project as their own. They are more likely to go the extra mile to solve problems, more likely to collaborate instead of saying "not my problem," and more likely to innovate. It creates a pride in workmanship that no amount of top-down pressure can replicate. As a final thought, consider that accountability also drives continuous learning, which keeps a team getting better. In a culture where mistakes aren't hidden, the team can have honest post-mortems and grow from them. Each person is motivated to improve their own performance because they know they own it and their teammates depend on it. This continuous improvement cycle is essentially what great construction management is all about – steadily building a learning organization that gets safer, faster, and more efficient with each project.

Fostering a culture of accountability – done the right way – is one of the greatest motivators in the field. It's not about cracking the whip, it's about creating an environment where everyone from project executive to apprentice feels responsible and proud to uphold high standards. When accountability is paired with clarity, empowerment, and real-time feedback, it stops being a dreaded word and becomes a source of engagement and excellence. People want to be held to a high standard when they know it's fair and aimed at growth. As one article succinctly put it, "Accountability isn't about blame – it's about ownership… A strong culture of accountability boosts engagement, sparks innovation, and drives long-term success." By embracing that mindset, construction leaders can turn accountability into a powerful engine that drives both people development and project performance, ensuring that the whole team is pulling together towards a successful outcome.

## Accountability Systems in Practice

Effective accountability systems in construction include:

Clear Expectations:

Roles and responsibilities must be defined so individuals understand what is expected

Measurable Outcomes:

Metrics provide an objective basis for evaluating performance

Regular Feedback:

Continuous evaluation allows for real-time improvement

Fair Consequences and Recognition:

Performance should be linked to both rewards and corrective actions

These elements create a system where accountability is consistent and transparent.

In practice, these elements must be connected, not isolated.

For example:

- Expectations without measurement lead to ambiguity
- Measurement without feedback leads to frustration
- Feedback without consequences leads to no change

The system only works when all four operate together.

A practical structure on a project might look like:

- Daily: Field-level expectations reinforced through pre-task planning
- Weekly: Metrics reviewed in short performance huddles

- Monthly: Formal evaluation tied to outcomes and trends
- Phase-end: Lookbacks that adjust expectations for the next phase

This cadence ensures accountability is continuous, not reactive.

## Balancing Accountability and Motivation

The challenge is balancing accountability with motivation. If accountability is too strict or punitive, it can discourage initiative and create fear. Workers may avoid reporting issues or taking responsibility.

On the other hand, if accountability is too loose, performance suffers because there are no consequences for poor execution.

The goal is a balanced system where:

- Good performance is recognized and rewarded
- Issues are addressed constructively
- Learning is prioritized over blame
- Standards remain clear and enforced

In the field, this balance is visible in how leaders respond to problems.

Two crews can make the same mistake and receive completely different outcomes depending on leadership:

- In a punitive system, the mistake leads to blame and silence
- In a balanced system, the mistake leads to correction and shared learning

The second system produces better long-term performance, even if it appears slower in the moment.

The key distinction is between:

- Negligence (willful disregard for standards)
- Error (mistake within a functioning system)

Negligence requires enforcement.

Error requires improvement.

High-performing organizations make this distinction clearly.

## Positive Accountability Culture

A positive accountability culture includes:

- Encouraging reporting of issues and near-misses
- Treating mistakes as learning opportunities
- Reinforcing desired behaviors through recognition

- Holding individuals responsible for commitments

Over time, this builds trust and engagement across the team.

Trust is the critical variable here. Without trust, accountability systems collapse into compliance behavior. People do the minimum required to avoid consequence.

With trust, accountability becomes proactive. People:

- Raise issues earlier
- Correct each other faster
- Take ownership without being forced

This directly improves safety, quality, and productivity.

## Linking Accountability to Performance Metrics

Accountability is closely tied to the metrics defined earlier in this chapter.

For example:

- Safety performance ties to incident rates and reporting behavior
- Cost performance ties to budget tracking and forecasting
- Schedule performance ties to milestone completion

When individuals understand how their actions affect these metrics, accountability becomes tangible.

The key is visibility. Metrics must not live only in reports. They must be:

- Seen in the field
- Discussed in meetings
- Used in decision-making

Otherwise, they lose connection to behavior.

## Shifting the Conversation

Over time, this reframing changes the tone of evaluation conversations. Instead of asking, "What went wrong and who is at fault?" leaders start from, "What did we commit to, what actually happened, and what support or system changes do you need to hit the mark next time?"

That shift makes accountability a source of motivation and pride. People know the standard, feel trusted to meet it, and understand how their effort on the ground shows up in how they are evaluated.

This shift moves evaluation from judgment to alignment.

## Weekly Performance Huddle

One simple way to make performance evaluation tangible is a short weekly performance huddle with key field leaders.

- Bring three or four key metrics for the job (safety, critical path activities, quality issues, change events)
- Ask each role what went well, what slipped, and what they need from others
- Capture one improvement action with a clear owner and due date

Keeping this brief and predictable builds a habit of using metrics as a tool for joint problem-solving instead of post-mortem criticism.

To make this effective, consistency matters more than duration. A 15-minute weekly cadence that is never skipped is more valuable than a longer meeting that happens irregularly.

### Operational Takeaways for Today's Job:

- Share key performance numbers with the team. Let everyone know at the end of the week how you're doing on safety (e.g., number of incidents or near-misses), schedule (tasks completed vs. planned), and quality (inspections passed). When the crew sees the results, they understand the impact of their work and where to focus.
- Implement a peer checking system. Encourage crew members to take a quick look at each other's work for critical tasks (like inspecting each other's harness setup or verifying measurements). Peer accountability builds a sense of shared responsibility and often catches mistakes that a formal inspection might miss.
- Set clear daily expectations and follow up. Each morning, outline what needs to be achieved and the safety/quality standards to meet. At day's end, gather the team for 5 minutes to review what was done well and what fell short. Consistent feedback loops ensure accountability – everyone knows they'll be asked to reflect on the day's performance.
- Lead by owning up to mistakes and expect others to do the same. If you as a manager slip on something (say, you forgot to arrange an inspection), admit it openly and fix it. This attitude trickles down, encouraging crew members to speak up about their own mistakes or delays so that issues can be addressed rather than hidden.
- Tie recognition to the right things. Praise a crew that finishes ahead of time and does it safely and correctly, not one that cuts corners to beat the clock. When promotions or rewards are given, base them on a mix of productivity, safety, and quality performance. This makes it clear that accountability isn't just about hitting numbers – it's about doing so the right way.

# Chapter 19

# Continuous Improvement in Practice

Continuous improvement (CI) means deliberately capturing what we learn on every project and using it to get better, day after day. Rather than assuming that current practices are "good enough," teams examine problems and mistakes systematically and adjust methods to raise performance. As quality pioneer W. Edwards Deming taught, organizations must "improve constantly and forever the system of production and service." In practice this requires building strong feedback loops and a culture of learning on the job. Each day's experiences – whether success, near miss, or failure – become inputs to the next day's plan and training. The reward for this discipline is higher safety, quality, and productivity. One design-build contractor that instituted daily improvement meetings, visual controls, and cross-team problem-solving saw on-time delivery jump by roughly 40% and procurement lead times drop by 25%.

By contrast, projects that fail to learn from their problems tend to "reinvent the wheel." Without CI, mistakes repeat, craftsmanship suffers, and cost overruns pile up. Skilled teams often produce one great outcome, only to watch new crews repeat the same error years later because the insight was never captured. In a high-turnover industry, hard-won knowledge must be preserved; otherwise each job becomes isolated, and every team must rediscover best practices from scratch.

This chapter explains how construction organizations can institutionalize continuous improvement in the field by building organizational memory – capturing and retrieving knowledge from past work. Developing learning loops – short cycles of work, review, and adjustment that turn daily work into real progress and cover how to treat failures (accidents, near misses, defects) as training opportunities so mistakes are not repeated. Toward the end of the chapter I provide practical tools and templates (forms, checklists, agendas) that teams can use on the job to support learning. Finally, I addresses sustaining a CI culture – ensuring that improvement efforts endure through metrics, leadership, and employee engagement. Throughout, the tone is practical and solution-oriented, aimed at superintendents, project managers, foremen, and field leaders who want to make improvement part of their daily routine. The focus is on application in real jobsite conditions – union or open-shop, large or small, civil or building work – not on abstract theory.

## Organizational Memory in Construction

Construction projects are often one-off or short-lived, and teams disperse quickly when a job ends. This makes it easy for hard-won lessons and institutional knowledge to vanish when personnel change or move on. Yet retaining that organizational memory is critical. For example, if a civil crew solves a complex excavation problem on one site, that insight should be available to crews on the next site who may face similar soil issues. Without memory, each project team "reinvents the wheel" because lessons learned on one project are not applied to the next.

Construction projects have a clear end date, but the lessons they generate do not have to disappear with the trailers. Organizational memory is the deliberate effort to capture what was learned on one job and make it easy to use on the next. Without a simple, disciplined way to store and retrieve those lessons, the same problems quietly reappear project after project, and the organization keeps paying the same tuition.

To break this cycle, companies must deliberately capture, organize, and share their knowledge. This means thinking of each project as a learning opportunity that contributes to a permanent library of good practices and warnings. Practical steps include:

- **After-Action Reviews (AARs) and Debriefs:** After completing a major phase (like finishing the foundation) or after any significant incident, gather the relevant team to discuss what happened. Document what went well and what didn't, why it happened, and what changes are needed. For example, you might use a simple AAR form that records the project name, date, event summary, contributing factors, and recommended actions. It only takes a few minutes to fill out, but these notes become a written memory that future teams can reference. The process of debriefing turns events into teachable insights.

- **Knowledge Management Systems:** Store lessons learned, best practices, and standard procedures in a central repository or digital "knowledge library." This might be a company intranet site, shared drive, or project database where team members can quickly find past reports, photos, and how-to guides. For instance, a superintendent could upload annotated photos showing the correct and incorrect way to install a framing detail, or a written report on a welding lesson learned. By indexing this information (by project name, trade, or issue type), crews ensure that knowledge survives staff turnover. Whenever a new project starts, reviewing the archive should be a standard kickoff activity.

- A practical approach is to maintain a lessons learned register that lives in the same ecosystem as drawings, RFIs, and submittals. Each entry briefly describes the issue, the root cause, and the fix, and is tagged by CSI division and project phase (design, procurement, construction, commissioning), so that future teams can search it when planning similar work. For example, a misaligned anchor bolt problem can be logged as "03 30 00 – cast in place concrete / construction," with a note that pre-pour checks must include independent verification of template location before concrete placement.

- **Standard Operating Procedures (SOPs) and Checklists:** Develop and update written procedures and checklists based on on-the-job experience. Whenever a recurring problem is identified, codify the solution into SOPs or toolbox talk material. These documents serve as a living memory of how work should be done.

The register should not be a dumping ground; it should be a short list of patterns that matter. At turnover, the project team can distill recurring issues into one or two pages, then route those into standard details, checklists, and training materials for the next job. When front-end planners or field leaders start a new project, reviewing these summaries alongside schedule and budget becomes as normal as reading the specs.

- **Mentoring and Peer Sharing:** Embed memory into people, not just paperwork. Pair less-experienced workers with veteran craftsmen so lessons can be passed on verbally. Encourage regular site "show-and-tell" sessions where techniques and lessons are demonstrated.

- **Project Debrief Reports:** At project close-out, compile a comprehensive lessons-learned report. Archive and circulate these reports so new teams can review them before starting similar work.

Using these mechanisms, construction organizations turn each project into a source of insight for the next one. One contractor, for example, started every new road project by reading lessons from past highway jobs; this helped them anticipate issues and adjust their plans ahead of time.

Building organizational memory is not automatic; it requires discipline and leadership support. Teams must actually use the stored information. It is not enough to file away a report; project leaders should regularly ask, "Does this lesson apply to our current work? How can we act on it?"

**Case Study:** Eagle Bridge Replacement. A heavy-civil contractor began a bridge replacement project by reviewing a detailed lessons-learned report from a nearly identical job a few years earlier. The old report noted that unexpected subsurface rock delays had cost extra hours. Acting on this memory, the new team ordered additional early geotechnical borings and contingency planning. When the same condition was encountered, they were prepared, avoiding delays.

## Learning Loops (Review → Adjust → Retrain)

At the heart of continuous improvement is an iterative cycle of work, review, and adjustment – a learning loop. After a short work cycle (for example, a shift, a week, or the completion of a specific task), teams step back to review what happened, decide on improvements, and retrain the crew on any changes. Then the cycle repeats. This is essentially the Plan-Do-Check-Act (PDCA) cycle made concrete on the jobsite.

Continuous improvement is only real if it shows up as a loop: review what happened, adjust the system, and retrain the people who use it. In construction, that loop must be fast, visible, and grounded in everyday tasks, or it will get buried under production pressure.

One effective model is a 48-hour learning loop for any incident, near miss, or major quality escape. Within 24–48 hours, the supervisor or manager gathers the people closest to the work for a short review focused on facts and contributing factors, not blame. The team then identifies at least one concrete change to an SOP, checklist, drawing note, or approval step that would reduce the chance of recurrence, and that change is implemented immediately rather than waiting for the next big policy update.

The loop closes when the affected crews are retrained, even if only in a 10–15 minute toolbox talk that uses the actual event as the example. The change and its rationale are also captured in the organizational memory system, tagged so future projects can find and use it. Over time, tools like checklists and procedures should evolve in visible ways; if they never change after incidents, they are forms, not learning devices. When crews see their feedback turn into real adjustments, they recognize that speaking up and participating in these loops is worth the effort.

- **Plan and Do (Work Execution):** Begin with a plan and carry out the work accordingly. As work proceeds (pouring concrete, welding pipe, installing drywall, etc.), observe and record any deviations or unexpected events. For example, note if the pouring crew fell behind schedule when needing extra vibration time, or if an electrical conduit had to be rerouted unexpectedly. Collect data whenever possible – such as daily progress photos, production counts (linear feet of pipe installed, number of fixtures wired), or even simple progress metrics on a whiteboard.

- **Check/Review:** At the end of the cycle, analyze the outcomes. Compare actual performance to the plan and targets. Look at safety logs (incidents or near misses), quality reports (rework or punch-list items), productivity numbers (e.g. cubic yards of concrete per day vs. planned), and budget/schedule variances. Also rely on direct observation and input: ask foremen and crew what went right and what didn't. This step pinpoints what succeeded and what failed, and why. For instance, a superintendent might discover that a framing crew consistently hit its wall targets each day, but a follow-up check finds defects in how door openings were cut – a gap not visible until paint – which explains why final inspections took longer than planned.

- **Act/Adjust:** Decide on changes needed based on the review. Prioritize the highest-impact issues first. Changes can involve work methods, sequencing, crew assignments, equipment, or scheduling. For example, if the review finds that an electrical change order was often missing information, the team might adjust by adding a step in their kickoff checklist to verify all drawings match the contract docs. If crews found that a scaffold setup at an unexpected elevation caused inefficiency, the adjustment might be to redesign the scaffold plan for future lifts.

- **Retrain/Rebrief:** Put the changes into practice by ensuring the crew understands them. Conduct a toolbox talk, quick demonstration, or a short drill to explain the new method. For example, if the update was a revised concrete placement sequence, gather the crew and walk them through the new sequence. If a safety improvement was identified (say, better harness attachment), physically show the correct technique or have a safety officer supervise practice runs. Even a brief reorientation ensures everyone adopts the improvements.

- **Repeat:** With the new changes in place, start the next work cycle (morning, week, or next task). Then do another review at the end of that cycle, and continue the loop. Each loop is an opportunity to learn more and refine further.

These loops can be formal (scheduled weekly meetings, monthly trainings) or informal (quick talks at the end of a shift). A useful practice is a weekly continuous improvement meeting on-site: a brief gathering (15–30 minutes) where the team reviews key metrics and incidents from the week and plans adjustments. A sample agenda for such a meeting could include: - Review Key Metrics: Compare actual vs. planned progress (e.g. work-hours expended, quantities installed) and review safety and quality figures (near misses, incidents, or rework hours).

- **Discuss Incidents and Issues:** Briefly recap any safety events or quality defects from the week, and confirm agreed lessons. For instance, if a subcontractor's rigging error delayed a lift, discuss how to avoid it next time (better pre-lift meeting, updated checklists, etc.).

- **Identify Improvement Ideas:** Ask crew members for suggestions to improve processes or eliminate waste. Prompt discussion with a question like, "How can we speed up the piping installation without cutting corners on safety?" Crew members often have valuable ideas, such as rearranging material staging to avoid double-handling.

- **Adjust Plan/Methods:** Decide on any changes to work methods, resource allocation, or sequencing based on the above reviews. Assign responsibility for implementing each change. For example, you might decide the superintendent will coordinate earlier with the concrete supplier on deliveries to prevent wait time.

- **Training and Briefing Needs:** Identify topics that need retraining or coaching. If the team noticed repeated errors in welding technique, plan a hands-on demo next week. If a permit renewal was overlooked, schedule a quick refresher on permit procedures.

- **Follow-Up:** Review action items from the last meeting, update their status, and close any completed tasks. This closure ensures that improvement items don't get forgotten.

Formatting this agenda as a one-page checklist or whiteboard helps leaders ensure each point is covered. Keeping the loop cycle short – ideally weekly or even daily – helps catch problems early and prevents small issues from compounding. Many contractors find that a 5–10 minute daily huddle each morning is equally powerful. In the morning meeting, crews can quickly run through yesterday's work, share one safety reminder or quality tip, and review the day's plan. These daily and weekly routines turn abstract CI ideas into habits on the site.

Several well-known methods embody this loop concept. The classic PDCA cycle (Plan-Do-Check-Act) is essentially what we described above. The military and healthcare use After-Action Reviews (AARs) in training and operations: participants answer questions like "What did we plan? What actually happened? Why the difference? What will we do next time?" Lean manufacturing uses Kaizen events (short focused workshops) and Gemba walks (manager visits to the work area) to achieve the same ends. Construction teams can borrow these ideas without needing fancy labels. For example, a foreman might shadow a crew for a few hours to spot inefficiencies, then lead a mini-plan session on how to fix them.

A key ingredient for learning loops is trust and openness. Workers must feel safe to share honest feedback without fear of punishment. When a superintendent consistently asks "What went wrong?" in a coaching tone – and then listens – crews learn that speaking up leads to solutions, not blame. Over time, this encourages a culture where improvements bubble up from the field.

On a large hospital renovation, the project team held a Monday morning meeting every week. In one such review, the concrete foreman reported that extra time was needed to clean out forms before each pour, causing delays. The team adjusted by adding a daily cleanup task to the schedule and assigning a worker specifically for that duty. They retrained the formwork crew on a faster cleaning technique. The next pour week they found progress was smoother and schedule adherence improved. This is a small but concrete example of a learning loop at work: observe an issue, adapt the plan, retrain, and reap the benefit in the next cycle.

## Core Learning Loop Process

At its core, a learning loop is a disciplined cycle that connects execution, reflection, adjustment, and retraining into a continuous system. The process is straightforward, but its effectiveness depends on consistency and follow-through.

The cycle begins with planning and execution. Work is carried out according to the established plan, but with deliberate attention to observation. As the work progresses, crews and supervisors should actively note deviations, inefficiencies, or unexpected conditions. These observations do not need to be complex; simple field notes, progress photos, or basic production tracking are often sufficient to capture meaningful insights.

At the end of the work cycle, the team shifts into review. This is where actual performance is compared against the plan. Safety records, quality reports, productivity data, and schedule progress are examined alongside direct feedback from the crews. The purpose is not to assign blame, but to identify gaps, patterns, and root causes.

Once the issues are understood, the team moves into adjustment. Decisions are made about what needs to change, whether that involves sequencing, crew assignments, equipment use, or planning assumptions. The focus should always be on the changes that will have the greatest impact with the least disruption.

These changes must then be translated into action through retraining. Crews need to understand not only what is changing, but why it is changing. This can take the form of a short toolbox talk, a demonstration, or a brief rebrief before work begins. Without this step, adjustments remain theoretical and never influence actual behavior.

The cycle then repeats, with each iteration building on the last. Over time, this simple loop transforms daily operations into a steady stream of incremental improvements. The strength of the system does not come from complexity, but from repetition and discipline.

## From Event to System Change

A common failure point in construction learning is stopping at discussion. Teams review an issue, agree that it should not happen again, and then return to work without changing the conditions that allowed it to occur.

For a learning loop to be complete, it must result in a system-level adjustment. That adjustment should be visible in how work is planned or executed going forward.

In practice, this means the outcome of a review must show up in at least one of the following:

- A revised checklist step that prevents the issue from being missed again
- A clarified drawing note or field instruction that removes ambiguity
- An added verification point or hold point before critical work proceeds
- A change in sequencing or crew assignment to improve workflow
- A clearly defined stop-work trigger that empowers crews to intervene

If none of these elements change, the loop has not truly closed. The conversation may have happened, but learning has not been embedded into the system.

The 48-hour learning loop reinforces this discipline by compressing the timeline between event and action. Instead of waiting for formal reports or periodic reviews, the system is adjusted while the conditions are still fresh and directly relevant to the crews involved.

## Weekly Learning Loop in Practice

Learning loops can be applied at multiple time scales, but one of the most effective formats is a weekly continuous improvement meeting. Typically lasting 15–30 minutes, this meeting provides a structured opportunity to review performance and make targeted adjustments.

A consistent agenda helps ensure that the meeting remains focused and productive. Teams review key metrics such as safety performance, schedule progress, quality issues, and cost indicators. They then discuss incidents or recurring problems, identifying both what happened and why it happened.

From there, the team identifies improvement opportunities and assigns responsibility for implementing changes. Training needs are also addressed, ensuring that any adjustments are reinforced with clear communication to the field. Finally, previous action items are reviewed and closed, maintaining accountability across cycles.

This meeting is not intended to be a report-out. It is a working session where decisions are made and actions are assigned. When conducted consistently, it becomes part of the project rhythm, ensuring that improvement is continuous rather than reactive.

## Daily Micro-Loops (Field-Level Application)

While weekly reviews provide structure, most improvement occurs at the daily level. Short, focused learning cycles embedded directly into field operations allow teams to respond to issues in real time.

A brief morning huddle, lasting five to ten minutes, can serve as a daily learning loop. During this time, crews review the previous day's performance, identify a specific issue or inefficiency, and adjust the day's plan accordingly. A single safety or quality focus is reinforced to maintain alignment.

These micro-loops are critical because they prevent small issues from compounding into larger problems. By addressing deviations immediately, teams maintain control over performance and reduce the likelihood of rework or incidents.

## Embedding Learning into Normal Work

Learning loops are not a separate activity layered on top of construction work. They are an extension of the work itself.

Many established methodologies reflect this same concept, including Plan-Do-Check-Act cycles, after-action reviews, Kaizen events, and Gemba walks. While the terminology may vary, the underlying behaviors remain consistent: work is reviewed regularly, adjustments are made quickly, crews are retrained immediately, and results are visible.

In practice, these behaviors appear in simple, everyday actions. A foreman may adjust crew positioning during a shift to improve efficiency. A superintendent may identify a coordination issue during a site walk and revise sequencing before the next work cycle begins. These are learning loops operating at their most practical level.

## Trust as the Foundation of the Loop

A learning loop cannot function without trust.

Workers must feel safe to:

- Report issues
- Admit mistakes
- Suggest improvements

If feedback is met with blame or punishment, the loop breaks. Problems go unreported, and opportunities for improvement are lost.

Conversely, when feedback leads to visible action, the loop strengthens. Crews begin to surface issues earlier, often before they develop into incidents or defects. This creates a proactive environment where improvement is driven from the field rather than imposed from above.

Leadership behavior is critical in establishing this trust. The response to early feedback sets the tone for the entire system. If leaders reinforce participation and act on input, the learning loop becomes self-sustaining.

## Key Principle

If a problem occurs and nothing changes in procedures, checklists, or training, then the organization has not learned.

Learning loops must result in visible, operational changes. Without that, they remain discussions rather than drivers of performance improvement.

## Turning Failures into Training

Most organizations have incident reports and case studies that are read once and then filed away. A more powerful approach is to turn selected failures—both internal events and external cases from earlier in this book—into short, repeatable training modules for crews and supervisors. This keeps hard earned lessons close to the work instead of trapped in binders and archives.

Each module can follow a simple three part structure. First, "What happened": a brief, factual overview of the event and its consequences, avoiding sensational detail. Second, "Why it happened": three to five bullet points on root causes and missed warning signs, written in plain language that connects to field decisions. Third, "What this means for us": specific behaviours, checks, or design and field practices the current team will adopt—for example, new verification points, clearer stop work criteria, or additional temporary works requirements. These modules can be delivered in 10–15 minutes during safety meetings, phase start huddles, or onboarding for new supervisors.

Failures only become useful when they change what people do on Monday morning. When a crew can point to a case study and say, "This is why we insist on independent verification before we remove shoring," the story has moved from a distant incident to an everyday habit. Over time, a library of such modules becomes part of the culture: a standing reminder that the organization is willing to look at its own mistakes and others' mistakes honestly, and is committed to building systems that do not repeat them.

## Integrating Continuous Improvement into Daily Operations

The elements discussed in this chapter—organizational memory, learning loops, and failure-based training—are not independent systems. They are parts of a single operating model. When they function together, continuous improvement becomes embedded in daily work rather than treated as a separate initiative.

Organizational memory ensures that lessons are not lost between projects. Learning loops ensure that new lessons are created and applied quickly. Training modules ensure that those lessons reach the field in a form that influences behavior.

When these systems are aligned, the flow of improvement becomes continuous:

- Work is performed
- Outcomes are reviewed
- Adjustments are made
- Crews are retrained
- Lessons are stored and reused

This cycle repeats across projects, crews, and phases of work.

The effectiveness of this system depends on consistency. If any part of the cycle is missing—if lessons are not captured, if adjustments are delayed, or if crews are not retrained—the system weakens and improvement slows.

Leaders play a critical role in maintaining this integration. They must ensure that:

- Lessons learned are actually reviewed at the start of new work
- Learning loops are executed consistently
- Training reflects current risks and recent events

Without leadership reinforcement, continuous improvement efforts tend to degrade into isolated activities rather than a sustained system.

## Failures as Training Modules

Failures come in many forms – accidents, near misses, quality defects, schedule overruns, cost overruns, and more. Rather than hiding or minimizing these events, high-performing teams treat every failure (and near miss) as a learning opportunity. Whenever something goes wrong, it should be analyzed for root causes and then incorporated into training or process changes so the same issue cannot happen again. This mindset turns painful lessons into valuable assets.

If a worker is injured because a step in a safety procedure was skipped, that incident should not be forgotten. Instead, conduct a brief investigation: what factors led to the violation? Was the hazard known but a control missed? Use the answers in a toolbox talk or training session to highlight the importance of each control measure. Even better, have the crew walk through the specific task under supervision to practice the safer method. In essence, the accident becomes a case study or role-play scenario. This turns experience into education.

Industry leaders echo this approach. Quality guru Deming said that "improvement is not a one-time effort," and systems must continually look for ways to reduce waste and improve. In safety, modern approaches like "Safety II" emphasize using near misses and successes as learning tools, not just focusing on failures. The critical point is: discovering what went wrong is useless unless it's shared and applied. A toolkit cannot stay hidden; it must be wielded to make next time safer.

Here are practical ways to use failures (and near misses) in training and improvement:

- Root Cause Analysis: After any significant event, gather the responsible team for a quick root-cause analysis. Techniques like the "5 Whys" or a simple fishbone diagram can help uncover why the error occurred. Was it a training gap, a plan deficiency, or just a one-off slip? Document the findings clearly. For example, if a scaffold collapsed, the analysis might reveal a missing locking pin and that the team had never been briefed on pin inspections.

- Post-Incident Briefings: Hold a structured debrief as soon as possible. Bring together the crew (and other relevant workers) and explain the incident in factual terms. Discuss the root cause and highlight what should be done differently. Encourage crew members to ask questions and share their perspective. This can be done immediately (on the spot) or in a short follow-up meeting. The goal is to ensure the lesson spreads beyond the individuals directly involved. For instance, after a near-miss with a suspended load, the site held a quick meeting on rigging safety, even though only one crew was involved.

- Scenario Drills: Turn a significant failure into a live training drill. For example, if a lift rigging error occurred on a jobsite, pause production and run a mock scenario the next day. Challenge crews to set up the lift correctly under supervision. If a concrete pour was rejected for a finishing defect, simulate the proper finishing technique in a controlled exercise. These drills force the crew to apply the "right way" under guidance, reinforcing the lesson. It's one thing to tell crews, "Don't repeat that mistake," and quite another to have them practice the improved method.

- Error-Proofing Training: Teach workers to recognize and avoid known failure modes using checklists, visual cues, or simple "poka-yoke" devices (error-proofing ideas). For instance, if material mix-ups have happened before, put samples of the correct mixtures on a wall chart, label bins, and train crews to double-check color or code labels before using. Show photos of what went wrong and what should have happened. If a quality control chart misinterpretation caused rework, walk through the chart logic step by step in a brief class.

- Lessons-Learned Files and Toolbox Talks: Maintain a brief "lessons learned" log on-site or company-wide where each incident (and success) is summarized in a few bullet points. Regularly review recent entries in toolbox talks. For example: "At Project Delta, we had a minor electrical shock last week because a cable was not properly grounded. Remember to verify all grounds before energizing." Repetition reinforces learning. Seeing the note in print gives it weight beyond the initial conversation.

- Failure Simulation and Role Play: For some complex problems, simulations can be enlightening. Consider using safety trainers or external facilitators who role-play a hazardous situation. For example, pretend a trench is about to collapse (using props or animation), and walk the crew through the correct emergency response. This immersive method makes lessons memorable.

- Celebrate Proper Responses: Don't just focus on what went wrong; also highlight what went right. If a worker correctly identifies a hazard or performs a rescue drill well, acknowledge it as a positive example. Over time, crews learn that reporting issues and responding properly are valued behaviors, not punishable ones.

Adopting this mindset – that every failure can be a free lesson – dramatically accelerates improvement. Workers learn that mistakes become learning moments, not just blame. Over time, the organization internalizes the lessons of the past. For instance, after investigating several incidents of weld defects, a team might realize all were due to unclear specifications. The solution: a quick mini-training on reading spec charts. By catching and correcting the root cause through training, future welds stay on target.

*After a major electrical arc-flash incident on one project, for example, the electrician's safety committee revamped their lockout/tagout training. They held a hands-on refresher for every crew, even demonstrating it live on one of the plant's breakers. That immediate application of learning from a real failure ensured the same mistake wasn't repeated elsewhere.*

In short, using failures as training modules creates a learning culture where practice continually adapts. Nothing goes into the incident file unanswered – every accident, near-miss, or defect prompts a question, "What will we do differently next time?" This approach makes improvement constant and concrete.

## Tools and Templates for Continuous Improvement

To make continuous improvement operational on the site, teams can use simple tools and templates. The idea is to embed CI into daily work through forms, checklists, and visual aids that prompt action and ensure consistency. Here are practical tools that can be readily deployed:

- *Lesson Learned Form:* A structured form to document each significant learning event (positive or negative). Key fields include: Project/Task Name, Date, Description of Event (what happened), Observed Cause(s), Consequences (safety, quality, schedule), Recommended Corrective Actions, Responsible Person, and Completion Status. For example, a form entry might read:
- *Event Title:* "Boom Lift Tipped During High-Wind Operation."
- *Description:* Brief narrative of what went wrong (e.g. "Lift was operated at wind speeds above the manufacturer's chart limit.").
- *Root Cause:* Wind thresholds were ignored due to production pressure.
- *Action Items:* Specific changes (e.g. "Revise lift plan to mandate shut-down at lower wind speeds; retrain signalers on limits.").
- *Owner & Due Date:* Who will implement each action and by when (e.g. "Project Engineer – revise lift plan by next Monday").
- *Status:* On review or completed.

By standardizing this form, every team captures lessons in the same way. Filled-out forms should be stored in a binder or di1me this builds an organizational memory of problems and fixes, which can be shared at kickoff meetings for new projects.

- *Weekly Learning Loop Agenda:* An agenda template for a short weekly meeting (15–30 min) to review what the team learned and plan adjustments. (This was outlined in section 19.2, and it can be formalized into a checklist or whiteboard agenda.) Sample items on the agenda include:
- *Review Key Metrics:* Actual vs. planned progress, safety stats (near misses, incidents), quality reports (rework hours), budget and schedule variances.

- *Discuss Incidents or Near Misses:* Summarize any safety or quality events from the last week and agreed lessons.
- *Identify Improvement Ideas:* Solicit crew suggestions for process improvements or waste elimination (e.g. "How can we speed up foundation prep without compromising safety?").
- *Adjust Plan/Methods:* Decide on changes to work methods, resource allocation, or sequencing. Assign who will implement each change.
- *Training Needs:* Identify topics for retraining or briefings (e.g. schedule a toolbox talk on a new hazard or skill).
- *Follow-Up:* Review action items from the last meeting; update status or close them.

Formatting this as a one-page checklist helps the meeting leader cover each point. Keep the meeting concise and focused on problem-solving, not extended lectures.

- *Daily Huddle (Crew Briefing) Agenda:* A checklist for a daily kickoff meeting. Even 5–10 minutes each morning can make a difference. Prompt items might include:
- *Yesterday's Summary:* What tasks were completed? Any surprises or hold-ups?
- *Today's Plan:* What work will we accomplish today, and who is responsible for each task?
- *Safety Moment:* Review one safety topic or near-miss from the site.
- *Coordination Needs:* Any interfaces with other trades or inspections planned?
- *Improvement Item:* Quick check – "Did anything not go well yesterday? How can we avoid it today?"
- *Logistics:* Materials, permits, equipment needed.

A written checklist or board (even a whiteboard posted at the toolbox) ensures consistency. It also serves as a record of what was covered each day, which is useful if someone missed the meeting.

- *Pre-Task Planning Checklist:* A pre-task planning form helps crews ask the right questions before starting any significant activity. Typical prompts include:
- *Have we done this before?* (If yes, what did we learn last time?)
- *What can go wrong?* (List hazards or failure modes.)
- *What measures will prevent each hazard?* (Controls, PPE, engineered safeguards.)
- *Are conditions different today?* (Weather, changes in personnel or site conditions.)
- *Do we have the right tools and materials?*
- *Emergency plan:* If something goes wrong, what is the immediate response?

By regularly completing this checklist, crews reinforce past lessons and ensure nothing obvious is overlooked. The completed forms can be signed off by supervisors or photographed and stored as proof of compliance.

- **Training Conversion Worksheet:** Use this when a site observation or evaluation reveals a training need. This form helps turn issues into training events. Columns might be: Date, Observed Issue (what deficiency was seen), Category (safety/quality/etc.), Proposed Training Topic, Delivery Format (toolbox talk, hands-on demo, video, etc.), Trainer/Lead, and Verification Method. For example, if several weld defects are found during an inspection, the worksheet entry might read: "Issue: Excess weld spatter on carbon steel joints. Proposed training: Refresher on welding technique and spatter control in next toolbox talk. Trainer: Welding foreman. Verification: Confirmed by QA inspector that subsequent welds meet standards." This ensures that real-world observations become documented training actions, and someone is accountable for delivering it.

- **5S Audit Checklist:** Regular workplace organization can prevent mistakes and speed work. A 5S audit (Sort, Set in Order, Shine, Standardize, Sustain) checklist ensures tools and materials are in their designated places, work areas are clean, and signage or labels are clear. For example, on a framing crew, the checklist might include verifying that safety tools (first-aid kit, fire extinguisher) are in place, unused materials are removed from the scaffold, and fasteners are stored by size. Conducting brief 5S checks (even weekly) helps teams spot small inefficiencies or hazards before they become problems.

- **Visual Management Board:** Create a visual control board or "mission control" in the site trailer or field office. This could be as simple as a whiteboard or as high-tech as a big screen with dashboards. It displays key information at a glance: the week's schedule vs. actual progress, safety stats (near misses/injuries), a list of open action items from improvement meetings, and any critical project alerts. Visual cues (red/yellow/green status lights, charts, photos) make issues obvious. For example, a chart of budget vs. actual can alert everyone if spending is running ahead. Posting current lesson-learned notes on the board reminds crews of lessons from other projects in real time. The board facilitates communication across the team – anyone can see where things stand and what needs attention.

- I**mprovement Suggestion Box or Log:** Encourage crew members to contribute ideas by providing a simple suggestion form or board. For example, post a clipboard with suggestion cards on site, or keep a running log where anyone can write an improvement idea (with contact info so you can follow up). Periodically review these suggestions in meetings and recognize those implemented. For instance, if a carpenter suggests reorganizing tool delivery based on floor levels, try it out and give credit. This formal channel shows that management values worker insight.

- **Supervisor Verification Checklist:** A simple daily or weekly checklist for supervisors to confirm that planned safety and quality measures are actually implemented. Example items:

- [ ] Site pre-task plan completed for each crew.

- [ ] Lessons Learned form from last incident has been reviewed and posted in the trailer.

- [ ] Toolbox talk on recent event or lesson was given to all crews.

- [ ] Corrective actions from last meeting have been started.

- [ ] All critical permits and procedures are up-to-date and being followed.

If an item cannot be checked off, the supervisor notes why and takes corrective action. This checklist serves as both a reminder and a feedback log – it reinforces that learning is being turned into action each day.

Each of these tools should be adapted to your organization's needs and style. The point is to embed continuous improvement into daily work, not rely on memory or informal whispers. Forms and checklists act as prompts: they ensure that whenever something noteworthy happens, it is captured and addressed. Over time, this consistency means no important lesson slips through the cracks.

Tip: Keep the tools simple and visible. For instance, laminate a pre-task checklist and leave it in each crew's toolbox, or integrate weekly agendas into your project management software. Even a wall chart of metrics updated by hand each week can have a big impact. The goal is to make improvement routine, not a special project.

## Sustaining a Continuous Improvement Culture

Getting started with CI is one challenge; keeping it going is another. Too often, projects kick off improvement activities with enthusiasm, only to see them fade after a few weeks. To sustain improvement over time, several factors are key:

- **Metrics and Feedback:** Track progress with meaningful indicators. Instead of arbitrary targets, use outcome-focused metrics. For example, track the project-level rework rate (hours of rework vs. total hours), number of safety observations closed per week, or percentage of suggested ideas implemented. Make these visible on a dashboard or at weekly meetings. Celebrate when numbers improve: post the trending graph of "Near Misses This Year" or announce, "We went from 4 OSHA-recordables last year to 1 this year." Seeing concrete evidence of progress (even small improvements) reinforces to everyone that CI is working. Conversely, declining or stagnant metrics signal when action is needed. Remember: what gets measured gets attention, so pick a few key indicators that reflect your team's main goals (safety, quality, schedule, morale).

- **Leadership Commitment:** Supervisors and managers must visibly support CI – not just talk about it. Field leaders should regularly review improvement metrics, attend daily and weekly learning meetings, and most importantly, act on the ideas generated. If leaders ignore suggestions or fail to follow up on action items, staff quickly lose interest. In practice, this means allocating time in the schedule for CI activities (don't treat them as extra work that can be dropped). For example, a project manager could carve out one hour every Monday morning for field walkthrough and CI huddle rather than letting it slide when deadlines loom. Leaders should also model the desired behavior: share a lesson they learned, participate in a drill, or help with a Kaizen event. When foremen and superintendents "walk the walk" of continuous learning, crews take notice. Visible commitment might also include rewarding team members who contribute ideas – for example, a monthly recognition of the "Best CI Idea" or a simple thank-you note.

- **Training and Competence:** Build CI skills among staff. Teach people how to conduct After-Action Reviews, use root-cause tools (like 5 Whys analysis), and apply new procedures. For example, you might spend a portion of a supervisor training class on how to lead a toolbox talk effectively or how to update a checklist. Cross-train crew members so they appreciate each other's roles; a plumber who understands an electrician's challenges can offer better input for process improvements. When everyone understands the improvement process, they will naturally look for enhancements. Consider pairing new foremen with experienced mentors who can coach them in continuous improvement techniques.

- **Empowerment and Recognition:** Create an environment where workers feel safe to voice ideas and raise issues. Encourage "Kaizen suggestions" by asking crews directly for feedback – during a meeting, put up an idea board, or even have an anonymous suggestion drop box. When someone fixes a wasteful practice or catches an error early, publicly acknowledge it. For example: "Yesterday Joe from the framing crew noticed a quality issue and stopped work to get it fixed. Great catch – that's exactly the kind of attention that keeps us safe and efficient." Simple shout-outs in morning meetings or a "CI Star" note on the bulletin board go a long way. Over time, people will proactively look for small wins when they know they will be heard.

- **Integration into Systems:** Make CI part of formal processes and responsibilities. For example, include a "lessons learned" agenda item in every milestone review or safety meeting. Require that someone present one improvement idea in each subcontractor coordination meeting. Tie participation in CI to performance reviews or project incentives. Some companies even build it into contracts: a contractor might agree to a sharing arrangement where cost savings from an improvement are split 50/50, giving everyone skin in the game. The idea is that CI is not a side activity, but woven into day-to-day operations and management structures.

- **Continuous Monitoring:** Treat your CI system itself like any other project and never let it grow stale. Just as you track schedules and budgets, track the health of your CI practices. If action-item lists start growing old with no resolution, or if the weekly meeting attendance drops off, take corrective action. Seasoned projects may implement internal "CI audits" – brief checks to ensure tools like the Lessons Learned form or Pre-Task checklist are actually being used. An example audit question could be, "Can anyone show me a lesson learned entry from this month?" or "Show me your updated project agenda with CI items." Use the same root-cause analysis on these problems: if people stop attending, ask why and fix the root cause (perhaps the meetings are too long or held at an inconvenient time).

By treating improvement as part of the job (not a one-off project), teams create a self-reinforcing cycle. Improved processes free up time and resources, which can be reinvested in further learning and innovation. As Deming said, organizations should be "continually looking for ways to reduce waste and improve quality" as an ongoing obligation.

In practice, CI report both cultural and measurable gains. One design-build firm found that engaging all 800 of its employees (site crews, office staff, subcontractors) in the process – through daily huddles, suggestion programs, and joint improvement events – was key to their success. Each person felt part of the improvement mission. The result was not just statistics (faster delivery, fewer safety incidents), but a sense of pride and ownership across the organization.

Leading an improvement culture can have other benefits too. Crews start to recognize and catch errors early, reducing rework. Workers become more engaged because they see their ideas taken seriously, which can help with retention. Even clients notice when a contractor is constantly improving: fewer delays and defects translate to higher client satisfaction.

Continuous improvement in construction is not a one-time fix but a permanent commitment. It requires: good record-keeping and sharing of lessons, disciplined short feedback loops of review and adjust, learning from every failure and near miss, practical tools to capture and act on ideas, and above all a culture that values learning and adaptation. By following the guidance in this chapter, field leaders can make their teams smarter and safer over time – turning every project into a stepping-stone for the next, rather than a blind alley.

### Operational Takeaways for Today's Job:

- Share key performance numbers with the team. Let everyone know at the end of the week how you're doing on safety (e.g., number of incidents or near-misses), schedule (tasks completed vs. planned), and quality (inspections passed). When the crew sees the results, they understand the impact of their work and where to focus.
- Implement a peer checking system. Encourage crew members to take a quick look at each other's work for critical tasks (like inspecting each other's harness setup or verifying measurements). Peer accountability builds a sense of shared responsibility and often catches mistakes that a formal inspection might miss.
- Set clear daily expectations and follow up. Each morning, outline what needs to be achieved and the safety/quality standards to meet. At day's end, gather the team for 5 minutes to review what was done well and what fell short. Consistent feedback loops ensure accountability – everyone knows they'll be asked to reflect on the day's performance.
- Lead by owning up to mistakes and expect others to do the same. If you as a manager slip on something (say, you forgot to arrange an inspection), admit it openly and fix it. This attitude trickles down, encouraging crew members to speak up about their own mistakes or delays so that issues can be addressed rather than hidden.
- Tie recognition to the right things. Praise a crew that finishes ahead of time and does it safely and correctly, not one that cuts corners to beat the clock. When promotions or rewards are given, base them on a mix of productivity, safety, and quality performance. This makes it clear that accountability isn't just about hitting numbers – it's about doing so the right way.
- **Capture and share knowledge:** Use AARs, SOPs, and reports to retain lessons learned. Create a searchable project archive so new teams can act on past insights rather than repeating mistakes.
- **Use short learning loops:** Hold daily or weekly review meetings (huddles, walkdowns) where crews compare actual results to plans. Adjust methods based on findings and retrain crews on changes immediately.
- **Learn from every failure:** Treat every accident, near miss, or defect as a training opportunity. Analyze root causes, train the crew on the fix, and update procedures so the mistake can't happen again.
- **Leverage simple tools:** Deploy checklists, forms, and visual boards to make improvement systematic. Examples include Lessons Learned forms, daily huddle agendas, pre-task planning sheets, and supervisor checklists. These tools prompt action and ensure no lesson is missed.
- **Measure and celebrate progress:** Track meaningful metrics (rework rate, suggestion count, safety trends) and make improvements visible. Celebrate small wins and improvements in meetings or via bulletin boards to keep the momentum going.
- **Engage people at all levels:** Continuous improvement must be more than a management slogan. Supervisors, foremen, and crews should all be involved. Empower workers to speak up with ideas, recognize them publicly, and connect CI activities to leadership support and training programs.

When these practices are ingrained into daily jobsite operations – supported by leadership and embraced by workers – even small steps forward can yield big leaps in project performance.

www.ingramcontent.com/pod-product-compliance
Ingram Content Group UK Ltd.
Pitfield, Milton Keynes, MK11 3LW, UK
UKHW052345280926
15472UKWH00009B/2769

9 798995 586104